Methods in Molecular Biology

Series Editor
John M. Walker
School of Life and Medical Sciences
University of Hertfordshire
Hatfield, Hertfordshire, AL10 9AB, UK

For further volumes:
http://www.springer.com/series/7651

High-Throughput Metabolomics

Methods and Protocols

Edited by

Angelo D'Alessandro

Department of Biochemistry and Molecular Genetics, University of Colorado Denver, Aurora, CO, USA

Humana Press

Editor
Angelo D'Alessandro
Department of Biochemistry and Molecular Genetics
University of Colorado Denver
Aurora, CO, USA

ISSN 1064-3745 ISSN 1940-6029 (electronic)
Methods in Molecular Biology
ISBN 978-1-4939-9235-5 ISBN 978-1-4939-9236-2 (eBook)
https://doi.org/10.1007/978-1-4939-9236-2

This Humana imprint is published by the registered company Springer Science+Business Media, LLC, part of Springer Nature.
The registered company address is: 233 Spring Street, New York, NY 10013, U.S.A.

Preface

High-Throughput Metabolomics: *So Much to Discover, So Little Time...*

Dum loquimur fugerit invida aetas... (Horace, Odes 1, 11, 8)

The post-genomic era and the bioinformatic revolution that accompanied it fostered new strides in the fields of metabolomics and lipidomics. These "omics" approaches are often referred to—rightfully so—as the "closest to the phenotype" and perceived by the scientific community as novel, especially in comparison to genomics, transcriptomics, and proteomics. Despite the aggressive and largely successful efforts to rebrand this discipline, metabolomics—defined as the comprehensive analysis of small molecule metabolites—is perhaps the oldest analytical tool mankind managed to harness. History is full of records describing symptoms and metabolic characteristics of metabolic diseases such as (*ante litteram*) diabetes: "the sweet taste" and "capacity to attract ants" of urine have been documented since the fifth century BCE in India and Greece, second century BCE in China. Centuries of advancements in the fields of chemistry and (clinical) biochemistry, recently accompanied by the introduction of tools like NMR and mass spectrometers, have simply provided a novel "magnifying lens" to expand our understanding of the small molecule world as a function of our attempts to "poke nature." From this perspective, metabolomics is nothing but the next iteration of a discipline that scientists have been investigating for decades with much less sophisticated tools, often compensating for the technological gap with incredible rigor and acumen. Building on decades of advancements and empowered by novel analytical and bioinformatics tools, scientists have embraced the "new" field of metabolomics to generate a wealth of data from laboratory studies, some of which are slowly transitioning into the clinics. This transition can be significantly sped up owing to the opportunity to perform large-scale studies in a high-throughput fashion both at the discovery phase (e.g., high-throughput screening of novel drugs) and clinical testing (e.g., in large-scale prospective studies). In this view, this entry of the Methods in Molecular Biology series focuses on recent technological, computational, and biostatistical advances in the field of high-throughput metabolomics. Chapters encompass methods, platforms, and analytical strategies for steady-state measurements and metabolic flux analysis with stable isotope-labeled tracers, in biological matrices of clinical relevance and model organisms. Mass spectrometry-based or orthogonal methods are discussed, along with computational and statistical methods to address data sparsity in high-throughput metabolomics approaches. Finally, a few representative applications are discussed, including biodosimetry, sports and wellness, and personalized metabolomics. The main take-home message we wish to share with the interested reader is that high-throughput metabolomics tools can bring about the next generation of clinical biochemistry in a cost-effective, but not necessarily less rigorous fashion than current analytical approaches, exponentially advancing our capacity to investigate nature while easing the advent of personalized medicine.

Prior to concluding this quick introduction to the contents of the book, I will take the chance to thank all the contributing authors for their support to this successful initiative and Dr. John Walker and David C. Casey (Springer Nature) and Julie Reisz Haines (University of Colorado Denver, Anschutz Medical Campus) for their invaluable editorial assistance.

Conflict of Interest

A.D. is founder and CSO of Omix Technologies, Inc.

Aurora, CO, USA ***Angelo D'Alessandro***

Contents

Contributors

RAJESH AGARWAL • *Department of Pharmaceutical Sciences, Skaggs School of Pharmacy and Pharmaceutical Sciences, University of Colorado Denver, Aurora, CO, USA*
SHUBHRA AGRAWAL • *Elucidata, Inc., Cambridge, MA, USA*
MIKAYLA ANDERSON • *Bloodworks Northwest Research Institute, Seattle, WA, USA*
LLUÍS AROLA • *Biochemistry and Biotechnological Department, Nutrigenomics Research Group, Universitat Rovira i Virgili, Tarragona, Spain; Biotechnological Area, EURECAT-Technological Center of Catalonia, Reus, Spain*
GIUSEPPE ASTARITA • *Department of Biochemistry and Molecular and Cellular Biology, Georgetown University, Washington, DC, USA*
CORAL BARBAS • *Facultad de Farmacia, Centro de Metabolómica y Bioanálisis (CEMBIO), Universidad CEU San Pablo, Madrid, Spain*
ROBERT M. BARKLEY • *Department of Pharmacology, University of Colorado Denver, Aurora, CO, USA*
BRYAN C. BERGMAN • *Division of Endocrinology, Diabetes, and Metabolism, School of Medicine, University of Colorado Anschutz Medical Campus, Denver, CO, USA*
MIRELA BERISA • *Donald B. and Catherine C. Cancer Metabolism Center, Memorial Sloan Kettering Cancer Center, New York, NY, USA*
KRISTEN E. BOYLE • *Department of Pediatrics, Section of Nutrition, University of Colorado Anschutz Medical Campus, Aurora, CO, USA*
EMILY C. BRITT • *Morgridge Institute for Research, Madison, WI, USA; Department of Nutritional Sciences, University of Wisconsin-Madison, Madison, WI, USA*
COREY D. BROECKLING • *Proteomics and Metabolomics Facility, Colorado State University, Fort Collins, CO, USA*
ERIC BUNKER • *Department of Chemistry and Biochemistry, University of Colorado, Boulder, CO, USA*
NÚRIA CANELA • *Technological Joint Unit of Omic Sciences, EURECAT-Technological Center of Catalonia, Universitat Rovira i Virgili, Reus, Spain*
PETER CARMELIET • *Laboratory of Angiogenesis and Vascular Metabolism, VIB Center for Cancer Biology (CCB), VIB, Leuven, Belgium; Laboratory of Angiogenesis and Vascular Metabolism, Department of Oncology and Leuven Cancer Institute (LKI), KU Leuven, Leuven, Belgium*
DOUGLAS A. CHAPNICK • *Department of Chemistry and Biochemistry, University of Colorado, Boulder, CO, USA*
THEKLA CORDES • *Department of Bioengineering, University of California San Diego, La Jolla, CA, USA*
JUSTIN R. CROSS • *Donald B. and Catherine C. Cancer Metabolism Center, Memorial Sloan Kettering Cancer Center, New York, NY, USA*
ANGELO D'ALESSANDRO • *Department of Biochemistry and Molecular Genetics, University of Colorado Denver, Aurora, CO, USA*
DENISE M. DAVIS • *Department of Radiology, School of Medicine, University of Colorado Denver, Aurora, CO, USA*

ANTONI DEL PINO • *Technological Joint Unit of Omic Sciences, EURECAT-Technological Center of Catalonia, Universitat Rovira i Virgili, Reus, Spain*

ANDREAS DRÄGER • *Center for Bioinformatics Tübingen (ZBIT), University of Tübingen, Tübingen, Germany; Department for Computer Science, University of Tübingen, Tübingen, Germany; German Center for Infection Research (DZIF), Tübingen, Germany*

ENRIQUE DURÁN-GUERRERO • *Instituto de Investigación Vitivinícola y Agroalimentario (IVAGRO), University of Cádiz, Puerto Real, Spain; Department of Analytical Chemistry, University of Cádiz, Puerto Real, Spain*

JING FAN • *Morgridge Institute for Research, Madison, WI, USA; Department of Nutritional Sciences, University of Wisconsin-Madison, Madison, WI, USA*

GIUSEPPINA FANELLI • *Department of Ecological and Biological Sciences (DEB), University of Tuscia, Viterbo, Italy*

SARAH-MARIA FENDT • *Laboratory of Cellular Metabolism and Metabolic Regulation, VIB Center for Cancer Biology (CCB), VIB, Leuven, Belgium; Laboratory of Cellular Metabolism and Metabolic Regulation, Department of Oncology and Leuven Cancer Institute (LKI), KU Leuven, Leuven, Belgium*

ÁNGELES FERNÁNDEZ-RECAMALES • *Department of Chemistry, Faculty of Experimental Sciences, University of Huelva, Huelva, Spain; International Campus of Excellence CeiA3, University of Huelva, Huelva, Spain*

IVO FEUSSNER • *Department of Plant Biochemistry, Albrecht-von-Haller-Institute for Plant Sciences, University of Goettingen, Goettingen, Germany; Service Unit for Metabolomics and Lipidomics, Goettingen Center for Molecular Biosciences (GZMB), University of Goettingen, Goettingen, Germany; Department of Plant Biochemistry, Goettingen Center for Molecular Biosciences (GZMB), University of Goettingen, Goettingen, Germany*

KIRSTIN FEUSSNER • *Department of Plant Biochemistry, Albrecht-von-Haller-Institute for Plant Sciences, University of Goettingen, Goettingen, Germany; Service Unit for Metabolomics and Lipidomics, Goettingen Center for Molecular Biosciences (GZMB), University of Goettingen, Goettingen, Germany*

XIAOYUN FU • *Bloodworks Northwest Research Institute, Seattle, WA, USA; Division of Hematology, Department of Internal Medicine, University of Washington School of Medicine, Seattle, WA, USA*

ANTONIA GARCÍA • *Facultad de Farmacia, Centro de Metabolómica y Bioanálisis (CEMBIO), Universidad CEU San Pablo, Madrid, Spain*

SARAH GEHRKE • *Department of Biochemistry and Molecular Genetics, University of Colorado Denver, Aurora, CO, USA*

SABU GEORGE • *Elucidata, Inc., Cambridge, MA, USA*

FEDERICA GEVI • *Department of Science and Technology for Agriculture, Forestry, Nature and Energy (DAFNE), University of Tuscia, Viterbo, Italy*

BART GHESQUIÈRE • *Metabolomics Expertise Center, VIB Center for Cancer Biology (CCB), VIB, Leuven, Belgium; Department of Oncology, Metabolomics Expertise Center, KU Leuven, Leuven, Belgium*

DEBASHIS GHOSH • *Department of Biostatistics and Informatics, Colorado School of Public Health, University of Colorado Anschutz Medical Campus, Aurora, CO, USA*

MIGUEL A. GIJÓN • *Department of Pharmacology, University of Colorado Denver, Aurora, CO, USA*

Joanna Godzien • *Facultad de Farmacia, Centro de Metabolómica y Bioanálisis (CEMBIO), Universidad CEU San Pablo, Madrid, Spain*
Álvaro González-Domínguez • *Department of Pediatrics, Hospital Universitario Puerta del Mar, Cádiz, Spain; Institute of Research and Innovation in Biomedical Sciences of the Province of Cádiz (INiBICA), Cádiz, Spain*
Raúl González-Domínguez • *Department of Chemistry, Faculty of Experimental Sciences, University of Huelva, Huelva, Spain; International Campus of Excellence CeiA3, University of Huelva, Huelva, Spain*
Rishabh Gupta • *Elucidata, Inc., Cambridge, MA, USA*
Kirk C. Hansen • *Department of Biochemistry and Molecular Genetics, University of Colorado Denver, Aurora, CO, USA*
Kathleen A. Harrison • *Division of Endocrinology, Diabetes, and Metabolism, School of Medicine, University of Colorado Anschutz Medical Campus, Denver, CO, USA*
Sarah Hayton • *Medical, Molecular and Forensic Sciences, Murdoch University, Murdoch, WA, Australia; Separation Science and Metabolomics Laboratory, Murdoch University, Murdoch, WA, Australia*
Pol Herrero • *Technological Joint Unit of Omic Sciences, EURECAT-Technological Center of Catalonia, Universitat Rovira i Virgili, Reus, Spain*
Sean Jacobson • *Center for Genes, Environment, and Health, National Jewish Health, Denver, CO, USA*
Neema Jamshidi • *University of California, Los Angeles, Los Angeles, CA, USA*
Rachel C. Janssen • *Department of Pediatrics, Section of Neonatology, University of Colorado Anschutz Medical Campus, Aurora, CO, USA*
Abhishek Jha • *Elucidata, Inc., Cambridge, MA, USA*
Joanna Kalucka • *Laboratory of Angiogenesis and Vascular Metabolism, VIB Center for Cancer Biology (CCB), VIB, Leuven, Belgium; Laboratory of Angiogenesis and Vascular Metabolism, Department of Oncology and Leuven Cancer Institute (LKI), KU Leuven, Leuven, Belgium*
Katerina Kechris • *Department of Biostatistics and Informatics, Colorado School of Public Health, University of Colorado Anschutz Medical Campus, Aurora, CO, USA*
Sahil Kumar • *Elucidata, Inc., Cambridge, MA, USA*
Evagelia C. Laiakis • *Department of Oncology, Lombardi Comprehensive Cancer Center, Georgetown University, Washington, DC, USA; Department of Biochemistry and Molecular and Cellular Biology, Georgetown University, Washington, DC, USA*
Alfonso María Lechuga-Sancho • *Department of Pediatrics, Hospital Universitario Puerta del Mar, Cádiz, Spain; Institute of Research and Innovation in Biomedical Sciences of the Province of Cádiz (INiBICA), Cádiz, Spain; Department of Mother and Child Health and Radiology, Faculty of Medicine, University of Cádiz, Cádiz, Spain*
Hongde Li • *Department of Biology, Indiana University, Bloomington, IN, USA*
Xuedong Liu • *Department of Chemistry and Biochemistry, University of Colorado, Boulder, CO, USA*
Ángeles López-Gonzálvez • *Facultad de Farmacia, Centro de Metabolómica y Bioanálisis (CEMBIO), Universidad CEU San Pablo, Madrid, Spain*
Garth L. Maker • *Medical, Molecular and Forensic Sciences, Murdoch University, Murdoch, WA, Australia; Separation Science and Metabolomics Laboratory, Murdoch University, Murdoch, WA, Australia*
David P. Marciano • *Department of Genetics, Stanford University School of Medicine, Stanford, CA, USA*

ROSA MARÍA MATEOS • *Department of Pediatrics, Hospital Universitario Puerta del Mar, Cádiz, Spain; Institute of Research and Innovation in Biomedical Sciences of the Province of Cádiz (INiBICA), Cádiz, Spain*

CHRISTIAN M. METALLO • *Department of Bioengineering, University of California San Diego, La Jolla, CA, USA; Moores Cancer Center, University of California San Diego, La Jolla, CA, USA; Diabetes and Endocrinology Research Center, University of California San Diego, La Jolla, CA, USA; Institute of Engineering in Medicine, University of California San Diego, La Jolla, CA, USA*

IÑIGO SAN-MILLÁN • *Division of Sports Medicine, University of Colorado School of Medicine, Aurora, CO, USA*

REIHANEH MOSTOLIZADEH • *Center for Bioinformatics Tübingen (ZBIT), University of Tübingen, Tübingen, Germany; Department for Computer Science, University of Tübingen, Tübingen, Germany; German Center for Infection Research (DZIF), Tübingen, Germany*

ROBERT C. MURPHY • *Department of Pharmacology, University of Colorado Denver, Aurora, CO, USA*

TRAVIS NEMKOV • *Department of Biochemistry and Molecular Genetics, University of Colorado Denver, Aurora, CO, USA*

WILLIAM M. OLD • *Department of Molecular, Cellular and Developmental Biology, University of Colorado, Boulder, CO, USA; Linda Crnic Institute for Down Syndrome, University of Colorado School of Medicine, Aurora, CO, USA*

GIUSEPPE PAGLIA • *Institute for Biomedicine, EURAC Research, Bolzano, Italy*

SWETABH PATHAK • *Elucidata, Inc., Cambridge, MA, USA*

HARRISON PIELKE-LOMBARDO • *Computational Bioscience Program, School of Medicine, University of Colorado Anschutz Medical Campus, Aurora, CO, USA*

SURBHI PODDAR • *Elucidata, Inc., Cambridge, MA, USA*

JESSICA E. PRENNI • *Proteomics and Metabolomics Facility, Colorado State University, Fort Collins, CO, USA*

MARIA ROSA RAS • *Technological Joint Unit of Omic Sciences, EURECAT-Technological Center of Catalonia, Universitat Rovira i Virgili, Reus, Spain*

DOMINIK REINHOLD • *PPD, Wilmington, NC, USA*

JULIE A. REISZ • *Department of Biochemistry and Molecular Genetics, University of Colorado Denver, Aurora, CO, USA*

SARA RINALDUCCI • *Department of Ecological and Biological Sciences (DEB), University of Tuscia, Viterbo, Italy*

MIGUEL ÁNGEL RODRIGUEZ • *Technological Joint Unit of Omic Sciences, EURECAT-Technological Center of Catalonia, Universitat Rovira i Virgili, Reus, Spain*

ANA SAYAGO • *Department of Chemistry, Faculty of Experimental Sciences, University of Huelva, Huelva, Spain; International Campus of Excellence CeiA3, University of Huelva, Huelva, Spain*

MÓNICA SCHWARZ • *Institute of Research and Innovation in Biomedical Sciences of the Province of Cádiz (INiBICA), Cádiz, Spain; "Salus Infirmorum" Faculty of Nursing, University of Cádiz, Cádiz, Spain; Instituto de Investigación Vitivinícola y Agroalimentario (IVAGRO), University of Cádiz, Puerto Real, Spain*

CARMEN SEGUNDO • *Institute of Research and Innovation in Biomedical Sciences of the Province of Cádiz (INiBICA), Cádiz, Spain; "Salus Infirmorum" Faculty of Nursing, University of Cádiz, Cádiz, Spain*

RAGHAV SEHGAL • *Elucidata, Inc., Cambridge, MA, USA*
GRETCHEN L. SEIM • *Morgridge Institute for Research, Madison, WI, USA; Department of Nutritional Sciences, University of Wisconsin-Madison, Madison, WI, USA*
NATALIE J. SERKOVA • *Department of Radiology, School of Medicine, University of Colorado Denver, Aurora, CO, USA*
AMY M. SHEFLIN • *Proteomics and Metabolomics Facility, Colorado State University, Fort Collins, CO, USA*
MICHAEL P. SNYDER • *Department of Genetics, Stanford University School of Medicine, Stanford, CA, USA*
JENNA STEINER • *Department of Radiology, School of Medicine, University of Colorado Denver, Aurora, CO, USA*
JASON M. TENNESSEN • *Department of Biology, Indiana University, Bloomington, IN, USA*
TIFFANY H. THOMAS • *Department of Pathology and Cell Biology, Columbia University Medical Center, New York, NY, USA*
ROBERT D. TRENGOVE • *Separation Science and Metabolomics Laboratory, Murdoch University, Murdoch, WA, Australia*
SARA VIOLANTE • *Donald B. and Catherine C. Cancer Metabolism Center, Memorial Sloan Kettering Cancer Center, New York, NY, USA*
YI WANG • *Bloodworks Northwest Research Institute, Seattle, WA, USA*
LINXING YAO • *Proteomics and Metabolomics Facility, Colorado State University, Fort Collins, CO, USA*
SIMONA ZARINI • *Department of Pharmacology, University of Colorado Denver, Aurora, CO, USA*
CONNIE ZHENG • *Department of Biochemistry and Molecular Genetics, University of Colorado Denver, Aurora, CO, USA*
JAMES C. ZIMRING • *Bloodworks Northwest Research Institute, Seattle, WA, USA; Division of Hematology, Department of Internal Medicine, University of Washington School of Medicine, Seattle, WA, USA; Department of Laboratory Medicine, University of Washington School of Medicine, Seattle, WA, USA*
LELLO ZOLLA • *Department of Science and Technology for Agriculture, Forestry, Nature and Energy (DAFNE), University of Tuscia, Viterbo, Italy*

Part I

Methods

Chapter 1

Sample Preparation and Reporting Standards for Metabolomics of Adherent Mammalian Cells

Sarah Hayton, Robert D. Trengove, and Garth L. Maker

Abstract

Metabolomics is an analytical technique that investigates the small molecules present within a biological system. Metabolomics of cultured cells allows profiling of the metabolic chemicals involved in a cell type-specific system and the response of that metabolome to external challenges, such as change in environment or exposure to drugs or toxins. The numerous benefits of in vitro metabolomics include a much greater control of external variables and reduced ethical concerns. There is potential for metabolomics of mammalian cells to uncover new information on mechanisms of action for drugs or toxins or to provide a more sensitive, human-specific early risk assessment in drug development or toxicology investigations. One way to achieve stronger biological outcomes from metabolomic data is via the use of these mammalian cultured cell models, particularly in a high-throughput context. With the sensitivity and quantity of data that metabolomics is able to provide, it is important to ensure that the sampling techniques have minimal interference when it comes to interpretation of any observed shifts in the metabolite profile. Here we describe a sampling procedure designed to ensure that the effects seen in metabolomic analyses are explained fully by the experimental factor and not other routine culture-specific activities.

Key words Cell culture, Mammalian cells, Adherent cells, Experimental design, Sample preparation, Quenching, Metabolite extraction, Extracellular, Intracellular, Metabolomics

1 Introduction

Metabolomics studies frequently state that the metabolome is a closer reflection of the phenotype of an organism, tissue, or cell than other "omics analyses," such as proteomics, transcriptomics, and genomics [1–3]. The small molecules that are shuffled around the vast network of metabolic pathways in a biological system are referred to as metabolites, with the "metabolome" made up of the many thousands of different metabolites and their relative levels of abundance. The goal of metabolomic profiling is to measure changes in the metabolome of a given system in response to a challenge to normal cellular homeostasis, whether physical, chemical, environmental, or other external stressor.

Angelo D'Alessandro (ed.), *High-Throughput Metabolomics: Methods and Protocols*, Methods in Molecular Biology, vol. 1978, https://doi.org/10.1007/978-1-4939-9236-2_1, © Springer Science+Business Media, LLC, part of Springer Nature 2019

Metabolomics has recently become an attractive application for untargeted, high-throughput screening analyses [4]. However, due to the data-rich, hypothesis-generating outcomes of untargeted metabolomics, it is understandable that such an approach might not be attractive to some investigators, particularly if a limited number of clinical or animal samples are available. In particular, handling the vast amounts of data can complicate meaningful biological interpretation of the data. This can be partly addressed by the use of cultured mammalian cells, as opposed to animal-based samples, to more easily accommodate re-visiting of the sample set if novel or previously unknown metabolites are highlighted by an untargeted study. The number of controls and replicates can be easily manipulated in design of cell culture-based experiments, benefiting the development, validation, and standardization of untargeted, high-throughput metabolomic studies. The importance of experimental design and standardized reporting requirements for these studies of cultured mammalian cells has been previously reviewed across a large number of published protocols [5, 6].

The focus of this protocol is to give the investigator a sound, high-throughput procedure to follow for handling and preparing cultured cell samples preceding instrumental analysis for metabolomics, which will have minimal possible interference on the detectable metabolome. It is designed so that any of the multiple platforms currently used for metabolomics can be applied to the prepared samples thereafter (e.g., gas chromatography-mass spectrometry, liquid chromatography-mass spectrometry). Here we provide a standardized protocol for untargeted metabolomic analysis applied to any cultured mammalian cell line, highlighting the importance of adequate reporting of culture conditions to allow for meaningful biological interpretation of metabolomic data and comparison of results across multiple studies.

2 Materials

2.1 Cell Lines

Cell lines should be sourced from reputable sources such as the European Collection of Cell Cultures (ECACC) or the American Type Culture Collection (ATCC). Examples of cell lines which have been studied using this protocol:

- B50 rat neuroblastoma (ECACC 85042302).
- HepG2 human hepatocarcinoma (ECACC 85011430; ATCC HB-8065).
- SH-SY5Y human neuroblastoma (ECACC 94030304; ATCC CRL-2266).

2.2 Growth Conditions

All cell culture media should be sterile-filtered (whether pre- or post-purchase) and medium and supplements within expiry periods. It is highly recommended to follow ECACC or ATCC recommendations of growth conditions for the specific cell line. The above cell lines were cultured at 37 °C and 5% CO_2 with the following media and supplements:

- Dulbecco's Modified Eagle's Medium (DMEM; 4.5 g/L glucose).
- Ham's F12 nutrient mixture.
- L-glutamine (2 mM, 1% v/v).
- Penicillin and streptomycin (combined 10^4 U/mL, 1% v/v).
- Fetal calf serum (heat-treated at 56 °C for 60 min; 5 or 10% v/v).

2.3 Cell Handling

All solvents used for cell culture treatments, PBS washes, and sample preparation (e.g., water, methanol) should be of the highest purity possible, preferably LC-MS grade. The use of lower-quality solvents may introduce impurities that can interfere with mass spectrometry data.

Suggested composition of solutions for cell handling:

- 0.25% trypsin-EDTA solution (2.5 g/L porcine trypsin and 0.2 g/L EDTA).
- Phosphate-buffered saline (PBS; 0.01 M phosphate buffer, 0.0027 M potassium chloride, and 0.137 M sodium chloride, pH 7.4).
- 0.4% w/v trypan blue dye in phosphate-buffered saline solution. A hemocytometer and microscope or an automated cell counter can be used for cell counting.

3 Methods

The entire workflow is summarized in Fig. 1.

3.1 Cell Culture Sample Setup

1. Seed cells at the required density (number of cells) in the required number of culture plates (*see* **Note 1**). The volume of medium should be determined based on the size of the culture plate chosen.
2. Cells should be left to grow and adhere for 24–48 h before treatment, depending on the specific cell type. Cells should not be allowed to become over-confluent, as the metabolome will be affected by the depletion of supplements in the medium or potentially by differentiation of the cells (*see* **Note 2**).

Reporting requirements:

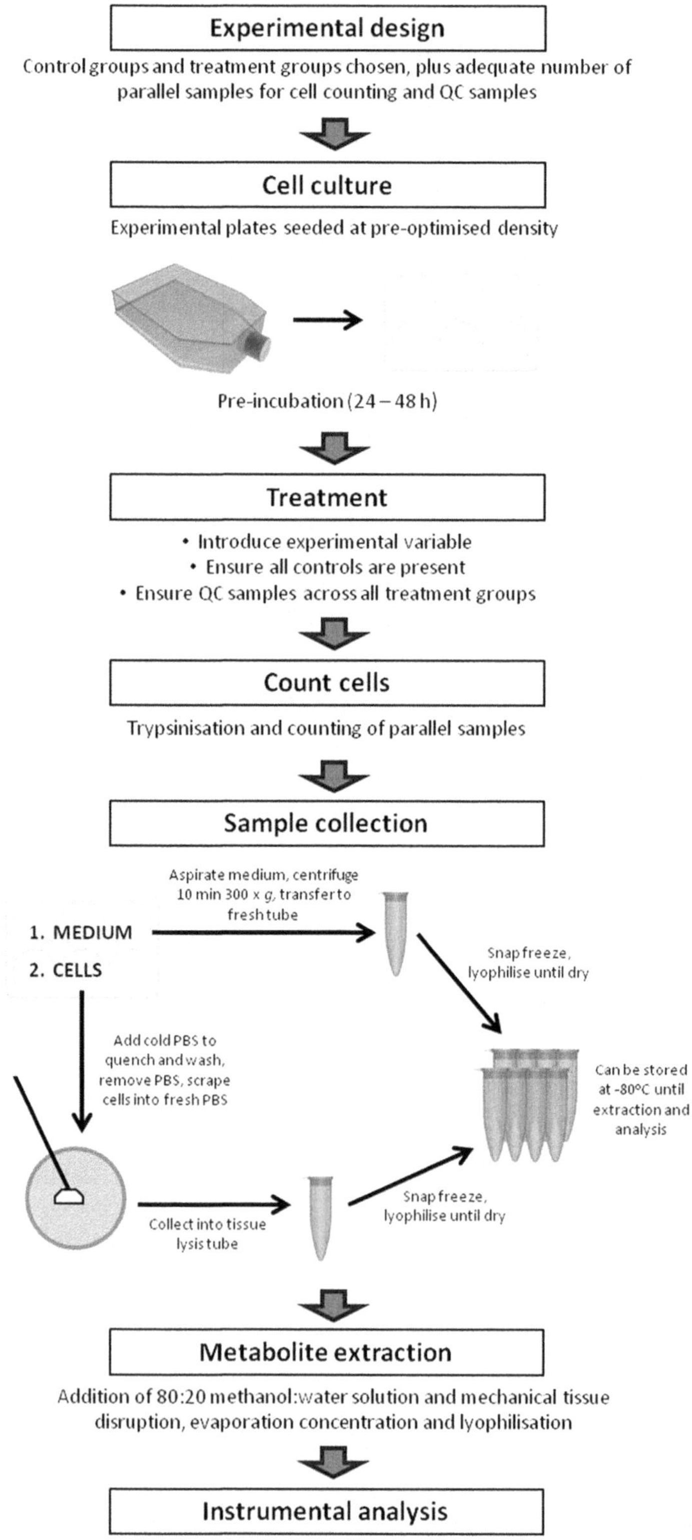

Fig. 1 Workflow schematic of sample preparation procedure for metabolomic analyses of adherent mammalian cells

Name and source of cell line, including results from genetic validation and mycoplasma testing.

Generation (passage) number of cells used to set up experimental samples (*see* **Note 3**).

Size and type of culture plates used and the seeding density as cell number.

Specifics of medium and supplements used, including type, supplier, concentration, and serum percentage.

Passaging procedures used during general cell culture.

Inoculation procedures for any treatments/exposures (nature of exposure, exposure time, doses).

Environmental conditions of cell incubation.

Setup of parallel samples for cell counting and quality control samples (*see* **Note 4**).

3.2 Counting of Cells in Duplicate Samples

1. Immediately before harvesting for metabolomics, trypsinize and count one representative (duplicate) sample per replicate for the number of viable cells, via addition of 0.4% *w/v* trypan blue dye (*see* **Note 5**). Cells can be counted manually or using an automated cell counter, which should be validated before use.
2. Immediately following cell counting, remove all culture plates from incubation, and place on ice. Care should be taken not to dislodge cells from the plate surface while transporting.

3.3 Sample Collection (Medium/Extracellular)

1. Remove medium from all culture plates into centrifuge tubes, and centrifuge at low rcf (300 × *g*) for 10 min to pellet any cells that may be present.
2. Transfer medium to fresh tubes, and store on ice before snap-freezing, lyophilization, and storage at −80 °C, prior to extraction and analysis of the extracellular metabolome. Freeze-drying unit used should be capable of reaching <−80 °C temperature and <0.002 mBar vacuum.

3.4 Quenching

1. Wash remaining adhered cells with cold phosphate-buffered saline (PBS; 4 °C) to remove any trace medium present. The volume used should be adequate to cover the cells in the culture plate on a level surface. The cold PBS also acts to quench the metabolism of the cells (*see* **Note 6**) and so should be added to all culture plates at the same time.
2. Remove and discard this wash PBS.

3.5 Sample Collection (Cells/Intracellular)

1. Add an equal volume of cold PBS to all plates for the collection of cells. Remove cells from the plate surface by scraping into the PBS, and collect into tubes appropriate for tissue lysis (normally screw cap, o-ring sealed). Store tubes on ice until snap-freezing and lyophilization.
2. Snap-freeze both cell and medium samples in liquid nitrogen, and lyophilize (freeze-dry) to minimize degradation of metabolites before extraction and analysis can take place.

3.6 Metabolite Extraction

1. Prepare metabolite extraction solution by the addition of required concentration of internal standard compounds (e.g., stable isotope-labeled chemicals) in an 80:20 methanol/water mixture (*see* **Note 7**).
2. Remove freeze-dried samples from −80 °C storage, and add the same volume of extraction solution to all samples. This required volume is determined by the amount of material collected and thus the amount of dried tissue present and the size of the storage tube. To prevent loss of sample during the concentration process (**step 6** below), no more than two-thirds of the maximum volume capacity of the tube should be used, e.g., a maximum of 1 mL total volume of extracted sample in a 1.5 mL microcentrifuge tube.
3. Extract metabolites from cell samples by vigorous sonication or tissue disruption to lyse any remaining intact cell membranes and release intracellular metabolites. Extracellular medium can be extracted by simple vortex mixing.
4. Centrifuge extracts at high rcf (e.g., $1.5 \times 10^4 \times g$) for 10 min to pellet the sample debris. Collect the same volume of supernatant into a fresh tube for every sample.
5. The remaining pellet can undergo a second extraction step (repeat of **steps 2–4**) to maximize metabolite recovery from the biological material. Following centrifugation of the repeat extraction, the supernatant can be added to the first collected extract.
6. Evaporate methanol in the extraction solution via vacuum concentration, and reconstitute the sample in the same volume of water (*see* **Note 8**).
7. Snap-freeze samples once again, and lyophilize by freeze-drying. Once completely dry, the samples can be stored at −80 °C until metabolomic analysis.
8. As water has been removed from the extracts, the samples can be reconstituted as desired, according to the chosen analytical platform. For example, for GC-based analysis the samples would be chemically derivatized, while for LC-based platforms,

the samples can be reconstituted in the required mobile phase solvent(s).

Reporting requirements:

Composition of extraction solution.

Description of extraction procedure (mechanical disruption, collection details, etc.)

If known, expected recovery rate and stability of extracted metabolites (*see* **Note 9**).

4 Notes

1. The quantity of cells that needs to be grown per sample extracted for metabolomic studies should be optimized specifically for the intended analytical instrument's detection capabilities and will vary depending on the specific cell being cultured. The exact quantity may depend on whether the study is an untargeted screen for metabolites or if a certain class or selection of metabolites is being targeted. For untargeted analysis, this is generally assessed by observing the total ion chromatogram (TIC) of a single sample, which is the combined signal of everything measured in that sample. An optimal number of cells per sample are chosen based on the maximum number of detectable metabolites that are confidently resolved from the background signal while also minimizing the occurrence of any high-abundance peaks that may cause overloading of the instrument detector, potentially obscuring any less abundant metabolites with similar chromatographic retention times. Once an optimal cell number for seeding density is obtained, this should remain consistent for the duration of the experiment, to minimize data manipulation in later interpretation stages.
2. If a change of medium is required during the incubation period after cell seeding, care should be taken not to disrupt the course of cell growth, as much as is practicable. All control samples should also have medium changed.
3. If multiple passages are required for repeated samples of the same experimental setup, then the range of passage generations used should be stated. Depending on the design for the specific metabolomic experiment, as well as the characteristics of the cells used, it may not be necessary to conduct experiments across multiple passages, especially for high-throughput cell metabolomics applications. This should be a part of the consideration during the researcher's experimental design [5].

4. The use of quality control samples within a metabolomic study, particularly for large sample sets, is crucially important for accurate correction and statistical interpretation of the data post-analysis [7, 8]. For an untargeted study, the QC samples are typically obtained from a pooled sample of all treatment groups, set up in parallel to the experimental samples, and collected simultaneously. If sample volume is sufficient, QC samples can be obtained by pooling an aliquot of the experimental samples. QC samples should be placed throughout the sequence during instrumental analysis, meaning that any drift in metabolite detection levels can be detected and modelled. This allows any variation observed from drift phenomena to be removed from statistical analysis of the data. The appropriate number of QC samples should be determined in the experimental design stages, to ensure that adequate sample material is prepared.
5. In cell culture experiments, especially when testing the response of a system to a physical challenge, the amount of material available for metabolomic analysis will likely differ between sample groups. It is therefore important to normalize metabolomic data post-analysis to remove this sample variation and allow for meaningful biological interpretation. A number of differing techniques have been used, including protein or DNA concentration, tissue weight, or cell number [9–12]. The most common method of normalization of sample variation is to the cell number upon sample harvesting, which should be considered as the standard approach when possible. A separate, parallel sample should be set up along with the sample for metabolite harvest and extraction, specifically for cell counting. This is necessary due to the destructive nature of harvesting for metabolomic analysis, as well as the leakage of metabolites encountered during trypsinization of cells for counting [13, 14].
6. An ideal extraction procedure for metabolomics of cultured cells should immediately quench metabolism in order to extract and collect all metabolites present at the end point of the experiment [15, 16]. PBS is a buffered, isotonic solution, so it causes less membrane leakage, thereby reducing loss of intracellular metabolites compared to other solvents that have been used in quenching, such as methanol.
7. The composition of extraction solution can be altered depending on the chemical properties of the specific metabolites targeted for analysis, to maximize recovery. The suggested extraction solution, methanol and water in an 80:20 ratio, gives a single-phase extraction supernatant and allows for the whole extract to be collected. Also common is the use of a dual-phase (polar and nonpolar) extraction using methanol, chloroform, and water (or acetonitrile). The ratio of these solvents can vary but is most commonly used at 1:1 methanol and chloroform,

with a smaller proportion of water or acetonitrile. This results in a separation of polar and nonpolar extraction phases, which can be collected separately. If multiple phase extractions and collections are carried out, this should be included in the reporting.

8. Extracts need to be reconstituted in a majority aqueous solution to allow for successful lyophilization, which itself enhances the conservation of metabolites for any extended period of deep-freeze storage.
9. If known, the expected recovery rate and stability of the extracted metabolites, dependent on the class of compound, should be included in basic reported information. Any such information that is known from previous metabolomic analyses using a particular extraction or instrumental protocol will benefit the researcher during the post-instrumental stages and ensure that accurate biological interpretation of the data is represented. It may benefit other researchers by allowing the comparison of multiple metabolomic studies.

References

1. Idle JR, Gonzalez FJ (2007) Metabolomics. Cell Metab 6:348–351
2. Blow N (2008) Biochemistry's new look. Nature 455:697–700
3. Fiehn O (2002) Metabolomics - the link between genotypes and phenotypes. Plant Mol Biol 48(1–2):155–171
4. Hayton S, Maker GL, Mullaney I, Trengove RD (2017) Untargeted metabolomics of neuronal cell culture: a model system for the toxicity testing of insecticide chemical exposure. J Appl Toxicol 37:1481–1492. https://doi.org/10.1002/jat.3498
5. Hayton S, Maker GL, Mullaney I, Trengove RD (2017) Experimental design and reporting standards for metabolomics of mammalian cell lines. Cell Mol Life Sci 74(24):4421–4441. https://doi.org/10.1007/s00018-017-2582-1
6. León Z, García-Cañaveras JC, Donato MT, Lahoz A (2013) Mammalian cell metabolomics: experimental design and sample preparation. Electrophoresis 34(19):2762–2775
7. Dunn WB, Broadhurst D, Begley P, Zelena E, Francis-McIntyre S, Anderson N, Brown M, Knowles JD, Halsall A, Haselden JN, Nicholls AW, Wilson ID, Kell DB, Goodacre R, Consortium THSMH (2011) Procedures for large-scale metabolic profiling of serum and plasma using gas chromatography and liquid chromatography coupled to mass spectrometry. Nat Protoc 6(7):1060–1083
8. Goodacre R, Broadhurst D, Smilde AK, Kristal BS, Baker JD, Beger R, Bessant C, Connor S, Capuani G, Craig A, Ebbels T, Kell DB, Manetti C, Newton J, Paternostro G, Somorjai R, Sjöström M, Trygg J, Wulfert F (2007) Proposed minimum reporting standards for data analysis in metabolomics. Metabolomics 3(3):231–241
9. Muschet C, Möller G, Prehn C, Hrabē de Angelis M, Adamski J, Tokarz J (2016) Removing the bottlenecks of cell culture metabolomics: fast normalization procedure, correlation of metabolites to cell number, and impact of the cell harvesting method. Metabolomics 12(10):151. https://doi.org/10.1007/s11306-016-1104-8
10. Hutschenreuther A, Kiontke A, Birkenmeier G, Birkemeyer C (2012) Comparison of extraction conditions and normalization approaches for cellular metabolomics of adherent growing cells with GC-MS. Anal Methods 4:1953–1963
11. Cao B, Aa J, Wang G, Wu X, Liu L, Li M, Shi J, Wang X, Zhao C, Zheng T, Guo S, Duan J (2011) GC-TOFMS analysis of metabolites in adherent MDCK cells and a novel strategy for identifying intracellular metabolic markers for use as cell amount indicators in data normalization. Anal Bioanal Chem 400(9):2983–2993
12. Silva LP, Lorenzi PL, Purwaha P, Yong V, Hawke DH, Weinstein JN (2013) Measurement of DNA concentration as a normalization strat-

egy for metabolomic data from adherent cell lines. Anal Chem 85(20):9536–9542

13. Dettmer K, Nürnberger N, Kaspar H, Gruber MA, Almstetter MF, Oefner PJ (2011) Metabolite extraction from adherently growing mammalian cells for metabolomics studies: optimization of harvesting and extraction protocols. Anal Bioanal Chem 399(3):1127–1139
14. Garcia-Canaveras JC, Lopez S, Castell JV, Donato MT, Laboz A (2016) Extending metabolome coverage for untargeted metabolite profiling of adherent cultured hepatic cells. Anal Bioanal Chem 408:1217–1230
15. Čuperlović-Culf M, Barnett DA, Culf AS, Chute I (2010) Cell culture metabolomics: applications and future directions. Drug Discov Today 15(15/16):610–621
16. Álvarez-Sánchez B, Priego-Capote F, Luque de Castro MD (2010) Metabolomics analysis II: preparation of biological samples prior to detection. Trends Analyt Chem 29(2):120–127

Chapter 2

High-Throughput Metabolomics: Isocratic and Gradient Mass Spectrometry-Based Methods

Travis Nemkov, Julie A. Reisz, Sarah Gehrke, Kirk C. Hansen, and Angelo D'Alessandro

Abstract

Metabolomics has emerged in the past decade as a highly attractive and impactful technique for phenotype-level profiling in diverse biological applications. Most recently, the dual developments of high-throughput analytical techniques along with dramatically increased sensitivity of high-resolution mass spectrometers have enabled the routine analysis of hundreds of unique samples per day. We have previously reported a robust 3 min isocratic metabolomics platform for the quantification of amino acids and the key pathways of central carbon and nitrogen metabolism. Building on this work, we describe here a 5 min reverse phase gradient followed by global, untargeted profiling of the hydrophilic metabolome. In addition to observing those metabolites measured in the 3 min run, the use of the longer gradient run here also allows for coverage of less polar compounds such as fatty acids and acylcarnitines, both key players in mitochondrial and lipid metabolism, without a significant sacrifice in throughput.

Key words Untargeted metabolomics, Mass spectrometry, Isocratic, Gradient, High-throughput

1 Introduction

In the last two decades, the cost of sequencing the human genome [1] has dropped from ~$300 million US dollars to ~$1000–1500 by use of whole-exome sequencing (www.genome.gov/sequencingcosts/). Technological advances have made the genome amenable to large-scale studies which foster personalized medicine initiatives around the globe. The introduction of high-throughput metabolomics technologies promises to achieve a similar goal.

Similarly to high-throughput genome sequencing tools, it is easy to anticipate how high-throughput metabolomics applications will soon become instrumental for the characterization of clinically relevant metabolic biomarkers requiring a final validation phase in large prospective cohorts, such as in the case of large-scale personalized medicine initiative studies. For example, high-throughput metabolomics approaches can be theoretically used to quantify

Angelo D'Alessandro (ed.), *High-Throughput Metabolomics: Methods and Protocols*, Methods in Molecular Biology, vol. 1978, https://doi.org/10.1007/978-1-4939-9236-2_2, © Springer Science+Business Media, LLC, part of Springer Nature 2019

metabolic markers of disease such as homocysteine levels in homocystinuria, an inborn error of metabolism [2], glucose levels in diabetes [3], and lactate [4] and succinate [5] levels as readouts for base deficit and tissue hypoxia in trauma. Historically, targeted metabolomics approaches that have been translated into clinical practice are based on multiple reaction monitoring-mass spectrometry (MRM-MS) [6], a highly sensitive method that requires preselection of molecular targets of interest, therefore affording detection and quantitation of only a handful of small molecules at a time. The introduction of metabolomics strategies based on high-resolution quadrupole-time of flight (QTOF) or quadrupole-Orbitrap-based instruments has enabled investigators to perform untargeted "discovery mode" metabolomics analyses while simultaneously quantifying metabolites of interest against external calibration curves or internal stable isotope-labeled standards.

Over the past 5 years, several approaches have been proposed to achieve high-throughput metabolomics workflows, either based on flow-injection mass spectrometry [7], ultra-high-pressure liquid chromatography (UHPLC) interfaced to MS [8, 9], or matrix-assisted laser desorption/ionization imaging mass spectrometry (IMS) [10]. These approaches have made the detection and quantification of over 100,000 features and thousands of named metabolites possible in less than 5 min through direct injection or isocratic/gradient-based chromatography or using in situ imaging approaches. Further advancements in terms of throughput and sensitivity are anticipated upon the recent introduction of microfluidic capillary electrophoresis coupled to MS for metabolomics applications [11, 12]. Optimization of rapid extraction protocols [13], automation [14], and, predictably, low-cost robotization of sample extraction protocols [15] would enable affordable processing of hundreds to a thousand samples per day in a cost-effective manner.

Recently, we described a high-throughput isocratic chromatography-based method for rapid quantitation of hydrophilic compounds from the central carbon and nitrogen pathways [8], including amino acids [16]. Below we describe an evolution of this method, where mobile phases are coupled with agents that improve chromatographic separation and ease anionic ionization of high-energy phosphate compounds (e.g., nucleoside triphosphates) to facilitate detection in negative ion mode. In addition, we describe a high-throughput gradient-based method that in part addresses the main limitation of the isocratic 3 min method, by making some classes of hydrophobic compounds (including acyl-conjugated carnitines, free fatty acids, and bile acids) amenable to detection and quantification. This method affords the rapid quantitation of up to 600,000 features from a clinically relevant sample (e.g., blood extracts), resulting in hundreds to thousands of named compounds (level of confidence based on high-resolution intact mass, transition fingerprints, and comparison of retention times against our in-house standard library). Given the rapidity of these approaches, >450 samples can be processed per day, making it realistic to anticipate the potential translation of such technology into the field of clinical biochemistry. These approaches are also compatible with the use of stable isotope-labeled internal standards, which allow for the correction of matrix-dependent ion suppression effects, as well as to monitor the stability of chromatographic peak shape and retention times and the quality of signal intensity at the MS and MS/MS level [8]. The use of stable isotope-labeled standards also streamlines quantitation of compounds of interest, thus normalizing matrix or batch effects as well as instrument-dependent variability across platforms, provided the concentrations of the labeled standards and the endogenous metabolite from the tested matrix are within the linearity range of the MS instrument. Stable isotope-labeled standards can be leveraged to quantify metabolic fluxes by determining ratios of spiked-in isotopologues against those deriving from alternative metabolism of stable isotope tracers in fluxomics experiments, in like fashion to what we described recently for lactate [17] and alanine [18] synthesis downstream to [1,2,3-$^{13}C_3$]glucose catabolism via glycolysis or the pentose phosphate pathway in red blood cells.

2 Materials

Prepare all solutions using highest-quality solvents and reagents. We recommend LC-MS grade solvents, such as Optima™ (Fisher). Solutions can be stored at room temperature unless otherwise stated.

2.1 General

1. Benchtop centrifuge capable of 18,213 × *g* with refrigeration to 4 °C.
2. Refrigeration space (4 °C) with power outlet(s) to accommodate vortex(es), often a cold room or chromatography refrigerator.
3. Vortex with foam microcentrifuge tube insert.

2.2 Mobile Phases for Ultra-High-Pressure Liquid Chromatography

1. Water.
2. Water containing 0.1% v/v formic acid.
3. Acetonitrile.
4. Acetonitrile containing 0.1% v/v formic acid.
5. Ammonium acetate.
6. Sterile disposable vacuum filtration system (0.22 μm membrane).

2.3 Extraction Solution

1. Water.
2. Acetonitrile.
3. Methanol.
4. 1 L glass bottle with cap, preferably new.
5. 1 L glass graduated cylinder, preferably new.

2.4 Tissue Sample Preparation

1. Bead beater (Next Advance, Storm 24).
2. Eppendorf™ Safe-lock tubes, 1.5 mL (catalog no. 022363204).
3. Glass beads, 1.0 mm diameter (NextAdvance SKU GB10).

2.5 Mass Spectrometry Data Acquisition and Analysis

1. Ultra-high-pressure liquid chromatography (UHPLC) system.
2. Analytical column: Kinetex® 1.7 μm C18 100 Å, UHPLC column 150 × 2.1 mm (Phenomenex catalog no. 00F-4475-AN).
3. Guard column: SecurityGuard™ ULTRA cartridge-UHPLC C18 for 2.1 mm ID columns (Phenomenex catalog no. AJ08782) with SecurityGuard™ ULTRA holder (catalog no. AJ09000).
4. Electrospray ionization (ESI) or heated ESI (HESI) source.
5. Mass spectrometer capable of high-resolution untargeted analysis with MS^1 acquisition. Instrument should be interfaced with a UHPLC system.
6. Data visualization software. We recommend Maven (freely available at http://genomics-pubs.princeton.edu/mzroll/index.php) preceded by use of RawConverter (http://fields.scripps.edu/rawconv) to transform Thermo. RAW files into the .mzXML format needed for Maven.
7. Drift and/or batch correction software. We utilize MetaboDrift, an Excel macro add-in.

3 Methods

3.1 Preparation of Chromatography Mobile Phases

1. Mobile phases for positive ion mode may be used as supplied. Phase A is water with 0.1% formic acid, and phase B is acetonitrile with 0.1% formic acid.
2. Prepare 25 mL of a 2 M stock of ammonium acetate in water. Vacuum filter resulting solution to remove any insoluble material.
3. Prepare 1 L of mobile phase A for negative mode (5% acetonitrile, 95% water, 1 mM ammonium acetate) by first diluting 2 M ammonium acetate (0.5 mL) with water (950 mL) and then adding acetonitrile (50 mL). Ensure thorough mixing and complete solubility.

4. Prepare 1 L of mobile phase B for negative mode (95% acetonitrile, 5% water, 0.5 mM ammonium acetate) by first diluting 2 M ammonium acetate (2.5 mL) with water (to the 50 mL mark) and then adding acetonitrile to the 1 L mark. Ensure thorough mixing and complete solubility.

3.2 Preparation of Stable Isotope-Labeled Standard Stock Solutions

1. Prepare stock solutions to 1000× the desired final concentration in extraction solution, which should be within 1–2 orders of magnitude of the typical biological range in the analyzed sample matrix. Experimentally determined reference ranges can be found by searching for the compound of interest in Human Metabolome Database (www.hmdb.ca) and clicking on the "Concentrations" tab. Stock solutions of stable isotope-labeled standards desired for analysis are weighed and dissolved in a solvent according to the chemical nature of the compound and manufacturer instructions. Small volumes (<10 μL) of concentrated formic acid or concentrated aqueous ammonium formate may be added as needed to neutralize pH. Typical stock concentrations (in mM) are given in Table 1.
2. Store stock solutions at −20 °C until use.

3.3 Preparation of Metabolite Extraction Solution Containing Stable Isotope-Labeled Standards

1. Prepare 1 L of extraction solution by adding methanol (0.5 L), acetonitrile (0.3 L), and water (0.2 L) to a clean glass graduated cylinder.
2. Pour into a clean glass bottle, cap, swirl to mix, and cool to subzero temperatures. We recommend overnight storage at −20 °C or, when time-restricted, at least 30 min at −80 °C.
3. Unused extraction solution may be stored capped at −20 °C.
4. Remove stable isotope-labeled standard stock solutions from −20 °C, and thaw on ice. Ensure that the solution is completely thawed and that no precipitation has occurred due to freeze-thaw. If precipitation has occurred, vortex the solution until compound has redissolved.
5. Pipette stable isotope-labeled standard into pre-cooled extraction solution at desired concentration. We recommend maintaining the total volume of internal standards at or less than 5% of the total extraction solution volume to avoid diluting the organic components and thereby potentially altering extraction efficiency.

3.4 Metabolite Extraction from Biofluid Samples

1. Thaw samples on ice or in refrigerator. Pre-cool labeled microcentrifuge tubes by immersing in ice. We utilize 1.7 mL tubes; however tube size can vary to accommodate volume of extraction and compatibility with rotor.
2. Briefly vortex thawed samples to homogenize; then transfer a 20 μL aliquot by micropipette to labeled and cooled

Table 1
Typical stock concentrations of standards

Compound	Vendor	Product no.	Conc. (mM)
Adenosine	CIL	CLM-3678-0.1	1
Amino acid standard mix	CIL	MSK-A2–1.2	1
Acylcarnitine standard mix	CIL	NSK-B-1	200×[a]
Betaine (*N*,*N*,*N*-trimethylglycine) HCl	CDN isotopes	D-3352	0.1–2
Choline	CDN isotopes	D-2464	0.1–2
Citric acid	CIL	DLM-3487-PK	1
N,*N*-Dimethylglycine HCl	CDN isotopes	D-3509	0.1–2
Fumaric acid	CIL	CLM-4454-PK	1
Glucose	CIL	CLM-1396-1	10
Glutathione (reduced)	CIL	CNLM-6245-10	1
α-Ketoglutarate	CIL	CLM-2411-PK	1
Palmitic acid	CIL	CLM-409-0.5	1
Sodium lactate	CIL	CLM-1578-0.5	40
Sphingosine 1-phosphate	Avanti	860659P	1
Succinic acid	CIL	CLM-1571-PK	1
Trimethylamine *N*-oxide	CIL	DLM-4779-1	1
Tryptophan	CIL	NLM-800-0.25	1
Uric acid	CIL	NLM-1697-0.5	1

CIL Cambridge Isotope Laboratories

[a]According to manufacturer instructions, acylcarnitine standards are to be diluted 200-fold into extraction solution. Concentrations vary for each compound

microcentrifuge tube. Remaining volumes may be refrozen and stored at −80 °C.

3. Add chilled extraction solution containing applicable standard(s) to each tube. For whole blood and red blood cell samples, add 180 μL of extraction solution so that the extraction ratio is 1:10. For plasma, serum, cell media, and other fluids, add 480 μL of extraction solution so that the extraction ratio is 1:25.
4. Vortex extraction samples vigorously for 30 min at 4 °C.
5. Centrifuge for 10 min at 18,213 × *g* (or maximum speed) at 4 °C to pellet insoluble materials such as proteins, nucleic acids, and less polar lipids, and ensure that the resulting supernatant is completely clear. If there are still particulates or cloudiness, transfer supernatants to a new tube, and repeat centrifugation.
6. Transfer ~100 μL of extract supernatant into labeled and cooled autosampler vials. Please note that, at this step, 100 μL of the extract could be dried down (e.g., under nitrogen flow or SpeedVac) prior to resuspension in equal volume of pure water (with or without 0.1% formic acid or 10 mM ammonium acetate) or methanol. Though time-consuming, this process can be automated in 96 well-plate format and ensures significantly improved chromatographic retention of hydrophilic compounds in both the isocratic and gradient-based separations.
7. For use as a quality control, prepare a pooled sample of each biological matrix being analyzed by combining 10 μL of each extract supernatant into an autosampler vial. Aliquot size may be adjusted so that the final QC sample volume is sufficient for QC runs at the beginning, middle, and end of the sequence for both positive and negative ion polarity modes or every 10–20 sample runs depending on the size of the sample set. Remaining supernatants may be stored at −80 °C (*see* **Note 1**).
8. Prepare a blank sample by adding 100 μL of extraction solution from **step 3** into a new, labeled autosampler vial.
9. Ensure that the sample vials do not contain air bubbles and are kept cold (in range of −20 °C in freezer to LC autosampler at 7 °C) prior to analysis. Perform MS analysis on samples.

3.5 Metabolite Extraction from Tissue Samples

1. Label new 1.5 mL bead-beater-safe tubes (such as Eppendorf™ Safe-lock tubes), and cool on tube rack immersed in dry ice.
2. Quickly transfer tissue samples from −80 °C freezer to rack on dry ice. Prevent tissue samples from thawing.

3. Transfer tissue to pre-chilled, tared tube. Quickly weigh and record the mass of each tissue to the nearest 0.1 mg; then immediately return tube to dry ice (*see* **Note 2**).
4. Add chilled extraction solution with applicable standard(s) to each tube to obtain a concentration of 15 mg tissue per mL of extraction solution (*see* **Note 3**).
5. For powderized tissue, proceed to **step 6**. For whole pieces of tissue, add one scoop of glass beads (~50–75 μL of volume) to each sample, and bead beat for 5 min at level 4 at 4 °C.
6. Vortex samples vigorously for 30 min at 4 °C.
7. Centrifuge for 10 min at 4 °C at 18,213 × *g* (or maximum speed) to pellet insoluble material (*see* **Note 4**).
8. If extraction was performed at a concentration greater than 15 mg/mL (in **step 4**), dilute an aliquot of supernatant with the same chilled extraction solution used for initial extraction (containing standards if applicable) to obtain a concentration of 15 mg/mL. Then repeat **step 7**.
9. Transfer ~100 μL extract supernatant into labeled and cooled autosampler vials.
10. For use as a quality control, prepare a pooled sample of each biological matrix being analyzed by combining 10 μL of each extract supernatant into an autosampler vial. Aliquot size may be adjusted so that the final QC sample volume is sufficient for QC runs at the beginning, middle, and end of the sequence for both positive and negative ion polarity modes or every 10–20 sample runs depending on the size of the sample set. Remaining supernatants may be stored at −80 °C (*see* **Note 1**).
11. Prepare a blank sample by adding 100 μL of extraction solution from **step 4** into a new, labeled autosampler vial.
12. Ensure that the sample vials do not contain air bubbles and are kept cold (in range of −20 °C in freezer to LC autosampler at 7 °C) prior to analysis.
13. Perform MS analysis on samples.

3.6 Metabolite Extraction from Cell Samples

1. Place sample tubes with frozen cells on dry ice.
2. Add ice cold extraction solution with applicable standard(s) to each tube such that the final extraction concentration is 2×10^6 cells per mL of solution (*see* **Notes 5** and **6**).
3. Ensure cell pellet is released from the tube into suspension, and vortex samples vigorously for 30 min at 4 °C.
4. Centrifuge for 10 min at 18,213 × *g* (or maximum speed) to pellet insoluble material.
5. Transfer ~100 μL extract supernatant into labeled and cooled autosampler vials.

6. For use as a quality control, prepare a pooled sample of each biological matrix being analyzed by combining 10 μL of each extract supernatant into an autosampler vial. Aliquot size may be adjusted so that the final QC sample volume is sufficient for QC runs at the beginning, middle, and end of the sequence for both positive and negative ion polarity modes or every 10–20 sample runs depending on the size of the sample set. Remaining supernatants may be stored at −80 °C (*see* **Note 1**).
7. Prepare a blank sample by adding 100 μL of extraction solution from **step 2** into a new, labeled autosampler vial.
8. Ensure that the sample vials do not contain air bubbles and are kept cold (in range of −20 °C in freezer to LC autosampler at 7 °C) prior to analysis.
9. Perform MS analysis on samples.

3.7 UHPLC-MS Data Acquisition: Isocratic Method

1. Prepare an isocratic UHPLC-MS positive ion method with the following chromatography conditions: Flow rate 0.25 mL/min; solvent composition 95% A, 5% B from time zero until 3 min; column temperature 25 °C; and sample compartment temperature 7 °C. Utilize a C18 column (method has been optimized for C18 column in Subheading 2.5) with a C18 guard column. Phases for positive mode should be supplemented with 0.1% (v/v) formic acid.
2. The mass spectrometry settings for this method are resolution 70,000 (Q Exactive) or 60,000 (Q Exactive HF), scan range 65–900 *m/z*, maximum injection time 200 ms, 2 microscans, automatic gain control (AGC) 3×10^6 ions, ESI source voltage 4.0 kV, capillary temperature 320 °C, and sheath gas 15, auxiliary gas 5, and sweep gas 0 (all nitrogen, measured in arbitrary units). Optimal ESI or HESI gas and voltage parameters may be determined by observing signal intensity as settings are changed.
3. If applicable, incorporate optimized gas and voltage settings into method.
4. Ensure polarity is set to positive, and save method.
5. Prepare a method for negative mode isocratic runs by repeating the above steps with the following changes: Utilize negative ionization solvents supplemented with 1 mM NH_4OAc instead of formic acid, and set solvent composition to 100%. A. Save negative mode method.
6. Prepare run sequence starting with several blank runs, 1–2 injections of QC sample, and then randomized analytical samples with QC samples inserted every 10–20 runs. End with several blank runs before changing polarities. We recommend collecting all data in one polarity mode and then switching to the other to minimize issues with column equilibration.

7. Before beginning, ensure stable column back pressure, stable MS background signal, and the absence of air bubbles or leaks in the LC system.

3.8 UHPLC-MS Data Acquisition: Gradient Method

1. Prepare a 5 min gradient UHPLC-MS positive ion method with the following chromatography conditions: Flow rate 0.45 mL/min, column temperature 45 °C, and sample compartment temperature 7 °C. Solvent gradient is as follows: 0–0.5 min 5% B, 0.5–1.1 min 5–95% B, 1.1–2.75 min hold at 95% B, 2.75–3 min 95–5% B, and 3–5 min hold at 5% B. Utilize a C18 column (method has been optimized for Kinetex C18 column) and a C18 guard column. Phases for positive mode should be supplemented with 0.1% (v/v) formic acid.
2. Add mass spectrometry settings to the gradient UHPLC-MS method: resolution 70,000 (Q Exactive) or 60,000 (Q Exactive HF), scan range 65–900 *m/z*, maximum injection time 200 ms, microscans 2, automatic gain control (AGC) 3×10^6 ions, source voltage 4.0 kV, capillary temperature 320 °C, and sheath gas 45, auxiliary gas 15, and sweep gas 0 (all nitrogen). Optimal ESI or HESI gas and voltage parameters may be determined by observing signal intensity as settings are changed.
3. If applicable, incorporate optimized gas and voltage settings into method.
4. Ensure polarity is set to positive, and save method.
5. Prepare a method for negative mode 5 min gradient runs by repeating the above steps with the following changes: Utilize solvents supplemented with 1 mM NH_4OAc (instead of formic acid) to favor negative ionization. Solvent gradient is as follows: 0–0.5 min 0% B, 0.5–1.1 min 0–100% B, 1.1–2.75 min hold at 100% B, 2.75–3 min 100–0% B, and 3–5 min hold at 0% B. Save negative mode method.
6. Prepare run sequence starting with several blank runs, 1–2 injections of the QC sample, and then randomized analytical samples with QC samples inserted every 10–20 runs. End with several blank runs before changing polarities. We recommend collecting all data in one polarity mode and then switching to the other to minimize issues with column equilibration. With a sufficient number of blanks to ensure column equilibration when switching modes and solvents, the instrumentation can run continuously under automation.
7. Before beginning, ensure stable column back pressure and MS background signal after the column temperature set point has been reached. Ensure also the absence of air bubbles or leaks in the LC system, then run sequence.

4 Data Analysis

1. Quantify metabolite peak areas using software of choice. We utilize Maven, which first requires file conversion from .raw to .mzXML (accomplished using RawConverter), and assign metabolite names using the KEGG database. Table 1 contains a list of observed retention times for acylcarnitines, fatty acids, and oxylipins observed on the 5 min gradient method. Reported times are an average of 3–10 measurements in various matrices (plasma, red blood cells, saliva, pancreatic tissue) and commercial standards prepared in the background of plasma or RBCs (*see* **Note 7**).
2. Utilize QC sample data to assess instrument performance and stability throughout analytical runs. Coefficients of variation (CV) for named metabolites should be <20%. If CV values are above this threshold while other readouts (pump pressure, total signal level, etc.) do not suggest technical issues with instrument performance, consider the use of a batch- or drift-correction software. When needed, we utilize the MetaboDrift normalization algorithms separately on positive and negative ion mode raw peak area data. Once the peak areas are appropriately normalized as needed for each polarity, the data can be merged for subsequent calculations and visualizations.
3. Use the ratio of peak areas for stable isotope-labeled standards and their unlabeled (light) counterparts to determine the absolute values of endogenous metabolites in the extract. Back-calculate the concentrations in the biological samples using the following equations:

For biofluids:

$$[\mathrm{Light}] = \left(\mathrm{Area}_{\mathrm{light}} / \mathrm{Area}_{\mathrm{heavy}}\right)[\mathrm{Heavy}] \times \mathrm{DF}$$

where DF is dilution factor—10 for RBCs and blood and 25 for serum and plasma as described above (*see* **Note 8**).

For tissues:

$$[\mathrm{Light}] = \left(\mathrm{Area}_{\mathrm{light}} / \mathrm{Area}_{\mathrm{heavy}}\right)[\mathrm{Heavy}] \times 1 / \mathrm{EC}$$

where EC is extraction concentration, described here as i15 mg/mL (*see* **Note 9**).

For cells:

$$[\mathrm{Light}] = \left(\mathrm{Area}_{\mathrm{light}} / \mathrm{Area}_{\mathrm{heavy}}\right)[\mathrm{Heavy}] \times 1 / \mathrm{EC}$$

where EC is extraction concentration, described here as 2×10^6 cells/mL (*see* **Note 10**).

5 Notes

1. If/when saved extracts are retrieved for use, spin at max speed for 10 min at 4 °C before aliquotting to AS vial.
2. Tissue extraction method is optimized for 2–20 mg pieces of tissue (wet weight). For powdered tissues, weigh and extract in similar manner, but skip the bead-beating step.
3. If the sample mass is too large to accommodate the volume for 15 mg/mL (e.g., >20 mg), all samples may be extracted at an identical higher concentration and extracts diluted in **step 8** of this section.
4. When centrifuging samples in the presence of glass beads, supernatant is not often fully clarified. A supernatant aliquot may be transferred to a new tube and respun to ensure complete isolation of insoluble material.
5. The cell extraction concentration of 2×10^6 cells/mL is optimized for average-sized eukaryotic cells. Extraction of metabolites from cells with significantly smaller diameter (e.g., T cells, neurons, bacteria) is best performed at higher concentrations.
6. If cell counts are approximated and/or imprecise, the insoluble protein pellet remaining after metabolite extraction may be dissolved in an appropriate chaotrope (e.g., 8 M urea) and quantified by the Bradford or BCA assay for post hoc data normalization.
7. Retention time drift and degradation of peak shape of the acylcarnitines sometimes occur during column age perhaps due to the presence of residual ammonium acetate. For robust retention time matching, use of a column naïve to negative mode additives, such as ammonium salts, is recommended.
8. The concentration of heavy standard utilized here is the final concentration in the extraction. For example, if the extraction solution contains 1 μM of standard and 20 μL of sample are mixed with 180 μL of extraction solution, the resulting concentration of heavy standard would be 0.9 μM.
9. If concentration of heavy standard is entered in μM units and the concentration of tissue extraction is measured in mg/mL, then the resulting concentration of light (endogenous metabolite) will be in units of μmol of metabolite per mg of tissue.
10. If concentration of heavy standard is entered in μM units and the concentration of cells in extraction is measured in cells/mL, then the resulting concentration of light (endogenous metabolite) will be in units of μmol of metabolite per cell.

Conflicts of Interest

A.D. and T.N. are part of Omix Technologies, Inc., and A.D. is a consultant for Hemanext, Inc.

References

1. Venter JC, Adams MD, Myers EW, Li PW, Mural RJ, Sutton GG, Smith HO, Yandell M, Evans CA, Holt RA, Gocayne JD, Amanatides P, Ballew RM, Huson DH, Wortman JR, Zhang Q, Kodira CD, Zheng XH, Chen L, Skupski M, Subramanian G, Thomas PD, Zhang J, Gabor Miklos GL, Nelson C, Broder S, Clark AG, Nadeau J, McKusick VA, Zinder N, Levine AJ, Roberts RJ, Simon M, Slayman C, Hunkapiller M, Bolanos R, Delcher A, Dew I, Fasulo D, Flanigan M, Florea L, Halpern A, Hannenhalli S, Kravitz S, Levy S, Mobarry C, Reinert K, Remington K, Abu-Threideh J, Beasley E, Biddick K, Bonazzi V, Brandon R, Cargill M, Chandramouliswaran I, Charlab R, Chaturvedi K, Deng Z, Di Francesco V, Dunn P, Eilbeck K, Evangelista C, Gabrielian AE, Gan W, Ge W, Gong F, Gu Z, Guan P, Heiman TJ, Higgins ME, Ji RR, Ke Z, Ketchum KA, Lai Z, Lei Y, Li Z, Li J, Liang Y, Lin X, Lu F, Merkulov GV, Milshina N, Moore HM, Naik AK, Narayan VA, Neelam B, Nusskern D, Rusch DB, Salzberg S, Shao W, Shue B, Sun J, Wang Z, Wang A, Wang X, Wang J, Wei M, Wides R, Xiao C, Yan C, Yao A, Ye J, Zhan M, Zhang W, Zhang H, Zhao Q, Zheng L, Zhong F, Zhong W, Zhu S, Zhao S, Gilbert D, Baumhueter S, Spier G, Carter C, Cravchik A, Woodage T, Ali F, An H, Awe A, Baldwin D, Baden H, Barnstead M, Barrow I, Beeson K, Busam D, Carver A, Center A, Cheng ML, Curry L, Danaher S, Davenport L, Desilets R, Dietz S, Dodson K, Doup L, Ferriera S, Garg N, Gluecksmann A, Hart B, Haynes J, Haynes C, Heiner C, Hladun S, Hostin D, Houck J, Howland T, Ibegwam C, Johnson J, Kalush F, Kline L, Koduru S, Love A, Mann F, May D, McCawley S, McIntosh T, McMullen I, Moy M, Moy L, Murphy B, Nelson K, Pfannkoch C, Pratts E, Puri V, Qureshi H, Reardon M, Rodriguez R, Rogers YH, Romblad D, Ruhfel B, Scott R, Sitter C, Smallwood M, Stewart E, Strong R, Suh E, Thomas R, Tint NN, Tse S, Vech C, Wang G, Wetter J, Williams S, Williams M, Windsor S, Winn-Deen E, Wolfe K, Zaveri J, Zaveri K, Abril JF, Guigó R, Campbell MJ, Sjolander KV, Karlak B, Kejariwal A, Mi H, Lazareva B, Hatton T, Narechania A, Diemer K, Muruganujan A, Guo N, Sato S, Bafna V, Istrail S, Lippert R, Schwartz R, Walenz B, Yooseph S, Allen D, Basu A, Baxendale J, Blick L, Caminha M, Carnes-Stine J, Caulk P, Chiang YH, Coyne M, Dahlke C, Mays A, Dombroski M, Donnelly M, Ely D, Esparham S, Fosler C, Gire H, Glanowski S, Glasser K, Glodek A, Gorokhov M, Graham K, Gropman B, Harris M, Heil J, Henderson S, Hoover J, Jennings D, Jordan C, Jordan J, Kasha J, Kagan L, Kraft C, Levitsky A, Lewis M, Liu X, Lopez J, Ma D, Majoros W, McDaniel J, Murphy S, Newman M, Nguyen T, Nguyen N, Nodell M, Pan S, Peck J, Peterson M, Rowe W, Sanders R, Scott J, Simpson M, Smith T, Sprague A, Stockwell T, Turner R, Venter E, Wang M, Wen M, Wu D, Wu M, Xia A, Zandieh A, Zhu X (2001) The sequence of the human genome. Science 291(5507):1304–1351
2. Scolamiero E, Cozzolino C, Albano L, Ansalone A, Caterino M, Corbo G, di Girolamo MG, Di Stefano C, Durante A, Franzese G, Franzese I, Gallo G, Giliberti P, Ingenito L, Ippolito G, Malamisura B, Mazzeo P, Norma A, Ombrone D, Parenti G, Pellecchia S, Pecce R, Pierucci I, Romanelli R, Rossi A, Siano M, Stoduto T, Villani GRD, Andria G, Salvatore F, Frisso G, Ruoppolo M (2015) Targeted metabolomics in the expanded newborn screening for inborn errors of metabolism. Mol BioSyst 11(6):1525–1535
3. Danaei G, Finucane MM, Lu Y, Singh GM, Cowan MJ, Paciorek CJ, Lin JK, Farzadfar F, Khang Y-H, Stevens GA, Rao M, Ali MK, Riley LM, Robinson CA, Ezzati M (2011) National, regional, and global trends in fasting plasma glucose and diabetes prevalence since 1980: systematic analysis of health examination surveys and epidemiological studies with 370 country-years and 2·7 million participants. Lancet 378(9785):31–40
4. Odom SR, Howell MD, Silva GS, Nielsen VM, Gupta A, Shapiro NI, Talmor D (2013) Lactate clearance as a predictor of mortality in trauma patients. J Trauma Acute Care Surg 74(4):999–1004
5. Lusczek ER, Muratore SL, Dubick MA, Beilman GJ (2017) Assessment of key plasma metabolites in combat casualties. J Trauma Acute Care Surg 82(2):309–316
6. D'Alessandro A, Gevi F, Zolla L (2011) A robust high resolution reversed-phase HPLC strategy to investigate various metabolic species

in different biological models. Mol BioSyst 7(4):1024–1032

7. Zampieri M, Sekar K, Zamboni N, Sauer U (2017) Frontiers of high-throughput metabolomics. Curr Opin Chem Biol 36:15–23
8. Nemkov T, Hansen KC, D'Alessandro A (2017) A three-minute method for high-throughput quantitative metabolomics and quantitative tracing experiments of central carbon and nitrogen pathways. Rapid Commun. Mass Spectrom 31(8):663–673
9. Lewis MR, Pearce JTM, Spagou K, Green M, Dona AC, Yuen AHY, David M, Berry DJ, Chappell K, Horneffer-van der Sluis V, Shaw R, Lovestone S, Elliott P, Shockcor J, Lindon JC, Cloarec O, Takats Z, Holmes E, Development NJK (2016) Application of ultra-performance liquid chromatography-TOF MS for precision large scale urinary metabolic phenotyping. Anal Chem 88(18):9004–9013
10. Miura D, Fujimura Y, Yamato M, Hyodo F, Utsumi H, Tachibana H, Wariishi H (2010) Ultrahighly sensitive in situ metabolomic imaging for visualizing spatiotemporal metabolic behaviors. Anal Chem 82(23):9789–9796
11. Gasilova N, Yu Q, Qiao L, Girault HH (2014) On-Chip spyhole mass spectrometry for droplet-based microfluidics. Angew Chem Int Ed 53(17):4408–4412
12. Wouters S, De Vos J, Dores-Sousa JL, Wouters B, Desmet G, Eeltink S (2017) Prototyping of thermoplastic microfluidic chips and their application in high-performance liquid chromatography separations of small molecules. J Chromatogr A 1523:224–233
13. Gehrke S, Reisz JA, Nemkov T, Hansen KC, D'Alessandro A (2017) Characterization of rapid extraction protocols for high-throughput metabolomics. Rapid Commun Mass Spectrom 31(17):1445–1452
14. Dias DA, Koal T (2016) Progress in metabolomics standardisation and its significance in future clinical laboratory medicine. EJIFCC 27(4):331–343
15. Liquid-handling Lego robots and experiments for STEM education and research. [date unknown]. http://journals.plos.org/plosbiology/article?id=10.1371/journal.pbio.2001413. Accessed 19 July 2017
16. Nemkov T, D'Alessandro A, Hansen KC (2015) Three-minute method for amino acid analysis by UHPLC and high-resolution quadrupole orbitrap mass spectrometry. Amino Acids 47(11):2345–2357
17. Reisz JA, Wither MJ, Dzieciatkowska M, Nemkov T, Issaian A, Yoshida T, Dunham AJ, Hill RC, Hansen KC, D'Alessandro A (2016) Oxidative modifications of glyceraldehyde 3-phosphate dehydrogenase regulate metabolic reprogramming of stored red blood cells. Blood 128(12):e32–e42
18. D'alessandro A, Dzieciatkowska M, Nemkov T, Hansen KC (2017) Red blood cell proteomics update: is there more to discover? Blood Transfus 15(2):182–187

Chapter 3

High-Throughput Metabolomics Based on Direct Mass Spectrometry Analysis in Biomedical Research

Raúl González-Domínguez, Álvaro González-Domínguez, Carmen Segundo, Mónica Schwarz, Ana Sayago, Rosa María Mateos, Enrique Durán-Guerrero, Alfonso María Lechuga-Sancho, and Ángeles Fernández-Recamales

Abstract

Metabolomics based on direct mass spectrometry analysis shows a great potential in biomedical research because of its high-throughput screening capability and wide metabolome coverage. This chapter contains detailed protocols to perform comprehensive metabolomic fingerprinting of multiple biological samples (serum, plasma, urine, brain, liver, spleen, thymus) by using complementary analytical platforms. The most important issues to be considered are discussed, including sample treatment, metabolomic analysis, raw data preprocessing, and data analysis.

Key words Metabolomics, Direct mass spectrometry analysis, Direct infusion, Flow injection, High-throughput

1 Introduction

1.1 The Potential of Direct Mass Spectrometry Analysis for High-Throughput Metabolomics

Nontargeted metabolomic analysis is very challenging because of the difficulty of simultaneously characterizing the entire set of metabolites present in biological systems in a comprehensive manner. To this end, nuclear magnetic resonance (NMR) and mass spectrometry (MS) are nowadays the most commonly employed metabolomic platforms, with complementary analytical performance and applicability [1, 2]. NMR is a rapid, nondestructive, and very reproducible technique, which requires a relatively simple sample pre-treatment step and provides important structural information. However, its low sensitivity and spectral resolution seriously compromise the detection of individual metabolites in complex samples. On the other hand, the combination of mass spectrometry with separation techniques, such as liquid chromatography (LC), gas chromatography (GC), or capillary

Angelo D'Alessandro (ed.), *High-Throughput Metabolomics: Methods and Protocols*, Methods in Molecular Biology, vol. 1978, https://doi.org/10.1007/978-1-4939-9236-2_3,

electrophoresis (CE), has demonstrated a great utility to perform both qualitative and quantitative metabolomic analyses. As a consequence of the inherent analytical bias introduced by these separation methods, the combination of complementary approaches has been postulated as the most suitable strategy to maximize metabolomic coverage [3, 4]. Nevertheless, the application of these hyphenated approaches significantly increases overall analysis times, which in turn can negatively affect technical stability (e.g., analytical drifts in mass accuracy and/or sensitivity).

To solve these limitations, the use of direct mass spectrometry analysis has emerged in recent years for high-throughput metabolomics [5, 6]. In this approach, sample extracts are directly introduced into the mass spectrometer without previous chromatographic or electrophoretic separation. For this purpose, the simplest instrumental configuration is direct infusion mass spectrometry (DIMS), which employs a syringe pump to constantly deliver samples into the MS system. A second alternative is flow injection mass spectrometry (FI-MS), based on the infusion of samples as a plug into a stream of solvent delivered by a LC pump. In general, DIMS-based metabolomics provides higher sensitivity and reproducibility, but as counterpart, FI-MS is a less sample volume consuming technique and enables automation. Direct mass spectrometry analysis exhibits multiple advantages over conventional MS hyphenated platforms thanks to the lack of a preceding separation step. First, it is noteworthy the higher technical simplicity of DI/FI-MS approaches, which are unaffected by common troubles usually found in chromatography such as column clogging and overpressure, changes in retention times and peak broadening due to stationary phase deterioration, or irregular baseline and signal drifts due to mobile phase contamination or leaks in the LC system, among other issues. Furthermore, direct introduction of samples without applying a previous chromatographic/electrophoretic step, which is inherently biased toward specific metabolite classes depending on the separation mode employed, usually provides more comprehensive metabolome coverage. Therefore, direct MS analysis has demonstrated huge potential for high-throughput metabolomic screening, of particular interest when dealing with large sample populations [6].

1.2 Application of Direct MS-Based Metabolomics in Biomedical Research

Metabolomics based on direct MS analysis has gained great popularity for mapping metabolic alterations associated with disease pathogenesis and progression in a holistic manner. For instance, various authors have previously described the application of MS fingerprinting to investigate the underlying pathology of different oncological disorders, including lung cancer [7–9], prostate cancer [10], kidney cancer [11], pancreatic cancer [12], and colorectal cancer [13–15]. Thereby, it has been demonstrated the occurrence of multiple metabolic dysregulations affecting to a wide range of pathways depending on the cancer type and stage, such as impaired energy metabolism, altered homeostasis of lipids, oxidative stress, and many others. DI/FI-MS-based metabolomics has also been employed to identify the characteristic metabotype associated with abnormal glucose metabolism in patients with impaired glucose tolerance [16] and insulin resistance [17] and in volunteers subjected to an oral sugar challenge [18]. Other studies performed on the last years dealt with the discovery of potential diagnostic biomarkers for Chagas disease [19], schistosomiasis [20], aspergillosis [21], Crohn's disease [22], induced stress [23], and preeclampsia [24]. However, it should be noted that the application of direct MS has mainly focused on the metabolomic study of Alzheimer's disease (AD). González-Domínguez et al. optimized a metabolomic approach based on a two-step extraction procedure and subsequent DIMS analysis of serum samples to elucidate pathological hallmarks associated with AD development and progression, in both human cohorts [25, 26] and transgenic APP×PS1 mouse models [27, 28]. Later, this methodology was adapted to get deeper insights into lipidomic [29] and phospholipidomic [30] alterations occurring in this neurodegenerative disorder. Furthermore, authors also described the implementation of this DIMS-based metabolomic platform for the analysis of other biological matrices, such as urine [31], brain [32], and other peripheral organs (i.e., liver, kidney, thymus, spleen) [33]. Thereby, a comprehensive characterization of AD-related metabolic disturbances was accomplished, comprising multiple impairments in the homeostasis of neurotransmitters, various lipid classes (e.g., phospholipids, glycerolipids, cholesterol, fatty acids), and antioxidants, among others. Interestingly, these findings were then validated by applying orthogonal hyphenated approaches, such as UHPLC-MS [34–38], GC-MS [35–39], and CE-MS [40], thus demonstrating the reliability of direct MS for metabolomic fingerprinting. In this line, Lin et al. reported the use of DIMS techniques to address metabolic alterations in the hippocampus [41] and cerebellum [42] of CRND8 transgenic mice, revealing significant perturbations in the regulation of neuroinflammatory processes. Finally, it is also noteworthy the recent development of a novel metabolomic approach based on flow injection analysis using an atmospheric pressure photoionization mass spectrometer (FI-APPI-MS) with the aim to complement the ionization capabilities of the electrospray source (ESI), usually employed in DIMS-based metabolomics [43]. This methodology was successfully applied to serum samples from AD patients [43] and APP×PS1 transgenic mice [27], which provided complementary findings to those obtained with DI-ESI-MS platforms.

2 Materials

2.1 Chemicals and Standards

Prepare all samples and solutions by using high-purity solvents and reagents, including methanol, ethanol, chloroform, dichloromethane, formic acid, ammonium formate, ammonium acetate, and toluene. Deionized water can be obtained from a Milli-Q Gradient system (Millipore, Watford, UK). For identification purposes, authentic metabolite standards must be purchased if available.

2.2 Sample Collection

2.2.1 Human Samples

Blood samples are extracted by venipuncture of the antecubital region after 8 h of fasting. All samples must be collected at the same time of the day in order to avoid the influence of the circadian rhythm. To obtain serum samples, blood is immediately cooled and protected from light for 30 min to allow clot retraction. On the contrary, plasma collection requires the addition of anticoagulants to the corresponding blood collection tube (*see* **Note 1**). Then, blood samples are centrifuged at 4000 × *g* for 10 min at 4 °C, and the resulting serum/plasma is divided into aliquots and frozen at −80 °C until analysis. On the other hand, urine samples are directly collected in sterile plastic containers known not to release plasticizers or other compounds into the sample (*see* **Note 2**). Before storing at −80 °C, it is recommended to perform a mild centrifugation step at 2000 × *g* for 10 min at 4 °C in order to remove human cells and bacteria, which may break upon sample freezing. The study population recruitment must be performed in accordance with the principles contained in the Declaration of Helsinki.

2.2.2 Animal Model Samples

After reception, mice must be acclimated for 3 days in rooms with a 12-h light/dark cycle at 20–25 °C, with water and food available ad libitum. Then, mice are individually anesthetized by isoflurane inhalation and exsanguinated by cardiac puncture. These blood samples are processed as previously described (Subheading 2.2.1) to obtain serum and plasma. Subsequently, brain and other peripheral organs (liver, kidney, spleen, and thymus) are rapidly removed and rinsed with saline solution (0.9% NaCl, w/v). Furthermore, brains are dissected into hippocampus, cortex, cerebellum, striatum, and olfactory bulbs. Urines are directly collected in sterile plastic containers and centrifuged at 2000 × *g* for 10 min at 4 °C (*see* **Note 2**). Finally, all these samples are snap-frozen in liquid nitrogen and stored at −80 °C until analysis. Handling of animals must be performed according to the directive 2010/63/EU stipulated by the European Community.

2.3 Instrumentation

Metabolomic methods described in this chapter have previously been validated in two high-resolution mass spectrometry systems: (1) quadrupole-time-of-flight (Q-TOF) mass spectrometer, model

QSTAR XL Hybrid system (Applied Biosystems), equipped with integrated syringe pump, Accela LC system (Thermo Fisher Scientific), and syringe pump model KDS 100 (KD Scientific); (2) Q-TOF mass spectrometer, model Xevo G2-S, coupled to an ACQUITY UPLC™ system (Waters). For accurate mass measurement, the QSTAR XL system is daily calibrated using renin and taurocholic acid in positive and negative ion modes, respectively. In the Xevo G2-S mass spectrometer, all spectra are acquired using a reference lock mass (leucine enkephalin) to ensure accuracy and reproducibility.

A cryogenic homogenizer model Freezer/Mill 6770 (SPEX SamplePrep) and a pellet mixer (VWR International) are used to perform metabolomic extraction of tissues. All samples are centrifuged in a centrifuge model Eppendorf 5804R. A vacuum manifold (Visiprep, Supelco) is employed to perform solid phase extraction (SPE) of urine samples.

3 Methods

3.1 Sample Treatment

3.1.1 Metabolomic Extraction of Serum/Plasma Samples

A two-step extraction protocol is applied to serum and plasma samples with the aim to fractionate the blood metabolome in two complementary extracts: (1) aqueous extract containing more polar compounds (i.e., low molecular weight metabolites and phospholipids) and (2) lipophilic extract, mainly composed of neutral lipids [25].

1. Thaw serum/plasma samples on an ice bath.
2. Add 400 μL of methanol/ethanol (1:1, v/v) to 100 μL of serum/plasma into an Eppendorf tube placed on an ice bath.
3. Stir the mixture during 5 min, using a vortex mixed or an orbital shaker (at 4 °C if possible), in order to precipitate proteins.
4. Centrifuge samples at 4000 × *g* for 10 min at 4 °C.
5. Transfer the supernatant to a new tube, take it to dryness under nitrogen stream (*see* **Note 3**), and reconstitute the resulting residue with 100 μL of methanol/water (80:20, v/v) containing 0.1% (v/v) formic acid (polar extract) (*see* **Note 4**).
6. Add 400 μL of chloroform/methanol (1:1, v/v) to the protein precipitate obtained in **step 3** placed on an ice bath.
7. Shake vigorously during 5 min using a vortex mixer (at 4 °C if possible).
8. Centrifuge samples at 10,000 × *g* for 10 min at 4 °C.

9. Take to dryness the supernatant under nitrogen stream (*see* **Note 3**) and reconstitute with 100 μL of dichloromethane/methanol (60:40, v/v) containing 0.1% (v/v) formic acid and 10 mM ammonium formate (lipophilic extract) (*see* **Note 4**).

3.1.2 Lipidomic Extraction of Serum/Plasma Samples

To get a deeper insight into the blood circulating lipidome, serum/plasma samples are treated following a modification of the Bligh-Dyer extraction protocol, which allows analyzing neutral lipids (e.g., diglycerides, triglycerides, cholesterol derivatives), usually not detected by means of conventional metabolomic approaches [29].

1. Thaw serum/plasma samples on an ice bath.
2. Mix 50 μL of serum/plasma with 150 μL of methanol containing 30 mM ammonium acetate in an Eppendorf tube placed on an ice bath.
3. Shake vigorously during 1 min using a vortex mixer (at 4 °C if possible).
4. Add 200 μL of pure chloroform.
5. Shake vigorously during 1 min using a vortex mixer (at 4 °C if possible).
6. Centrifuge samples at 10,000 × *g* for 10 min at 4 °C.
7. Transfer the organic layer to a new tube and keep for further analysis.

3.1.3 Metabolomic Extraction of Urine Samples

Direct MS fingerprinting of urine samples requires the application of a pre-treatment step with the aim to reduce the high salinity of this biological fluid, which may compromise further analysis due to matrix effects (e.g., ion suppression). For this purpose, two different approaches can be applied after thawing urine samples on an ice bath, as describe elsewhere [31].

1. Sample dilution.
   (a) Centrifuge urine samples at 4000 × *g* for 10 min at 4 °C.
   (b) Dilute tenfold with methanol/water (1:1, v/v).
2. Mixed mode solid phase extraction (MM-SPE).
   (a) Condition SPE cartridges (Isolute Multimode, 500 mg of sorbent) with 2 mL of pure methanol.
   (b) Load 1.5 mL of urine, previously centrifuged at 4000 × *g* for 10 min at 4 °C.
   (c) Clean the SPE cartridge with 2 mL of deionized water.
   (d) Carry out the elution with: (1) 0.5 mL of methanol, (2) 0.5 mL of 10 mM ammonium acetate (pH = 3) in methanol, and (3) 0.5 mL of 5% (v/v) ammonia in methanol.

All eluents are pumped through the SPE cartridges by using a vacuum manifold (Visiprep, Supelco).

3.1.4 Metabolomic Extraction of Tissue Samples

Extraction of tissue samples, including the hippocampus, brain cortex, cerebellum, striatum, olfactory bulbs, liver, kidney, spleen, and thymus, is carried out in accordance with the methodology previously optimized by González-Domínguez et al. [33, 36].

1. Cryo-homogenize tissues to obtain a fine powder using a Freezer/Mills 6770 homogenizer during 30 s at rate of 10 strokes per second (*see* **Note 5**).
2. Weigh 30 mg of homogenized tissue (or the entire sample for smaller organs) in 1.5 mL Eppendorf tubes. Samples are kept frozen until the addition of the extraction solvent.
3. Add 10 μL mg^{-1} of pre-cooled 0.1% (v/v) formic acid in methanol (−20 °C).
4. Use a pellet mixer (VWR international) to disrupt cells during 2 min. Perform this extraction step inside an ice bath to avoid sample heating by friction.
5. Centrifuge at 10,000 × *g* for 10 min at 4 °C.
6. Transfer the supernatant to a new tube and keep for further analysis (polar extract).
7. Add 10 μL mg^{-1} of pre-cooled chloroform/methanol (2:1, v/v), containing 0.1% (v/v) formic acid and 10 mM ammonium formate (−20 °C), to the pellet obtained in **step 5** (*see* **Note 6**).
8. Repeat **steps 4–6** to obtain the corresponding lipophilic extracts.

3.1.5 Preparation of Quality Control Samples

For each biological matrix, prepare quality control (QC) samples by pooling equal volumes of each individual sample. The analysis of these QC samples allows monitoring the stability and performance of the system along the analysis period [44]. These samples must be analyzed at the start of the run in order to equilibrate the analytical system as well as at intermittent points throughout the sequence to monitor system stability.

3.2 Direct MS Metabolomic Fingerprinting

3.2.1 Direct Infusion Electrospray Ionization Mass Spectrometry (DI-ESI-MS)

Sample extracts are directly infused into the QTOF-MS system (QSTAR XL, Applied Biosystems) by using an integrated syringe pump operating at 5 μL min^{-1} flow rate. Mass spectra are obtained by electrospray ionization (ESI) in positive and negative ionization modes in separate runs, as described elsewhere [25, 27, 32, 33]. Briefly, full-scan spectra are acquired for 0.2 min in the *m/z* range 50–1100 Da, with 1.005 s scan time. The source temperature is maintained at 60 °C, and high-purity nitrogen is used as curtain and nebulizer gas at flow rates of 1.13 L min^{-1} and 1.56 L min^{-1}, respectively. The ion spray voltage (IS), declustering potential (DP), and focusing potential are fixed at 3300/−4000 V, 60/−100 V, and 250/−250 V in positive and negative ion modes, respectively.

3.2.2 Flow Injection Electrospray Ionization Mass Spectrometry (FI-ESI-MS)

For FI-ESI-MS analysis, sample extracts are introduced into the mass spectrometer (Xevo G2-S) by flow injection using an ACQUITY UPLC™ system [18]. Methanol containing 0.1% (v/v) formic acid and 10 mM ammonium formate is delivered at 200 μL min^{-1} as flow injection solvent. Mass spectra are obtained in positive and negative ion modes, by injecting 5 μL of sample. Full-scan spectra are acquired in the *m/z* range 50–1100 Da during 1.5 min (scan time 0.5 s). Capillary and cone voltages are set at 3.0 kV and 30 V, respectively, and source temperature is maintained at 120 °C. Desolvation gas flow (high-purity nitrogen) is fixed at 250 L h^{-1}, at 150 °C of temperature.

3.2.3 Flow Injection Atmospheric Pressure Photoionization Mass Spectrometry (FI-APPI-MS)

The application of FI-APPI-MS fingerprinting requires the coupling of a high-resolution mass spectrometer equipped with an atmospheric pressure photoionization source (QSTAR XL), a LC system for flow injection of samples (Accela), and a syringe pump to deliver the dopant reagent for photospray ionization. Following the previously optimized protocol, methanol is used as flow injection solvent at 50/100 μL min^{-1} in positive and negative ion modes, respectively [43]. On the other hand, toluene is delivered at 20/40 μL min^{-1} as photoionization dopant, in both ionization modes. Mass spectra are obtained in positive and negative ion modes by injecting 5 μL of sample. Full-scan spectra are acquired in the *m/z* range 50–1100 Da, with 1.005 s of scan time. The ion spray voltage (IS), declustering potential (DP), and focusing potential are fixed at 1500/−2300 V, 50/−50 V, and 250/−250 V in positive and negative ion modes, respectively. The source temperature is maintained at 400 °C, and the gas flows (high-purity nitrogen) are fixed at 1.13 L min^{-1} for curtain gas, 1.50 L min^{-1} for nebulizer gas, 3.0 L min^{-1} for heater gas, and 1 L min^{-1} for lamp gas.

3.3 Raw Data Processing

In direct MS-based metabolomics, data preprocessing only implies the application of a single peak detection step in order to convert original files into a two-dimensional data matrix of spectral peaks and their intensities. For this purpose, different software must be employed depending on the MS system employed to acquire metabolomic fingerprints.

1. Metabolomic data obtained by DI-ESI-MS and FI-APPI-MS analysis are submitted to peak detection by using the MarkerView™ software (Applied Biosystems). For this, all peaks above the noise level (10 counts, determined empirically

from experimental spectra) are selected and binned in intervals of 0.1 Da. For FI-MS fingerprints, this processing step is limited to scans within the apex of infusion profiles [43].

2. Metabolomic data obtained by FI-ESI-MS analysis are submitted to peak detection by using the MarkerLynx™ software (Waters). To this end, all peaks above the noise level (200 counts, determined empirically from experimental spectra) are selected and binned in intervals of 0.02 Da. This processing step is limited to scans within the apex of infusion profiles [18].

3.4 Data Analysis

Multiple statistical packages can be used to analyze metabolomic data. In this section, we focus on the MetaboAnalyst 3.0 web tool (http://www.metaboanalyst.ca/), the SIMCA-P™ software (version 11.5, UMetrics AB, Umeå, Sweden), and the STATISTICA 8.0 software (StatSoft, Tulsa, USA). Thereby, a conventional pipeline for metabolomic data analysis involves the following steps:

1. Data filtering by choosing masses present in at least 50% of samples.
2. Estimation of missing values by means of the *k*-nearest neighbor (KNN) algorithm.
3. Data filtering based on the interquartile range (IQR) in order to remove variables showing little variance within the study population.
4. Data normalization according to the total area sum.
5. Pareto scaling to reduce the relative importance of larger values, and logarithmic transformation in order to approximate a normal distribution [45].
6. Application of multivariate statistical techniques with the aim to visualize general trends and detect discriminant patterns among the study groups: principal component analysis (PCA), partial least squares discriminant analysis (PLS-DA).
7. Application of univariate statistical techniques to identify discriminant metabolites among the study groups: *t*-test or one-way analysis of variance (ANOVA) (*see* **Note 7**), with Bonferroni or false discovery rate (FDR) correction for multiple testing.

4 Notes

1. Various anticoagulants can be used to obtain plasma samples, including citrate, ethylenediaminetetraacetic acid (EDTA), and heparin. Advantages and drawbacks of using each anticoagulant have recently been reviewed by Hernandes et al. [46].

2. Preservatives can be added to enhance the stability of urine samples (e.g., borate, sodium azide), which are prone to bacterial contamination during collection and storage. However, the general recommendation of the European Consensus Expert Group is to avoid the use of additives [47].
3. The evaporation of sample extracts can also be carried out by using vacuum concentrators (e.g., SpeedVac), if available.
4. Depending on the sensitivity of the MS system employed, sample extracts can be directly analyzed without performing a preconcentration step.
5. Smaller organs (e.g., hippocampus, striatum, olfactory bulbs) can be directly extracted without prior cryo-homogenization.
6. This second extraction step is of great interest for studying peripheral organs (i.e., liver, kidney, spleen, thymus), due to the high content of neutral lipids in these tissues. However, **steps 7** and **8** can be omitted when brain tissue is analyzed, as previously reported [32, 36].
7. Nonparametric methods must be used when variables do not show normal distribution (checked by normal probability plots) and variances are not homogeneous (checked by Levene's test).

References

1. Emwas AHM, Salek RM, Griffin JL, Merzaban J (2013) NMR based metabolomics in human disease diagnosis: applications, limitations, and recommendations. Metabolomics 9:1048–1072
2. Theodoridis G, Gika HG, Wilson ID (2011) Mass spectrometry-based holistic analytical approaches for metabolite profiling in systems biology studies. Mass Spectrom Rev 30:884–906
3. González-Domínguez R, Sayago A, Fernández-Recamales A (2017) Metabolomics in Alzheimer's disease: the need of complementary analytical platforms for the identification of biomarkers to unravel the underlying pathology. J Chromatogr B Analyt Technol Biomed Life Sci 1071:75–92
4. Gonzalez-Dominguez A, Duran-Guerrero E, Fernandez-Recamales A, Lechuga-Sancho AM, Sayago A, Schwarz M, Segundo C, Gonzalez-Dominguez R (2017) An overview on the importance of combining complementary analytical platforms in metabolomic research. Curr Top Med Chem 17:3289–3295
5. Draper J, Lloyd AJ, Goodacre R, Beckmann M (2013) Flow infusion electrospray ionisation mass spectrometry for high throughput, non-targeted metabolite fingerprinting: a review. Metabolomics 9:S4–S29
6. González-Domínguez R, Sayago A, Fernández-Recamales A (2017) Direct infusion mass spectrometry for metabolomic phenotyping of diseases. Bioanalysis 9:131–148
7. Guo Y, Wang X, Qiu L, Qin X, Liu H, Wang Y, Li F, Wang X, Chen G, Song G, Li F, Guo S, Li Z (2012) Probing gender-specific lipid metabolites and diagnostic biomarkers for lung cancer using Fourier transform ion cyclotron resonance mass spectrometry. Clin Chim Acta 414:135–141
8. Cameron SJ, Lewis KE, Beckmann M, Allison GG, Ghosal R, Lewis PD, Mur LA (2016) The metabolomic detection of lung cancer biomarkers in sputum. Lung Cancer 94:88–95
9. Lokhov PG, Kharybin ON, Archakov AI (2012) Diagnosis of lung cancer based on direct-infusion electrospray mass spectrometry of blood plasma metabolites. Int J Mass Spectrom 309:200–205
10. Lokhov PG, Dashtiev MI, Moshkovskii SA, Archakov AI (2010) Metabolite profiling of blood plasma of patients with prostate cancer. Metabolomics 6:156–163

11. Lin L, Yu Q, Yan X, Hang W, Zheng J, Xing J, Huang B (2010) Direct infusion mass spectrometry or liquid chromatography mass spectrometry for human metabonomics? A serum metabonomic study of kidney cancer. Analyst 135:2970–2978
12. Ritchie SA, Akita H, Takemasa I, Eguchi H, Pastural E, Nagano H, Monden M, Doki Y, Mori M, Jin W, Sajobi TT, Jayasinghe D, Chitou B, Yamazaki Y, White T, Goodenowe DB (2013) Metabolic system alterations in pancreatic cancer patient serum: potential for early detection. BMC Cancer 13:416
13. Ritchie SA, Ahiahonu PW, Jayasinghe D, Heath D, Liu J, Lu Y, Jin W, Kavianpour A, Yamazaki Y, Khan AM, Hossain M, Su-Myat KK, Wood PL, Krenitsky K, Takemasa I, Miyake M, Sekimoto M, Monden M, Matsubara H, Nomura F, Goodenowe DB (2010) Reduced levels of hydroxylated, polyunsaturated ultra long-chain fatty acids in the serum of colorectal cancer patients: implications for early screening and detection. BMC Med 8:13
14. Li F, Qin X, Chen H, Qiu L, Guo Y, Liu H, Chen G, Song G, Wang X, Li F, Guo S, Wang B, Li Z (2013) Lipid profiling for early diagnosis and progression of colorectal cancer using direct-infusion electrospray ionization Fourier transform ion cyclotron resonance mass spectrometry. Rapid Commun Mass Spectrom 27:24–34
15. Williams MD, Zhang X, Park JJ, Siems WF, Gang DR, Resar LM, Reeves R, Hill HH Jr (2015) Characterizing metabolic changes in human colorectal cancer. Anal Bioanal Chem 407:4581–4595
16. Lokhov PG, Trifonova OP, Maslov DL, Balashova EE, Archakov AI, Shestakova EA, Shestakova MV, Dedov II (2014) Diagnosing impaired glucose tolerance using direct infusion mass spectrometry of blood plasma. PLoS One 9:e105343
17. Renner S, Römisch-Margl W, Prehn C, Krebs S, Adamski J, Göke B, Blum H, Suhre K, Roscher AA, Wolf E (2012) Changing metabolic signatures of amino acids and lipids during the prediabetic period in a pig model with impaired incretin function and reduced β-cell mass. Diabetes 61:2166–2175
18. González-Domínguez R, Mateos RM, Lechuga-Sancho AM, González-Cortés JJ, Corrales-Cuevas M, Rojas-Cots JA, Segundo C, Schwarz M (2017) Synergic effects of sugar and caffeine on insulin-mediated metabolomic alterations after an acute consumption of soft drinks. Electrophoresis 38:2313–2322
19. Antunes LC, Han J, Pan J, Moreira CJ, Azambuja P, Borchers CH, Carels N (2013) Metabolic signatures of triatomine vectors of *Trypanosoma cruzi* unveiled by metabolomics. PLoS One 8:e77283
20. Ferreira MS, de Oliveira DN, de Oliveira RN, Allegretti SM, Catharino RR (2014) Screening the life cycle of *Schistosoma mansoni* using high-resolution mass spectrometry. Anal Chim Acta 845:62–69
21. de Francisco TMG, Zaramella IF, Gasparetto JC, Cerqueira LB, Piantavini MS, Pontarolo R, Campos FR (2009) Rapid detection of aspergillosis in immunocompromised patients using DIMS and chemometric analysis. Anal Methods 7:6346–6351
22. Jansson J, Willing B, Lucio M, Fekete A, Dicksved J, Halfvarson J, Tysk C, Schmitt-Kopplin P (2009) Metabolomics reveals metabolic biomarkers of Crohn's disease. PLoS One 4:e6386
23. Lorenzo-Tejedor M, de la Cámara C, Lopez-Anton R, Bailon R, Aguiló J, Bernal-Ruiz ML (2015) Direct infusion electrospray mass spectrometry as a new non-invasive tool for serum metabolomics in induced-stress subjects. Eur J Psychiat 29:259–275
24. Anand S, Young S, Esplin MS, Peaden B, Tolley HD, Porter TF, Varner MW, D'Alton ME, Jackson BJ, Graves SW (2016) Detection and confirmation of serum lipid biomarkers for preeclampsia using direct infusion mass spectrometry. J Lipid Res 57:687–696
25. González-Domínguez R, García-Barrera T, Gómez-Ariza JL (2014) Using direct infusion mass spectrometry for serum metabolomics in Alzheimer's disease. Anal Bioanal Chem 406:7137–7148
26. González-Domínguez R, García-Barrera T, Gómez-Ariza JL (2012) Metabolomic approach to Alzheimer's disease diagnosis based on mass spectrometry. Chem Papers 66:829–835
27. González-Domínguez R, García-Barrera T, Vitorica J, Gómez-Ariza JL (2015) Application of metabolomics based on direct mass spectrometry analysis for the elucidation of altered metabolic pathways in serum from the APP/PS1 transgenic model of Alzheimer's disease. J Pharm Biomed Anal 107:378–385
28. González-Domínguez R, García-Barrera T, Vitorica J, Gómez-Ariza JL (2015) Metabolomic research on the role of interleukin-4 in Alzheimer's disease. Metabolomics 11:1175–1183
29. González-Domínguez R, García-Barrera T, Gómez-Ariza JL (2014) Metabolomic study of

lipids in serum for biomarker discovery in Alzheimer's disease using direct infusion mass spectrometry. J Pharm Biomed Anal 98:321–326
30. González-Domínguez R, García-Barrera T, Gómez-Ariza JL (2014) Combination of metabolomic and phospholipid-profiling approaches for the study of Alzheimer's disease. J Proteome 104:37–47
31. González-Domínguez R, Castilla-Quintero R, García-Barrera T, Gómez-Ariza JL (2014) Development of a metabolomic approach based on urine samples and direct infusion mass spectrometry. Anal Biochem 465:20–27
32. González-Domínguez R, García-Barrera T, Vitorica J, Gómez-Ariza JL (2015) Metabolomic screening of regional brain alterations in the APP/PS1 transgenic model of Alzheimer's disease by direct infusion mass spectrometry. J Pharm Biomed Anal 102:425–435
33. González-Domínguez R, García-Barrera T, Vitorica J, Gómez-Ariza JL (2015) High throughput multi-organ metabolomics in the APP/PS1 mouse model of Alzheimer's disease. Electrophoresis 36:2237–2249
34. González-Domínguez R, Rupérez FJ, García-Barrera T, Barbas C, Gómez-Ariza JL (2016) Metabolomic-driven elucidation of pathological mechanisms associated with Alzheimer's disease and mild cognitive impairment. Curr Alzheimer Res 13:641–653
35. González-Domínguez R, García-Barrera T, Vitorica J, Gómez-Ariza JL (2015) Deciphering metabolic abnormalities associated with Alzheimer's disease in the APP/PS1 mouse model using integrated metabolomic approaches. Biochimie 110:119–128
36. González-Domínguez R, García-Barrera T, Vitorica J, Gómez-Ariza JL (2014) Region-specific metabolic alterations in the brain of the APP/PS1 transgenic mice of Alzheimer's disease. Biochim Biophys Acta 1842:2395–2402
37. González-Domínguez R, García-Barrera T, Vitorica J, Gómez-Ariza JL (2015) Metabolomics reveals significant impairments in the immune system of the APP/PS1 transgenic mice of Alzheimer's disease. Electrophoresis 36:577–587
38. González-Domínguez R, García-Barrera T, Vitorica J, Gómez-Ariza JL (2015) Metabolomic investigation of systemic manifestations associated with Alzheimer's disease in the APP/PS1 transgenic mouse model. Mol BioSyst 11:2429–2440
39. González-Domínguez R, García-Barrera T, Gómez-Ariza JL (2015) Metabolite profiling for the identification of altered metabolic pathways in Alzheimer's disease. J Pharm Biomed Anal 107:75–81
40. González-Domínguez R, García A, García-Barrera T, Barbas C, Gómez-Ariza JL (2014) Metabolomic profiling of serum in the progression of Alzheimer's disease by capillary electrophoresis-mass spectrometry. Electrophoresis 35:3321–3330
41. Lin S, Liu H, Kanawati B, Liu L, Dong J, Li M, Huang J, Schmitt-Kopplin P, Cai Z (2013) Hippocampal metabolomics using ultrahigh-resolution mass spectrometry reveals neuroinflammation from Alzheimer's disease in CRND8 mice. Anal Bioanal Chem 405:5105–5117
42. Lin S, Kanawati B, Liu L, Witting M, Li M, Huang J, Schmitt-Kopplin P, Cai Z (2014) Ultrahigh resolution mass spectrometry-based metabolic characterization reveals cerebellum as a disturbed region in two animal models. Talanta 118:45–53
43. González-Domínguez R, García-Barrera T, Gómez-Ariza JL (2015) Application of a novel metabolomic approach based on atmospheric pressure photoionization mass spectrometry using flow injection analysis for the study of Alzheimer's disease. Talanta 131:480–489
44. Sangster T, Major H, Plumb R, Wilson AJ, Wilson ID (2006) A pragmatic and readily implemented quality control strategy for HPLC-MS and GC-MS-based metabonomic analysis. Analyst 131:1075–1078
45. van den Berg RA, Hoefsloot HCJ, Westerhuis JA, Smilde AK, van der Werf MJ (2006) Centering, scaling, and transformations: improving the biological information content of metabolomics data. BMC Genomics 7:142
46. Hernandes VV, Barbas C, Dudzik D (2017) A review of blood sample handling and pre-processing for metabolomics studies. Electrophoresis 38:2232–2241
47. Yuille M, Illig T, Hveem K, Schmitz G, Hansen J, Neumaier M, Tybring G, Wichmann E, Ollier B (2010) Laboratory management of samples in biobanks: European consensus expert group report. Biopreserv Biobank 8:65–69

Chapter 4

Traveling Wave Ion Mobility Mass Spectrometry: Metabolomics Applications

Giuseppe Paglia and Giuseppe Astarita

Abstract

Ion mobility (IM) spectrometry can separate gas-phase ions according to their charge, molecular shape, and size. In recent years, several IM technologies have been integrated with mass spectrometry (MS) and launched as commercially available instrumentation for metabolomics analysis. The addition of IM to MS-based metabolomics workflows provides an additional degree of separation to chromatography and MS resolving power, improving peak capacity and signal-to-noise ratio. Moreover, it makes possible to experimentally derive collision cross section (CCS), which can be used as an additional coordinate for metabolite identification, together with accurate mass and fragmentation information. The addition of CCS to current metabolome database would allow to filter and score molecules based on their CCS values, adding more confidence in the identification process during metabolomics experiments.

In this chapter, we present procedures for the integration of travelling-wave (TW)-IM into traditional MS-based metabolomics workflows.

Key words Ion mobility, Collision cross section, Drift time, Travelling-wave ion mobility, Metabolomics, Mass spectrometry

1 Introduction

Mass spectrometry (MS)-based metabolomics has been increasingly used for biological research and biomarker discovery. To maximize the number of measured metabolites, chromatographic separation is often used before MS detection. Such an approach, however, often incurs into three main limitations: (1) biological samples contain several thousands of metabolites, which represent a challenge even when using long chromatographic separations to completely resolve them; (2) many metabolites are isobars having the same exact mass but different chemical structure and cannot be distinguished using only exact mass values; and (3) large-scale metabolomics experiments often require high-throughput chromatographic runs, which might decrease chromatographic peak capacity, in turn decreasing the number of features detected [1].

Angelo D'Alessandro (ed.), *High-Throughput Metabolomics: Methods and Protocols*, Methods in Molecular Biology, vol. 1978, https://doi.org/10.1007/978-1-4939-9236-2_4,

To offset some of these limitations (increasing throughput while maximizing peak capacity), in recent years, ion mobility (IM) has been used in combination with MS for metabolomics applications. IM is a technique that allows gas-phase electrophoresis separation of ions within a chamber filled with a buffer gas and subjected to an electrical field. Ions are separated based on molecular shape, size, and charge as well as to the nature of the buffer gas [2–4].

IM occurs on a timescale of milliseconds, making it suitable for combining with MS, where mass separation is typically obtained in the microsecond scale [3].

The integration of IM separation in a MS-based metabolomics workflows significantly increases the number of features detected due to the separation of co-eluting compounds, noise reduction, and resolution of isobaric components. Because chromatographic separation occurs in seconds, IM-MS can be further coupled to chromatography, providing an additional degree of separation to increase peak capacity, distinguish metabolites from matrix interferences, or even resolve structural isomers. The additional dimension of ion mobility separation increased the peak capacity of a traditional MS experiment by over fivefold [5, 6] (Fig. 1), and it compensates the reduction of peak capacity due to shorter chromatographic run in high-throughput metabolomics [1]. Thus, IM provides cleaner mass spectral data and increased signal-to-noise ratio for complex mixtures of metabolites in biological samples [7–11].

The combination of IM with traditional MS-based analytical approaches offers other major advantages for metabolomics. The ion mobility measurement (drift time) can be converted to a collision cross section (CCS) value, which provides an orthogonal physicochemical parameter that increases the confidence of metabolite identification. CCS values are comparable across different laboratories and various CCS databases, containing both empirical and theoretical values, and are now publicly available [12–20].

Furthermore, IM-derived CCSs are more reproducible than retention times, which may shift in values due to sample loading and matrix effect; thus, they provide a useful additional coordinate for metabolite identification in metabolomics [19, 20].

Standardized procedure for including IM in metabolomics workflows is necessary at this stage, also considering that CCS values can be obtained using different types of IM technologies [21].

Here, we present experimental protocols for integrating travelling-wave (TW)-IM with a traditional liquid chromatography (LC)-MS-based metabolomics workflow.

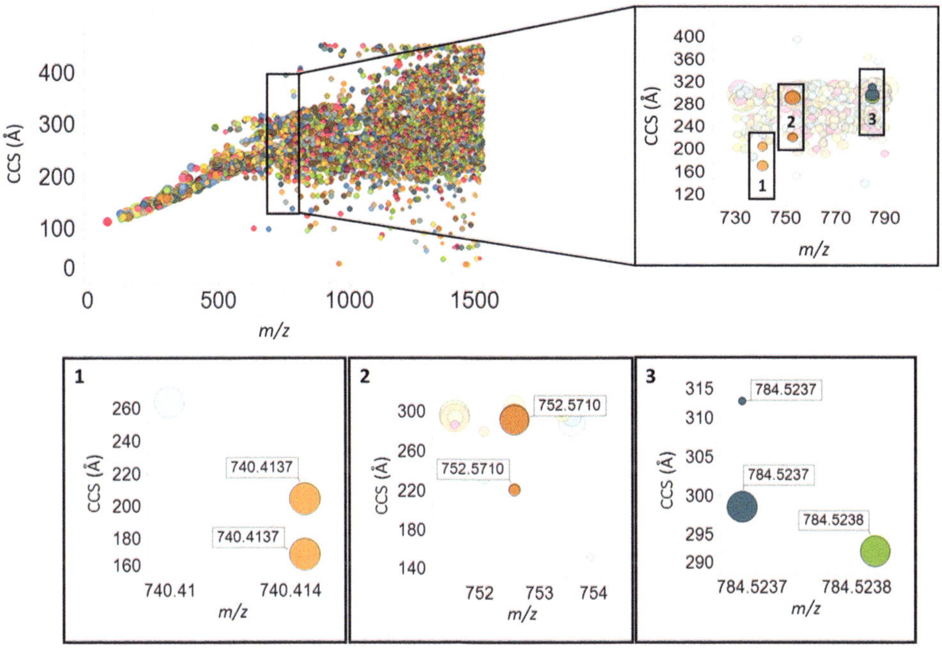

Fig. 1 The additional dimension of ion mobility separation increased the peak capacity and allows to separate and to identify near isobaric compounds during a typical metabolomics/lipidomics experiment

2 Materials

2.1 Equipment

1. Liquid chromatography system: ACQUITY UPLC (Waters Corporation, Milford, Massachusetts, USA).
2. TWIM-MS system: SYNAPT High Definition Mass Spectrometry (HDMS) (Waters Corporation, Milford, Massachusetts, USA).

2.2 Solvents and Chemicals

1. Water, methanol, chloroform, isopropanol, acetonitrile (ACN) (MS grade) (Sigma-Aldrich).
2. Formic acid and ammonium formate (Sigma-Aldrich).
3. System-suitability standard solution for polar metabolites: water/acetonitrile 50:50 (vol/vol) containing selected metabolites at the concentration of 10 mg/mL. Select metabolites considering the mass range that will be used during the metabolomics analysis. For analysis of small polar metabolites, we suggest to use the following metabolites: niacinamide (HMDB0001406), pyroglutamic acid (HMDB0000267), phenylalanine (HMDB0000159), succinic acid (HMDB0000254), hypoxanthine (HMDB0000157), arginine (HMDB0000517), inosine (HMDB0000195), S-adenosylhomocysteine (SAH) (HMDB0000939), and raffinose (HMDB0003213). Niacinamide, pyroglutamic acid,

phenylalanine, succinic acid, hypoxanthine, arginine, inosine, SAH, and raffinose can be purchased from Sigma-Aldrich.

4. System-suitability standards solution for lipids: 1 mg/mL of standard lipids in isopropanol. We suggest using commercially available lipid extracts as system-suitability standards to confirm visualization and separation of several lipid classes that might be present in biological samples. Natural lipids and lipid extracts can be purchased from Avanti Polar Lipids, Alabaster, Alabama, USA.
5. Poly-DL-alanine solution: water/acetonitrile 50:50 (vol/vol) containing poly-DL-alanine (Sigma-Aldrich. cat # p9003) at the concentration of 10 μg/mL.
6. Mobile phase A polar metabolites: ACN plus 0.1% of formic acid.
7. Mobile phase B polar metabolites: water plus 0.1% of formic acid.
8. Weak wash solvent polar metabolites: acetonitrile/water (10:90, vol/vol).
9. Strong wash solvent polar metabolites: acetonitrile/water (40:60, vol/vol).
10. Mobile phase A lipids: 10 mM ammonium formate with 0.1% formic acid in ACN/water (40:60 vol/vol).
11. Mobile phase B lipids: 10 mM ammonium formate with 0.1% formic acid in ACN/isopropanol (10:90 vol/vol).
12. Weak wash solvent lipids: ACN/water/isopropanol (30:30:40, vol/vol/vol).
13. Strong wash solvent lipids: isopropanol/water/formic acid/dichloromethane (95:5:2:1, vol/vol/vol/vol).
14. Leucine enkephalin (Sigma-Aldrich).
15. Sodium formate (Sigma-Aldrich).

2.3 Supplies

1. ACQUITY CSH C18 column (2.1 × 100 mm ID, 1.7 μm).
2. ACQUITY UPLC BEH Amide column (2.1 × 150 mm ID, 1.7 μm).
3. Glass Vials (8 mL, 40 mL, 1.5 mL for autosampler and LC analysis).
4. Glass pipettes (5 mL, 10 mL).
5. Glass Pasteur pipettes.
6. Caps with Teflon liner.
7. Conical insert for reducing the volume of the autosampler vials.

8. Vial racks.
9. Vial trays.

2.4 Software

1. MassLynx (Waters Corporation, Milford, Massachusetts, USA).
2. Progenesis® QI (Nonlinear Dynamics, Newcastle, UK).
3. DriftScope™ software (Waters Corporation, Milford, Massachusetts, USA).

3 Methods

3.1 Preparation of IM-MS Experiment Instrument Setting and Optimization

1. Perform routine maintenance and MS instrument calibration according to the instrument manufacturer's guidelines.
2. Set the mass spectrometer on Mobility-TOF mode using the settings provided in Table 1 or 2 that were previously optimized using a mixture of small metabolites and lipids, respectively.
3. Infuse the system-suitability standards solutions in the mass spectrometer at the rate of 5 μL/min in both ES^+ and ES^-. During the infusion optimize the IM parameters. Make sure that all molecules included in your system-suitability standards solution fit in the IM separation window. In order to optimize the TWIMS separation, adjust the following parameters:
   (a) *T*-wave velocity. Expand/narrow the drift time distribution profile.
   (b) *T*-wave height. Expand/narrow the drift time distribution profile.
   (c) IM separation gas flow rate. Shift the drift time values.
4. Data can be visualized using DriftScope software. By selecting the 2D plot (m/z vs. drift time), check that all masses included in your system-suitability standard solution are detected (*see* **Note 1**).

3.2 Preparation of IM-MS Experiment Calibration

1. Perform routine maintenance and MS instrument calibration according to the instrument manufacturer's guidelines (*see* **Note 2**).
2. Infuse the freshly prepared solution of poly-DL-alanine at 10 μL/min using the experimental settings previously optimized for both positive and negative ionization modes. Ensure that the beam is stable and that you clearly see the complete range of m/z that you expect to use for the calibration. Inspect the peaks of the singly charged oligomers reported in Table 3.
3. Annotate the drift time of the standards determined at the apex of the IM peak in both ES^+ and ES^- ionization modes.

Table 1
Suggested TWIM-MS settings for small polar metabolites

		ES⁻	ES⁻
MS			
Capillary		2.2	2.2
Cone		40 V	30 V
Source		110 (°C)	110 (°C)
Desolvation		450 (°C)	500 (°C)
Desolvation		450 (L/h)	500 (L/h)
Cone		20	20
EDC delay		1.58 V	1.58 V
Optic		Resolution	Resolution
MS		0.2 scan/s	0.2 scan/s
Lock mass		Leu enk (2 μg/mL)	Leu enk (2 μg/mL)
Lock mass		15 μL/min	15 μL/min
Triwave DC			
Trap DC	Entrance	3	3
	Bias	60	60
	Trap DC	0	0
	Exit	3	3
IMS DC	Entrance	25	25
	Helium cell DC	35	35
	Helium exit	–5	–5
	Bias	3	3
	Exit	0	0
Transfer DC	Entrance	3	3
	Exit	15	15
Triwave			
Trap	Wave velocity	311 (m/s)	311 (m/s)
	Wave height	6 (V)	6 (V)

(continued)

Table 1
(continued)

		ES^-	ES^-
IMS	Wave velocity	650 (m/s)	650 (m/s)
	Wave height	40 (V)	40 (V)
Transfer	Wave velocity	220 (m/s)	220 (m/s)
	Wave height	4 (V)	4 (V)
Gas control			
IMS gas (nitrogen)		90 (mL/min)	90 (mL/min)
Helium cell		180 (mL/min)	180 (mL/min)

4. Annotate the enhanced duty cycle delay coefficient (EDC delay) in use in the mass spectrometer. This is a correction factor (c) that can be found in the MassLynx Tune windows. Click System > Acquisition Settings. Then, on the Acquisition Setup tab, click EDC delay (mass < 5000 Da). The EDC delay correction factor © is then used to calculate the corrected drift time:

$t_d' = t_d - (c\,(\sqrt{(m/z)}/1000)$ ms.

t_d' = corrected drift time.

t_d = drift time.

c = EDC delay

5. To obtain calibration coefficients, use cross-section data (Ω) for singly charged poly-DL-alanine oligomers (Table 3) (*see* **Note 3**).
6. Correct published cross sections by accounting for reduced mass and charge state. Reduced mass:

$m = (M_{ion} \times m_{gas}/M_{ion} + m_{gas})$

m = reduced mass.

M_{ion} = molecular weight ion.

m_{gas} = molecular weight gas.

7. Calculate the Normalized Cross-Section: $\Omega' = [\Omega \cdot \text{Ö} \cdot (m)]/z$

Ω' = normalized collision cross section.

Ω = reference collision cross section.

z = charge.

Table 2
Suggested TWIM-MS settings for lipids

		ES⁻	ES⁻
MS			
Capillary voltage		2.5 kV	2.5 kV
Cone voltage		40 V	30 V
Source temperature		120 (°C)	120 (°C)
Desolvation temperature		500 (°C)	500 (°C)
Desolvation gas flow		600 (L/h)	600 (L/h)
Cone gas flow		50 (L/h)	50 (L/h)
EDC delay coefficient		1.58 V	1.58 V
Optic mode		Resolution	Resolution
MS scan rate		0.2 scan/s	0.2 scan/s
Lock mass solution		Leu enk (2 μg/mL)	Leu enk (2 μg/mL)
Lock mass flow rate		15 μL/min	15 μL/min
Triwave DC			
Trap DC	Entrance	3	3
	Bias	60	60
	Trap DC	0	0
	Exit	3	3
IMS DC	Entrance	25	25
	Helium cell DC	35	35
	Helium exit	−5	−5
	Bias	3	3
	Exit	0	0
Transfer DC	Entrance	3	3
	Exit	3	3
Triwave			
Trap	Wave velocity	311 (m/s)	311 (m/s)
	Wave height	4 (V)	4 (V)

(continued)

Table 2
(continued)

		ES⁻	ES⁻
IMS	Wave velocity	900 (m/s)	900 (m/s)
	Wave height	40 (V)	40 (V)
Transfer	Wave velocity	191 (m/s)	191 (m/s)
	Wave height	4 (V)	4 (V)
Gas control			
IMS gas (nitrogen)		90 (mL/min)	90 (mL/min)
Helium cell		180 (mL/min)	180 (mL/min)

Table 3
Collision cross section (CCS) in nitrogen for singly charged protonated and deprotonated oligomers of poly-DL-alanine [22]

		ES⁺		ES⁻	
Sequence poly-DL-alanine	**Molecular weight**	**$[M+H]^+$**	**CCS_{N2} (Å2)**	**$[M-H]^-$**	**CCS_{N2} (Å2)**
A_3	231.122	232.13	151	230.114	150
A_4	302.159	303.167	166	301.151	165
A_5	373.196	374.204	181	372.188	179
A_6	444.233	445.241	195	443.225	195
A_7	515.27	516.278	211	514.262	209
A_8	586.307	587.315	228	585.3	223
A_9	657.345	658.352	243	656.337	238
A_{10}	728.382	729.39	256	727.374	253
A_{11}	799.419	800.427	271	798.411	267
A_{12}	870.456	871.464	282	869.448	279
A_{13}	941.493	942.501	294	940.485	294
A_{14}	1012.53	1013.538	306	1011.522	308

8. Plot t_d' vsersus Ω', and fit a power-trend line of the form $y = Ax^b$ to the Data, to Obtain a calibration curve. Annotate the equation (*see* **Note 4**).

3.3 Preparation of IM-MS Experiment Build CCS Library

CCS values can be used as additional orthogonal measurement, in addition to m/z, in searchable database to increase the confidence of metabolite identification. $^{TW}CCS_{N2}$ reference values have been reported in the literature [19, 20]. Nevertheless, the number of CCS_{N2} values available for metabolites and lipids is still limited and does not cover the complexity of the metabolome and lipidome. Therefore, there is the necessity to extend and populate existing metabolomics and lipidomics databases with CCS_{N2} values.

Building a $^{TW}CCS_{N2}$ database requires experimental procedures based on calibration with compounds having known CCS_{N2} such as poly-DL-alanine (Fig. 2).

1. Prepare a solution containing standard metabolites or lipids of interest at the concentration of 10 μg/mL.
2. After performing calibration with poly-DL-alanine, directly infuse the solution containing the standard metabolites at 5 μL/min (in both ES^+ and ES^- ionization modes). Annotate the drift time of the standard compounds determined at the apex of the IM peak in both ES^+ and ES^- ionization modes and convert experimental drift times in collision cross sections using the equation of the calibration curve obtained for poly-alanine. IM parameter settings used during poly-DL-alanine calibration and those used to obtain $^{TW}CCS_{N2}$ for standard compounds and unknown metabolites must be identical (*see* **Note 5**).
3. Repeat the infusion of standard metabolites three times and calculate $^{TW}CCS_{N2}$ and standard deviation.
4. Annotate experimental condition used for deriving experimental $^{TW}CCS_{N2}$.
5. To validate the new $^{TW}CCS_{N2}$ values, repeat the whole procedure using a different IM settings (e.g., change *T*-wave velocity and *T*-wave height).
6. Perform calibration with poly-DL-alanine using a new IM setting (repeat the procedure reported in Subheading 3.2).
7. Derive $^{TW}CCS_{N2}$ values of standard metabolites or lipids with the new IM setting, and compare with the previous values.

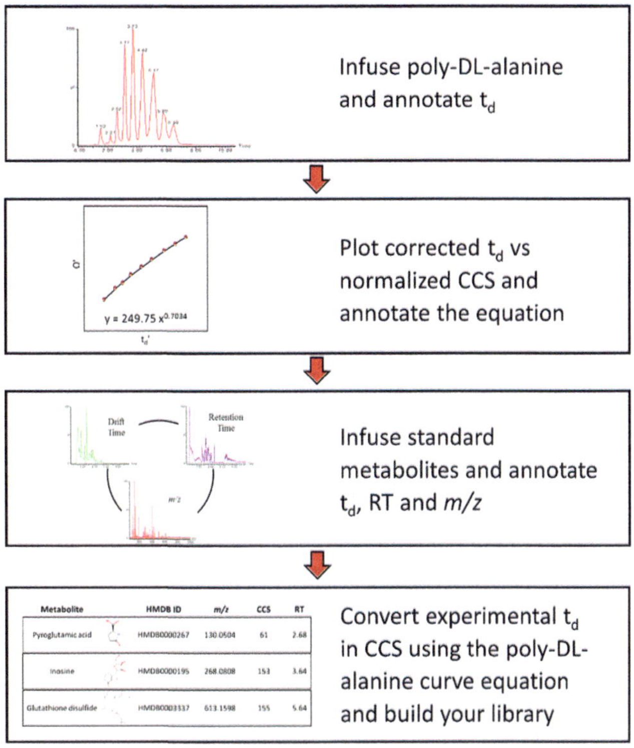

Fig. 2 Experimental workflow for building $^{TW}CCS_{N2}$ database

3.4 LC-TWIM-MS Analysis of Polar Metabolites

Metabolic profiling of the small polar metabolites can be achieved by coupling hydrophilic interaction liquid chromatography (HILIC) with TWIM-MS. We suggest to use the following column for the separation of small polar compounds. ACQUITY UPLC BEH Amide column (2.1 × 150 mm, particle size 1.7 μm) [20]:

1. Prepare QC samples by pooling together small aliquots of your samples.
2. Calibrate the mass spectrometer using sodium formate in both positive and negative ionization modes. Set the mass spectrometer to mobility-TOF using the setting reported in Table 1.
3. Perform ion mobility calibration as described in Subheading 3.2.
4. Set the UPLC using a column temperature at 45 °C, the flow rate at 0.4 mL/min, the injection volume at 5 μL, and the autosampler temperature at 4 °C. The separation of polar metabolites is achieved by using the following linear elution gradient: 0 min, 99% of A; 6 min, 40% of A; 8 min, 99% of A;

10 min, 99% of A. Equilibrate the chromatographic system by running the gradient four times without injecting any sample.

5. Run the system-suitability standard solution for polar metabolites, and ensure that the LC-TWIM-MS system is working fine by checking mass accuracy, retention time, area, and $^{TW}CCS_{N2}$ of the polar metabolites.
6. Equilibrate the LC-TWIM-MS system by injecting at least ten QC samples, and then analyze samples in a randomized order including a QC sample every eight injections (*see* **Note 6**).

3.5 LC-TWIM-MS Analysis of Lipids

LC-TWIM-MS analysis of lipid can be achieved using a reversed-phase chromatography. We suggest to use the Charged Surface Hybrid (CSH™) C18 (2.1 × 100 mm) 1.7 μm.

1. Prepare QC samples by pooling together small aliquots of your samples.
2. Calibrate the mass spectrometer using sodium formate in both positive and negative ionization modes. Set the mass spectrometer to mobility-TOF using the setting reported in Table 2.
3. Perform ion mobility calibration as described in Subheading 3.2.
4. Prepare the UPLC system by setting the column temperature at 55 °C, the flow rate at 0.4 mL/min, the injection volume at 5 μL, and the autosampler temperature at 10 °C.
5. The separation of lipids is achieved by using the following elution gradient: 0 min, 40% of B; 2 min, 43% of B; 2.1 min, 50% of B; 12 min, 54% of B; 12.1 min, 70% of B; 18 min, 99% of B; 18.1 min, 40% of B; 22 min, 40% of B.
6. Equilibrate the chromatographic system by running the gradient four times without injecting any sample.
7. Run the system-suitability standard solution for polar metabolites, and ensure that the LC-TWIM-MS system is working fine by checking mass accuracy, retention time, area, and $^{TW}CCS_{N2}$ of the polar metabolites (*see* **Note 7**).
8. Equilibrate the LC-TWIM-MS system by injecting at least ten QC samples and then analyze samples in a randomized order including a QC sample every eight injections (*see* **Note 6**).

3.6 Data Processing and Analysis

During metabolomics experiment, it is possible to automatically generate a list of features, and each feature is associated with a specific retention time and mass values. Specific software, as Progenesis QI, allow a similar approach during UPLC-TWIM-MS experiments. In this case each feature will be associated with three molecular identifier, $^{TW}CCS_{N2}$, retention time, and *m/z* [19, 20]. Indeed,

Progenesis QI software is coded to automatically convert drift time data into CCS values using the calibration curve previously acquired.

1. Import the UPLC-TWIMS-MS data into Progenesis® QI software.
2. Align ion maps in the retention time direction, and perform peak picking.
3. Convert drift time data into $^{TW}CCS_{N2}$ values by providing the equation for the calibration curve acquired using poly-DL-alanine.
4. Perform statistical analysis and select features of interest according to significance.
5. Search the features in databases containing $^{TW}CCS_{N2}$ information by matching experimentally derived $^{TW}CCS_{N2}$ values against $^{TW}CCS_{N2}$ databases [19, 20].
6. Select a ΔCCS less than that of a given threshold as tolerance parameters. The difference between reference $^{TW}CCS_{N2}$ (contained in the database) and experimental $^{TW}CCS_{N2}$ values (ΔCCS) contributes to the identification score, in addition to accurate masses, isotopic pattern, and retention times.

4 Notes

1. While adjusting IM parameters might occur that the same ion peak is detected at the beginning and at the end of the driftogram. This phenomenon occurs because a new ion packet is released from the Trap region before the previous ion packet has been delivered to the pusher region. To fix this artifact, you can increase *T*-wave height and decrease *T*-wave velocity in the IMS cell.
2. Allow the mass spectrometer to remain in Operate mode for at least 1 h prior to CCS calibration to ensure a stable calibration. This procedure is necessary only before performing a CCS calibration.
3. These $^{TW}CCS_{N2}$ values were obtained using nitrogen. If you plan to use helium, you will need CCS for singly charged poly-DL-alanine oligomers obtained by using helium (CCS_{He}) [22].
4. CCS calibration can be performed automatically using MassLynx. Using the instrument's IntelliStart feature, select CCS calibration. Infuse the poly-DL-alanine into the mass spectrometer and acquire calibration data automatically via IntelliStart. Accept the calibration and use the instrument's

LockCCS function to set up. Immediately after the CCS calibration, perform the LockCCS by infusing leucine enkephalin at 10 μL/min. Use IntelliStart, and select LockSpray source setup. Start flow, and verify that $[M+H]^+$ or $[M-H]^-$ ions of leucine enkephalin can be seen in the Tune window. IntelliStart then acquires data automatically.

5. Metabolites can provide multiple ions, such as adducts, multimers, multiple charged ions, and fragment ions. Pay attention during the procedure for deriving $^{TW}CCS_{N2}$, since every ion produced from a single metabolite will have its unique $^{TW}CCS_{N2}$.
6. During untargeted metabolomics do not include blank samples during sample. Include blank sample at the beginning and at the end of the batch.
7. More specific calibrants can be used for deriving $^{TW}CCS_{N2}$ values for phospholipids [23].

References

1. Rainville PD, Wilson ID, Nicholson JK et al (2017) Ion mobility spectrometry combined with ultra performance liquid chromatography/mass spectrometry for metabolic phenotyping of urine: effects of column length, gradient duration and ion mobility spectrometry on metabolite detection. Anal Chim Acta 982:1–8
2. Kanu AB, Dwivedi P, Tam M et al (2008) Ion mobility-mass spectrometry. J Mass Spectrom 43:1–22
3. Lapthorn C, Pullen F, Chowdhry BZ (2013) Ion mobility spectrometry-mass spectrometry (IMS-MS) of small molecules: separating and assigning structures to ions. Mass Spectrom Rev 32:43–71
4. May JC, McLean JA (2015) Ion mobility-mass spectrometry: time-dispersive instrumentation. Anal Chem 87:1422–1436
5. Dwivedi P, Schultz AJ, Hill HH Jr (2010) Metabolic profiling of human blood by high-resolution ion mobility mass spectrometry (IM-MS). Int J Mass Spectrom 298:78–90
6. Dwivedi P, Puzon G, Tam M et al (2010) Metabolic profiling of Escherichia coli by ion mobility-mass spectrometry with MALDI ion source. J Mass Spectrom 45:1383–1393
7. Shah V, Castro-Perez JM, McLaren DG et al (2013) Enhanced data-independent analysis of lipids using ion mobility-TOFMS E to unravel quantitative and qualitative information in human plasma. Rapid Commun Mass Spectrom 27:2195–2200
8. Hart PJ, Francese S, Claude E et al (2011) MALDI-MS imaging of lipids in ex vivo human skin. Anal Bioanal Chem 401:115–125
9. Ahonen L, Fasciotti M, af Gennäs GB et al (2013) Separation of steroid isomers by ion mobility mass spectrometry. J Chromatogr A 1310:133–137
10. Malkar A, Devenport NA, Martin HJ et al (2013) Metabolic profiling of human saliva before and after induced physiological stress by ultra-high performance liquid chromatography-ion mobility-mass spectrometry. Metabolomics 9:1192–1201
11. Kaplan K, Dwivedi P, Davidson S et al (2009) Monitoring dynamic changes in lymph metabolome of fasting and fed rats by electrospray ionization-ion mobility mass spectrometry (ESI-IMMS). Anal Chem 81:7944–7953
12. Zhang X, Romm M, Zheng X et al (2016) SPE-IMS-MS: an automated platform for sub-sixty second surveillance of endogenous metabolites and xenobiotics in biofluids. Clin Mass Spectrom 2:1–10
13. Stow SM, Causon TJ, Zheng X et al (2017) An interlaboratory evaluation of drift tube ion mobility-mass spectrometry collision cross section measurements. Anal Chem 89:9048–9055
14. Zheng X, Aly NA, Zhou Y et al (2017) A structural examination and collision cross section database for over 500 metabolites and xenobiotics using drift tube ion mobility spectrometry. Chem Sci 8:7724–7736

15. Zhou Z, Tu J, Zhu ZJ (2018) Advancing the large-scale CCS database for metabolomics and lipidomics at the machine-learning era. Curr Opin Chem Biol 42:34–41
16. Zhou Z, Tu J, Xiong X et al (2017) LipidCCS: prediction of collision cross-section values for lipids with high precision to support ion mobility-mass spectrometry-based lipidomics. Anal Chem 89:9559–9566
17. Righetti L, Bergmann A, Galaverna G et al (2018) Ion mobility-derived collision cross section database: application to mycotoxin analysis. Anal Chim Acta 1014:50–57
18. Pacini T, Fu W, Gudmundsson S et al (2015) Multidimensional analytical approach based on UHPLC-UV-ion mobility-MS for the screening of natural pigments. Anal Chem 87:2593–2599
19. Paglia G, Angel P, Williams JP et al (2015) Ion mobility-derived collision cross section as an additional measure for lipid fingerprinting and identification. Anal Chem 87:1137–1144
20. Paglia G, Williams JP, Menikarachchi LC et al (2014) Ion mobility-derived collision cross-sections to support metabolomics applications. Anal Chem 86:3985–3993
21. May JC, Morris CB, McLean JA (2017) Ion mobility collision cross section compendium. Anal Chem 89:1032–1044
22. Bush MF, Campuzano IDG, Robinson CV (2012) Ion mobility mass spectrometry of peptide ions: effects of drift gas and calibration strategies. Anal Chem 84:7124–7130
23. Hines KM, May JC, McLean JA et al (2016) Evaluation of collision cross section calibrants for structural analysis of lipids by traveling wave ion mobility-mass spectrometry. Anal Chem 88:7329–7336

Chapter 5

Capillary Electrophoresis Mass Spectrometry as a Tool for Untargeted Metabolomics

Ángeles López-Gonzálvez, Joanna Godzien, Antonia García, and Coral Barbas

Abstract

Although capillary electrophoresis (CE) coupled to mass spectrometry (MS) is a separation technique not extensively implemented, it offers differential possibilities in the study of polar and ionic metabolites in complex matrices with minimum sample treatment. However, in order to get successful results, some efforts at early stages and following specific recommendations are necessary.

In this chapter, we describe our updated and well-tested methods for untargeted metabolomics using CE-MS-TOF for common biological samples: urine, serum or plasma, feces, tissues, and cells. Sample treatment, as well as separation and detection conditions are described in detail and other steps in the workflow for untargeted metabolomics are also explained. Special attention is paid to instrumental setup and advices for daily practice.

Characteristic electropherograms obtained with each type of sample are depicted as well as groups of metabolites easily measured by this technique. Their global or individual comparisons have been given undoubtedly important information to unveil altered metabolic pathways, diagnosis, and prognosis or biomarker discovery in the study of diseases or conditions over decades.

Key words Fingerprinting, Biological samples, Polar metabolites, Untargeted metabolomics, CE-MS

1 Introduction

Capillary electrophoresis, as a separation technique mainly used for polar and ionic compounds, able to deal with complex matrices and based on different principles of separation than chromatographic techniques has found a place in untargeted metabolomics when coupled with mass spectrometry as a detector [1].

It is not a mainstream instrumentation for several reasons: (1) there are not many people trained to use it; (2) it is not very robust. It requires careful set up of conditions and ideally one person working with the instrument instead of working in "open access" mode. In addition, migration time shifts are higher than in LC and data pretreatment parameters should be carefully considered.

Angelo D'Alessandro (ed.), *High-Throughput Metabolomics: Methods and Protocols*, Methods in Molecular Biology, vol. 1978, https://doi.org/10.1007/978-1-4939-9236-2_5,

Despite of it, for those taking the necessary work to deal with the limitations, it offers a lot of opportunities to uncover compounds that are not easy to analyze with other techniques [2]. CE as compared to GC does not require derivatization, which makes easier sample handling and avoids problems related to multiderivatization. When compared to HILIC chromatography, CE permits the dissolution of the sample in water which is the proper medium for polar and ionic substances.

Capillary zone electrophoresis (CZE), the most common mode of CE when coupled to MS, separates first in CE on the basis of charge-to-size ratio of the analytes and then in the MS on the basis of its mass-to-charge ratio (m/z). In this way, it offers an alternative selectivity to LC that can be of interest for broader metabolite coverage. The coupling to MS is important in untargeted metabolomics although it restricts the substances that can be added to the separation buffer to those that can be made volatile.

The applications of CE-MS in untargeted metabolomics expand not only to any biological fluid but also to any cell culture or tissue from which metabolites can be extracted. In this chapter we will describe the applications in which our laboratory has a broad experience such as urine, plasma, serum, feces, tissues, and cell cultures.

2 Materials

2.1 Samples

1. Urine samples
   First-void urine samples are preferred compared to spot urine samples because the influence of lifestyle factors (such as diet, physical exertion, and stress) on the urinary metabolic profile is minimized and normalized.
2. Plasma and serum samples
   Fresh EDTA anticoagulated blood must be quickly centrifuged and then frozen at −80 °C.
3. Feces samples
   Fresh stool samples collected from each subject, must be frozen immediately, and stored at −80 °C until they are processed.
4. Tissue samples
   Tissue sample after dissection must be weighted and frozen at once in liquid nitrogen, and stored at −80 °C until they are processed.
5. Cell culture samples
   Parasites are grown in RPMI-1640 medium supplemented with 10% heat inactivated fetal calf serum at 26 °C. Large-scale cultures are carried out in a roller apparatus (Gibco, Cell Culture) with an inoculum of 4×10^5 promastigotes × mL^{-1}.

Biological replicates are collected in the late log phase of promastigote culture. Before metabolite extraction, metabolism is quenched by shaking the 10 mL culture bottle in an ethanol/dry-ice bath for 40 s. Finally, a volume equivalent to 4×10^7 promastigotes or amastigotes are centrifuged at $2000 \times g$ for 10 min at 4 °C, washed three times in cold (4 °C) PBS (137 mM NaCl, 8 mM Na_2HPO_4, 2.7 mM KCl, and 1.5 mM KH_2PO_4; pH 7.0) and pellet stored at −80 °C until they are processed.

2.2 Chemicals for Metabolic Fingerprinting in Biological Samples

All reagents and solvents are purchased from Sigma-Aldrich (Steinheim, Germany) (unless indicated otherwise). Individual 1000 mg/L (ppm) stock solutions of the reference compounds are prepared in Milli-Q water (unless indicated otherwise). Prepare all solutions using ultrapure Milli-Q® water (Millipore, Billerica, MA, USA) and MS grade reagents. Store all prepared solutions at 4 °C and −20 °C (according to prescription). Internal standard methionine sulfone is prepared in ultrapure water at a concentration of 25 mM. Conduct all volume and weight measurements in volumetric flasks and analytical balances, respectively. Diligently follow all waste disposal regulations when disposing of waste materials. Before the analysis, the solutions must be thawed and diluted appropriately to obtain a working concentration of 10 ppm of each standard and 0.2 mM of L-methionine sulfone.

2.3 Solvents and Buffer Reagents

1. Acetonitrile (ACN), methanol (MeOH) and isopropanol (IPA), all of them MS quality, came from Sigma (St. Louis, MO, USA).
2. Sodium hydroxide (NaOH) 97% is from Panreac (Barcelona, Spain).
3. Formic acid analytical grade came from Sigma (St. Louis, MO, USA).
4. Background electrolyte (BGE): 1.0 M formic acid diluted from concentrated formic acid in ultrapure water with 10% MeOH.
5. CE sheath liquid used for detection: 100 mL MeOH, 100 mL ultrapure water, 4 μL concentrated formic acid, and the reference mass standards from Agilent Technologies (10 μL 5 mM purine and 10 μL 2.5 mM HP922).

2.4 CE-MS Equipment and Setup

1. Metabolomics studies are performed using a capillary electrophoresis (7100 Agilent Technologies, Waldbronn, Germany) coupled to an Accurate-Mass Time of Flight mass spectrometer (TOF-MS) system (6224 Agilent Technologies) and controlled by MassHunter.
2. Fused-silica capillary (Agilent Technologies) 125 cm total length, 50 μm internal diameter. The capillary is cut to a length of 1 m eliminating the window of UV detection (*see* **Note 1**).

The capillary tip that becomes the sprayer side (MS side) is a key point in the system. It must be smooth in order to achieve a stable nebulization (*see* **Note 2**). Observe the end of the capillary under a microscope or a magnifying glass by rotating it up to 360° (*see* **Note 2**). Finally, sonicate both ends for 1 or 2 min in IPA–water (1:1, v/v) to clean them. Longer lifetime is reached if both ends of the capillary are maintained in water to keep them wet after being used.

3. Check the assembly of the needle and the sprayer tip with the nebulizer adjustment kit to detect any damage that could cause bad performance. The cone must be symmetrical. If the tip does not appear smooth and even or the cone isn't symmetrical, the sprayer needle needs to be replaced (Fig. 1). In addition, it is necessary to confirm the protrusion of the capillary. The protrusion must be about ½ to ¼ of the capillary diameter as depicted in Fig. 1.
4. The CE system is coupled to the MS by an electrospray source (ESI). Connect the capillary and the sheath liquid pipe to the sprayer kit. Care should be taken not to damage the capillary end. The tip of the inlet electrode and the tip of the MS sprayer must be at the same level (*see* **Note 3**). A 1200 series ISO Pump from Agilent Technologies is used to supply the sheath liquid.

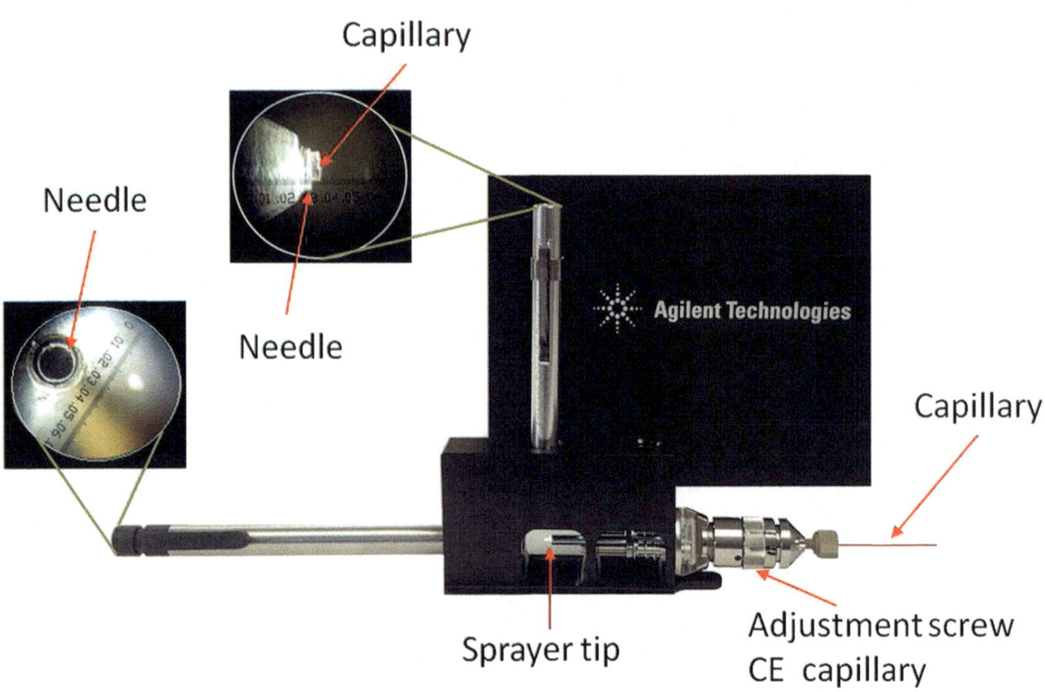

Fig. 1 Looking through the nebulizer adjustment kit: needle and protrusion. ESI sprayer with the capillary, the adjustment screw, and the spray tip

5. Optimization of the sensitivity and System suitability: Verify the sensitivity of the detection by analyzing two times a solution with three standards at 10 mg/L (test solution) and compare the results with a test electropherogram. We recommend optimizing the sensitivity always by enlarging or shortening the protrusion of the capillary. That could be done easily by rotating clockwise o counter clockwise the knob in the spray holder (directly, do not disassemble it). Analyze the same standard solution again (3–4 times) until the best sensitivity in the MS detection is achieved. Afterward, evaluate the system suitability based on the repeatability of multiple analysis (≥5) of the same standard solution and evaluate the CV of the peak abundances and migration times of the compounds before running the study sequence.
6. Data acquisition is performed by Mass Hunter WorkStation Software (version B.06.01) (Agilent Technologies, Santa Clara, CA, USA).

2.5 Other Equipment

1. Centrifuge Megafuge 1.0R (Heraeus Instruments, Langenselbold).
2. Centrifuge 5415 R (Eppendorf, Hamburg, Germany).
3. Speedvac concentrator: SpeedVac SPD121P (Thermo Electron Corporation, Milford, MA, USA).
4. Centrifree® ultracentrifugation device (Millipore Ireland Ltd., Ireland) 30 kDa protein cutoff (PN: 4104).
5. Polypropylene vials (Agilent Technologies, Santa Clara, USA) with 0.250 mL.
6. 3 mL Vacuette plastic tubes containing 4.5 mg of EDTA·K3 for plasma collection (Greiner Bio-One, Brazil).
7. Eppendorf tubes 0.5 and 1.5 mL (Eppendorf).
8. Glass beads, acid-washed: 425–600 μm and 5–7 mm (Sigma-Aldrich, St. Louis, MO, USA).
9. Liquid nitrogen (Air Liquide, Class Alphagaz™ 1).
10. Tissuelyzer LT (Qiagen, Hilden, Germany).
11. Vortex mixer.

2.6 Software

1. Mass Hunter WorkStation Software (version B.06.01) (Agilent Technologies, Santa Clara, CA, USA).
2. MassHunter Qualitative Analysis Software (version B.06.00) (Agilent Technologies, Santa Clara, CA, USA).
3. Mass Hunter Profinder (version B.08.00) (Agilent Technologies, Santa Clara, CA, USA).
4. SIMCA-P + version 12.0.1 (Umetrics, Umeå, Sweden) for multivariate data analysis.

5. CEU Mass Mediator (http://ceumass.eps.uspceu.es/). This is a tool to search measured accurate masses against known metabolite databases (KEGG, Metlin, LipidMaps and HMDB).

3 Methods

3.1 Urine Sample Treatment for Metabolite Fingerprinting

Urine is a biofluid that contains metabolic signatures of many biochemical pathways. It is commonly used for metabolomics investigation and clinical assays due to noninvasive collection, the complex nature, and the ability to collect multiple samples over a period of time. However, it is important to determine appropriate sampling time (24 h collection versus random or timed).

Urine is a main carrier of hydrophilic waste compounds from the bloodstream filtered by the kidney. The main drawback is the variability in its concentration that makes necessary the use of different types of normalization.

Urine contains mostly water (>90%) and small molecules as urea, creatinine, electrolytes, etc. with very low level of proteins. Even though the presence of blood in urine is highly unlikely, based on our experience, when collected from newborns it is more common, and a method is optimized for those samples that is included below.

CE has great potential capabilities for the study of polar and ionic compounds in aqueous samples and enormous simplicity in sample treatment of urine samples.

3.1.1 Urine Preparation

1. Thaw samples on ice.
2. Vortex for 1 min.
3. Centrifuge for 20 min at 4 °C at 13,200 × *g*.
4. Place 20–100 μL (depending on the sample volume available) of the urine supernatant in an Eppendorf tube.
5. Add in proportion 1:1 (sample/solutions of 0.2 M formic acid containing 0.2 mM of the methionine sulfone with internal standard).
6. Vortex for 1 min.
7. Centrifuge for 15 min at 4 °C at 13,200 × *g*.
8. Place 50 μL of the supernatant in the CE vial (polypropylene vials). Only a few nanoliters will be injected.
9. For long sequences (more than 2 days of analysis), close the vials hermetically with crim caps and keep them at low temperature in the fridge (≤3 days). The day of the analysis, take them out of the fridge, replace caps by CE caps, and centrifuge the vials (*see* **Note 4**).

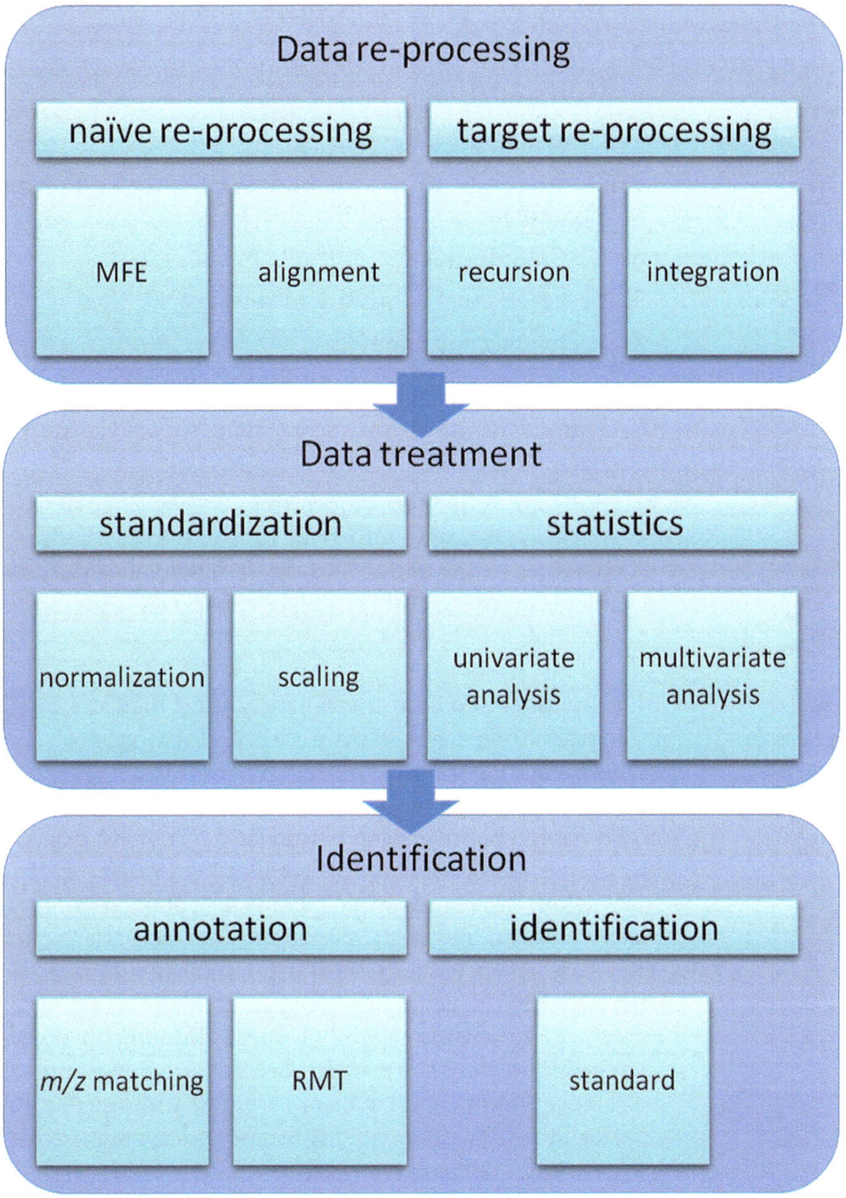

Fig. 2 Scheme of main steps in data processing for CE-MS based untargeted metabolic phenotyping

3.1.2 Urine Preparation with Proteins by the Presence of Blood

1. Thaw samples on ice.
2. Vortex-mix samples for 1 min.
3. On ice, transfer 50 μL of urine to an Eppendorf tube (0.5 mL).
4. Add 50 μL 0.2 M formic acid with 0.4 mM methionine sulfone (IS).
5. Vortex-mix 5 min.
6. Centrifuge for 10 min at 13,200 × *g* at 4 °C.

Deproteinization

1. Transfer 60 μL of supernatant.
2. Add 180 μL MeOH (cold).
3. Vortex-mix 5 min.
4. Centrifuge for 15 min at 13,200 × *g* at 4 °C.
5. Transfer 200 μL of the supernatant to an Eppendorf tube (0.5 mL).
6. Dry in Speedvac for approx 2 h.
7. Resuspend in 100 μL 0.1 M formic acid.
8. Vortex-mix for 5 min.
9. Centrifuge for 5 min at 13,200 × *g* at 4 °C.
10. Transfer 80 μL of the supernatant to CE vial (polypropylene vials).
11. For long sequences (more than 2 days of analysis), close the vials hermetically with crim caps and keep them at low temperature in the fridge (≤3 days). The day of the analysis, take them out of the fridge, replace caps by CE caps, and centrifuge the vials (*see* **Note 4**).

3.2 Plasma and Serum Sample Treatment for Metabolite Fingerprinting

Blood plasma and serum are widely used matrices in clinical and biological studies as both can give a global view of the instantaneous metabolic status.

Blood is composed of cells and approximately 50% (v/v) of liquid named plasma. Both plasma and serum are derived from full blood; serum comes from blood allowed to clot, centrifuged, and supernatant collected, while plasma is obtained when anticoagulants are added and the supernatant collected after centrifugation. Therefore, even though both are aqueous solutions (more than 90% water) plasma contains between 6% and 8% of proteins and this percentage in serum is slightly lower. Albumin constitutes approximately 60% of all of the protein. Before the analysis, proteins should be removed to prevent capillary clogging and signal noise.

Serum and plasma contain also peptides, amino acids, nitrogen compounds, electrolytes, carbohydrates, lipids and waste substances collected from the body more or less dissolved [3]. There is a comprehensive study that describes a higher sensitivity of metabolites in serum but better reproducibility in plasma [4].

Hemolyzed samples always should be avoided [5] and if they are present, they will be likely outliers.

3.2.1 Plasma Preparation

1. Collect fresh fasting blood into EDTA-treated tubes.
2. Shake gently and thoroughly for about 15 min.
3. Centrifuge anticoagulated blood at 1000 × *g* for 10 min at 4 °C.
4. Collect the supernatant and freeze the plasma at −80 °C in 100 μL aliquots.

3.2.2 Serum Preparation

1. Collect fresh fasting blood into serum-separating tubes.
2. Invert gently and let rest for about 45 min at room temperature for complete coagulation.
3. Centrifuge at 1000 × *g* for 10 min at 4 °C.
4. Collect the supernatant and freeze the serum at −80 °C in 100 μL aliquots.

3.2.3 Deproteinization of Plasma or Serum

1. Thaw samples (either plasma or serum) on ice.
2. Place 100 μL of serum or plasma in Eppendorf tubes.
3. Add 100 μL of 0.2 M formic acid with 5% ACN and 0.4 mM methionine sulfone, IS.
4. Mix by vortexing for approximately 1 min.
5. Transfer the mixed samples to centrifree Millipore (30 kDa protein cutoff) filters (*see* **Note 5**).
6. Centrifuge for 70 min at 4 °C with 2000 × *g*.
7. Transfer the filtrate directly to a polypropylene vial for the analysis and close the vial with aluminum crimp top-seal.
8. Before the analysis it is necessary to change caps and centrifuge the vials (*see* **Note 4**).

3.3 Feces Sample Treatment for Metabolite Fingerprinting

Feces are samples directly related to gut microbiota. The gut metabolome regulates many processes, such as immunity and energy-harvesting metabolism of dietary components [6]. It may be associated with the incidence and progression of numerous diseases [7].

Fecal metabolites reflect complex interactions between dietary inputs, intestinal bacteria, and host metabolism [8], and risk, or progression, of diseases such as colorectal cancer. According to Rojo et al. in a review from 2017 [9] even though CE is able to give information as far as around 23% of the total number of metabolites found to be altered in person's colon, CE as separation technique for gut microbiota metabolomics analysis is used by less than 5% of the studies.

Feces are made up of 75% water and 25% solid matter that consists of dead bacteria (30%), indigestible food matter (30%), fats (10–20%), inorganic substances (10–20%), and protein (2–3%). Sample treatment includes extraction directly with water or PBS and the lyophilization of the feces, to avoid a bias due to the material consistence.

3.3.1 Fecal Debris Removal

1. Weight with accuracy approximately 0.4 g of feces.
2. Add 1.2 mL PBS.
3. Vortex-mix for 1 min (resuspension).
4. Centrifuge at 1000 × *g* for 1 min at 4 °C.

3.3.2 Prokaryotic Cell Isolation

1. Transfer supernatant to an Eppendorf tube 1.5 mL.
2. Centrifuge at 13,200 × *g* for 5 min at 4 °C.
3. Remove supernatant (keeping bacterial pellets).

3.3.3 Metabolite Double Solvent Extraction

1. Add 1.2 mL of cold MeOH (−80 °C) to the bacterial pellets.
2. Vortex-mix 10 s.
3. Sonicate 30 s, 15 W on ice bath (−20 °C).
4. Repeat three times, 5 min on ice bath in between.
5. Centrifuge at 12,000 × *g*, for 5 min at 4 °C, 10 min (Debris pelleting).
6. Repeat **steps 1–5** and collect all the supernatant.
7. Transfer 100 μL supernatant to an Eppendorf tube.
8. Dry in Speedvac for 1 h approximately.
9. Resuspend in 100 μL of 0.1 M formic acid with 0.2 mM methionine sulfone (IS).
10. Centrifuge for 20 min at 16,000 × *g* at 4 °C before analysis.
11. Transfer 80 μL of the supernatant to a polypropylene vial for analysis.
12. For long sequences (more than 2 days of analysis), close the vials hermetically with crim caps and keep them at low temperature in the fridge (≤3 days). The day of the analysis, take them out of the fridge, replace caps by CE caps and centrifuge the vials (*see* **Note 4**).

3.4 Tissue Sample Treatment for Metabolite Fingerprinting

Tissues are complex structures that can early reflect metabolomics modification in pathological conditions as cancer much sooner than mainstream biofluids. They could come from biopsies (human) or the whole organ (mouse) and could comprise myocardium, liver, kidney, colon, adipose tissue, prostate, lung, aorta, stomach, etc. in human and animal models. Tissue collection and homogeneity generate intrinsic challenges in nontargeted metabolomics especially in tumor sample collection [10]. Sample treatment with tissue involves a prestep with optimized solvents and tools for mechanical disruption that allow us to obtain the tissue homogenate from which metabolites will be extracted. For complete cell disruption of hard tissues, a cell disrupter can be a useful [1]. Checking the process performance with a microscope is recommended.

3.4.1 Tissue Preparation

1. Immerse the samples in liquid nitrogen.
2. Weight frozen tissue in precooled 2 mL tubes.
3. Add cold MeOH (−20 °C) 50% in the ratio sample–solvent 1:10 (w/v). That is, for example for 20 mg of tissue, 200 μL of the mixture must be added.

3.4.2 Disruption and Homogenization of the Tissue

1. Add one 5 mm or 7 mm (very hard tissue) stainless steel bead to each tube.
2. Place each tube in TissueLyser LT Adapter. Perform 3 cycles of 5 min at 50 Hz.
3. Take 100 μL of homogenate to an Eppendorf tube.
4. Add 100 μL of 0.2 M formic Acid with 5% ACN.
5. Vortex-mix for 20 min.
6. Centrifuge for 10 min at 16,000 × *g* at 4 °C.
7. Transfer supernatant to a Millipore filter (30 kDa protein cutoff).
8. Centrifuge for 70 min at 2000 × *g* at 4 °C.
9. Transfer the filtrate to an Eppendorf tube.
10. Dry in Speedvac for 2 h.
11. Resuspend in 100 μL of 0.1 M formic Acid with 0.2 mM methionine sulfone (IS).
12. Centrifuge for 20 min at 16,000 × *g* at 4 °C before analysis.
13. Transfer 80 μL of the supernatant to a polypropylene vial for analysis.
14. For long sequences (more than 2 days of analysis), close the vials hermetically with crim caps and keep them at low temperature in the fridge (≤3 days). The day of the analysis, take them out of the fridge, replace caps by CE caps, and centrifuge the vials (*see* **Note 4**).

3.5 Cell Culture Sample Treatment for Metabolite Fingerprinting (e.g., Metabolite Extraction of Leishmania)

Metabolite fingerprinting of cells can offer a snapshot of cell activity changes and endogenous biochemical reaction products occurring within a living cell [11]. Cell metabolome can be defined as the set of all the metabolites present in cells and considered the best indicator of an organism's phenotype [12].

Metabolomics offers new opportunities to unveil new potential diagnostic biomarkers, in drug discovery as well as in the study of the final effect of a drug and how it challenges cell or parasite metabolism. From the study of the intracellular or endometabolome (all the metabolites present inside the cell) and extracellular or exometabolome (all the metabolites in the surrounding extracellular medium) of a specific disease complementary information can be obtained and uncovered.

Many applications of CE-MS in CEMBIO can be highlighted in the study of human and rodent stem cells for therapeutic intervention [13] along with in host-pathogen interactions that are stablished after viral, bacterial, and parasitic infections.

Cell density is a crucial factor that contributes to possible alterations in the cell metabolome as the quantity of metabolites is directly proportional to the number of cells. In a metabolomic study is recommended to adjust the number of cells of each sample to 1×10^7 cells whenever possible.

3.5.1 Metabolite Extraction

1. Add 450 μL of cold (−20 °C) MeOH/H_2O (4/1, v/v) to the cell pellet and vortex vigorously.
2. The cells are mechanically disrupted by three freeze–thaw cycles in liquid N_2, followed by lysis for 10 min at 50 Hz in a TissueLyser LT with glass beads (50 mg, 425–600 μm).
3. The cellular debris is removed by centrifugation at 13,200 × *g* at 4 °C for 10 min.
4. Transfer 150 μL of the clarified supernatant to a 1.5 mL Eppendorf tube and evaporated to dryness in a SpeedVac at 35 °C.
5. The metabolite extracts are resuspended in 150 μL of 0.1 M formic Acid with 0.2 mM methionine sulfone (IS).
6. Centrifuge for 20 min at 13,200 × *g* at 4 °C before analysis.
7. Transfer 100 μL of the supernatant to a polypropylene vial for analysis and cap the vial.
8. For long sequences (more than 2 days of analysis), close the vials hermetically with crim caps and keep them at low temperature in the fridge (≤3 days). The day of the analysis, take them out of the fridge, replace caps by CE caps, and centrifuge the vials (*see* **Note 4**).

3.6 Quality Control Samples

Quality control (QC) samples are required at the beginning of the sequence to stabilize the system and throughout the analytical runs at periodic intervals of time to monitor variations in signal across the time. It is recommended to start a new sequence with 10 injections of the QC and reinject after every 5–10 samples (depending on the length of the sequence) analyzed in a randomized order. Analyze the QC throughout the run to provide a measurement not only of the system's stability and performance but also of the reproducibility of the sample treatment procedure (*see* **Note 6**).

3.6.1 QC Preparation

1. Take equal aliquots of each sample and mix them all together in an Eppendorf tube.
2. Treat the pool of sample following the same procedure as the study samples simultaneously. For some types of sample such as a cell culture, feces, or tissues, quality controls are prepared once the samples are treated.

3.7 CE Conditioning and Sample Injection

1. All separations are performed in normal polarity with a background electrolyte containing 1.0 M formic acid in Milli-Q® water with 10% MeOH (v/v) at a temperature of 20 °C.
2. New capillaries are preconditioned with a flush (950 mbar) of 1.0 M NaOH for 30 min, followed by Milli-Q® water for 30 min and the background electrolyte for 30 min.

3. Before each analysis, the background electrolyte in the inlet vial is refreshed by automatic refill.
4. Before each analysis, the capillary is conditioned by flushing with the background electrolyte for 5 min at 950 mbar.
5. The samples are hydrodynamically injected at 50 mbar for 17–50 s depending on the type of sample. Stacking effect is achieved by injecting a plug of background electrolyte at 100 mbar afterward for 20 s.
6. The separation conditions are conducted at +30 kV voltage and an internal pressure of 25 mbar for a run time of 30 min. The current observed under these conditions is +23 μA.
7. The sheath liquid with two reference masses allows the correction of mass accuracy in the MS. It is continuously infused at 6 μL/min to the ESI source. This flow is achieved by a pump that works at 0.6 mL/min with a 1:100 split.
8. The MS is operated in positive mode and nitrogen is used both as a nebulizing and a drying gas.
9. The MS parameters are: Drying gas temperature 200 °C, nebulizer pressure 10 psi, flow rate 10 L/min, capillary voltage 3500 V, fragmentor 125 V, skimmer 65 V, and octopole 750 V. Data are acquired in positive mode with a full scan from m/z 85 to 1000 at a rate of 1.41 scan per second (*see* **Note 7**).

3.8 Data Treatment

Metabolomics data treatment is a complex process with multiple steps. For a comprehensive review on the topic, *see* [14]. A brief overview of the main steps is depicted in Fig. 2. There are minor differences in this procedure compared to other techniques employed in metabolomics; therefore, it will be discussed briefly.

3.8.1 Data Reprocessing

1. Check the quality of individual electropherograms of all samples (MassHunter Qualitative software) by comparing their total ion chromatogram (TIC) and their current to detect instrumental errors.
2. Extract ion $[M+H]^+$ of methionine sulfone (m/z = 182.0487 Da) from the electropherograms of all samples for checking the consistency of the internal standard signal (in terms of intensity and MT) in all samples.
3. Acquired files are processed through MassHunter Profinder software applying two levels feature extraction procedure.
   (a) To remove background noise and extract compounds firstly naive molecular feature extraction (MFE) is applied. It is quick method which does not use exhaustive EIC extraction. Based on total ion count abundance, the

algorithm will assign to the same feature related coeluting signals (isotopes, adducts, charges, multimers) and merge them into a single value called feature. Since it can yield several false negatives, list of ions created during this procedure is used for target data mining called recursion.

(b) Recursive feature extraction reduces significantly false positives searching the data for already found signals and performing EIC extraction and integration.

(c) Extracted features are aligned across all the samples to match signals corresponding to the same compound across all samples. For successful alignment, attention should be paid to the MT window which should be adjusted to the observed MT shift. This shift can be determined based on the internal standard peak. Here it is defined as absolute (min) and relative (%) value. Absolute value should correspond to the maximum observed shift in the acquired data.

(d) Results of all reprocessing are visualized and the quality of each data integration should be inspected and if needed, manually corrected. This plays very important role for CE-MS data due to the MT shift. The software creates a list of all possible components as represented by the full TOF mass spectral data. Each compound is described by its neutral mass, migration time, and abundance.

3.8.2 Normalization and Scaling

Normalization refers to the reduction of biological and/or analytical variation between samples. Scaling aims to reduce the differences related to the magnitude of the signal.

Normalization for Biological Variation

This type of normalization is obligatory when the amount of biological sample or concentration of metabolites in the sample is not the same across the samples. This includes, for example, cells, tissues, or urine. Cells can be normalized by the number of cells, protein or DNA content, tissues buy the weight of the sample or protein amount, while urine can be normalized to its total volume (in the case of 24-h collection), amount of creatinine, specific gravity or osmolality. Sophisticated computational solutions also can be applied such as medium fold change normalization or probabilistic quotient normalization.

Normalization Related to Analytical Variation

Analytical variation can be reduced by normalization to the internal standard(s) signal or total signal (response). If signal drift is not observed over the sequence, this normalization can be skipped.

Scaling

Scaling is a mandatory process for multivariate data analysis to reduce the impact of the absolute signal. Otherwise, high abundant signals contribute more to the model than the low abundant ones, and that is not correct from biological perspective. Most common scaling ie Pareto and UV, (auto scaling) and that should be optimized in each experiment.

3.8.3 Statistical Analysis

Statistical analysis for the selection of potential markers can be performed either by multivariate or univariate analysis:

1. Univariate statistical analysis (i.e., the Student's *t*-test) can be performed using MS Excel (Microsoft), where type of comparison (paired, unpaired), variance assumption (equal, unequal), and tile (one, two) should be studied. Significance is expressed through *p* value with a common cutoff of 0.05.
2. Multivariate analysis can be applied to simplify the dimensionality of the sample and thus it permits an easier representation of data. Using SIMCA-P + 12.0 (Umetrics, Sweden) for example clean data matrices can be loaded for multivariate data analysis such as principal component analysis (PCA), principal least squares discriminant analysis (PLS-DA), or orthogonal projection to latent structure-discriminant analysis (OPLS-DA). These models illustrate the degree and tidiness of the sample separation either spontaneously (PCA) or following used classification (PLS and OPLS). Based on supervised models, significant metabolites are selected by means of S-plot or VIP score (variable importance in projection).

3.8.4 Identification

There are four levels of confidence in the metabolite annotation [15]:

(Level 1) Identified compounds: by comparing two orthogonal characteristics with standard analyzed under identical analytical conditions, e.g., m/z and MT or m/z and fragmentation pattern. (Level 2) Putatively annotated compounds: without chemical standards as reference, based the m/z matching or spectral similarity with the libraries or databases. (Level 3) Putatively characterized compounds: chemical class assignment based on similarity of physicochemical properties. (Level 4) Unknowns: no identification or classification, although they can be differentiated and quantified based on spectral data.

The easiest and the most common way to putatively identify a metabolic feature is by linking its accurate mass with a certain structure using public or in-housedatabases. For a review about public databases in MS Metabolomics [16, 17]. The use of the CEU Mass Mediator is highly recommend.

After obtaining a list of possible candidates, identification by authentic standards must be done whenever it is possible. The compounds detected are show in Table 1 in increasing neutral exact mass order.

Table 1
Compounds identified with standards by CE-MS ordered by increasing neutral exact mass

Name	Formula	Monoisotopic mass (Da)	Migration time (min)
Urea	CH_4N_2O	60.0324	25.22
Glycine	$C_2H_5NO_2$	75.0320	9.42
Putrescine	$C_4H_{12}N_2$	88.1000	7.39
Alanine	$C_3H_7NO_2$	89.0477	12.89
Sarcosine	$C_3H_7NO_2$	89.0477	14.03
Cadaverine	$C_5H_{14}N_2$	102.1157	7.40
2-Aminobutyric acid	$C_4H_9NO_2$	103.0633	12.74
4-Aminobutyric acid	$C_4H_9NO_2$	103.0633	10.61
N,*N*-Dimethylglycine	$C_4H_9NO_2$	103.0633	14.18
N-Methylalanine	$C_4H_9NO_2$	103.0633	13.19
Choline	$C_5H_{13}NO$	103.0997	10.08
Serine	$C_3H_7NO_3$	105.0426	10.51
Hypotaurine	$C_2H_7NO_2S$	109.0197	22.88
Creatinine	$C_4H_7N_3O$	113.0589	12.19
Proline	$C_5H_9NO_2$	115.0633	10.83
Guanidineacetic acid	$C_3H_7N_3O_2$	117.0537	9.15
Betaine	$C_5H_{11}NO_2$	117.0790	16.24
Valine	$C_5H_{11}NO_2$	117.0790	10.32
Threonine	$C_4H_9NO_3$	119.0582	15.16
Cysteine	$C_3H_7NO_2S$	121.0198	11.08
Taurine	$C_2H_7NO_3S$	125.0147	25.89
3-Methylhistamine	$C_6H_{11}N_3$	125.0953	12.53
Pipecolic acid	$C_6H_5NO_2$	129.0790	20.53
4-Hydroxyproline	$C_5H_9NO_3$	131.0582	12.16
Creatine	$C_4H_9N_3O_2$	131.0695	9.60
Isoleucine	$C_6H_{13}NO_2$	131.0946	14.25
Leucine	$C_6H_{13}NO_2$	131.0946	14.37
Norleucine	$C_6H_{13}NO_2$	131.0946	10.47
Asparagine	$C_4H_8N_2O_3$	132.0535	10.82
Ornithine	$C_5H_{12}N_2O_2$	132.0899	10.48

(continued)

Table 1
(continued)

Name	Formula	Monoisotopic mass (Da)	Migration time (min)
Aspartic acid	$C_4H_7NO_4$	133.0375	11.18
Malic acid	$C_4H_6O_5$	134.0215	27.00
Homocysteine	$C_4H_9NO_2S$	135.0354	15.36
Hypoxanthine	$C_5H_4N_4O$	136.0385	17.03
3-Aminobenzamide	$C_7H_8N_2O$	136.0637	10.53
N-Methylnicotinamide	$C_7H_8N_2O$	136.0637	11.43
Acetylcholine	$C_7H_{15}NO_2$	145.1103	11.28
Spermidine	$C_7H_{19}N_3$	145.1579	7.03
Glutamine	$C_5H_{10}N_2O_3$	146.0691	10.78
Lysine	$C_6H_{14}N_2O_2$	146.1055	10.48
Glutamic acid	$C_5H_9NO_4$	147.0532	15.66
Methionine	$C_5H_{11}NO_2S$	149.0511	14.89
N-Methyladenine	$C_6H_7N_5$	149.0701	11.74
Histidine	$C_6H_9N_3O_2$	155.0695	8.65
β-Imidazolelactic acid	$C_6H_8N_2O_3$	156.0535	12.30
Allantoin	$C_4H_6N_4O_3$	158.0440	24.54
Carnitine	$C_7H_{15}NO_3$	161.1052	12.60
Phenylalanine	$C_9H_{11}NO_2$	165.0790	11.16
3-Methylhistidine	$C_7H_{11}N_3O_2$	169.0851	11.29
Arginine	$C_6H_{14}N_4O_2$	174.1117	8.36
Citrulline	$C_6H_{13}N_3O_3$	175.0957	10.97
Hippuric acid	$C_9H_9NO_3$	179.0582	25.69
p-Aminophenylalanine	$C_9H_{12}N_2O_2$	180.0899	11.05
Methionine sulfone (IS)	$C_5H_{11}NO_4S$	181.0409	16.69
Tyrosine	$C_9H_{11}NO_3$	181.0739	16.40
Phosphocholine	$C_5H_{14}NO_4P$	183.0660	26.45
o-Phosphoserine	$C_3H_8NO_6P$	185.0089	28.38
Glycyl leucine	$C_8H_{16}N_2O_3$	188.1161	12.70
N2-Acetyllysine	$C_8H_{16}N_2O_3$	188.1161	13.32
NGNG-dimethylarginine	$C_8H_{18}N_4O_2$	202.1430	8.80

(continued)

Table 1 (continued)

Name	Formula	Monoisotopic mass (Da)	Migration time (min)
Spermine	$C_{10}H_{26}N_4$	202.2157	6.94
Acetylcarnitine	$C_9H_{17}NO_4$	203.1158	13.21
Tryptophan	$C_{11}H_{12}N_2O_2$	204.0899	15.69
Carnosine	$C_9H_{14}N_4O_3$	226.1066	9.77
Cystine	$C_6H_{12}N_2O_4S_2$	240.0239	15.89
Thiamine	$C_{12}H_{16}N_4OS$	264.1045	9.63
Adenosine	$C_{10}H_{13}N_5O_4$	267.0968	13.60
Octanoylcarnitine	$C_{15}H_{29}NO_4$	287.2096	10.50
Decanoylcarnitine	$C_{19}H_{37}NO_4$	315.2410	12.99

This confirmation includes injection of the corresponding standards, a pool of sample, and the spiked pool of sample. These analyses must be repeated with electrophoretic conditions identical to the former analysis. Comparison of the migration time and the MS spectra for the sample, standards, and spiked sample comprise the final confirmation under identical conditions. Sample spiking is critical for CE due to the MT shift.

To minimize MT shift on the annotation process its relative MT value versus the IS can be used. In this way, even very drastic changes in the relative MT can be reduced and MT library can be used for annotation.

Novel identification method for CE-ESI-TOF-MS [18] includes induced in-source fragmentation to obtain structural information about metabolites. Precursor ions and their product ions are linked based on the correlation analysis ($r \geq 0.9$) while free available MS/MS spectra are used for spectral matching (0–10 eV).

Figure 3 shows the total ion electropherograms of the 5 types of samples described in this chapter. The list of fingerprinted compounds includes organic amines, short-chain organic acids, amino acids, amino acid derivatives, and carnitines. Figure 4 shows the percentage of each group of compounds in plasma.

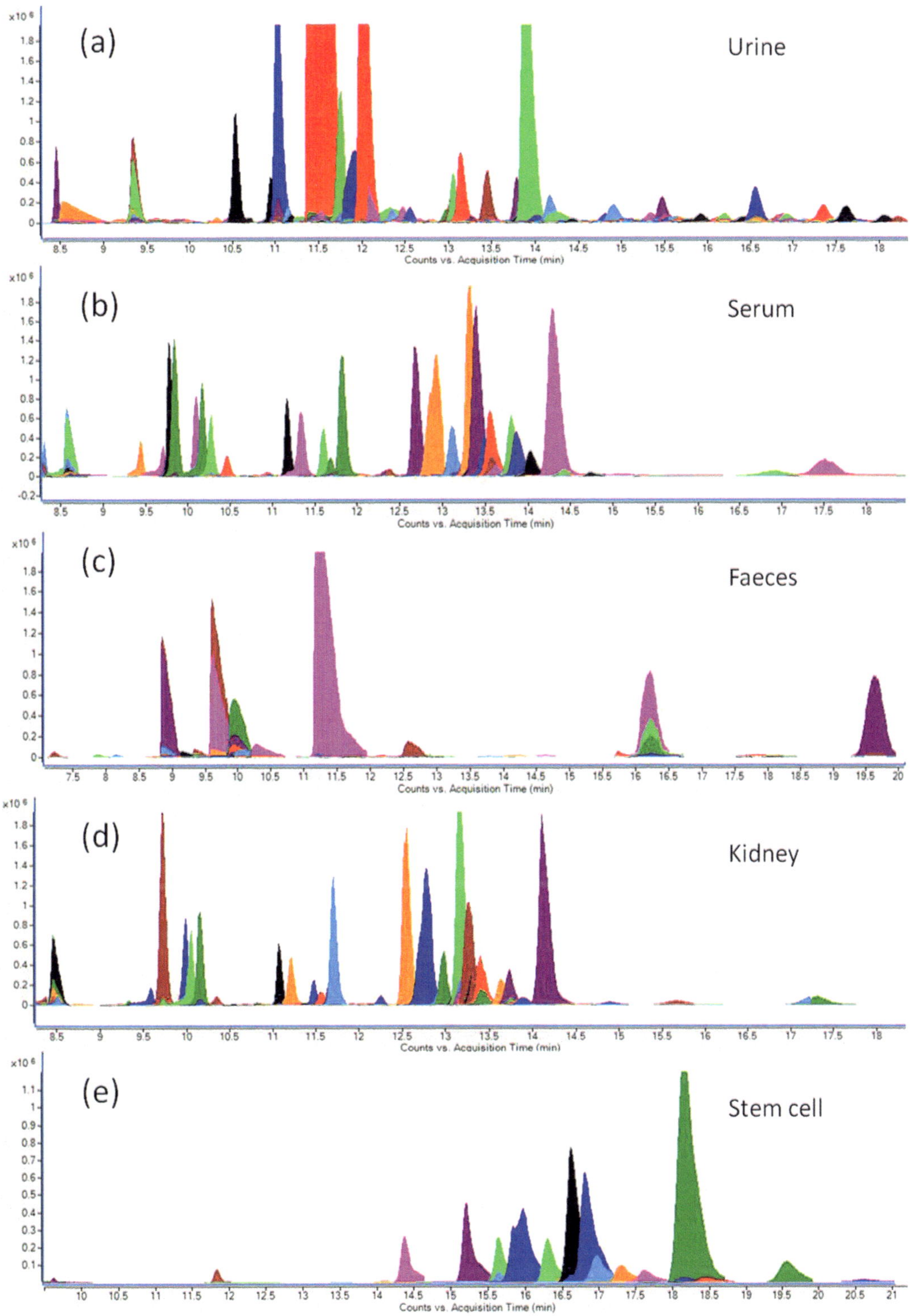

Fig. 3 Representative extracted compounds chromatograms (ECC) of different types of samples: (**a**) urine extract; (**b**) serum extract; (**c**) faeces extract; (**d**) kidney extract; (**e**) stem cell extract

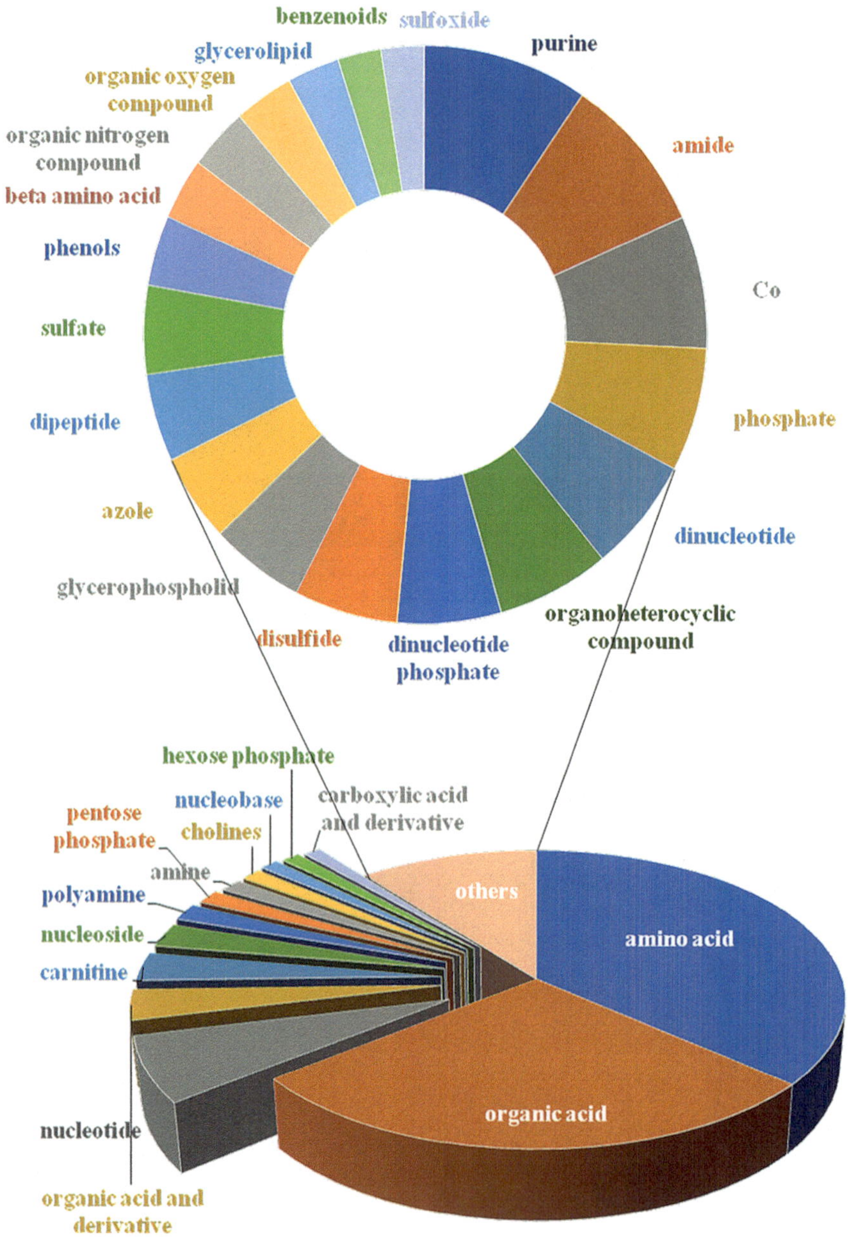

Fig. 4 Metabolites reported by CE-MS based metabolic phenotyping

4 Notes

1. For cutting the capillary to the required length, use a diamond blade capillary cutter (e.g., Beckman Coulter, Fullerton, CA, USA). Burn both ends using a window maker or a lighter to remove the polyamide coating. Wipe off the burnt ends with

IPA–water (1:1, v/v) until they are clean. Care should be taken while handling the uncoated capillary section as its strength is decreased. If the analysis is done without removing the polyamide coating at the sprayer (MS) side, the sheath liquid causes the polyamide coating to swell and expand at the sprayer tip. The nebulization will be disrupted, which will decrease the sensitivity and produce abnormalities on the current intensity.

2. TIP: If the capillary looks smooth, it is ready to use; otherwise file it using the cutter until it is smooth and clean the particles that have been able to enter the capillary by dipping it in isopropanol–water (1:1) in an ultrasound bath.
3. If the height difference is <1 cm the siphoning effect is negligible. In case of larger differences, siphoning increases, causing an introduction of air into the capillary, which results in problems such as unstable current and inaccurate sample injection volume.
4. As the vials have a very narrow insert in the final part, it is necessary to centrifuge them to remove bubbles that may occurs when transferring the sample.
5. Better repeatability and sensitivity in plasma/serum are obtained with ultrafiltration than with classical precipitation [19].
6. Samples in the sequence of analysis must be randomized to analyze samples from all the different groups each day. This allows checking the instrument performance each day.
7. Most robust methods are achieved if two segments in the MS nebulizer are programmed: 0 psi (no spray) in the conditioning and injection steps and at min 1 of analysis, apply 10 psi [20].

Acknowledgments

This work was supported by a grant from the Spanish Ministerio de Economía y Competitividad (CTQ2014-55279-R).

References

1. Garcia A, Godzien J, Lopez-Gonzalvez A, Barbas C (2017) Capillary electrophoresis mass spectrometry as a tool for untargeted metabolomics. Bioanalysis 9(1):99–130. https://doi.org/10.4155/bio-2016-0216
2. Canuto GA, Castilho-Martins EA, Tavares MF, Rivas L, Barbas C, López-Gonzálvez Á (2014) Multi-analytical platform metabolomic approach to study miltefosine mechanism of action and resistance in Leishmania. Anal Bioanal Chem 406(14):3459–3476. https://doi.org/10.1007/s00216-014-7772-1
3. Psychogios N, Hau DD, Peng J, Guo AC, Mandal R, Bouatra S, Sinelnikov I, Krishnamurthy R, Eisner R, Gautam B, Young N, Xia J, Knox C, Dong E, Huang P, Hollander Z, Pedersen TL, Smith SR, Bamforth F, Greiner R, McManus B, Newman JW, Goodfriend T,

Wishart DS (2011) The human serum metabolome. PLoS One 6(2):e16957. https://doi.org/10.1371/journal.pone.0016957

4. Yu Z, Kastenmüller G, He Y, Belcredi P, Möller G, Prehn C, Mendes J, Wahl S, Roemisch-Margl W, Ceglarek U, Polonikov A, Dahmen N, Prokisch H, Xie L, Li Y, Wichmann HE, Peters A, Kronenberg F, Suhre K, Adamski J, Illig T, Wang-Sattler R (2011) Differences between human plasma and serum metabolite profiles. PLoS One 6(7):e21230. https://doi.org/10.1371/journal.pone.0021230
5. Dudzik D, Barbas-Bernardos C, García A, Barbas C (2018) Quality assurance procedures for mass spectrometry untargeted metabolomics. A review. J Pharm Biomed Anal 147:149–173. https://doi.org/10.1016/j.jpba.2017.07.044
6. Le Chatelier E, Nielsen T, Qin J, Prifti E, Hildebrand F, Falony G, Almeida M, Arumugam M, Batto JM, Kennedy S, Leonard P, Li J, Burgdorf K, Grarup N, Jørgensen T, Brandslund I, Nielsen HB, Juncker AS, Bertalan M, Levenez F, Pons N, Rasmussen S, Sunagawa S, Tap J, Tims S, Zoetendal EG, Brunak S, Clément K, Doré J, Kleerebezem M, Kristiansen K, Renault P, Sicheritz-Ponten T, de Vos WM, Zucker JD, Raes J, Hansen T, Bork P, Wang J, Ehrlich SD, Pedersen O, MetaHIT consortium (2013) Richness of human gut microbiome correlates with metabolic markers. Nature 500(7464):541–546. https://doi.org/10.1038/nature12506
7. Loftfield E, Vogtmann E, Sampson JN, Moore SC, Nelson H, Knight R, Chia N, Sinha R (2016) Comparison of collection methods for fecal samples for discovery metabolomics in epidemiologic studies. Cancer Epidemiol Biomark Prev 25(11):1483–1490. https://doi.org/10.1158/1055-9965.EPI-16-0409
8. Yano JM, Yu K, Donaldson GP, Shastri GG, Ann P, Ma L, Nagler CR, Ismagilov RF, Mazmanian SK, Hsiao EY (2015) Indigenous bacteria from the gut microbiota regulate host serotonin biosynthesis. Cell 161(2):264–276. https://doi.org/10.1016/j.cell.2015.02.047
9. Ferrer M, Raczkowska BA, Martínez-Martínez M, Barbas C, Rojo D (2017) Phenotyping of gut microbiota: focus on capillary electrophoresis. Electrophoresis 38(18):2275–2286. https://doi.org/10.1002/elps.201700056
10. Naz S, Moreira dos Santos DC, García A, Barbas C (2014) Analytical protocols based on LC-MS, GC-MS and CE-MS for nontargeted metabolomics of biological tissues. Bioanalysis 6(12):1657–1677. https://doi.org/10.4155/bio.14.119
11. Wang TJ, Larson MG, Vasan RS, Cheng S, Rhee EP, McCabe E, Lewis GD, Fox CS, Jacques PF, Fernandez C, O'Donnell CJ, Carr SA, Mootha VK, Florez JC, Souza A, Melander O, Clish CB, Gerszten RE (2011) Metabolite profiles and the risk of developing diabetes. Nat Med 17(4):448–453. https://doi.org/10.1038/nm.2307
12. Buescher JM, Liebermeister W, Jules M, Uhr M, Muntel J, Botella E, Hessling B, Kleijn RJ, Le Chat L, Lecointe F, Mäder U, Nicolas P, Piersma S, Rügheimer F, Becher D, Bessieres P, Bidnenko E, Denham EL, Dervyn E, Devine KM, Doherty G, Drulhe S, Felicori L, Fogg MJ, Goelzer A, Hansen A, Harwood CR, Hecker M, Hubner S, Hultschig C, Jarmer H, Klipp E, Leduc A, Lewis P, Molina F, Noirot P, Peres S, Pigeonneau N, Pohl S, Rasmussen S, Rinn B, Schaffer M, Schnidder J, Schwikowski B, Van Dijl JM, Veiga P, Walsh S, Wilkinson AJ, Stelling J, Aymerich S, Sauer U (2012) Global network reorganization during dynamic adaptations of Bacillus subtilis metabolism. Science 335(6072):1099–1103. https://doi.org/10.1126/science.1206871
13. Mastrangelo A, Panadero MI, Pérez LM, Gálvez BG, García A, Barbas C, Rupérez FJ (2016) New insight on obesity and adipose-derived stem cells using comprehensive metabolomics. Biochem J 473(14):2187–2203. https://doi.org/10.1042/BCJ20160241
14. Godzien J, Ciborowski M, Angulo S, Barbas C (2013) From numbers to a biological sense: how the strategy chosen for metabolomics data treatment may affect final results. A practical example based on urine fingerprints obtained by LC-MS. Electrophoresis 34(19):2812–2826. https://doi.org/10.1002/elps.201300053
15. Sumner LW, Amberg A, Barrett D, Beale MH, Beger R, Daykin CA, Fan TW, Fiehn O, Goodacre R, Griffin JL, Hankemeier T, Hardy N, Harnly J, Higashi R, Kopka J, Lane AN, Lindon JC, Marriott P, Nicholls AW, Reily MD, Thaden JJ, Viant MR (2007) Proposed minimum reporting standards for chemical analysis chemical analysis working group (CAWG) metabolomics standards initiative (MSI). Metabolomics 3(3):211–221. https://doi.org/10.1007/s11306-007-0082-2
16. Vinaixa M, Schymanski EL, Neumann S, Navarro M, Salek RM, Yanes O (2015) Mass spectral databases for LC/MS and GC/MS-based metabolomics: state of the field and future prospects. TrAC Trends Anal Chem 78:23–35
17. Gil de la Fuente A, Grace Armitage E, Otero A, Barbas C, Godzien J (2017) Differentiating signals to make biological sense—a guide through databases for MS-based non-targeted metabolomics. Electrophoresis 38(18):2242–2256. https://doi.org/10.1002/elps.201700070

18. Godzien J, Armitage EG, Angulo S, Martinez-Alcazar MP, Alonso-Herranz V, Otero A, Lopez-Gonzalvez A, Barbas C (2015) In-source fragmentation and correlation analysis as tools for metabolite identification exemplified with CE-TOF untargeted metabolomics. Electrophoresis 36(18):2188–2195. https://doi.org/10.1002/elps.201500016
19. Dunn WB, Broadhurst D, Begley P, Zelena E, Francis-McIntyre S, Anderson N, Brown M, Knowles JD, Halsall A, Haselden JN, Nicholls AW, Wilson ID, Kell DB, Goodacre R, Consortium HSMH (2011) Procedures for large-scale metabolic profiling of serum and plasma using gas chromatography and liquid chromatography coupled to mass spectrometry. Nat Protoc 6(7):1060–1083. https://doi.org/10.1038/nprot.2011.335
20. Domínguez-Álvarez J, Rodríguez-Gonzalo E, Hernández-Méndez J, Carabias-Martínez R (2011) Programmed nebulizing gas pressure for efficient and stable capillary electrophoresis-mass spectrometry analysis of anionic compounds in positive separation mode. Anal Chem 83(7):2834–2839. https://doi.org/10.1021/ac1032897

Part II

Lipidomics

Chapter 6

Overview of Lipid Mass Spectrometry and Lipidomics

Simona Zarini, Robert M. Barkley, Miguel A. Gijón, and Robert C. Murphy

Abstract

Mass spectrometry has played a critical role in the identification and quantitation of lipids present in biological extracts. Various strategies have emerged in order to carry out lipidomic studies. These include both shotgun approaches as well as those engaging liquid chromatographic separation of lipid species prior to mass spectrometric analysis. Nonetheless challenges remain at every level of the lipidomic experiment, including extraction of lipids, identification of specific species, and quantitation of the vast array of lipids present in the sample extract. New strategies have emerged to address some of these issues; however, precise quantitation remains a significant challenge. The use of the ratio of the abundance of the molecular ion species to that of an internal standard enables quite accurate assessment of fold changes within complex lipid species without the need for exact quantitation. Challenges continue to remain in terms of availability of reference standard material as well as relevant internal standards.

Key words Lipidomics, Mass spectrometry, Identification, Quantitation

1 Introduction to Lipid Biochemistry

Lipids are hydrophobic and amphipathic molecules derived from biochemical processes by either condensation of carbanion thioesters or carbon-carbon bond formation from carbocations of isoprene units [1]. In spite of this seemingly simple definition, a complex array of lipids are observed in all living systems. These include fatty acyls and their derivatives, polyketides from the plant kingdom, glycerolipids (GL), glycerophospholipids (the traditional term phospholipids, abbreviated as PL, will be used) of several different classes, sphingolipids (SL), and saccharolipids from the carbanion-based mechanism of formation. The abundant lipids derived by carbocation-based chemistry include prenols and steroids/sterols (ST). For each of these seven distinct classes of lipids, hundreds to thousands of individual molecular species can exist especially for those lipids that incorporate fatty acyl groups to form the GL, PL, and SL classes of lipids. In part this is due to the biosynthetic pathway of fatty acid biosynthesis involved in generating numerous fatty acids. Nevertheless, in mammalian cells this

Angelo D'Alessandro (ed.), *High-Throughput Metabolomics: Methods and Protocols*, Methods in Molecular Biology, vol. 1978, https://doi.org/10.1007/978-1-4939-9236-2_6, © Springer Science+Business Media, LLC, part of Springer Nature 2019

pathway leads only to saturated fatty acids such as myristate, palmitate, or stearate and monounsaturated fatty acids such as oleate and palmitoleate following the action of stearoyl-CoA desaturase (SCD). There are three mammalian fatty acid desaturases, which act on CoA esters of carboxylic acids and introduce double bonds between methylene groups at specific distances from the carboxyl moiety [2]: SCD is a Δ^9-desaturase, while FADS2 is a Δ^6-desaturase and FADS1 is a Δ^5-desaturase. These enzymes can work in concert with each other so that many different polyunsaturated fatty acids can be generated that are homoconjugated with *cis* double bonds (Fig. 1).

Seven classes of elongases are known to exist that add additional 2-carbon units from acetate and are responsible for the formation of very long-chain fatty acyl groups (even up to 36 carbon atom chains). They are known as elongases of very long-chain fatty acids, or ELOVLs 1–7 [3]. The basis of complexity for fatty acyl groups in complex lipids within the mammalian cell is due also to the dietary intake of linoleic and linolenic acid that have two and three double bonds, respectively, in a homoconjugated arrangement. These polyunsaturated fatty acids can be elongated and desaturated by the abovenamed enzymes (Fig. 1) to lead to important fatty acids such as arachidonic acid, eicosapentaenoic acid, and docosahexaenoic acid.

Polyunsaturated fatty acids occur in mammalian systems and have double bonds in various positions, designated by a nomenclature system that indicates the location of the last double bond (in a homoconjugated series) from the methyl terminus ($n-x$) rather than the carboxyl terminus. The desaturases and elongases described above only result in fatty acid species (including polyunsaturated fatty acids) that are either nine carbons removed from the terminal methyl group ($n-9$) if the initial desaturation occurred with stearic acid (18:0) or seven carbon atoms ($n-7$) if the first step of desaturation occurred when the fatty acid was palmitate (16:0). Yet common polyunsaturated fatty acids, the most abundant species observed when liberated from complex lipids such as GL or PL by hydrolysis or as free acids in plasma (Fig. 2), include linoleic acid (18:2, $n-6$), arachidonic acid (20:4, $n-6$), and docosahexaenoic acid (22:6, $n-3$).

There are a number of enzymatic steps involved in the biosynthesis of GLs (triglycerides, TG; diglycerides, DG; and monoglycerides) and PLs (phosphatidic acid, PA; phosphatidylinositol, PI; phosphatidylglycerol, PG; cardiolipin, CL; bis(monoacylglycerol) phosphate, BMP; phosphatidylcholine, PC; phosphatidylethanolamine, PE; and phosphatidylserine, PS). A somewhat simplified pathway is illustrated in Fig. 3 which is initiated by the covalent binding of fatty CoA esters to glycerol-3-phosphate. Covalent reaction of the fatty CoAs with a free hydroxyl group of the glycerol is catalyzed by various acyltransferases including GPATs 1–4

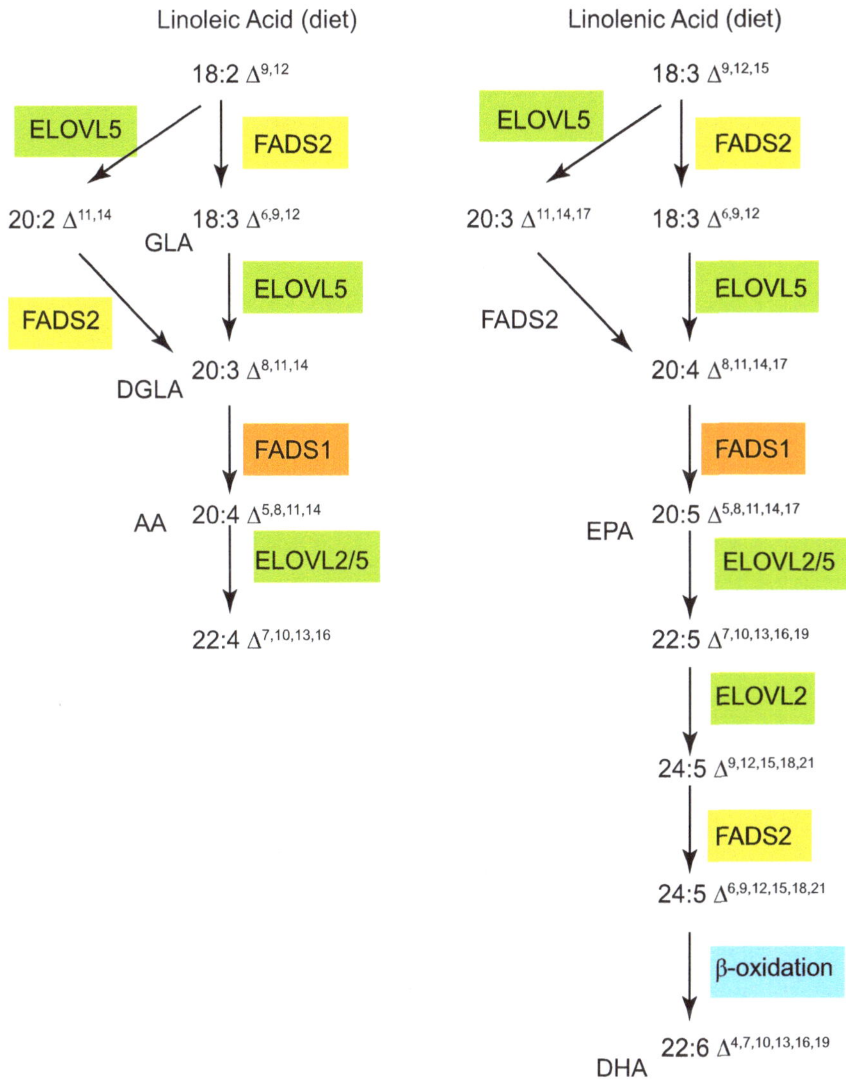

Fig. 1 Biochemical steps in the conversion of linoleic acid ($n-6$) and linolenic acid ($n-3$) into polyunsaturated fatty acids found in mammalian tissues. Enzymes engaged at various steps include very long-chain elongases (ELOVL), fatty acid desaturases (FADS), and the enzyme complex of β-oxidation

which are located at the endoplasmic reticulum. This first lipid product, termed lysophosphatidic acid, can be further reacted with a second fatty CoA ester, catalyzed by an acyltransferase of the AGPAT1–5 family to form diacyl-phosphatidic acids. Phosphatidic acid (PA) is converted to several phospholipids including PI, PG, and CL by the esterification of the phosphate monoester with glycerol, inositol, or another PG. PA can also be hydrolyzed by a phospholipase C-like activity termed phosphatidic acid phosphatase (PAP) to form a diacylglycerol [4]. Diacylglycerols are the immediate precursors of the triglyceride family of lipids after a third fatty

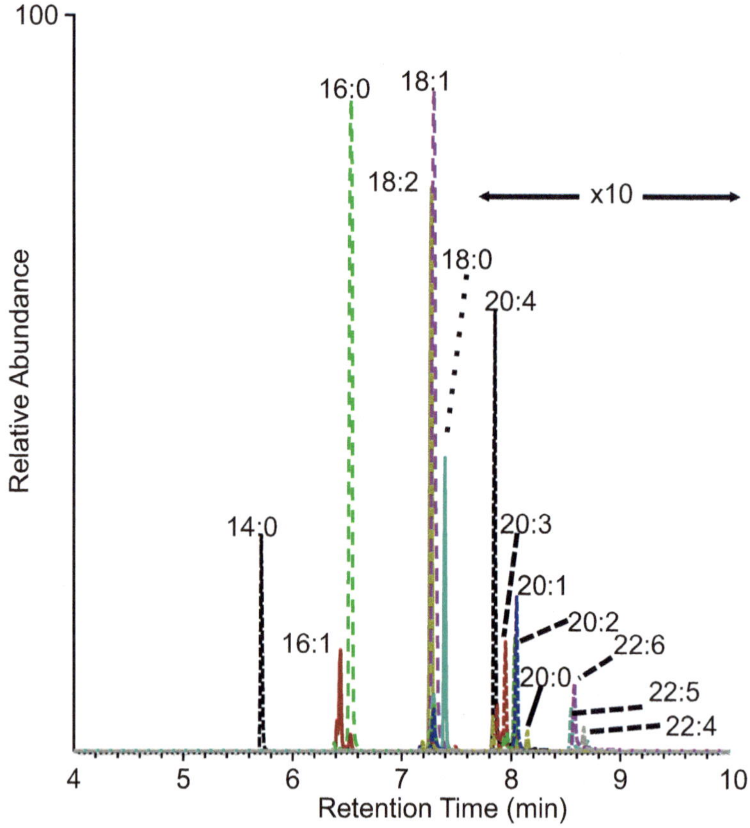

Fig. 2 Analysis of fatty acids present in human plasma by capillary GC/MS of pentafluorobenzyl ester derivatives and negative-ion chemical ionization. Carbon chain length: total number of unsaturations is indicated for each fatty acid. For clearer visualization, the signals for the indicated fatty acids have been amplified tenfold

acid acylation catalyzed by DGAT1 or DGAT2 [5]. Diglycerides can also be converted into the PC, PE, and PS phospholipids through the CDP phospholipid activation pathway. Thus molecular species of lipids observed in a lipidomic study are a result of numerous enzymatic reactions that incorporate fatty acyl CoA with little specificity or control of the possible 20 common fatty acid CoA esters found in cells. The combinatorial nature of these enzymatic steps leads to thousands of different molecular species in the GL and PL families.

The synthesis of sphingolipids is equally complex and is initiated by condensation of palmitoyl CoA with the amino acid serine, catalyzed by serine-palmitoyl CoA transferase (SPT) to form 3-ketosphinganine (Fig. 4). Recent studies have also revealed that other amino acids such as glycine and alanine can participate in this reaction, and furthermore other fatty acid CoA esters can lead to sphingolipids of unique structure [6]. After a series of reduction

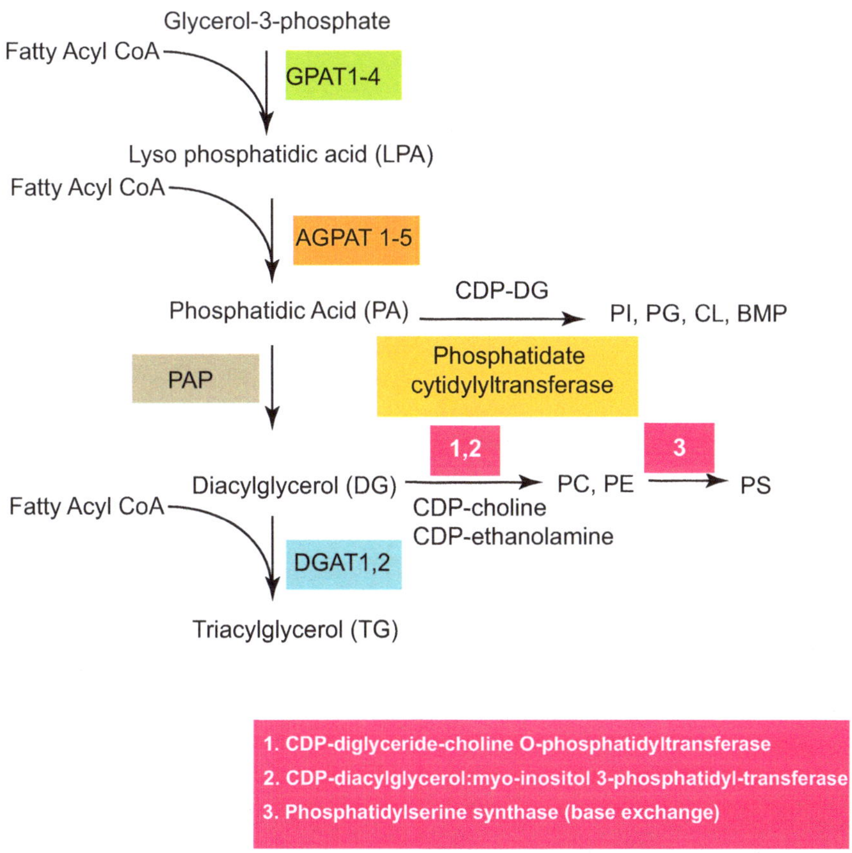

Fig. 3 Pathway of phospholipid biosynthesis from glycerol-3-phosphate and the attachment of fatty acids as their CoA esters catalyzed by various acyltransferases including glycerol-phosphate acyl transferase (GPAT) and acyl-glycerophosphate acyl transferase (AGPAT) isoforms. Conversion of phosphatidic acid into CDP-diglyceride and reaction with various polar head groups to form PI, PG, cardiolipin, or BMP primary phospholipids. PA can also be hydrolyzed to a free diglyceride catalyzed by phosphatidate phosphatase (PAP) and then condensated with CDP-activated polar head groups to form PC, PE, and PS. Also shown is the conversion of diacylglycerol into triacylglycerols by the reaction with another fatty acyl CoA ester, catalyzed by diacylglycerol acyl transferase (DGAT)

reactions sphinganine is formed. Ceramides are generated by N-acylation of the free amino group of sphinganine with various fatty acyl CoA esters, catalyzed by ceramide synthase and followed by introduction of a double bond on the long-chain base. Sphingosine is formed by N-deacylation of a ceramide, which is a reversible reaction. This sphingolipid unit can also be phosphorylated to form sphingosine-1-phosphate (S-1-P), a rather important signaling molecule [7]. Ceramides are the precursors of a series of products after reaction of the primary alcohol residue of ceramide with activated forms of phosphocholine resulting in sphingomyelin (SM), or activated sugar moieties resulting in hexosyl-ceramides. Sphingolipids can have extended carbohydrate chains attached to the ceramide backbone as very complex sphingolipid families,

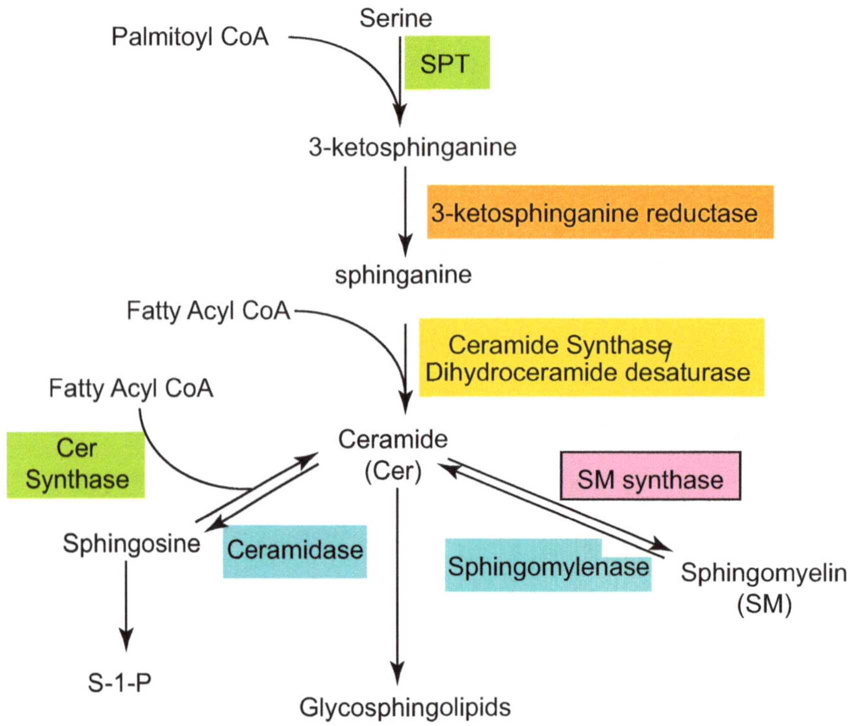

Fig. 4 Biosynthetic pathway leading to the synthesis of sphingolipids starting from the condensation of palmitoyl CoA with serine, catalyzed by serine-palmitoyl CoA transferase (SPT). Various sphingolipid intermediates are formed culminating in the production of the ceramide structure. Ceramides are then converted into sphingolipids such as sphingomyelin, S-1-P, sphingosine, and multiple families of glycosphingolipids

including gangliosides, cerebrosides, or sulfatides, where the carbohydrate portion of the molecule greatly contributes to the molecular species diversity [8].

The steroid/sterol (ST) family of molecules are derived from initial acetyl-CoA condensation with acetoacetyl-CoA catalyzed by HMG-CoA synthase (Fig. 5). This trihydroxy product is reduced by HMG-CoA reductase in a rate-limiting step of sterol biosynthesis. The resulting five-carbon acid, mevalonate, is then multiply phosphorylated to form isopentyl pyrophosphate. Carbocation-based condensation reactions lead to farnesyl-CoA and geranyl-CoA and finally to form squalene in a reaction catalyzed by the enzyme squalene synthase. The pyrophosphate intermediates of this complex prenol biosynthetic pathway can also be transformed into dolichols and ubiquinols. Squalene is interesting in that it is a branched-chain, polyunsaturated hydrocarbon yet is highly abundant in certain localized sites such as in hair follicles [9]. The hydrocarbon squalene must be functionalized by squalene monooxygenase to form 2,3-oxidosqualene which is required for cyclization by squalene epoxygenase to form the first steroid structure of this biochemical pathway,

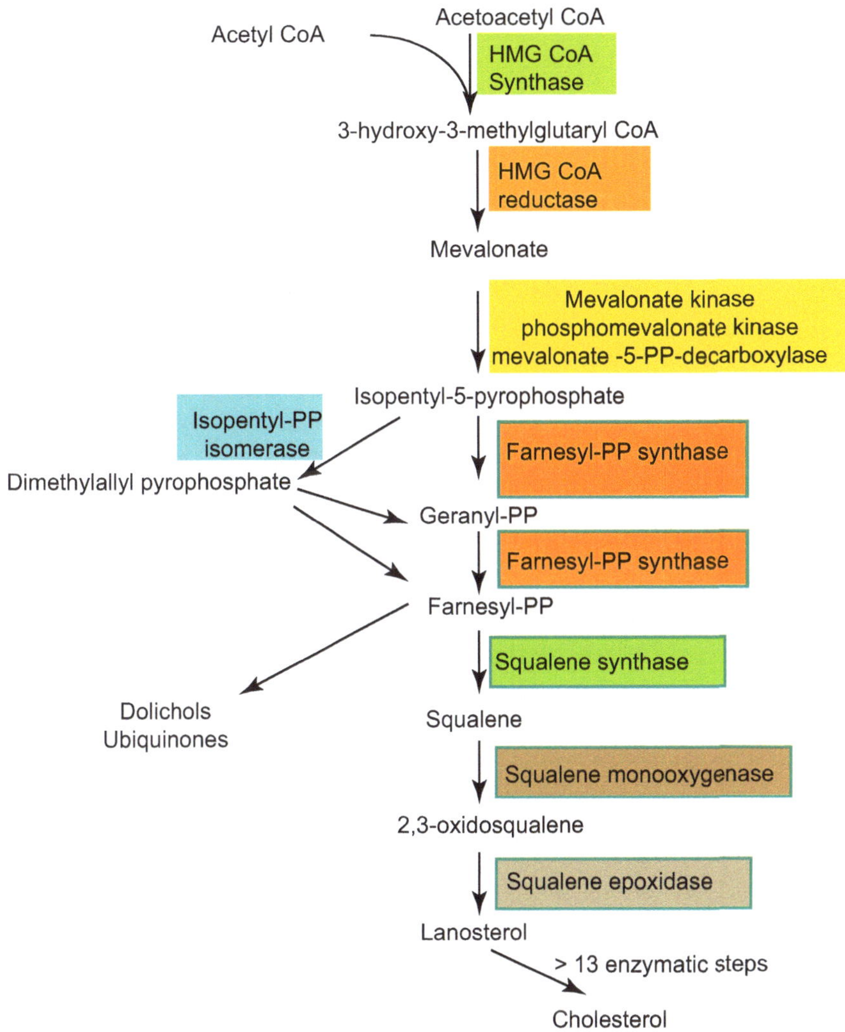

Fig. 5 Biosynthetic pathway leading to cholesterol formation, starting from the catalyzed condensation of acetyl-CoA with acetoacetyl-CoA to form 3-hydroxy-3-methylglutaryl CoA (HMG CoA)which is then reduced to mevalonic acid. Subsequent carbocation condensations of 5-carbon-membered units lead to prenol lipids such as dolichols and ubiquinols. Finally squalene is converted to cholesterol. Cholesterol is the starting substrate for all sterols and steroids

lanosterol. Lanosterol has the steroid nucleus trimmed and then various functional groups introduced in at least 15 to 18 enzymatic steps to finally form cholesterol which is the most abundant single molecular species of lipids present in mammalian cells. Cholesterol is the precursor of the steroid hormones including progesterone, corticosterone, aldosterone, cortisol, androstenedione, or testosterone and the estrogens estrone, estradiol, or estriol, as well as many other steroids. This sterol family of lipids includes many members that function by regulating intra- and intercellular events. The synthesis of bile acids such as cholic acid and chenodeoxycholic acid and their conjugates with glycine and taurine uses cholesterol as the starting substrate. Recently, 147 bile acids have been analyzed in human urine in a lipidomic study [10].

2 Lipidomics

The qualitative and quantitative analysis of lipid substances in a biological extract is perhaps the most generic definition of the term *lipidomics.* In most cases, analysis involves the use of mass spectrometry because of the inherent sensitivity, specificity, and wide dynamic range for quantitative analysis by the modern mass spectrometer. While lipids themselves are metabolites of enzymatic processes formed within cells, lipidomic analysis is typically separated from metabolomics studies in that only lipid species are the target of analysis. Nevertheless there is a very large difference between the fundamental features of lipidomic studies and metabolomics studies. While metabolomics may cover approximately 1000 to 2000 different small molecule metabolites which appear in cells, lipidomics is challenged by the assessment of lipids for which there may be greater than 50,000 molecular species occurring within a particular cell extract. The coverage of metabolites has been estimated to be 80% to 90% by LC-MS/MS techniques [11], but the coverage of lipids within an extract may be only 2% to 5% by LC-MS/MS.

Another challenging feature is the dynamic range in lipid concentration that is presented for analysis. In a recent study of the analysis of lipids in standard human plasma prepared by the National Institutes of Standards and Technology [12], over 2500 lipids were reported and quantitated. The absolute amounts of identified lipids in plasma included very high quantities of cholesterol esters, triacylglycerols, and phospholipids (several millimoles per liter) to femtomolar concentrations of biologically active prostaglandins. This is a range of 12 orders of magnitude.

An additional unique challenge to lipidomic analysis of biological samples is the variety of physical properties of lipids. Metabolomics studies aim at small molecules that are water-soluble and can therefore be extracted simultaneously by relatively simple methods. In contrast, lipids range from extremely hydrophobic (e.g., TGs) to mostly water-soluble (e.g., CoA esters), with many of them displaying amphipathic properties (e.g., lysophospholipids). Therefore, common extraction techniques are limited to certain groups of lipids with the exclusion of others, and a wide lipidomic coverage involves multiple extractions, as exemplified in the abovementioned study of the lipids in NIST plasma.

Thus the challenge of lipidomic analysis is not only the large number and variety of lipids which may be present but also the choice of extraction procedures and the extraordinary range of concentrations which demands the most sensitive techniques to be employed.

There have been many different strategies described to carry out lipidomic studies with the most popular techniques being the shotgun approach and targeted techniques engaging chromatographic separation and tandem mass spectrometry. None of these techniques is ideal for all cases, and each of the approaches has advantages and disadvantages. One way to organize these approaches is presented as an overview in Fig. 6.

Virtually all lipidomic studies begin in an attempt to provide mechanistic clues about disease processes, to assess the role of a protein or enzyme in cellular biochemistry, or to discover lipids present in biological systems. Careful planning of the exact lipidomic experiment and consideration of the most appropriate workflow are necessary to maximize the success of the endeavor. The initial step in all lipidomic studies involves extraction of the lipids of interest away from the biological matrix. In some cases this might involve solvent extraction, differential solubility in immiscible solvents, or solid-phase extraction. Organic extractions can be very generic such as the Bligh and Dyer strategy (chloroform/methanol/water extraction) [13], the Folch extraction [14], or rather specific extraction protocols. For example, the Bligh and Dyer extract can isolate approximately 90% to 92% of GLs in plasma as well as efficiently extract PLs. But if only neutral lipids were the target of analysis, it would be useful to separate the very abundant PL from the GL before analysis. In this case, one might employ isooctane/ethyl acetate (25/70: v/v) and obtain 93% extraction of GL with very poor extraction of PL in plasma [15]. More recently, solid-phase extraction techniques that employ either polar separations (normal phase) or reverse-phase separations with elution of specific lipid classes by lipophilicity have become quite popular. A very simple approach for abundant lipid analysis is simply a protein crash using ethanol or methanol to obtain the alcohol-soluble lipid components, but this extraction retains abundant salts in the ethanolic solution. However, it must be stated that not all lipids can be extracted by an immiscible solvent strategy. For example, while metabolites of arachidonic acid are often extracted using immiscible solvents, one cannot extract the sulfidopeptide leukotriene LTC_4 since it is not soluble in any organic solvent that is also immiscible with water.

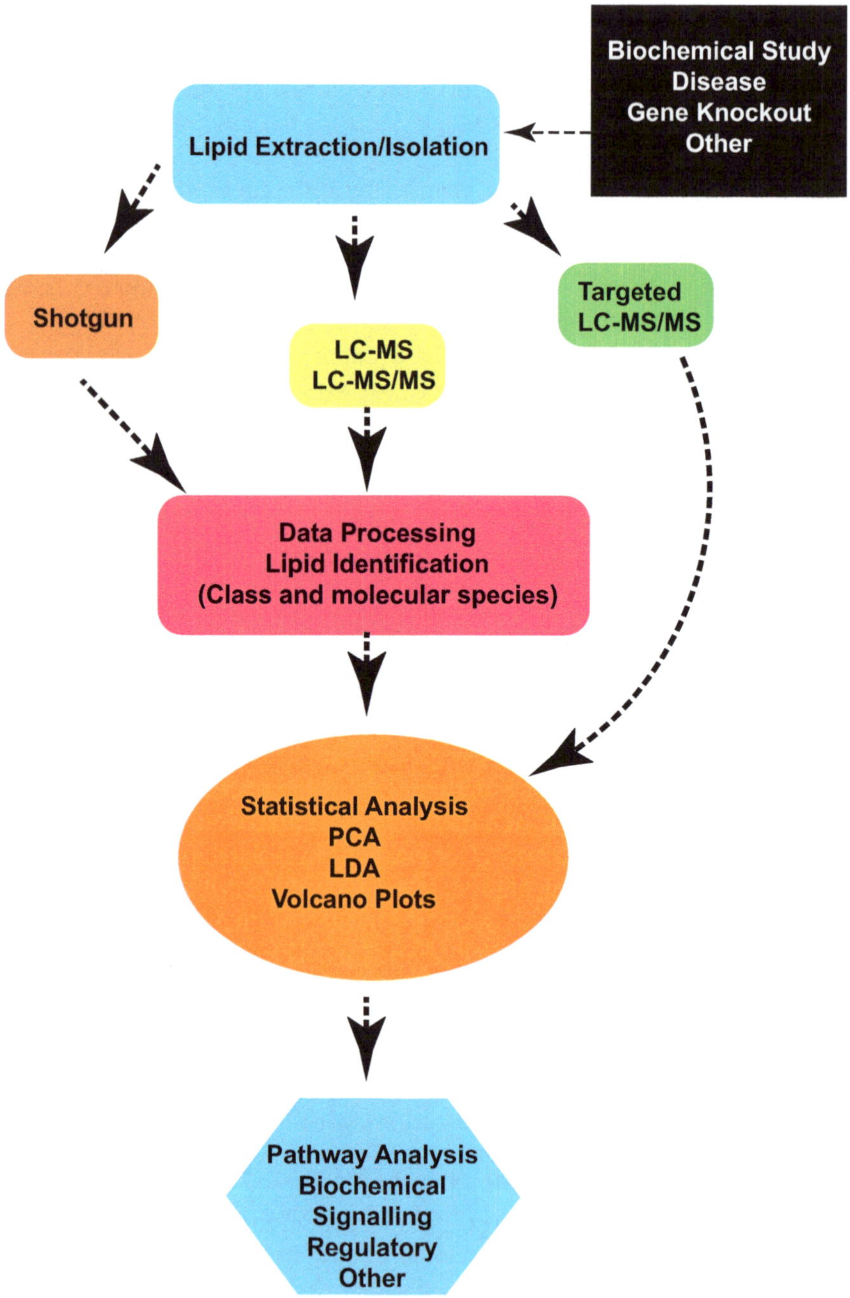

Fig. 6 A lipidomic workflow that embodies techniques of lipid extraction through mass spectrometric analysis, data processing, and statistical analysis

3 Lipid Mass Spectrometry

Mass spectrometry is the predominant analytical strategy engaged for lipidomic studies. In most cases electrospray ionization (ESI) is employed to generate ions. Lipids can form either positive or negative ions in the electrospray process. Positive ions are formed by proton attachment to yield the $[M+H]^+$ ion or adduct formation such as ammonium adducts $[M+NH_4]^+$ or alkali metal adducts $[M+Li]^+$, $[M+Na]^+$, or $[M+K]^+$. However, negative ions are the predominant form of ionization of many lipids, and these involve a proton loss $[M-H]^-$.

While some lipidomic approaches measure only molecular ion species, most studies engage tandem mass spectrometry as a means to generate more specific information about the lipid and improve identification of the molecular species. Therefore knowledge of the tandem mass spectrometric behavior of these different lipid classes is a critical component of lipidomic studies. There is a relatively large literature available on the collision-induced decomposition mechanisms of each lipid class, and these have been recently summarized [16]. The collisional activation of neutral lipids such as TGs and DGs [17] typically requires ESI formation of a charged ion by ammonium ion attachment or alkali metal ion attachment. The collision activation of an ammonium attachment ion of neutral lipids proceeds by loss of one of the acyl groups as a neutral fatty acid and free ammonia. This is the case for TG, DG, monoglycerides, as well as cholesteryl esters [18]. These pathways of product ion formation are illustrated in Fig. 7, with the formation of the primary product ions (diglyceride-like ion structures) after CID of the asymmetric TG (16:0/18:1/18:2) $[M+NH_4]^+$ and observed as the loss of each ester as a neutral carboxylic acid and neutral ammonia (NH_3).

For the case of a pure TG, three abundant product ions appear (Fig. 8), each corresponding to the loss of one of the acyl groups as a free carboxylic acid plus ammonia, at *m/z* 575, 577, and 601 corresponding to the loss of 18:1 (282 Da) plus NH_3 (17 Da), the loss of 18:2 (280 Da) plus NH_3 (17 Da), and the loss of 16:0 (256 Da) plus NH_3 (17 Da), respectively. When biological extracts are analyzed, each $[M+NH_4]^+$ ion may contain multiple molecular species that are isobaric. This complication often results in the appearance of more than three fatty acid neutral loss ions [17]. In order to obtain more complete information of the TG molecular species present, additional tandem mass spectrometry could be used, such as MS^3 for each of the initial product ions (diglyceride-like ions) which subsequently yield their own product ions following CID that correspond to the neutral loss of a second free fatty acid or an acylium ion. The third acyl group of the TG species can be calculated from the original mass of the starting TG [17].

Collisional activation of PLs results in very characteristic product ions for each phospholipid class depending upon whether the positive-ion molecular ion species $[M+H]^+$ or the negative-ion species $[M-H]^-$ is activated in the collision cell (Fig. 9). The CID of $[M+H]^+$ primarily leads to information about the polar head group. In the case of PC, the major product ion is observed at *m/z* 184

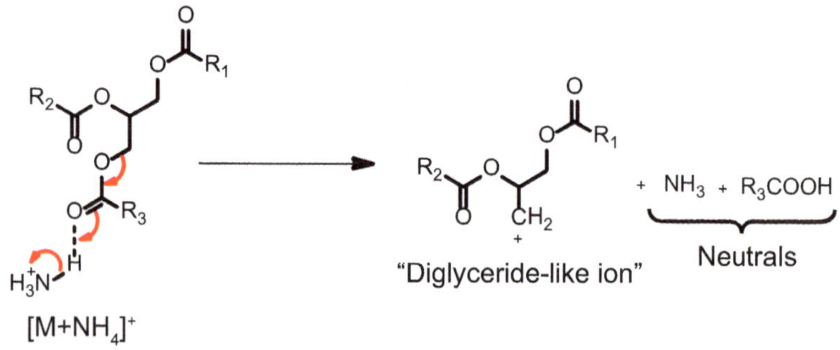

Fig. 7 Collisional activation of the adduct $[M+NH_4]^+$ derived from triacylglycerols to form diglyceride-like ions and neutral ammonia and one of the three RCOOH as a neutral free acid

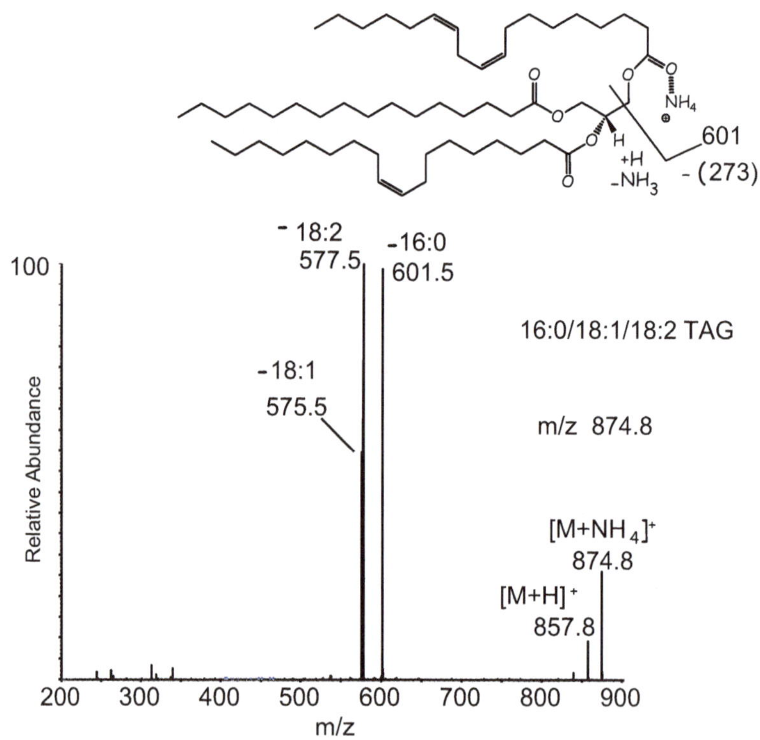

Fig. 8 Electrospray ionization of TG (16:0/18:1/18:2) generated the ammoniated molecular ion $[M+NH_4]^+$ at *m/z* 874.8. Collisional activation of this ion $[M+NH_4]^+$ leads to the abundant product ions at *m/z* 575.5 corresponding to the loss of 18:1 acyl chain plus ammonia, *m/z* 577.5 corresponding to the loss of 18:2 acyl chain plus ammonia, and *m/z* 601.5 corresponding to the loss of 16:0 acyl chain plus ammonia

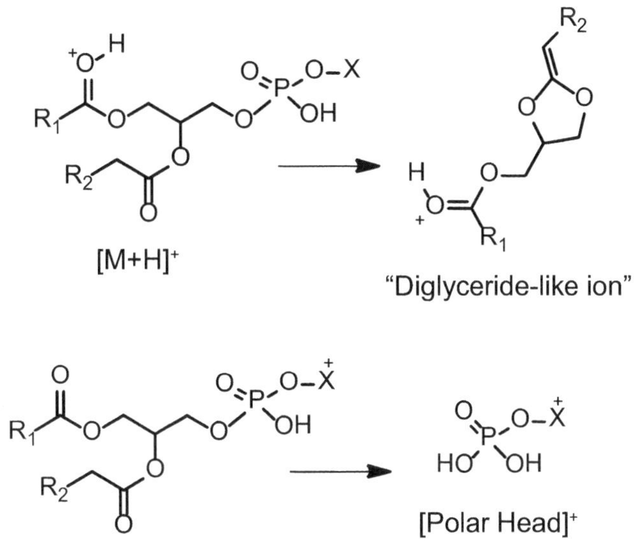

Fig. 9 Collisional activation of the protonated $[M+H]^+$ derived from a phospholipid to form either a diglyceride-like ion or fragment ion corresponding to the polar head group, depending on the location of the charge site

(Fig. 10a). Other phospholipid classes such as PG, PI, and PS lead to the formation of abundant diglyceride-like cations with a neutral mass loss corresponding to the polar head group. In Fig. 10b this is observed as $[M-141]^+$ for PE molecular species. On the other hand, CID of negative ions $[M-H]^-$ from phospholipids results in a number of different ions, but typically the most abundant product ions correspond to the carboxylate anions (Fig. 10c, d) as illustrated in Fig. 11.

Collisional activation of sphingolipids is rather complex in terms of product ion formation and whether positive and negative molecular ions are activated in the collision cell for a specific sphingolipid family member. A number of cleavage reactions are centered on the functional groups of the sphingolipid [16]. One of the common product ions derived from $[M+H]^+$ is the ion *m/z* 264 for the abundant dihydroxy sphingolipids termed SL(d18:1) derived from palmitoyl CoA and serine condensation (Fig. 12). Some sphingolipid classes such as the sulfatides readily form abundant negative ions $[M-H]^-$ since this molecule contains a sulfuric acid ester moiety. CID of these ions yield a product ion at *m/z* 97 corresponding to HSO_4^-. The glycosphingolipids are dominated by product ions resulting from cleavage centered around the sugar bonds. These product ions are of great use in structural characterization of such molecules. In lipidomic studies, a useful way to observe bioactive ceramides is to carry out a precursor ion scan for *m/z* 264, which is independent of the N-acyl group.

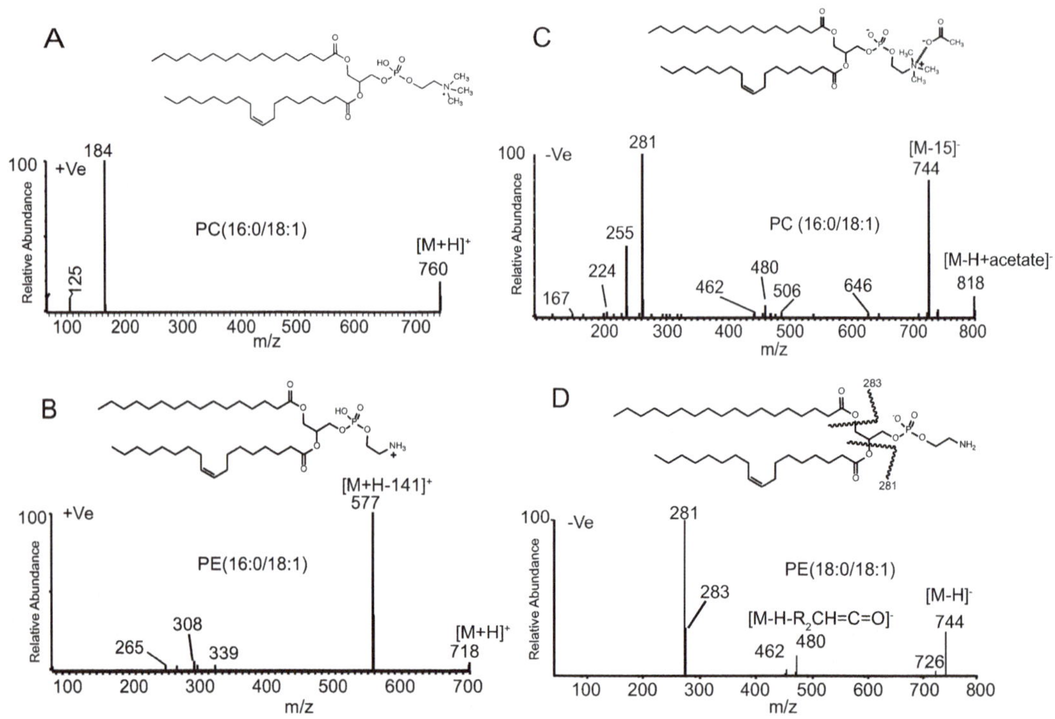

Fig. 10 Tandem mass spectrometry of PC and PE phospholipid molecular species as both positive- and negative-ion molecular species. (**a**) Collisional activation of the $[M+H]^+$ at *m/z* 760 from PC (16:0/18:1). (**b**) Collisional activation of the positive molecular ion $[M+H]^+$ at *m/z* 718 from PE (16:0/18:1). (**c**) Collisional activation of the acetate adduct negative ion at *m/z* 818 derived from PC (16:0/18:1) revealing abundant carboxylate anions at *m/z* 255 and *m/z* 281. (**d**) Collisional activation of the $[M-H]^-$ ion derived from PE (18:0/18:1) revealing the carboxylate anions at *m/z* 281 and *m/z* 283

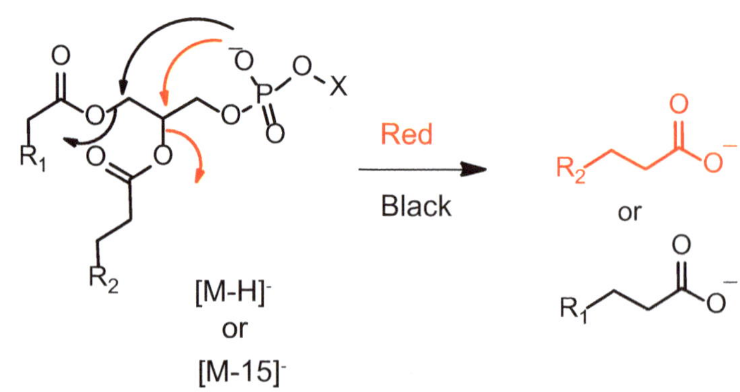

Fig. 11 Collisional activation of the protonated $[M-H]^-$ derived from a phospholipid to form carboxylate anions from either the *sn*−1 (black arrow) or *sn*−2 (red arrow). In the case of phosphatidylcholine species, the ion undergoing this reaction is $[M-15]^-$ corresponding to loss of a methyl group from the choline residue

The sterol class of lipids (SL) often have unique structures for each family member, but species such as cholesterol esters can be detected by precursor ion scanning for *m/z* 369 (positive ions) [15]. Most sterols and steroids are present in biological extracts as single entities, and therefore targeted analysis using specific precursor and product ions is typically employed in lipidomic studies. An example is the analysis of testosterone where the $[M+H]^+$ ion at *m/z* 289 yields *m/z* 97 (Fig. 13). The ion *m/z* 97 has a very distinct structure and is a characteristic product ion from steroids retaining an α,β-unsaturated moiety in the steroid A-ring [19]. The estrogens such as estradiol and estriol are typically analyzed as negative ions $[M-H]^-$ and a complex set of product ions result from CID [20]. Nonetheless very useful ion transitions are available for lipidomic analyses of these molecules.

The analysis of eicosanoids and in particular prostaglandins and leukotrienes is carried out by targeted analysis because of the very low abundance of these bioactive lipids present in a typical biological extract. In order to obtain maximum sensitivity of the mass spectrometer targeted analysis is performed with the $[M-H]^-$ and a specific product ion for the LC-MS/MS assay. It should be further pointed out that for eicosanoid analysis, commercially available stable isotope-labeled internal standards and reference

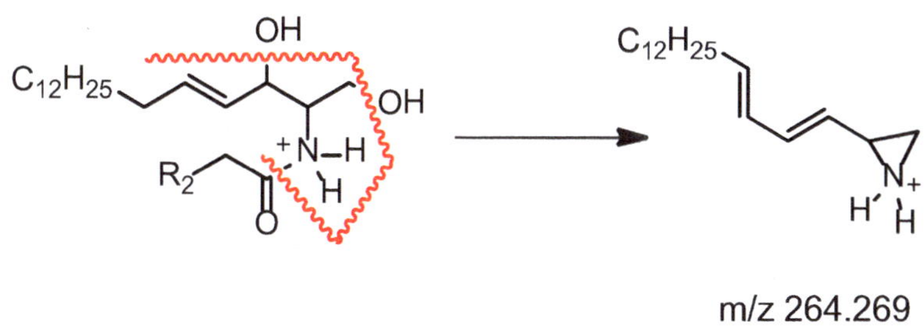

Fig. 12 Collisional activation of a protonated ceramide $[M+H]^+$ to form the common ion *m/z* 264.269

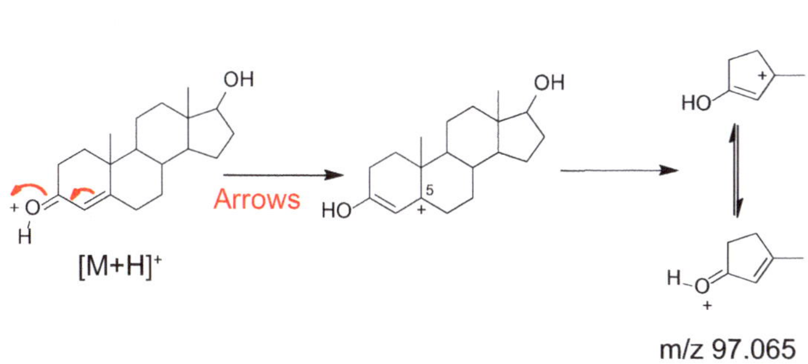

Fig. 13 Collisional activation of protonated testosterone $[M+H]^+$ to form the ion *m/z* 97.065

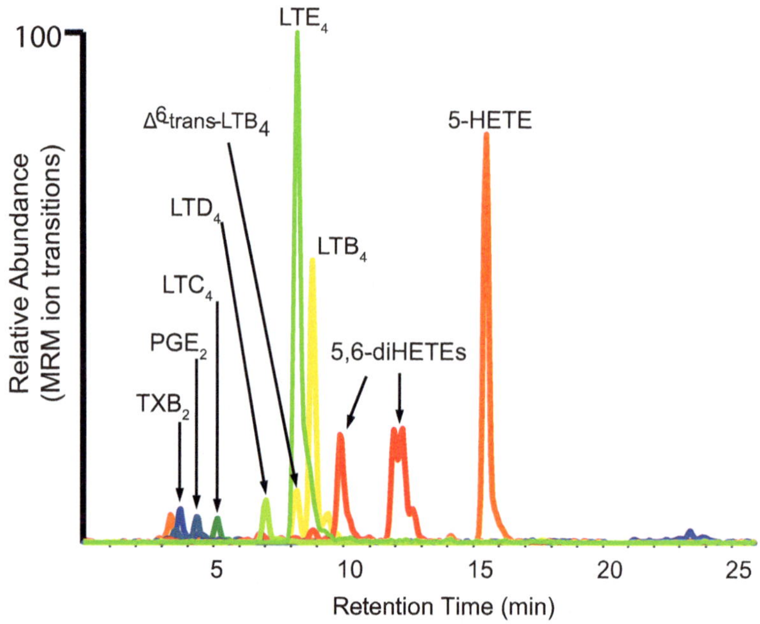

Fig. 14 Targeted analysis of eicosanoids generated in a murine inflammation model where mice were treated intraperitoneally with 1 mg zymosan for 2 h. Eicosanoids were isolated from the peritoneal exudate by solid-phase extraction after adding deuterated internal standards. The extracts were subjected to LC-MS/MS using a multiple-reaction monitoring strategy. Each of the individual eicosanoid labels corresponds to an individual ion transition monitored in this LC-MS/MS run [24]

standards enable accurate, precise, and sensitive assays to be developed [21]. There have been a large number of lipidomic studies of eicosanoids carried out over the past 30 years using a targeted approach [22, 23]. In fact, many of the strategies developed to carry out eicosanoid studies led to strategies of lipidomics currently in use. Figure 14 shows the analysis of eicosanoids generated from peritoneal macrophages stimulated with calcium ionophore [24]. Each of the colored traces corresponds to one separate precursor-product ion transition specific to one eicosanoid.

4 Lipidomic Strategies-Shotgun Analysis

Shotgun analysis is a rather unbiased approach that involves no additional separation of lipids into specific classes by chromatographic means, but rather the analysis of the major components by direct mass spectrometric detection of molecular ions. Tandem mass spectrometry and high-resolution techniques play an important role in the identification of observed molecular ions [25] as does the use of specific gas-phase ion chemistry and electrospray ionization chemistry to introduce lipids into the mass spectrometer [26].

As originally developed, shotgun lipidomics took advantage of unique features of electrospray ionization chemistry. Analysis of neutral pH lipid extracts using negative-ion electrospray voltages yielded signals predominantly from anionic lipid species. After raising the pH of the solution, negative-ion electrospray ionization produced signals from mildly basic lipid components. Finally, positive ion electrospray ionization was used to detect neutral, polar species. The anionic lipids which carry a negative charge at or close to physiological pH included CL as well as PI, PG, PS, and PA. The weakly anionic lipids included PE, ceramides, and free fatty acids. The neutral polar lipids, detected as cationic species, included PC, SM, certain glycosphingolipids, and neutral lipids such as CE and TG.

Tandem mass spectrometric techniques were employed to increase specificity in this shotgun approach by measuring specific product ions such as *m/z* 241 and 153 (PI), carboxylate anions (negative ions) from the fatty acyl groups of PL, and the positive ion *m/z* 184 for PC molecular species (*see* Tables in references 25, 26). Additionally, specific neutral losses could be used to detect lipids such as PE $[M+H-141]^+$ and the neutral loss of fatty acids plus the charging species for TG. The PE lipids from mouse heart were detected using product ion scanning for carboxylate anions from the PE fatty acyl groups, and TG species from mouse heart were detected as neutral loss ions corresponding to loss of specific fatty acids (plus charging group). The advantage of shotgun lipidomics is that it can be easily implemented in the mass spectrometry laboratory equipped with a triple quadrupole or quadrupole-TOF instrument, and a large number of lipid molecular species could be detected. However, lipid species analyzed are often only the most abundant species. The use of high-resolution mass measurement significantly enhanced the reliability of lipid identifications when using shotgun lipidomics. Disadvantages of shotgun lipidomics included the difficulty in detecting minor lipid species as well as the overlap of molecular weights of lipids from different classes since there was no prior separation. A rather specific example is the identification of BMP, bis(monoacylglycerol) phosphate, versus PG, which are isomeric species but readily separated by chromatography [27]. There are excellent reviews available describing the shotgun lipidomic approach [26, 28].

One of the rather interesting applications of shotgun lipidomics is the use of MALDI-based tissue imaging mass spectrometry. In these studies lipids are detected as they exist in situ in a tissue slice. Imaging mass spectrometry has become a robust technology, and in large part identification of lipids is based upon shotgun lipidomic techniques [29].

5 Lipidomic Strategies-Chromatography-Based Analysis

The second major approach engages chromatographic separation in order to better define the class of lipids being analyzed by mass spectrometry. LC-MS and the molecular ion mass-to-charge ratio are used to identify the lipid. LC-MS/MS adds the additional feature of gas-phase ion chemistry to better define the lipid species. Of particular value are chromatographic separations of GL and PL based on polarity such as using normal-phase silica [15] and HILIC [30] columns, since these separations place lipids of the same subclass of polarity and not lipophilicity as is the case for reversed-phase chromatography.

A variant of this approach, often termed *targeted* LC-MS/MS, typically employs isotope dilution as well as the power of chromatographic separation to enhance sensitivity for quantitative analysis. For such separations, reversed-phase columns are typically, but not exclusively, employed for maximum reproducibility and separation efficiency. It should be noted that once tandem mass spectrometry (MS/MS) is engaged, biases in ion chemistry complicate the relationship between a measured ion abundance and concentration of the molecular species of the lipid molecular species being analyzed. This is due to the fact that gas-phase reactions are induced in the collision cell and the rates of these reactions (product ion abundance and response factor) are dependent upon instrument conditions and exact lipid molecular species. In many targeted LC-MS/MS assays, precise standard curves using reference standard material are generated in order to establish the exact response factors of the target molecular species being measured [31]. The isotope dilution techniques are highly quantitative [32]. The difficulty with targeted LC-MS/MS is that if applied to the very complex mixtures, such as that found for GL and PL in biological extracts, only a limited number of internal standards can be used and the tandem mass spectrometry only yields relative quantitative numbers.

6 Lipid Identification

Irrespective of whether one engages an unbiased approach to detect and measure lipids or a targeted approach to define specific lipids, a major step of lipidomics is the processing of the mass spectrometric data. This is important in two aspects: (1) a valid means by which one can identify the lipid molecular species and (2) a quantitative assessment of lipid amount in the sample. Identification of lipids might be somewhat more difficult with shotgun and LC/MS approaches that employ database matching from large archives of lipid mass spectrometric data. Database searching alone can lead to erroneous identification of lipids. For example, the database may contain entries of plant or archaeal lipids, and the computer search identifies this lipid in a mammalian cell extract. It is important to be critical in the interpretation of database search results and carefully peruse the lists of automatically identified lipids. A molecular ion mass, even at high resolution, may not be sufficient for unique lipid identification. GL and PL may have numerous molecular species with the exact same elemental composition. If the instrumental measurement error in mass analysis is not near 1 ppm, additional elemental compositions could possibly be correct, increasing the number of lipids that should be considered as responsible for that ion abundance. An important nomenclature shorthand has been proposed to define the level at which lipids are identified by mass spectrometry [33]. For example, a positive ion observed between *m/z* 764.4 and 764.7 could be any of 58 lipid molecular ions if alkali metal salts are included. If the mass was measured at *m/z* 764.5230 at 1 ppm mass accuracy and considering only 1 phosphorus and 1–3 nitrogen atoms (present in several phospholipid classes), only 1 empirical formula fits within the 1 ppm error limit ($C_{43}H_{75}NO_8P$) corresponding to PE (38:6). Note that this level of PL identification does not reveal fatty acyl constituents but total number of acyl carbon atoms and double bonds. Normal-phase LC-MS separation can remove these salts and allows one to define the lipid as a PE from the HPLC retention time that accounts for the defined elements of the glycerophosphoethanolamine backbone and the remaining atoms assigned to the fatty acyl carbon atoms. Still the specific fatty acyl groups present are not known from this mass spectrometric data. Tandem mass spectrometry of PE species as $[M-H]^-$ yields abundant carboxylate anions that might be *m/z* 283.2637 and 303.2324 defining this molecular species as PE(18:0_20:4). Additional information is needed to place the fatty acyl groups at *sn*−1 or *sn*−2 on the glycerol backbone, the carbon position of the double bonds on the fatty acyl group, or their geometry being either *cis* or *trans*.

Assignment of double bond positions along the fatty acyl chain in an intact phospholipid can be accomplished by the use of ozone chemistry in a technique called Oz ID [34] or use of the Paternò-Büchi reaction which yields photochemical adducts of double bonds with acetone [35]. An alternative technique would be to isolate the phospholipid of interest, hydrolyze to yield free fatty acids, and carry out capillary GC/MS to assign fatty acids double bond positions by comparing retention times of authentic standard fatty acids. This approach can even distinguish *cis* from *trans* fatty acids by retention times. However, in this GC/MS strategy, information relevant to the intact molecule present in the biological extract is lost.

7 Absolute Quantitation

The second major objective of lipidomic studies is to obtain information related to abundance of each lipid. While many biochemical studies seek measures of absolute molar amounts, this can be difficult to achieve in many lipidomic studies. The reason for this is due to the large number of unique molecular species in the classes GL, PL, and SL present in biological systems. Few internal standards or, more importantly, lipid reference standards are available for the internal standard dilution method.

Lipid molecular ion species are typically formed by ESI with an ion abundance in proportion to the concentration presented to the mass spectrometer. However, the proportionality constant is different for each lipid class, and this is often referred to as response factor. In order to convert this ESI signal into a quantitative measure of the lipid, one has to have a means by which one can standardize the sensitivity of the mass spectrometer to that particular lipid species during the analysis. This is typically carried out by adding an internal standard of the same lipid class and measuring the internal standard molecular ion abundance as well as the target lipid molecular ion abundance. The ratio of the target lipid molecular ion abundance to that of the internal standard molecular ion abundance is calculated as the basic quantitative measure. This conversion to a ratio measurement is at the heart of all mass spectrometric quantitation rather than the absolute signal yield. In one common strategy, a standard curve is generated from this abundance ratio from a separate experiment where the internal standard is maintained constant and various amounts of reference standard are added, followed by mass spectrometric measurement of the abundance ratio of ESI molecular ions. The standard curve-based quantitative information is strictly valid for only that individual lipid molecular species for which there is an internal standard. Other molecular species have errors depending on unique species response factors.

8 Relative Quantitation

An emerging strategy is to measure the abundance of an ion and calculate the ratio of its abundance, relative to the same ion derived from the internal standard (i.e., molecular ion species or product ion in a targeted experiment). This is used when one internal standard is used for each lipid class such as PE or PC lipids and a large number of molecular species are present in that class grouping. In most cases, no reference material is available for each of the hundreds of molecular species being measured, and therefore strict calibrations curves cannot be generated to report data in a specific number of molar units. Yet the ratio calculated remains as a precise measure of the quantity of that molecular species in the biological extract. The exact response factors cannot be measured, but if operating parameters are not changed during the execution of the lipidomic experiment, the response factors should be constant for each member of the family. One cannot precisely compare one species to another, i.e., cannot compare abundance of PC(18:0_20:4) to the abundance of PC(18:0_16:1) or certainly across lipid classes, but one can compare the same lipid such as PC(18:0_20:4) in group A to the same species PC(18:0_20:4) in group B of the lipidomic study. We suggest that reporting the ratios of target molecule to internal standard is a precise measure of that amount of a lipid species relative to the amount of that same lipid species in other samples within that specific lipidomic study that use the same internal standard. In many ways this is similar to the way in which gene expression differences are reported in high-throughput chip experiments or to comparative proteomics data. A major advantage of this technique is the ability to use targeted analysis using LC-MS/MS and yield a large set of data for statistical analysis of lipid molecular species changes in lipidomic studies with common statistical approaches. For example, this relative quantitation method has been used to generate abundance information for over 900 molecular species of phospholipids in a recent lipidomic study [36].

9 Statistical Analysis of Lipidomic Data Sets

One of the unique features of most lipidomic studies is the generation of a large amount of data that must be perused in order to extract relevant information. This has naturally led to the use of statistical tools to assist in the data analysis and achieve the goals originally set up upon initiating the lipidomic studies. The type of information that is sought in these studies determines to large extent the depth at which statistical analysis is employed. In order to visualize data from hundreds of lipid species, graphical tools enable one to identify lipids that are relevant to the mechanism by which lipids are responsible for a particular phenotype in cells or complex organism. Differing groups of biological samples could include disease (pathology) versus normal, animals with genes knocked out versus wild type, or drug-treated versus untreated. Multiple measurements are required in order to calculate the population mean and apply a Student's *t*-test or some of its variants depending upon whether or not equal variance in the populations is present [37, 38].

Such graphical tools can be quite valuable to visualize enormous amounts of data. For example, volcano plots that reveal fold changes in lipid abundance as a function of significance (Student's *t*-test) are becoming widely employed. An example in Fig. 15

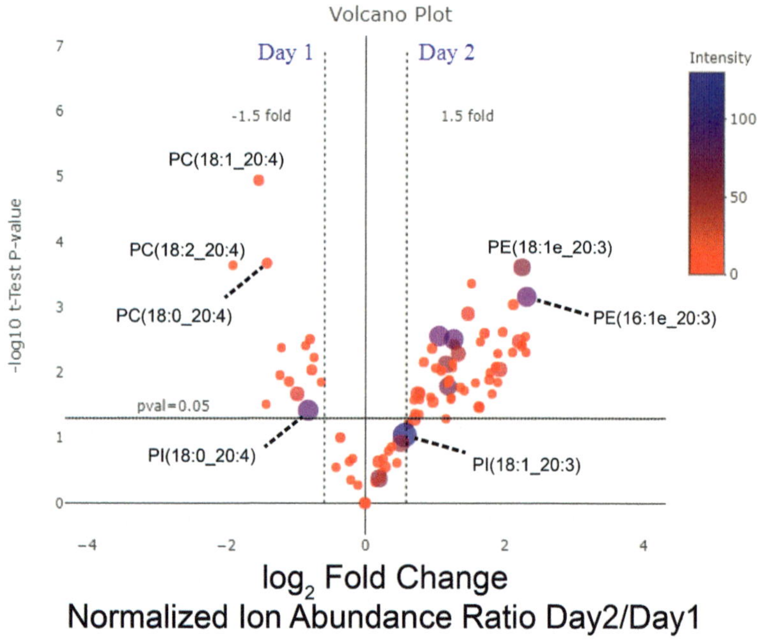

Fig. 15 Volcano plot of the relative ion intensities for phospholipids containing polyunsaturated 20-carbon fatty acyl groups from RAW 264.7 cells kept in culture for 1 versus 2 days [39]. The measured ion intensity (color bar) of approximately 100 molecular species of phospholipids containing polyunsaturated fatty acyl groups were normalized to the added internal standard ion intensity added to the extracted PLs from an equal number of cells from Day 1 culture and Day 2 culture. Those PL species with a ratio of more than twofold in abundance at Day 2 relative to Day 1 are in the upper right sector, with a *p*-value (from Student's *t*-test) lower than 0.05. Those species significantly more abundant at Day 1 compared to Day 2 are in the left upper sector. The size of the circles and color represent the relative abundance ratios of the individual phospholipid molecular species. Note the higher abundance of those PL containing 20:4 at Day 1 and those containing 20:3 at Day 2

shows the difference in phospholipid molecular species containing polyunsaturated fatty acids in RAW 264.7 cells carried in culture for 2 days as the only treatment. In this case the appearance of PLs containing a 20:3 fatty acyl chain (identified in additional experiments as Mead acid, 20:3 $n-9$) readily reveals phospholipid remodeling and fatty acid synthesis to allow the cell to adjust to the loss of critical $n-6$ fatty acids [39].

More sophisticated analyses are required if multiple groups are to be compared by informatics techniques including ANOVA and MANOVA [38]. Unfortunately, many of the reports of the value of these techniques deal little with the accuracy of the lipid identification, nevertheless the use of principal component analysis (PCA) and discriminant analysis methods including linear discriminant analysis (LDA) are becoming more widely employed. Yet the value of accurate lipid identification is of prime importance in understanding lipidomic data. While many of the informatics approaches focus on a desire to just divide populations of lipids into group A and group B in order to identify biomarkers or to facilitate understanding of the effects of drugs in human populations on lipid biochemistry, mass spectrometric data obtained in a careful lipidomic strategy can reveal complex biochemistry relevant to understanding important cellular mechanisms.

Acknowledgments

Studies that led to the development of these ideas were supported in part by grants from the National Institutes of Health (HL117798, HL34303, and ES022172) and a grant from the Global Alliance to Prevent Prematurity and Stillbirth (GAPPS).

References

1. Fahy E, Subramaniam S, Brown HA, Glass CK, Merrill AH Jr, Murphy RC, Raetz CRH, Russell DW, Seyama Y, Shaw W, Shimizu T, Spener F, van Meer G, VanNieuwenhze MS, White SH, Witztum JL, Dennis EA (2005) A comprehensive classification system for lipids. J Lipid Res 46:839–862
2. Chilton FH, Murphy RC, Wilson BA, Sergeant S, Ainsworth H, Seeds MC, Mathias RA (2014) Diet-gene interactions and PUFA metabolism: a potential contributor to health disparities and human diseases. Nutrients 6:1993–2022
3. Jakobsson A, Westerberg R, Jacobsson A (2006) Fatty acid elongases in mammals: their regulation and roles in metabolism. Prog Lipid Res 45:237–249
4. Fagone P, Jackowski S (2009) Membrane phospholipid synthesis and endoplasmic reticulum function. J Lipid Res 50:S311–S316
5. Yamashita A, Hayashi Y, Nemoto-Sasaki Y, Ito M, Oka S, Tanikawa T, Waku K, Sugiura T (2014) Acyltransferases and transacylases that determine the fatty acid composition of glycerolipids and the metabolism of bioactive lipid mediators in mammalian cells and model organisms. Prog Lipid Res 53:18–81
6. Duan J, Merrill AH Jr (2015) 1-Deoxysphingolipids encountered exogenously and made de novo: dangerous mysteries inside an enigma. J Biol Chem 90:15380–15389
7. Maceyka M, Harikumar KB, Milstien S, Spiegel S (2012) Sphingosine-1-phosphate signaling and its role in disease. Trends Cell Biol 22:50–60
8. Merrill AH Jr (2011) Sphingolipid and glycosphingolipid metabolic pathways in the era of sphingolipidomics. Chem Rev 111:6387–6422
9. Fitzgerald M, Murphy RC (2007) Electrospray mass spectrometry of human hair wax esters. J Lipid Res 48:1231–1246
10. Sarafian MH, Lewis MR, Pechlivanis A, Ralphs S, McPhail MJ, Patel VC, Dumas ME, Holmes E, Nicholson JK (2015) Bile acid profiling and quantification in biofluids using ultra-performance liquid chromatography tandem mass spectrometry. Anal Chem 87:9662–9670
11. Lu W, Su X, Klein MS, Lewis IA, Fiehn O, Rabinowitz JD (2017) Metabolite measurement: pitfalls to avoid and practices to follow. Annu Rev Biochem 20:277–304
12. Quehenberger O, Armando AM, Brown AH, Milne SB, Myers DS, Merrill AH, Bandyopadhyay S, Jones KN, Kelly S, Shaner RL, Sullards CM, Wang E, Murphy RC, Barkley RM, Leiker TJ, Raetz CR, Guan Z, Laird GM, Six DA, Russell DW, McDonald JG, Subramaniam S, Fahy E, Dennis EA (2010) Lipidomics reveals a remarkable diversity of lipids in human plasma. J Lipid Res 51:3299–3305
13. Bligh EG, Dyer WJ (1959) A rapid method of total lipid extraction and purification. Can J Biochem Physiol 37:911–917
14. Lebaron FN, Folch J (1959) The effect of pH and salt concentration on aqueous extraction of

brain proteins and lipoproteins. J Neurochem 4:1–8

15. Hutchins PM, Barkley RM, Murphy RC (2008) Separation of cellular non-polar neutral lipids by normal phase chromatography and analysis by electrospray ionization mass spectrometry. J Lipid Res 49:804–813
16. Murphy RC (2015) Tandem mass spectrometry of lipids: molecular analysis of complex lipids. In: Gaskell S (ed) New developments in mass spectrometry. Royal Society of Chemistry, London
17. McAnoy AM, Wu CC, Murphy RC (2005) Direct qualitative analysis of triacylglycerols by electrospray mass spectrometry using a linear ion trap. J Am Soc Mass Spectrom 16:1498–1509
18. Hutchins PM, Murphy RC (2011) Peroxide bond driven dissociation of hydroperoxy-cholesterol esters following collision induced dissociation. J Am Soc Mass Spectrom 22:867–874
19. Williams TM, Kind AJ, Houghton E, Hill DW (1999) Electrospray collision-induced dissociation of testosterone and testosterone hydroxy analogs. J Mass Spectrom 34:206–216
20. Wooding KM, Barkley RM, Hankin JA, Johnson CA, Bradford AP, Santoro N, Murphy RC (2013) Mechanism of formation of the major estradiol product ions following collisional activation of the molecular anion in a tandem quadrupole mass spectrometer. J Am Soc Mass Spectrom 24:1451–1455
21. Kita Y, Tokuoka SM, Shimizu T (2017) Mediator lipidomics by liquid chromatography-tandem mass spectrometry. Biochim Biophys Acta 1862:777–781
22. Murphy RC, Barkley RM, Zemski Berry K, Hankin JA, Harrison K, Johnson C, Krank J, McAnoy A, Uhlson C, Zarini S (2005) Electrospray ionization and tandem mass spectrometry of eicosanoids. Anal Biochem 346:1–42
23. Björkhem I (1979) Selective ion monitoring in clinical chemistry. CRC Crit Rev Clin Lab Sci 11:53–105
24. Zarini S, Gijón MA, Ransome AE, Murphy RC, Sala A (2009) Transcellular biosynthesis of leukotrienes in vivo during mouse peritoneal inflammation. Proc Natl Acad Sci U S A 106:8296–8301
25. Sales S, Knittelfelder O, Shevchenko A (2017) Lipidomics of human blood plasma by high-resolution shotgun mass spectrometry. Methods Mol Biol 1619:203–212
26. Han X, Gross RW (2003) Global analyses of cellular lipidomes directly from crude extracts of biological samples by ESI mass spectrometry: a bridge to lipidomics. J Lipid Res 44:1071–1079
27. Hankin JA, Murphy RC, Barkley RM, Gijón MA (2015) Ion mobility and tandem mass spectrometry of phosphatidylglycerol and bis(monoacylglycerol)phosphate (BMP). Intl J Mass Spectrom 378:255–263
28. Han X, Yang K, Gross RW (2012) Multi-dimensional mass spectrometry-based shotgun lipidomics and novel strategies for lipidomic analyses. Mass Spectrom Rev 31:134–178
29. Norris JL, Caprioli RM (2013) Analysis of tissue specimens by matrix-assisted laser desorption/ionization imaging mass spectrometry in biological and clinical research. Chem Rev 113:2309–2342
30. Hájek R, Jirásko R, Lísa M, Cífková E, Holčapek M (2017) Hydrophilic interaction liquid chromatography-mass spectrometry characterization of gangliosides in biological samples. Anal Chem 89:12425–12432
31. Bowden JA, Heckert A, Ulmer CZ, Jones CM, Koelmel JP, Abdullah L, Ahonen L, Alnouti Y, Armando AM, Asara JM, Bamba T, Barr JR, Bergquist J, Borchers CH, Brandsma J, Breitkopf SB, Cajka T, Cazenave-Gassiot A, Checa A, Cinel MA, Colas RA, Cremers S, Dennis EA, Evans JE, Fauland A, Fiehn O, Gardner MS, Garrett TJ, Gotlinger KH, Han J, Huang Y, Neo AH, Hyötyläinen T, Izumi Y, Jiang H, Jiang H, Jiang J, Kachman M, Kiyonami R, Klavins K, Klose C, Köfeler HC, Kolmert J, Koal T, Koster G, Kuklenyik Z, Kurland IJ, Leadley M, Lin K, Maddipati KR, McDougall D, Meikle PJ, Mellett NA, Monnin C, Moseley MA, Nandakumar R, Oresic M, Patterson R, Peake D, Pierce JS, Post M, Postle AD, Pugh R, Qiu Y, Quehenberger O, Ramrup P, Rees J, Rembiesa B, Reynaud D, Roth MR, Sales S, Schuhmann K, Schwartzman ML, Serhan CN, Shevchenko A, Somerville SE, St John-Williams L, Surma MA, Takeda H, Thakare R, Thompson JW, Torta F, Triebl A, Trötzmüller M, Ubhayasekera SJK, Vuckovic D, Weir JM, Welti R, Wenk MR, Wheelock CE, Yao L, Yuan M, Zhao XH, Zhou S (2017) Harmonizing lipidomics: NIST interlaboratory comparison exercise for lipidomics using SRM 1950-metabolites in frozen human plasma. J Lipid Res 58:2275–2288
32. Hinz C, Liggi S, Griffin JL (2017) The potential of ion mobility mass spectrometry for high-throughput and high-resolution lipidomics. Curr Opin Chem Biol 42:42–50
33. Liebisch G, Vizcaíno JA, Köfeler H, Trötzmüller M, Griffiths WJ, Schmitz G,

Spener F, Wakelam MJ (2013) Shorthand notation for lipid structures derived from mass spectrometry. J Lipid Res 54:1523–1530

34. Poad BLJ, Zheng X, Mitchell TW, Smith RD, Baker ES, Blanksby SJ (2018) Online Ozonolysis combined with ion mobility-mass spectrometry provides a new platform for lipid isomer analyses. Anal Chem 90:1292–1300
35. Ma X, Chong L, Tian R, Shi R, Hu TY, Ouyang Z, Xia Y (2016) Proc Natl Acad Sci U S A 113:2573–2578
36. Wall VZ, Barnhart S, Kramer F, Kanter JE, Vivekanandan-Giri A, Pennathur S, Bolego C, Ellis JM, Gijón MA, Wolfgang MJ, Bornfeldt KE (2017) Inflammatory stimuli induce acyl-CoA thioesterase 7 and remodeling of phospholipids containing unsaturated long (≥C20)-acyl chains in macrophages. J Lipid Res 58:1174–1185
37. Schwudke D, Shevchenko A, Hoffmann N, Ahrends R (2017) Lipidomics informatics for life-science. J Biotechnol 261:131–136
38. Checa A, Bedia C, Jaumot J (2015) Lipidomic data analysis: tutorial, practical guidelines and applications. Anal Chim Acta 885:1–16
39. Okuno T, Gijon MA, Zarini S, Martin SA, Barkley RM, Johnson CA, Ohba M, Yokomizo T, Murphy RC (2018) Altered eicosanoid production and phospholipid remodeling during cell culture. J Lipid Res 59:542–549

Chapter 7

LC-MS/MS-MRM-Based Targeted Metabolomics for Quantitative Analysis of Polyunsaturated Fatty Acids and Oxylipins

Xiaoyun Fu, Mikayla Anderson, Yi Wang, and James C. Zimring

Abstract

LC-MS/MS with multiple reaction monitoring (MRM) is a powerful tool for targeted metabolomics analysis including screening and quantification of known metabolites. Given the complexity of biological samples, the difference in ionization efficiency, and signal intensity of each metabolite, isotopically labeled internal standards are often used for accurate quantification. In this chapter, we describe a detailed protocol for the quantitative analysis of polyunsaturated fatty acids (PUFAs) and their oxidized products (oxylipins) by LC-MS/MS-MRM with isotope dilution. PUFAs are very susceptible to oxidation by both enzymatic and nonenzymatic pathways. Free PUFAs and corresponding oxylipins, known as bioactive lipids, are involved in many processes with varying biological functions depending on their chemical structure and concentration. Accurate quantification is thus becoming crucial to understanding the role of these bioactive lipids in health, disease(s), and other settings.

Key words LC-MS, MRM, Polyunsaturated fatty acid, Oxylipins, Eicosanoids, Docosanoids, Lipid mediators, Lipid oxidation, Isotope dilution, Hydroxyeicosatetraenoic acids (HETEs), Hydroxyoctadecadienoic acids (HODEs)

1 Introduction

Unlike untargeted metabolomics [1–5], which aims to profile as many metabolites as possible in a given set of samples, targeted metabolomics focuses on screening and providing absolute quantitation of known metabolites [6–9]. Untargeted metabolomics assays require high-resolution mass spectrometers for accurate mass identification; in contrast, targeted metabolomics can be performed on a triple quadrupole mass spectrometer. Selected reaction monitoring (SRM) is the most common mode of tandem mass spectrometry used for targeted analysis. The first quadrupole (Q1 or MS1) selects the precursor molecule with an *m/z* of given metabolite(s). The precursor ion is then fragmented via collision-induced dissociation (CID) in the second quadrupole (Q2), and

Angelo D'Alessandro (ed.), *High-Throughput Metabolomics: Methods and Protocols*, Methods in Molecular Biology, vol. 1978, https://doi.org/10.1007/978-1-4939-9236-2_7, © Springer Science+Business Media, LLC, part of Springer Nature 2019

finally, the fragmented product ion with a specified m/z is monitored in the third quadrupole (Q3 or MS2). Specific precursor-product ion pair(s) are selected for analysis of each metabolite. The term multiple reaction monitoring (MRM) refers to monitoring multiple different precursor-product ion pairs in a single analysis.

When coupled with modern separation techniques such as liquid chromatography (LC), gas chromatography (GC), or capillary electrophoresis (CE), MRM-based analyses provide high sensitivity, selectivity, reproducibility, and a broad dynamic range, optimal for the analysis of complex biological samples [10–12]. Numerous applications of targeted metabolomics analyses include early disease detection, therapy prediction, prognosis, biomarker identification, and insight into specific metabolic pathways and fluxes [13–15].

Accurate quantification of metabolites in complex biological samples is a significant challenge for MRM-based approaches. Many factors including sample preparation, extraction efficiency, matrix and ionization effects, and interference of other metabolites with the same m/z of precursor-product ion pairs (transitions) can affect the results. Therefore, it is crucial to optimize the conditions of both separation (e.g., LC column and solvents) and mass spectrometer parameters (e.g., collisional energy). Ideally, stable isotope-labeled internal standards should be used for accurate quantification. Because the internal standards have the same molecular formula and chemical structure as the analytes of interest, with the exception of having heavier elements such as ^{13}C, ^{2}H, and ^{15}N in the molecule, isotopically labeled internal standards have identical extraction recovery, ionization efficiency, chromatographic characteristics, and fragmentation patterns. This approach, named "isotope dilution," is commonly used in quantitative analysis by mass spectrometry. Here we describe a detailed protocol for quantification of polyunsaturated fatty acids (PUFAs) and their corresponding oxidized products (oxylipins) by LC-MS-MRM with isotope dilution.

Polyunsaturated fatty acids (PUFAs) including omega-3 and omega-6 fatty acids are essential building blocks of animal cell membranes and important lipid precursors for a wide array of biologically active molecules. PUFA-derived oxylipins, including eicosanoids, docosanoids, and other oxidation products, are endogenous metabolites generated by the action of lipoxygenase, cyclooxygenase, and cytochrome P450 enzymes, as well as a non-enzymatic peroxidation pathway [16]. PUFAs and their derived oxylipins are diverse molecules with multiple isomeric forms and present at a wide range of concentrations in biological samples. We developed a comprehensive LC-MS/MS-MRM-based targeted metabolomics platform to quantify a panel of PUFAs and corresponding oxylipins using isotopically labeled internal standards [17, 18]. The assay allows for the simultaneous quantification of 50 related analytes in biological samples with high sensitivity, reproducibility, and throughput.

2 Materials

2.1 Samples

1. Whole blood, plasma, packed red blood cells or platelets (50–100 μL), and other biological samples (*see* **Note 1**).
2. Pooled biological quality control sample (BioQC) (50 μL aliquot) (*see* **Note 2**).

2.2 Chemicals and Standards

1. Chemicals: LC-MS grade acetonitrile (CH_3CN), methanol (MeOH), 2-propanol, and ammonium acetate (NH_4OAc). MilliQ water (18.7 MΩ).
2. Butylated hydroxytoluene (BHT): 10 mM in MeOH is made fresh (*see* **Note 3**).
3. Standards: A list of PUFA and oxylipin standards (available from Cayman Chemicals, Ann Arbor, MI, USA) (*see* Table 1 and **Note 4**).
4. Internal standards (IS): Isotopically labeled PUFAs and oxylipins (available from Cayman Chemicals) with concentrations of stocking and final working solution of IS (1 × IS) listed in Table 2 (*see* **Note 5**).

2.3 LC-MS/MS Instrumentation

1. HPLC system with auto sampler: Here we utilize Waters ACQUITY UPLC I-Class system, including ACQUITY binary solvent manager, ACQUITY sample manager, and ACQUITY column manager.
2. LC column: ACQUITY UPLC™ HSS T3 C18, 1.8 μm, 2.1 × 100 mm.
3. Triple quadrupole mass spectrometer with an electrospray ionization source (ESI): Here we utilize AB Sciex QTRAP™ 6500 linear ion trap quadrupole mass spectrometer, equipped with standard ESI. Low mass range is configured for MRM data acquisition.
4. Software: AB Sciex analyst® version 1.6.2 is used to control all LC-MS/MS system. AB Sciex MultiQuant™ version 2.1.1 is used to process the data.

2.4 Other Equipment

1. Gas generator or high-pressure gas tank for the mass spectrometer. Here we use a genius 3031 (peak scientific) nitrogen and dry gas generator.
2. Eppendorf™ 5810R centrifuge.
3. Ultrasonic baths (Branson).
4. Vortex mixer.
5. Freezers: −20 °C and −80 °C freezers.

3 Methods

3.1 Sample Preparation

1. Biological samples: Biological samples are aliquoted into 1.5 mL Eppendorf tubes in volumes suitable for extraction (50 or100 μL), snap-frozen using liquid nitrogen, and stored at −80 °C until analysis (*see* **Note 6**).
2. BioQC sample: The BioQC is composed of pooled biological samples depending on the identity of samples for analysis. Pooled samples are mixed well and aliquoted in volumes mimicking samples for analysis (50 or 100 μL). These samples are snap-frozen using liquid nitrogen and stored at −80 °C until use.
3. Mixture of standards (QC): Individual stock solutions of each non-isotopically labeled standard are made in MeOH (*see* **Note 7**). A mixture of standards (QC) made from individual stock solutions serve as quality control. QC standards are diluted with 10 mM BHT in MeOH to the concentration listed in Table 1 and aliquoted in volumes mimicking samples for analysis (50 or 100 μL) and stored at −80 °C until use (*see* **Note 8**).
4. Mixture of internal standards (IS) (*see* **Note 9**): Individual isotopically labeled standard stock solutions are prepared in

Table 1
Composition and concentration of QC

Name	Abbreviation	Chemical formula	MW	Conc. (nM)
Prostaglandin E2	PGE2	$C_{20}H_{32}O_5$	352.5	20
Prostaglandin D2	PGD2	$C_{20}H_{32}O_5$	352.5	20
13,14-Dihydro-15-keto prostaglandin E2	13,14-dh-15-k PGE2	$C_{20}H_{32}O_5$	352.5	20
13,14-Dihydro-15-keto prostaglandin D2	13,14-dh-15-k PGD2	$C_{20}H_{32}O_5$	352.5	20
Leukotriene B4	LTB4	$C_{20}H_{32}O_4$	336.5	20
13-Hydroxy-octadecatrienoic acid	13-HOTrE	$C_{18}H_{30}O_3$	294.4	20
13-Hydroxy-octadecatrienoic acid	13-HOTrE (gamma)	$C_{18}H_{30}O_3$	294.4	20
9-Hydroxy-octadecatrienoic acid	9-HOTrE	$C_{18}H_{30}O_3$	294.4	20
12,13-Dihydroxy-octadecenoic acid	12,13-diHOME	$C_{18}H_{34}O_4$	314.5	50
9,10-Dihydroxy-octadecenoic acid	9,10-diHOME	$C_{18}H_{34}O_4$	314.5	50
13-Hydroxyoctadecadienoic acid	13-HODE	$C_{18}H_{32}O_3$	296.5	50
9-Hydroxyoctadecadienoic acid	9-HODE	$C_{18}H_{32}O_3$	296.5	50
12(13)epoxy-octadecenoic acid	12,13-EpOME	$C_{18}H_{32}O_3$	296.5	50
9(10)epoxy-octadecenoic acid	9,10-EpOME	$C_{18}H_{32}O_3$	296.5	50

(continued)

Table 1 (continued)

Name	Abbreviation	Chemical formula	MW	Conc. (nM)
13-Oxo-octadecadienoic acid	13-oxoODE	$C_{18}H_{30}O_3$	294.4	50
9-Oxo-octadecadienoic acid	9-oxoODE	$C_{18}H_{30}O_3$	294.4	50
12-Hydroxy-eicosapentaenoic acid	12-HEPE	$C_{20}H_{30}O_3$	318.5	50
15-Hydroxy-eicosatetraenoic acid	15-HETE	$C_{20}H_{32}O_3$	320.5	50
11-Hydroxy-eicosatetraenoic acid	11-HETE	$C_{20}H_{32}O_3$	320.5	50
9-Hydroxy-eicosatetraenoic acid	9-HETE	$C_{20}H_{32}O_3$	320.5	50
8-Hydroxy-eicosatetraenoic acid	8-HETE	$C_{20}H_{32}O_3$	320.5	50
12-Hydroxy-eicosatetraenoic acid	12-HETE	$C_{20}H_{32}O_3$	320.5	50
5-Hydroxy-eicosatetraenoic acid	5-HETE	$C_{20}H_{32}O_3$	320.5	50
15-Oxo-eicosatetraenoic acid	15-OxoETE	$C_{20}H_{30}O_3$	318.5	50
12-Oxo-eicosatetraenoic acid	15-OxoETE	$C_{20}H_{30}O_4$	319.5	50
5-Oxo-eicosatetraenoic acid	15-OxoETE	$C_{20}H_{30}O_5$	320.5	50
16-Hydroxy-docosahexaenoic acid	16-HDoHE	$C_{22}H_{32}O_3$	344.5	50
14-Hydroxy-docosahexaenoic acid	14-HDoHE	$C_{22}H_{32}O_3$	344.5	50
17-Hydroxy-docosahexaenoic acid	17-HDoHE	$C_{22}H_{32}O_3$	344.5	50
α-Linolenic acid (18:3 *n*−3)	ALA	$C_{18}H_{30}O_2$	278.4	500
γ-Linolenic acid (18:3 *n*−6)	GLA	$C_{18}H_{30}O_2$	278.4	250
Linoleic acid (18:2 *n*−6)	LA	$C_{18}H_{32}O_2$	280.5	2000
Eicosapentaenoic acid (20:5 *n*−3)	EPA	$C_{20}H_{30}O_2$	302.5	250
Arachidonic acid (20:4 *n*−6)	AA	$C_{20}H_{32}O_2$	304.5	1500
Dihomo-γ-linolenic acid (20:3 *n*6)	DGLA	$C_{20}H_{34}O_2$	306.5	250
Docosahexaenoic acid (22:6 *n*−3)	DHA	$C_{22}H_{32}O_2$	328.5	500
Docosapentaenoic acid (22:5 *n*−3)	DPA	$C_{22}H_{34}O_2$	330.5	500
Adrenic acid (22:4 *n*−6)	FA22:4	$C_{22}H_{36}O_2$	332.5	250

MeOH. A high concentration stock solution (20×) of mixed internal standards (*see* Table 2) is made from individual stock solutions with 10 mM BHT. The IS stock solution is aliquoted and stored at −80 °C until use (*see* **Note 10**).

3.2 Extraction of Analytes

1. Make 1× IS working solution: 20× IS stock solution is diluted to 1× with MeOH with 10 mM BHT. Concentrations of 1× IS are listed in Table 2 (*see* **Note 11**).

Table 2
Concentrations of isotopically labeled internal standards (IS)

Name	Abbreviation	20× stock solution (μM)	1× (nM)
Prostaglandin E2-d4 (3,3,4,4)	PGE2-d4	0.2	10
Prostaglandin D2-d4 (3,3,4,4)	PGD2-d4	0.2	10
12,13-Dihydroxy-octadecenoic acid-d4 (9,10,12,13)	12,13-diHOME-d4	1.0	50
9,10-Dihydroxy-octadecenoic acid-d4 (9,10,12,13)	9,10-diHOME-d4	1.0	50
13S-Hydroxyoctadecadienoic acid-d4 (9,10,12,13)	13-HODE-d4	1.0	50
9S-Hydroxyoctadecadienoic acid-d4 (9,10,12,13)	9-HODE-d4	1.0	50
15S-Hydroxy-eicosatetraenoic acid-d8 (5,6,8,9,11,12,14,15)	15-HETE-d8	1.0	50
12S-Hydroxy-eicosatetraenoic acid-d8 (5,6,8,9,11,12,14,15)	12-HETE-d8	1.0	50
5S-Hydroxy-eicosatetraenoic acid-d8 (5,6,8,9,11,12,14,15)	5-HETE-d8	1.0	50
13-Oxo-octadecadienoic acid-d3 (9,10,12)	13-OxoODE-d3	0.4	20
Eicosapentaenoic acid-d5 (19,19,20,20,20)	EPA-d5	5.0	250
Docosahexaenoic acid-d5 (21,21,22,22,22)	DHA-d5	10.0	500
Arachidonic acid-d8 (5,6,8,9,11,12,14,15)	AA-d8	30.0	1500
α-Linolenic acid-d14 (2,2,2,2,4,4,5,5,6,6,7,7,8,8)	ALA-d14	10.0	500
Dihomo-γ-linolenic acid-d6 (8,9,11,12,14,15)	DGLA-d6	5.0	250
Linoleic acid-d4 (9,10,12,13)	LA-d4	40.0	2000

2. Mix a 50 or 100 μL aliquot of each sample (water blank, QC, BioQC, and samples) with water and 1× IS mixture (1:1:1, vol/vol).
3. Add methanol to a final concentration of 80% methanol [vol/vol] (*see* **Note 12**).
4. Vortex for 1 min, sonicate for 10 min in a water bath chilled to 4 °C, and incubate at −20 °C for 1 h.
5. Centrifuge samples for 20 min at 20,000 rcf and 4 °C. Transfer supernatant into a new Eppendorf tube and discard pellet.
6. Supernatants are either immediately analyzed or stored at −80 °C for future analysis.

3.3 LC-MS/MS Analysis

1. Dilute extracted supernatant 1:1 vol/vol with 10 mM NH_4OAc in water for LC/MS/MS analysis. 25 μL of samples is injected for each analysis.

2. UPLC conditions:
   (a) Autosampler temperature: 8 °C.
   (b) Column temperature: 40 °C.
   (c) Mobile phase A: water/acetonitrile (75/25, vol/vol) with 5 mM ammonium acetate.
   (d) Mobile phase B: 2-propanol/acetonitrile/water (50/45/5, vol/vol/vol) with 5 mM ammonium acetate (*see* **Note 13**).
   (e) Flow rate: 0.3 mL/min.
   (f) Solvent gradient: 0–4 min, 1–50% B; 4–10 min, 50–90% B; 10–12 min, 90–99% B; 12–14 min, 99% B.
3. QTRAP 6500 mass spectrometer settings (*see* **Note 14**):
   (a) ESI in negative mode, unit resolution for both Q1 and Q3. Pause 3 ms between mass ranges.
   (b) Source parameter: Ionspray voltage (IS), −4500 V; temperature (TEM), 520 °C; curtain gas (Cur), 35.0 psi; collision gas (CAD), high; ion source gas (GS1), 48 psi; ion source gas 2 (GS2), 60 psi.
   (c) Compound parameter: Entrance potential (EP), −10 V; collision cell exit potential (CXP), −11 V; declustering potential (DP) and collision energy (CE), *see* Table 3.
   (d) Divert valve settings (*see* **Note 15**): 1.5–16 min to source.

Table 3
Optimized MRM transitions and conditions

Group	Abbreviation_product ion mass	Precursor (*m/z*)	Product (*m/z*)	DP (V)	CE (V)
	PGF2α_291*	353.2	291.2	-50	-30
	PGF2α_309	353.2	309.2	-50	-25
	PGE2_271*	351.2	271.2	-50	-20
	PGE2_315	351.2	315.2	-50	-15
Prostaglandins	PGD2_271*	351.2	271.2	-50	-20
	PGD2_315	351.2	315.2	-50	-15
	PGE2-d4_275*	355.2	275.2	-50	-20
	PGD2-d4_275*	355.2	275.2	-50	-20
	13,14-dh-15-k PGE2_207*	351.2	207.2	-50	-26
	13,14-dh-15-k PGE2_175	351.2	175.2	-50	-30
	13,14-dh-15-k PGD2_207*	351.2	207.2	-50	-26
	13,14-dh-15-k PGD2_175	351.2	175.2	-50	-30
Leukotriene	LT-B4_195*	335.2	195.1	-50	-22
	LT-B4_129	335.2	129.1	-50	-26
	12,13-diHOME-d4_185*	317.2	185.1	-50	-30
	12,13-diHOME_183*	313.1	183.1	-50	-30
	12,13-diHOME_129	313.1	129.1	-50	-30
	9,10-diHOME-d4_203*	317.2	203.1	-50	-30
	9,10-diHOME_201*	313.1	201.1	-50	-30
	9,10-diHOME_171	313.1	171.1	-50	-30

(continued)

Table 3
(continued)

Group	Abbreviation_product ion mass	Precursor (*m/z*)	Product (*m/z*)	DP (V)	CE (V)
Oxylipins	13-HODE-d4_198*	299.3	198.0	-60	-25
From LA	13-HODE-d4_114	299.3	114.1	-60	-30
	13-HODE_195	295.2	195.1	-60	-25
	13-HODE_113	295.2	113.1	-60	-30
	9-HODE-d4_172*	299.3	172.1	-50	-25
	9-HODE_171*	295.2	171.1	-50	-25
	HODE/EpOME_277	295.2	277.2	-50	-25
	12,13-EpOME_195*	295.2	195.0	-60	-25
	9,10-EpOME_171*	295.2	171.0	-50	-25
	13-OxoODE-d3_198	296.2	198.2	-50	-30
	13-OxoODE-d3_114*	296.2	114.1	-50	-30
	13-oxoODE_113*	293.2	113.1	-50	-30
	13-oxoODE_195	293.2	195.2	-50	-30
	9-OxoODE-d3_186	296.2	186.1	-50	-30
	9-oxoODE_185*	293.2	185.1	-50	-30
	9-oxoODE_125	293.2	125.1	-50	-30
	9-HOTrE_171	293.2	171.1	-35	-22
	9-HOTrE_121	293.2	121.1	-35	-23
Oxylipins	13-HOTrE_195	293.2	195.1	-50	-23
From ALA/GLA	13-HOTrE_223	293.2	223.1	-50	-23
	13-HOTrE(gamma)_193	293.2	195.1	-35	-23
	13-HOTrE(gamma)_149	293.2	223.1	-35	-23
	HOTrE_275	293.2	275.2	-50	-23
	18-HEPE_215	317.2	215.2	-50	-20
	18-HEPE_259	317.2	259.2	-50	-20
	15-HEPE_219	317.2	219.1	-50	-18
	15-HEPE_121	317.2	121.1	-50	-24
	12-HEPE_179*	317.2	179.1	-50	-19
	12-HEPE_208	317.2	208.1	-50	-18
Oxylipins	11-HEPE_167	317.2	167.1	-50	-20
From EPA	11-HEPE_195	317.2	195.1	-50	-18
	9-HEPE_123	317.2	123.1	-50	-20
	9-HEPE_167	317.2	167.1	-50	-20
	8-HEPE_155	317.2	155.1	-50	-20
	8-HEPE_127	317.2	127.1	-50	-26
	5-HEPE_115	317.2	115.0	-50	-19
	HEPE_255	317.2	255.1	-50	-19
	15-HETE_219*	319.2	219.1	-40	-18
	15-HETE_175	319.2	175.1	-40	-20
	15-HETE-d8_226*	327.3	226.1	-40	-20
	15-HETE-d8_182	327.3	182.2	-40	-20
	12-HETE_179*	319.2	179.1	-30	-20
	12-HETE_208	319.2	208.1	-30	-20
	12-HETE-d8_184*	327.3	184.1	-30	-20
	12-HETE-d8_214	327.3	214.1	-30	-20
	11-HETE_167*	319.2	167.1	-30	-20

(continued)

Table 3
(continued)

Group	Abbreviation_product ion mass	Precursor (*m/z*)	Product (*m/z*)	DP (V)	CE (V)
Oxylipins	11-HETE_149	319.2	149.1	–30	–20
From AA	9-HETE_167*	319.2	167.0	–30	–20
	9-HETE_123	319.2	123.1	–30	–23
	9-HETE_179	319.2	179.1	–30	–20
	8-HETE_155*	319.2	155.1	–30	–21
	8-HETE_127	319.2	127.1	–30	–25
	5-HETE_115*	319.2	115.0	–30	–19
	5-HETE_203	319.2	115.0	–30	–22
	5-HETE-d8_116*	327.3	116.0	–30	–19
	HETE/EET_257	319.2	257.2	–30	–20
	14,15-EET_219	319.2	219.1	–40	–18
	14,15-EET_175	319.2	175.1	–40	–20
	11,12-EET_167	319.2	167.1	–30	–20
	11,12-EET_179	319.2	179.1	–30	–18
	8,9-EET_155	319.2	155.1	–30	–20
	8,9-EET_167	319.2	167.1	–30	–20
	5,6-EET_191	319.2	191.1	–30	–16
	5,6-EET_115	319.2	115.1	–30	–20
	14,15-DHET_207	337.2	207.1	–40	–25
	14,15-DHET_129	337.2	129.1	–40	–25
	11,12-DHET_167	337.2	167.1	–40	–26
	11,12-DHET_149	337.2	149.1	–40	–32
	8,9-DHET_127	337.2	127.1	–40	–28
Oxylipins	8,9-DHET_185	337.2	185.1	–40	–24
From AA	5,6-DHET_145	337.2	145.1	–40	–25
	5,6-DHET_191	337.2	191.1	–40	–28
	15-oxoETE_113	317.2	113.1	–45	–23
	15-oxoETE_139	317.2	139.1	–45	–28
	12-oxoETE_153	317.2	153.1	–45	–22
	12-oxoETE_179	317.2	179.1	–45	–25
	5-oxoETE_203	317.2	203.2	–45	–25
	5-oxoETE_163	317.2	163.1	–45	–23
	oxoETE_273	317.2	273.2	–45	–20
Oxylipins	16-HDoHE_233*	343.2	233.0	–50	–19
From DHA	14-HDoHE_205*	343.2	205.0	–50	–18
	17-HDoHE_201*	343.2	201.2	–50	–18
	EPA-d5*	306.3	306.3	–50	–5
	EPA*	301.2	301.2	–50	–5
	DHA-d5*	332.3	332.3	–50	–5
	DHA*	327.2	327.2	–50	–5
	AA-d8*	311.3	311.3	–50	–5
	AA*	303.2	303.2	–50	–5

(continued)

Table 3 (continued)

Group	Abbreviation_product ion mass	Precursor (*m/z*)	Product (*m/z*)	DP (V)	CE (V)
PUFAs	ALA-d14*	291.3	291.3	-60	-5
	ALA*	277.2	277.2	-60	-5
	DGLA-d6*	311.3	311.3	-70	-5
	DGLA*	305.3	305.3	-70	-5
	LA-d4*	283.3	283.3	-90	-5
	LA*	279.2	279.2	-90	-5
	DPA*	329.3	329.3	-50	-5
	FA22:4*	331.3	331.3	-80	-5

*Transitions used for quantification

4. MRM transitions and parameters: For MRM analysis, it is essential to select the appropriate transitions and optimize DP and CE for each transition of the metabolites (*see* **Note 16**). Table 3 lists optimized DP and CE of transitions for the analytes of interest. The transitions with "*" are used for quantification [17] (*see* **Notes 17–19**).
5. Set up batch file and perform sample analyses (*see* **Note 20**).

3.4 Data Analysis

1. MultiQuant 2.1 software is used to process LC-MS/MS-MRM data, and the peak area of the extracted chromatogram is used for quantifying the concentration or relative abundance of metabolites of interest (*see* **Note 21**).
2. Concentrations of metabolites are calculated by the following formula (*see* **Note 22**):

$$[X] = \frac{\text{Peak Area}X}{\text{Peak Area IS}} \times [\text{IS}]$$

3. Data evaluation and quality control: Relative standard deviations of peak areas from each internal standard are used to evaluate reproducibility of assays, including extractions, dilutions, and the analysis. The variations of metabolite concentrations in QC and BioQC are used for evaluating intra-batch and batch-to-batch variations (*see* **Note 23**).

4 Notes

1. Other biological samples can be analyzed with this assay; however sample preparation and extraction procedures may vary. For example, tissue samples may need to be homogenized prior to extraction.
2. The BioQC sample is important to assess batch-to-batch and day-to-day reproducibility of the assay including extractions,

LC separation, and mass spectrometer detection. It is made from a pool of multiple individual samples (*see* Subheading 3.1 sample preparation) and is treated exactly the same as other biological samples for extraction and analysis. The BioQC should be injected every 10–15 samples to monitor instrument stability, especially for large sets of samples.

3. Butylated hydroxytoluene (BHT) is an antioxidant that is added to the internal standard solution to prevent oxidation that may occur during storage and sample processing.
4. A number of HPLC standard mixtures are also available from Cayman Chemicals.
5. Stable isotope-labeled internal standards most commonly have heavy hydrogen (D or ^{2}H), heavy carbon (^{13}C), or heavy nitrogen (^{15}N). We choose to use deuterated internal standards based on cost and availability. A number of HPLC isotopically labeled standard mixtures are also available from Cayman Chemicals.
6. For the samples that were previously frozen, thaw and aliquot samples at 4 °C. Please note that multiple freeze-thaw cycles may affect the results.
7. In general, higher concentrations of stock solutions are preferred for long-term storage stability. We aim for concentrations at least 5 mg/mL based on solubility. For QC and internal standard preparation, a serological pipette, glass syringe, or other suitable pipette for organic solvents is used to ensure accurate volumes.
8. The QC sample is used for monitoring batch-to-batch variations in instrument sensitivity and internal standard concentrations. In general, aliquots of the QC mixture can be stored at −80 °C for 1 year.
9. When choosing an internal standard for purchase, the labeled analog should be at least 3 Da different in molecular weight. Please note that the concentrations of internal standards provided by the manufacturer may vary from batch to batch. You can confirm and correct the concentrations of the IS using a QC mixture.
10. The internal standard stock solution is made at a concentration (usually 20×) higher than the concentration desired for extractions. The 20× stock IS is aliquoted to a volume needed to extract a batch of samples and stored at −80 °C to avoid multiple freeze-thaw cycles. For example, 30 samples for extraction at 50 μL will require a 20× IS stock aliquot of 80 μL. Aliquots of the 20× IS mixture can be stored at −80 °C for 1 year.
11. It is important to dilute the extraction such that the concentrations of all analytes and internal standards fall within the linear dynamic range. The 1× IS concentration listed in this

protocol is based on the analysis of RBC units [17] and is a good starting point for other sample types, though the concentration of the IS solution may need to be adjusted.

12. 80% MeOH accounts for the methanol in the IS and QC. Other organic solvents or solid phase extraction [16, 19] may also be used. Please keep in mind that the concentration of organic solvent and the extraction method have a significant effect on the extraction efficiency and need to be optimized based on the metabolites of interest.
13. We include 50% 2-propanol in mobile phase B to elute highly hydrophobic compounds from biological samples that may otherwise remain in the column, altering column performance.
14. Mass spectrometer settings may vary for different instruments; please refer to the manufacturer's instruction for tuning the instruments.
15. For analysis of biological samples, the divert valve is used to exclude unnecessary salts (elute earlier) and other lipids (elute later) for maintaining mass spectrometer performance.
16. In order to optimize mass MRM parameters, a standard solution (500 nM or 1 μM) of metabolite standards is used for acquiring MS^2, and two to four product ions are chosen for further optimization based on abundance and uniqueness. If metabolites have precursors with the same m/z (e.g., isomers) and similar retention time, a unique product ion should be used for quantification. In general, instruments have a specific setting for compound optimization. For example, on an AB Sciex QTRAP 6500 in "optimization" mode, declustering potential (DP), collision energy (CE), and other parameters can be automatically tuned to yield the most sensitive detection parameters for any given transitions.
17. For isotopically labeled internal standards, we use the same mass spectrometer parameters and transitions that correspond to the unlabeled metabolites. It is necessary to confirm the transitions of internal standards by MS^2, especially for deuterated internal standards. The product ions that may result from fragmentation involving either labeled or unlabeled element(s) in the molecule should not be used for quantification.
18. For your reference, Table 3 lists all the transitions we optimized, but only transitions with "*" are used for quantification in our protocol. It is important to limit the number of transitions per period in an MRM method in order to have optimal acquisition cycle times for quantification. You can use multiple periods with different MRM methods based on the retention times of the analytes. Many isomers share the same transitions. In order to conserve cycle time, a shared transition can be used for quantification if the isomers are able to be

completely separated by HPLC. For example, transition 295.2 → 195.1 can be used to monitor both 13-HODE and 12,13-EpOME.

19. You may find MRM conditions for additional oxylipins in ref. 19 and by searching the Metlin-MRM database (https://metlin.scripps.edu).
20. For quality control, we set the batch injections in following order: blank, blank with IS, QC, BioQC, samples, and BioQC. Blank with IS is used for monitoring the assay blank and may be used for blank subtraction.
21. Target analytes are quantified relative to their deuterium-labeled analogs. For quantification of analytes without available isotopically labeled internal standards, either report the relative abundance or use a correction factor for "real concentration." For example, isotopically labeled 8-, 9-, and 11-HETEs are not commercially available. We use 15-HETE-d8 as an internal standard, so the concentration calculated by the formula is the relative abundance, which can be used for relative comparison among the samples.
22. In order to report "real concentration" for the analytes without isotopically labeled analogs, a correction factor should be added in the formula to account for matrix effects and the transition intensity difference among the internal standard and analytes. Ideally, the correction factor should be determined using the addition of standards with known concentrations of analytes to pooled samples (or BioQC). Alternatively, a standard calibration curve can be used.
23. Assessing the reproducibility of the overall analysis includes analyzing the standard QC and BioQC that are injected multiple times throughout the course of a batch of injections. The relative standard deviation of the calculated concentrations from both the QC and BioQC should not exceed 15%. However, in general, metabolites that are low-abundance in biological samples tend to have a higher relative standard deviation. In this case, alterations to the protocol such as a higher injection volume or less dilution of extracted sample are recommended for increased quantification accuracy.

References

1. Guma M, Tiziani S, Firestein GS (2016) Metabolomics in rheumatic diseases: desperately seeking biomarkers. Nat Rev Rheumatol 12:269–281
2. Hill CB, Bacic A, Roessner U (2014) LC-MS profiling to link metabolic and phenotypic diversity in plant mapping populations. Methods Mol Biol 1198:29–41
3. Qi Y, Song Y, Gu H et al (2014) Global metabolic profiling using ultra-performance liquid chromatography/quadrupole time-of-flight mass spectrometry. Methods Mol Biol 1198:15–27
4. Ruiz-Canela M, Hruby A, Clish CB et al (2017) Comprehensive Metabolomic profiling

and incident cardiovascular disease: a systematic review. J Am Heart Assoc 6

5. Woodside JV, Draper J, Lloyd A et al (2017) Use of biomarkers to assess fruit and vegetable intake. Proc Nutr Soc 76:308–315
6. Begou O, Gika HG, Wilson ID et al (2017) Hyphenated MS-based targeted approaches in metabolomics. Analyst 142:3079–3100
7. Lu W, Bennett BD, Rabinowitz JD (2008) Analytical strategies for LC-MS-based targeted metabolomics. J Chromatogr B Analyt Technol Biomed Life Sci 871:236–242
8. Wei R, Li G, Seymour AB (2014) Multiplexed, quantitative, and targeted metabolite profiling by LC-MS/MRM. Methods Mol Biol 1198:171–199
9. Wei R, Li G, Seymour AB (2010) High-throughput and multiplexed LC/MS/MRM method for targeted metabolomics. Anal Chem 82:5527–5533
10. Savolainen OI, Sandberg AS, Ross AB (2016) A simultaneous metabolic profiling and quantitative multimetabolite Metabolomic method for human plasma using gas-chromatography tandem mass spectrometry. J Proteome Res 15:259–265
11. Maclennan MS, Kok MGM, Soliman L et al (2018) Capillary electrophoresis-mass spectrometry for targeted and untargeted analysis of the sub-5kDa urine metabolome of patients with prostate or bladder cancer: a feasibility study. J Chromatogr B Analyt Technol Biomed Life Sci 1074-1075:79–85
12. Pannkuk EL, Laiakis EC, Authier S et al (2016) Targeted metabolomics of nonhuman primate serum after exposure to ionizing radiation: potential tools for high-throughput Biodosimetry. RSC Adv 6:51192–51202
13. Zhu J, Djukovic D, Deng L et al (2014) Colorectal cancer detection using targeted serum metabolic profiling. J Proteome Res 13:4120–4130
14. Gasperotti M, Masuero D, Guella G et al (2014) Development of a targeted method for twenty-three metabolites related to polyphenol gut microbial metabolism in biological samples, using SPE and UHPLC-ESI-MS/MS. Talanta 128:221–230
15. Datta PK, Deshmane S, Khalili K et al (2016) Glutamate metabolism in HIV-1 infected macrophages: role of HIV-1 Vpr. Cell Cycle 15:2288–2298
16. Masoodi M, Volmer DA (2014) Comprehensive quantitative determination of PUFA-related bioactive lipids for functional lipidomics using high-resolution mass spectrometry. Methods Mol Biol 1198:221–232
17. Fu X, Felcyn JR, Odem-Davis K et al (2016) Bioactive lipids accumulate in stored red blood cells despite leukoreduction: a targeted metabolomics study. Transfusion 56:2560–2570
18. Zimring JC, Slichter S, Odem-Davis K et al (2016) Metabolites in stored platelets associated with platelet recoveries and survivals. Transfusion 56:1974–1983
19. Wang Y, Armando AM, Quehenberger O et al (2014) Comprehensive ultra-performance liquid chromatographic separation and mass spectrometric analysis of eicosanoid metabolites in human samples. J Chromatogr A 1359:60–69

Chapter 8

Untargeted and Semi-targeted Lipid Analysis of Biological Samples Using Mass Spectrometry-Based Metabolomics

Julie A. Reisz, Connie Zheng, Angelo D'Alessandro, and Travis Nemkov

Abstract

Liquid chromatography coupled to mass spectrometry (LC-MS)-based metabolomics and lipidomics offers invaluable tools to qualitatively and quantitatively study biological systems. Historically, unbiased (or discovery) analysis has been performed independently of targeted, quantitative analysis such as multiple reaction monitoring (MRM). These practices have been aptly carried out based on technical limitations of each assay. The wide mass scanning ranges typical of discovery approaches limit assay sensitivity, while targeted methods that improve analyte detection do not acquire data on ions not included in the targeted assay design. Recent improvements to quadrupole-Orbitrap technology have improved both scan speed as well as sensitivity, thus making these instruments more robust. By combining the improved robustness and coverage with stable isotope dilution (SID) techniques, advantages of the separate assays can now be realized in a single run, thereby improving the throughput of this type of analysis.

Key words Untargeted metabolomics, Mass spectrometry, Oxylipids, Bile acids, Gradient, High-throughput, Absolute quantification

1 Introduction

Technological developments in mass spectrometers over the past decade have enabled these instruments to rapidly scan through wide mass ranges to profile compounds with largely varying abundances, all while maintaining high mass accuracy and resolving power [1–3]. While tandem in-space instruments such as the triple quadrupole (QqQ) have served as a standard instrument for quantitative analyses due to high sensitivity and specificity [2], recent improvements to the tandem in-time quadrupole-Orbitrap, such as the Q Exactive HF, now achieve scanning up to 18 Hz with a resolving power up to 240,000 (FWHM) at 200 *m/z*, thus providing enhanced spectral quality and femtogram-level sensitivity over a dynamic concentration range of more than four orders of magnitude [4]. These advancements provide researchers with not only a

Angelo D'Alessandro (ed.), *High-Throughput Metabolomics: Methods and Protocols*, Methods in Molecular Biology, vol. 1978,
https://doi.org/10.1007/978-1-4939-9236-2_8,

robust tool to study complex samples from a systems-wide perspective [5] but also the opportunity to achieve both a discovery analysis—where data on features within a particular mass range are collected in an unbiased fashion—as well as a quantitative, semi-targeted analysis all within the same analytical run without significant sacrifice of analytical fidelity. Using a stable-isotope dilution (SID) approach, samples are extracted with exogenously added isotopologues of compounds of interest, which, by compensating for extraction efficiency and matrix effects, can then be subsequently used for quantification by peak area comparison [6, 7]. As the data from these runs are acquired in an unbiased fashion (e.g., scanning within a mass range of 90–1250 *m/z*), data can be interrogated post hoc to provide information on the lipidome at large, in addition to the quantitative information provided by SID, thereby enabling a high-throughput robust method to quantitatively measure systems. This approach thus maintains the hypothesis-generating capabilities of discovery analysis while providing quantitative information that is important for assessing the magnitude of a biological effect and enabling the direct comparison of datasets generated on varying technological platforms [8].

In this chapter, we demonstrate examples of this approach to quantify bile acids and oxylipids, while gathering additional data in a wider mass range capable of detecting species from cholesterol, cholesteryl ester, monoacylglyceride, diacylglyceride, triacylglyceride, sphingomyelin, phosphatidylcholine, phosphatidylethanolamine, phosphatidylserine, phosphatidylinositol, and lysophospholipid compound classes for follow-up discovery analyses. These lipid classes offer important information regarding the energy, inflammatory, and stress status of the system. Produced in the liver, bile acids are involved in complex signaling pathways that link gut microbiota with gastrointestinal and overall systemic health [9, 10]. Oxylipids are also acutely involved in inflammatory signaling pathways [11] and can serve as markers of oxidative stress [12, 13]. While profiling the levels of these compounds undoubtedly offers valuable information about the status of a studied system, the unbiased acquisition of additional analytes using a discovery approach affords the ability to expand analysis post hoc to look at precursors, catabolites, and modified species that may not have garnered interest during experimental design.

2 Materials

Prepare all solutions using Optima solvents (Thermo Fisher) and LC-MS grade reagents. Solutions can be stored at room temperature unless otherwise stated.

2.1 General

1. Benchtop centrifuge capable of 16,000 × *g* with refrigeration to 4 °C.
2. Vortex.
3. Freezer, −20 °C.

2.2 Extraction Solution

1. Methanol.

2.3 Sample Preparation

1. Extraction solution, 80 μL per sample.
2. 10× stock solution in methanol of stable isotope-labeled lipid standards, 10 μL per sample.
3. SPLASH® LIPIDOMIX® Mass Spec Standard (Avanti Polar Lipids, Inc.).
4. Microcentrifuge tubes (1.5–2.0 mL), cooled on ice.

2.4 Mobile Phases for Ultra-High-Pressure Liquid Chromatography

1. Water.
2. Acetonitrile (ACN).
3. Isopropanol.
4. 1 L bottles [2] with caps, preferably new.
5. 1 L graduated cylinder.
6. 2 M ammonium acetate solution prepared in highest quality water and sterile filtered (0.22 μm membrane).

2.5 Mass Spectrometry Data Acquisition and Analysis

1. UHPLC.
2. UHPLC column—we recommend the Waters ACQUITY HSS T3 2.1 × 150 mm, 1.8 μm particle size.
3. HESI or ESI source.
4. Mass spectrometer capable of high-resolution untargeted analysis with MS^1 acquisition. Instrument should be interfaced with a LC system, preferably UHPLC.
5. Data analysis and visualization software.

3 Methods

3.1 Preparation of Mobile Phases for Chromatography

1. Prepare 1 L of Phase A by adding to a clean, dry 1 L graduated cylinder 2.5 mL of 2 M ammonium acetate (via micropipette) followed by water to the 750 mL mark. Finally, add acetonitrile to the 1 L mark.
2. Carefully pour into a clean, dry 1 L glass bottle. Swirl to mix and ensure the absence of any insoluble material.
3. Prepare 1 L of Phase B1 by adding to a clean, dry 1 L graduated cylinder 2.5 mL of 2 M ammonium acetate followed by

water to the 50 mL mark. Next add 500 mL of isopropanol and 450 mL of acetonitrile.

4. Carefully pour into a clean, dry 1 L glass bottle. Swirl to mix and ensure the absence of any insoluble material.
5. Prepare 1 L of Phase B2 by adding to a clean, dry 1 L graduated cylinder 2.5 mL of 2 M ammonium acetate followed by isopropanol to the 900 mL mark. Next add 100 mL of acetonitrile.
6. Carefully pour into a clean, dry 1 L glass bottle. Swirl to mix and ensure the absence of any insoluble material.

3.2 Preparation of Extraction Solution Containing Stable Isotope-Labeled Standards

1. Prepare stock solutions to 2× the desired final concentration in extraction solution, which should be within 1–2 orders of magnitude of the typical biological range in the analyzed sample matrix. Experimentally determined reference ranges can be found by searching for the compound of interest in Human Metabolome Database (www.hmdb.ca) and clicking on the "Concentrations" tab. Stock solutions of stable isotope-labeled standards desired for analysis are weighed and dissolved in a solvent according to the chemical nature of the compound and manufacturer instructions. Typical stock concentrations (in μM) are given in Table 1, and analyte retention times are given in Table 2.
2. Store stock solutions at −20 °C until use.

3.3 Extraction of Lipids from Biofluid Samples

1. Thaw samples on ice or in refrigerator. Pre-cool labeled microcentrifuge tubes by immersing in ice. We recommend 1.7–2.0 mL high-quality tubes to minimize introduction of plasticizers or contaminants into the mass spectrometer.
2. Briefly vortex thawed samples to homogenize, and then transfer a 10 μL aliquot by micropipette to labeled and cooled microcentrifuge tube. Remaining volumes may be refrozen and stored at −80 °C.
3. *If utilizing stable isotope-labeled standards for oxylipid or bile acid quantitation*: Add 80 μL of chilled extraction solvent (methanol) and 10 μL of chilled 2× extraction solution containing applicable standard(s) to each tube. *If using stable isotope-labeled standards for phospholipid or cholesterol* quantitation: Add 1–10 μL of chilled SPLASH® LIPIDOMIX® Mass Spec Standard mixture (depending on compound class to be quantified, *see* manufacturer instructions) to each tube, and add chilled extraction solvent (methanol) for a final volume of 90 μL. If *performing relative quantitation only*: Add 90 μL of chilled extraction solvent (methanol) to each tube.
4. Vortex samples vigorously and briefly at room temperature to mix.

Table 1
Typical standard stock concentrations (2× desired final concentration)

Compound	Vendor	Product no.	Conc. (μM)
Oxylipids			
Prostaglandin E2-d4	Cayman Chemical	314010	0.020
Arachidonic acid-d8	Cayman Chemical	390010	0.128
Docosahexaenoic acid-d5	Cayman Chemical	10005057	0.048
Eicosapentaenoic acid-d5	Cayman Chemical	10005056	0.488
15(S)-HETE-d8	Cayman Chemical	334720	0.024
12(S)-HETE-d8	Cayman Chemical	334570	0.049
5(S)-HETE-d8	Cayman Chemical	334230	0.018
9(S)-HODE-d4	Cayman Chemical	338410	0.050
13(S)-HODE-d4	Cayman Chemical	338610	0.100
(±)9(10)-DiHOME-d4	Cayman Chemical	10009993	0.050
(±)12(13)-DiHOME-d4	Cayman Chemical	10009994	0.047
13-OxoODE-d3	Cayman Chemical	338620	0.027
Dihomo-g-linolenic acid-d6	Cayman Chemical	10458	1.067
Linoleic acid-d4	Cayman Chemical	390150	1.054
α-Linolenic acid-d14	Cayman Chemical	9000433	1.026
Bile acids			
Taurodeoxycholic acid-d4	Cambridge Isotope Labs	DLM-9568-0.01	0.1
Taurochenodeoxycholic acid-d4	Cambridge Isotope Labs	DLM-9562-0.01	0.1
Tauroursodeoxycholic acid-d4	Cambridge Isotope Labs	DLM-9882-0.01	0.1
Taurocholic acid-d4	Cambridge Isotope Labs	DLM-9572-0.01	0.1
Taurolithocholic acid-d4	Cambridge Isotope Labs	DLM-9570-0.01	0.1
Cholic acid-d4	Cambridge Isotope Labs	DLM-2611-0.05	0.1
Deoxycholic acid-d4	Cambridge Isotope Labs	DLM-2824-0.01	0.1
Chenodeoxycholic acid-d4	Cambridge Isotope Labs	DLM-6780-0.05	0.1
Ursodeoxycholic acid-d4	Cambridge Isotope Labs	DLM-9574-0.05	0.1
Glycocholic acid-d4	Cambridge Isotope Labs	DLM-2742-0.01	0.1
Glycochenodeoxycholic acid-d4	Cambridge Isotope Labs	DLM-7804-0.01	0.1

Table 2
Retention times of oxylipids and bile acids

Compound	Retention time (ave, min)
Oxylipids	
9(10)-DiHOME-d4	4.08
9(10)-DiHOME	4.09
12(13)-DiHOME-d4	3.94
12(13)-DiHOME	3.95
9(10)-EpOME	5.91
12(13)-EpOME	5.83
9-HODE	5.12
9-HODE-d4	5.10
13(S)-HODE	4.93
13(S)-HODE-d4	5.08
9-oxoODE	5.32
13-oxoODE	5.07
13-oxoODE-d3	5.06
5-HETE	5.75
5-HETE-d8	5.66
8-HETE	5.54
9-HETE	5.47
11-HETE	5.21
12-HETE	5.38
12-HETE-d8	5.36
15-HETE	5.18
15-HETE-d8	5.11
Adrenic acid	9.23
Arachidonic acid	8.12
Arachidonic acid-d8	8.03
Dihomo-γ-linolenic acid	8.79
Dihomo-γ-linolenic acid-d6	8.71
Docosahexaenoic acid	7.71
Docosahexaenoic acid-d5	7.64

(continued)

Table 2
(continued)

Compound	Retention time (ave, min)
Docosapentaenoic acid	8.45
Eicosapentaenoic acid	7.25
Eicosapentaenoic acid-d5	7.17
Linoleic acid	8.35
Linoleic acid-d4	8.28
α-Linolenic acid	7.40
α-Linolenic acid-d14	7.35
Bile acids	
Chenodeoxycholic acid	4.82
Cholic acid	3.69
Deoxycholic acid	4.98
Glycochenodeoxycholic acid	3.57
Glycolic acid	2.76
Taurochenodeoxycholic acid	3.66
Taurocholic acid	2.93
Taurodeoxycholic acid	3.85
Taurolithocholic acid	4.48
Tauroursodeoxycholic acid	2.53
Ursodeoxycholic acid	3.66

5. Incubate for 30 min in the −20 °C freezer. Samples are kept stationary during this time.
6. Centrifuge for 10 min at 12,000 × *g* at 4 °C to pellet insoluble materials such as proteins and nucleic acids, and ensure that the resulting supernatant is completely clear. If there are still particulates or cloudiness, transfer supernatants to a new tube and repeat centrifugation.
7. For analysis of oxylipids and bile acids, add 40 μL of supernatant and 40 μL of 10 mM ammonium acetate to an LC-compatible autosampler vial. Briefly vortex to mix, and then check vials individually to ensure the absence of air bubbles at the bottom of the vial. This is particularly likely to occur when vials contain a conical insert. For analysis of cholesterol, cholesteryl esters, and phospholipids, this dilution step can be

skipped or alternatively a separate set of autosampler vials prepared containing 80 μL of undiluted extract.

8. For use as a quality control, prepare a pooled sample by combining 10 μL of each extract supernatant into an autosampler vial and diluting the final volume 1:1 (v/v) with 10 mM ammonium acetate for oxylipid/bile acid analysis or no dilution of pool for analysis of cholesterol, cholesteryl esters, and phospholipids. Aliquot size may be adjusted so that the final QC sample volume is sufficient for QC runs at the beginning, middle, and end of the sequence or every 10–20 sample runs depending on the size of the sample set. Remaining supernatants may be stored at −80 °C (*see* **Note 1**).
9. Store autosampler vials containing samples in the temperature range of −20 to +4 °C until analysis. We recommend analyzing within 24 h, preferably as soon as possible after extracts are prepared.

3.4 Extraction of Lipids from Cell Samples

1. Place tubes containing frozen cell pellets on dry ice.
2. Add chilled extraction solvent (methanol) containing stable isotope-labeled lipid standards, if desired, to a final concentration of 2e6 cells per mL of solvent (*see* **Note 5**).
3. Vortex samples vigorously at 4 °C for 15 min.
4. Incubate for 15 min in the −20 °C freezer. Samples are kept stationary during this time.
5. Centrifuge for 10 min at 12,000 × *g* at 4 °C to pellet insoluble materials such as proteins and nucleic acids, and ensure that the resulting supernatant is completely clear (*see* **Note 6**). If there are still particulates or cloudiness, transfer supernatants to a new tube, and repeat centrifugation.
6. For analysis of oxylipids and bile acids, add 40 μL of supernatant and 40 μL of 10 mM ammonium acetate to an LC-compatible autosampler vial. Briefly vortex to mix, and then check vials individually to ensure the absence of air bubbles at the bottom of the vial. This is particularly likely to occur when vials contain a conical insert. For analysis of cholesterol, cholesteryl esters, and phospholipids, this dilution step can be skipped or alternatively a separate set of autosampler vials prepared containing 80 μL of undiluted extract.
7. For use as a quality control, prepare a pooled sample by combining 10 μL of each extract supernatant into an autosampler vial and diluting the final volume 1:1 with 10 mM ammonium acetate for oxylipid/bile acid analysis or no dilution of pool for analysis of cholesterol, cholesteryl esters, and phospholipids. Aliquot size may be adjusted so that the final QC sample volume is sufficient for QC runs at the beginning, middle, and

end of the sequence or every 10–20 sample runs depending on the size of the sample set. Remaining supernatants may be stored at −80 °C (*see* **Note 1**).

8. Store autosampler vials containing samples in the temperature range of −20 to +4 °C until analysis. We recommend analyzing within 24 h, preferably as soon as possible after extracts are prepared.

3.5 Extraction of Lipids from Tissue Samples

1. Label new 1.5 mL bead-beater-safe tubes (such as Eppendorf Safe-lock tubes), and cool on tube rack immersed in dry ice.
2. Quickly transfer tissue samples from −80 °C freezer to rack on dry ice. Prevent tissue samples from thawing.
3. Transfer tissue to pre-chilled, tared tube. Quickly weigh and record the mass of each tissue to the nearest 0.1 mg, and then immediately return tube to dry ice (*see* **Note 2**).
4. Add chilled extraction solution with applicable standard(s) to each tube to obtain a concentration of 15 mg tissue per mL of extraction solution (*see* **Note 3**).
5. For powderized tissue: Proceed to **step 6**. For whole pieces of tissue: Add one scoop of glass beads (~50–75 μL of volume) to each sample, and bead beat for 5 min at level 4 at 4 °C.
6. Vortex samples vigorously for 30 min at 4 °C.
7. Centrifuge for 10 min at 4 °C at 18,213 × *g* (or maximum speed) to pellet insoluble material (*see* **Note 4**).
8. If extraction was performed at a concentration greater than 15 mg/mL (in **step 4**), dilute an aliquot of supernatant with the same chilled extraction solution used for initial extraction (containing standards if applicable) to obtain a concentration of 15 mg/mL. Then repeat **step 7**.
9. For analysis of oxylipids and bile acids, add 40 μL of supernatant and 40 μL of 10 mM ammonium acetate to an LC-compatible autosampler vial. Briefly vortex to mix, and then check vials individually to ensure the absence of air bubbles at the bottom of the vial. This is particularly likely to occur when vials contain a conical insert. For analysis of cholesterol, cholesteryl esters, and phospholipids (Table 3), this dilution step can be skipped or alternatively a separate set of autosampler vials prepared containing 80 μL of undiluted extract.
10. Transfer ~100 μL extract supernatant into labeled and cooled autosampler vials.
11. For use as a quality control, prepare a pooled sample of each biological matrix being analyzed by combining 10 μL of each extract supernatant into an autosampler vial. Aliquot size may be adjusted so that the final QC sample volume is sufficient for QC runs at the beginning, middle, and end of the sequence

Table 3
Retention times of phospholipids in SPLASH® LIPIDOMIX® mass spec standard kit

Compound	Retention time (ave, min)
15:0-18:1(d7) PC	9.96
15:0-18:1(d7) PE	10.02
15:0-18:1(d7) PS	9.17
15:0-18:1(d7) PG	9.07
15:0-18:1(d7) PI	8.99
15:0-18:1(d7) PA	9.53
18:1(d7) LPC	6.70
18:1(d7) LPE	6.83
18:1(d7) Chol Ester $[M+NH_4]^+$	12.50
18:1(d7) MG	7.83
15:0-18:1(d7) DG	10.61
15:0-18:1(d7)-15:0 TG	11.82
18:1(d9) SM	9.79
Cholesterol (d7) $[M–H_2O+H]^+$	10.11

for both positive and negative ion polarity mode or every 10–20 sample runs depending on the size of the sample set. Remaining supernatants may be stored at −80 °C (*see* **Note 1**).

12. Prepare a blank sample by adding 100 μL of extraction solution from **step 4** into a new, labeled autosampler vial.
13. Ensure that the sample vials do not contain air bubbles and are kept cold (in range of −20 °C in freezer to LC autosampler at 7 °C) prior to analysis.
14. Perform MS analysis on samples.

3.6 UHPLC-MS Data Acquisition

1. Prepare a gradient UHPLC-MS negative ion global oxylipid method with the following chromatography conditions: Column temperature 45 °C and sample compartment temperature 7 °C. Using Phase A and B1, solvent gradient: 0–1 min 25% B1 and 0.3 mL/min, 1–2 min 25–50% B1 and 0.3 mL/min, 2–8 min 50–90% B1 and 0.3 mL/min, 8–10 min 90–99% B1 and 0.3 mL/min, 10–14 min hold at 99% B1 and 0.3 mL/min, 14–14.1 min 99–25% B1 and 0.3 mL/min, 14.1–16.9 min hold at 25% B1 and 0.4 mL/min, and 16.9–17 min hold at 25% B1 and resume flow of 0.3 mL/min.

Utilize a C18 column (method has been optimized for 2.1 × 150 mm, 1.8 μm column), and if desired, a C18 guard column (*see* **Note 2**).

2. Prepare a gradient UHPLC-MS positive ion global lipid method with the following chromatography conditions: Column temperature 45 °C and sample compartment temperature 7 °C. Using Phase A and B2, solvent gradient: 0–10 min 30–100% B2 and 0.325 mL/min; 10–12 min 100% B2 and 0.325 mL/min; 12–12.5 min 100–30% B2 and 0.4 mL/min; 12.5–14.9 min 30% B2 and 0.4 mL/min; 14.9–15 min 30% B2 and 0.325 mL/min.
3. Add mass spectrometry settings to the global UHPLC-MS method. The following are our routine settings: resolution 70,000 (Q Exactive) or 60,000 (Q Exactive HF), scan range 150–1500 *m/z*, maximum injection time 200 ms, microscans 2, automatic gain control (AGC) 3e6.
4. Determine optimal ESI (or HESI) gas and voltage settings by observing signal intensity as settings are changed. It is critical to use the flow rate of the analytical method during this test. We use 4.0 kV source voltage, 320 °C capillary temperature, and 45 sheath gas, 15 aux gas, and 0 sweep gas (all nitrogen) for this method.
5. Incorporate optimized gas and voltage settings into method, ensure polarity is set to negative, and save method.
6. Prepare run sequence starting with several blank runs, 1–2 injections of QC sample, then randomized analytical samples with QC samples inserted every 5–10 runs. End with several blank runs before changing polarities. We recommend collecting all data in one polarity mode, then switching to the other to minimize issues with column equilibration.
7. Before beginning, ensure stable back pressure and background signal after the column temperature set point has been reached. Ensure also the absence of air bubbles or leaks in the LC system, and then run sequence.

4 Data Visualization and Semi-targeted Analysis

4.1 Data Processing and Absolute Quantification

1. Utilize peak integration software to quantify MS^1 peak areas for the calculated intact masses of interest ±15 ppm. We recommend the freely available tools MAVEN (http://genomics-pubs.princeton.edu/mzroll/index.php) for visualization and integration of mzXML files preceded by RawConverter which converts .raw to the .mzXML format.
2. Utilize QC sample data to assess instrument performance and stability throughout analytical runs. Coefficients of variation

(CV) for named metabolites should be <20%. If CV values are above this threshold while other readouts (pump pressure, total signal level, etc.) do not suggest technical issues with instrument performance, consider the use of a batch- or drift-correction software. When needed, we utilize the MetaboDrift normalization algorithms.

3. Use the ratio of peak areas for stable isotope-labeled standards and their unlabeled (light) counterparts to determine the absolute values of endogenous metabolites in the extract (*see* **Note 7**). Back-calculate the concentrations in the biological samples using the following equations:

   For biofluids:

   $$[\text{Light}] = \left(\text{Area}_{\text{light}} / \text{Area}_{\text{heavy}}\right)[\text{Heavy}] \times \text{DF}$$

   where DF is dilution factor, described here as 10.

   For tissues:

   $$[\text{Light}] = \left(\text{Area}_{\text{light}} / \text{Area}_{\text{heavy}}\right)[\text{Heavy}] \times 1 / \text{EC}$$

   where EC is extraction concentration, described here as 15 mg/mL (*see* **Note 8**).

   For cells:.

   $$[\text{Light}] = \left(\text{Area}_{\text{light}} / \text{Area}_{\text{heavy}}\right)[\text{Heavy}] \times 1 / \text{EC}$$

   where EC is extraction concentration, described here as 2×10^6 cells/mL (*see* **Note 9**).

4.2 Data Visualization and Untargeted Discovery Analysis

1. While this chapter outlines the major workflow parameters used for Compound Discoverer 2.1 (ThermoFisher), multiple data analysis packages are available for chromatographic alignment, feature detection, compound identification, and pathway analysis including XC-MS [14–16] and MetaboAnalyst [17, 18].
2. A standard workflow for discovery analysis using Compound Discoverer 2.1 is shown in Fig. 1. For an in-depth explanation of parameter functions and optimization, please refer to the software manual and tutorials (https://mycompounddiscoverer.com/tutorials/).
3. All search parameter settings are maintained at default except for the following modifications:
   - Detect unknown compounds—min. peak intensity to 10,000.
   - Search ChemSpider—databases to BioCyc; Human Metabolome Database; KEGG; LipidMAPS; NIST. Set mass tolerance to 5 ppm.

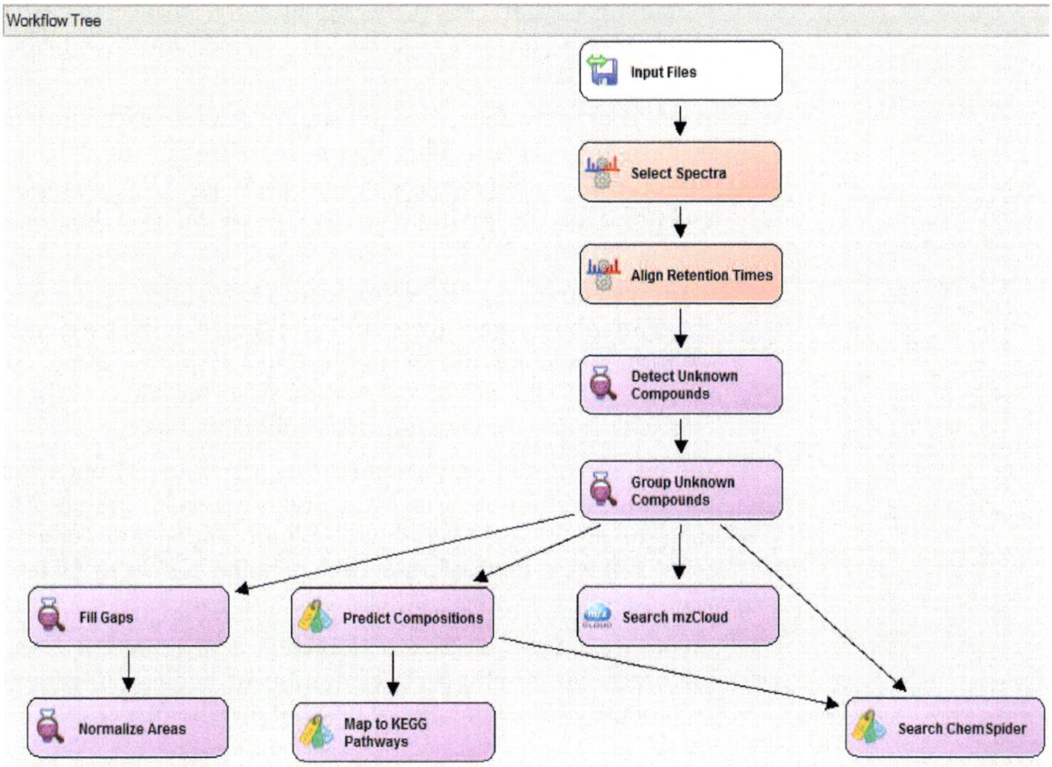

Fig. 1 Compound discoverer workflow tree

5 Notes

1. If/when the samples are retrieved from the freezer for future analysis, vortex briefly to mix, and centrifuge 10 min at max speed, 4 °C to ensure the absence of insoluble material in the supernatant.
2. For the described method, we routinely utilize a Waters ACQUITY HSS T3 column (2.1 × 150 mm, 1.8 μm particle size) without a guard column.
3. If the sample mass is too large to accommodate the volume for 15 mg/mL (e.g., >20 mg), all samples may be extracted at an identical higher concentration and extracts diluted in **step 8** of this section.
4. When centrifuging samples in the presence of glass beads, supernatant is not often fully clarified. A supernatant aliquot may be transferred to a new tube and respun to ensure complete isolation of insoluble material.
5. The cell extraction concentration of 2 × 10^6 cells/mL is optimized for average-sized eukaryotic cells. Extraction of

metabolites from cells with significantly smaller diameter (e.g., T cells, neurons, bacteria) is best performed at higher concentrations.

6. If cell counts are approximated and/or imprecise, the insoluble protein pellet remaining after metabolite extraction may be dissolved in an appropriate chaotrope (e.g., 8 M urea) and quantified by the Bradford or BCA assay for post hoc data normalization.
7. The concentration of heavy standard utilized here is the final concentration in the extraction. For example, if the extraction solution contains 1 μM of standard and 20 μL of sample are mixed with 180 μL of extraction solution, the resulting concentration of heavy standard would be 0.9 μM.
8. If concentration of heavy standard is entered in μM units and the concentration of tissue extraction is measured in mg/mL, then the resulting concentration of light (endogenous metabolite) will be in units of μmol of metabolite per mg of tissue.
9. If concentration of heavy standard is entered in μM units and the concentration of cells in extraction is measured in cells/mL, then the resulting concentration of light (endogenous metabolite) will be in units of μmol of metabolite per cell.

Acknowledgments

The authors are grateful to Sarah Gehrke, B.S. (University of Colorado Denver), for contributions to Table 1.

Conflicts of Interest

The authors disclose that A.D. and T.N. are part of Omix Technologies, Inc.

References

1. Dettmer K, Aronov PA, Hammock BD (2007) Mass spectrometry-based metabolomics. Mass Spectrom Rev 26:51–78
2. Lu W, Bennett BD, Rabinowitz JD (2008) Analytical strategies for LC-MS-based targeted metabolomics. J Chromatogr B Analyt Technol Biomed Life Sci 871:236–242
3. Gowda GAN, Djukovic D (2014) Overview of mass spectrometry-based metabolomics: opportunities and challenges. Methods Mol Biol 1198:3–12
4. Scheltema RA, Hauschild J-P, Lange O et al (2014) The Q Exactive HF, a Benchtop mass spectrometer with a pre-filter, high-performance quadrupole and an ultra-high-field Orbitrap analyzer. Mol Cell Proteomics 13:3698–3708
5. Fiehn O (2002) Metabolomics--the link between genotypes and phenotypes. Plant Mol Biol 48:155–171
6. Nemkov T, Hansen KC, D'Alessandro A (2017) A three-minute method for high-throughput quantitative metabolomics and quantitative tracing experiments of central carbon and nitrogen pathways. Rapid Commun Mass Spectrom 31:663–673
7. Dietmair S, Timmins NE, Gray PP et al (2010) Towards quantitative metabolomics of mammalian cells: development of a metabolite extraction protocol. Anal Biochem 404:155–164

8. Kapoore RV, Vaidyanathan S (2016) Towards quantitative mass spectrometry-based metabolomics in microbial and mammalian systems. Philos Transact A Math Phys Eng Sci 374
9. Jia W, Xie G, Jia W (2018) Bile acid-microbiota crosstalk in gastrointestinal inflammation and carcinogenesis. Nat Rev Gastroenterol Hepatol 15:111–128
10. Schaap FG, Trauner M, Jansen PLM (2014) Bile acid receptors as targets for drug development. Nat Rev Gastroenterol Hepatol 11:55–67
11. Dennis EA, Norris PC (2015) Oxylipid storm in infection and inflammation. Nat Rev Immunol 15:511–523
12. Fu X, Felcyn JR, Odem-Davis K et al (2016) Bioactive lipids accumulate in stored red blood cells despite leukoreduction: a targeted metabolomics study. Transfusion (Paris) 56:2560–2570
13. de Wolski K, Fu X, Dumont LJ et al (2016) Metabolic pathways that correlate with post-transfusion circulation of stored murine red blood cells. Haematologica 101:578–586
14. Smith CA, Want EJ, O'Maille G et al (2006) XCMS: processing mass spectrometry data for metabolite profiling using nonlinear peak alignment, matching, and identification. Anal Chem 78:779–787
15. Tautenhahn R, Patti GJ, Rinehart D et al (2012) XCMS online: a web-based platform to process untargeted metabolomic data. Anal Chem 84:5035–5039
16. Gowda H, Ivanisevic J, Johnson CH et al (2014) Interactive XCMS online: simplifying advanced metabolomic data processing and subsequent statistical analyses. Anal Chem 86:6931–6939
17. Xia J, Wishart DS (2011) Web-based inference of biological patterns, functions and pathways from metabolomic data using MetaboAnalyst. Nat Protoc 6:743–760
18. Chong J, Soufan O, Li C et al (2018) MetaboAnalyst 4.0: towards more transparent and integrative metabolomics analysis. Nucleic Acids Res 46:W486–W494

Chapter 9

HPLC-MS/MS Methods for Diacylglycerol and Sphingolipid Molecular Species in Skeletal Muscle

Kathleen A. Harrison and Bryan C. Bergman

Abstract

HPLC-MS/MS has enabled the quantitative analysis of complex mixtures of lipid molecular species. Several separate analyses, using methods that have been optimized for individual lipid classes, provide good lipidomic profiles, but may not be desirable for laboratories constrained by available instrumentation and wanting a higher throughput. Here we describe two methods using binary gradient HiLiC HPLC and triple quadrupole MS that together provide a lipidomic profile for lipids of interest in type 2 diabetes research. Methods for analysis of molecular species of diacylglycerol, ceramide, dihydroceramide, sphingosine, glucosyl- and lactosylceramide, sphingomyelin, and acylcarnitine from skeletal muscle and primary culture cells are described.

Key words Lipidomics, Diacylglycerols, Ceramides, Dihydroceramides, Sphingosine, Glucosylceramides, Lactosylceramides, Acylcarnitines, Sphingomyelins, HPLC-MS/MS, HiLiC

Abbreviations

C:DB	Fatty acyl groups are indicated by total carbon number and number of carbon-carbon double bonds
TG	Triacylglycerides
1,2-DG or 1,3-DG	1,2- or 1,3-Diacylglycerols
Cer	N-acyl sphingosines or ceramides
dhCer	N-acyl-sphinganines or dihydroceramides
SO	Sphingosine
GluCer	1-O-glucosylceramides
LacCer	1-O-lactosylceramides
SPM	Sphingomyelins
PC	Phosphatidylcholine
AC	Acylcarnitines
S1P	Sphingosine-1-phospate

Angelo D'Alessandro (ed.), *High-Throughput Metabolomics: Methods and Protocols*, Methods in Molecular Biology, vol. 1978, https://doi.org/10.1007/978-1-4939-9236-2_9,

1 Introduction

The intracellular accumulation of diacylglycerol (DG) and sphingolipids are potential mechanisms explaining skeletal muscle insulin resistance in the pathogenesis of type 2 diabetes. DG and sphingolipids are greater in insulin-resistant humans and animal models [1–10] and increase after high-fat feeding and lipid infusions that acutely induce insulin resistance [7–9]. Importantly, not all lipids appear to be equal, as specific isomers and species of DG [10–15] and species of sphingolipids appear to be more related to insulin resistance than others [10, 16–21]. Thus, monitoring multiple classes of lipids and their molecular species is essential to discern the roles of specific ectopic lipids in promoting insulin resistance.

The use of HPLC-MS/MS has enabled quantitative analysis of complex biological mixtures of lipid molecular species. The vast differences in the concentrations, solubilities, and polarities of lipid classes within a typical sample, as well as the sheer number of molecular species within each class, present a challenge even when using HPLC-MS/MS. Thus quantitating all lipid classes within a single LCMS run, from the nonpolar TGs to very polar sphingomyelins, is still problematic [22]. Instead, multiple analyses per sample using multiple methods may be used to obtain a lipidomic profile. Despite elegant methods for the different lipid classes [23–25], our requirement for lipidomic profiles from large numbers of precious samples of limited size, and the reality of limited equipment, necessitated development of methods to maximize the classes of lipids that can be measured using a minimal number of analyses with minimal sample prep. Two methods were developed using a HPLC-triple quadrupole MS that rely on normal-phase HiLiC columns to separate lipid classes based on their polarities. Multiple time periods of MRMs centered around the retention times of the lipid classes are used. Sample prep consists of simple lipid extraction to avoid complications posed by hydrolysis or derivatization (*see* **Note 1**). These methods present a suitable compromise that allow us to obtain quantitative data for the classes of lipids of interest in type 2 diabetes research with only two analyses of the tissue lipid extract.

The first method described is used to analyze the more nonpolar lipid classes. This nonaqueous method employs a HiLiC column in a simple normal-phase capacity, as no water is used in the mobile phases. The abundant TGs, the 1,3- and 1,2-DG isomers, ceramides, and dhCer are separated according to lipid class polarity. The injection solvent was chosen so that these nonpolar lipids are dissolved, but the more polar lipids are only sparingly dissolved and thus do not build up on the front of the column appreciably. TGs are best analyzed separately [26] as the quantity of TGs is often orders of magnitude greater than DGs and ceramides (*see* **Note 2**).

The second method for analysis of the more polar sphingolipids and acylcarnitines employs a separate HiLiC column. The column separates in hydrophilic interaction mode, with a percentage of water in the mobile phase throughout the separation. A more polar injection solvent is used for this method so that all classes of lipids are solubilized. Nonpolar lipids elute quickly, while the glucosylceramides (GluCer), sphingosine (SO), lactosylceramides (LacCer), sphingomyelins (SPM), acylcarnitines (AC), and sphingosine-1-phosphate (S1P) are retained longer and separate according to their polarities. MRM transitions for phosphatidylcholine (PC) molecular species are also included to aid proper peak assignment of SPM species (*see* **Note 3**).

2 Materials

2.1 Reagents and Solvents

Solvents of the highest available quality (Fisher Optima LCMS grade) were used for stock solutions of reference standards and internal standards, lipid extractions, mobile phases, and injection solvents. Milli-Q water was produced on site. LCMS grades of acetic acid (HOAc) and ammonium acetate (NH_4OAc) and dichloromethane (DCM) stabilized with amylene were used.

2.1.1 Mobile Phase Solutions

1. Post-column ionization buffer: isopropanol, water (95:5 v/v), 30 mM ammonium acetate (NH_4OAc) (*see* **Note 4**).
2. "Nonpolar lipid" nonaqueous normal-phase method: A = isooctane, B = isooctane, acetonitrile, isopropanol (705:20:75).
3. "Polar lipid" aqueous HiLiC method: A = hexane, isopropanol, acetic acid (300:200:0.28), 10 mM NH_4OAc, B = hexane, isopropanol, water, acetic acid (300:400:84:0.784), 10 mM NH_4OAc (*see* **Note 5**).

2.1.2 Injection Solvents

1. Nonpolar lipid method: hexane, dichloromethane (95:5, v/v).
2. Polar lipid method: mobile phases A and B (85:15, v/v).

2.2 Equipment

2.2.1 HPLC-MS

Agilent 1100 HPLC system with binary pump 1312A, autosampler 1329A, and capillary pump 1376A used for post-column addition of ionization buffer, interfaced with an API 2000 triple quadrupole mass spectrometer (ABSciex) operated in MRM mode. Unless noted otherwise API2000 parameters are ISV 5000v, FP 350v, EP 10v, GS1 30 (40 for polar lipid method), GS2 20 (30 for polar lipid method), TIS temp 100C, CAD gas 12, and CUR gas 12. Declustering potentials (DP) and collision energies (CE) were optimized individually for each lipid class while infusing reference standards with a syringe pump. Autosampler sample tray temperature set to 15 °C.

2.2.2 HPLC Columns

1. Nonpolar lipid method: Kinetex HiLiC, 2.6 μm, 100 Å, 2.1 × 100mm, fitted with a Security Guard silica 2 × 4 mm cartridge pre-column (Phenomenex).
2. Polar lipid method: Luna HiLiC, 3 μm, 200 Å, 3 × 100 mm, fitted with a Security Guard silica 2 × 4 mm cartridge pre-column (Phenomenex).

2.2.3 General Lab Equipment

16 mm glass culture tubes, 13 mm glass screw cap tubes and PTFE-lined screw caps, ALS vials with 500 μl flat-bottom inserts, and PTFE-lined silicone septa screw caps. Tissue homogenizer. Repeat pipettor for addition of internal standard solutions.

2.3 Standards

Either stable isotope-labeled or non-endogenous molecular species serve as internal standards. Quantitation is based on standard curves made from as many reference standards as available, using the internal standard for each for each lipid class (*see* **Note 6**). High-quality standards were purchased from Avanti Polar Lipids, Cayman Chemical, Cambridge Isotope Laboratories, or Sigma.

Separate internal standard mixtures of the polar and nonpolar lipids are created by combining aliquots from individual standard stock solutions. The mixtures are dried in a screw cap bottle under N2 and redissolved in DCM and methanol (80:20) and then diluted to the concentrations indicated in Table 1. One ml aliquots of the combined mixtures were placed into ALS vials, capped with PTFE-lined caps, and stored −80 until use. Reference standards are combined in 20 ml of methanol and dichloromethane (80:20, v/v) in the concentrations listed below in Table 2. Aliquots (430 μl) of the mixtures are placed in ALS vials, capped and stored at −80 °C until use.

Table 1
Internal standard mixtures

1. Nonpolar lipid IS mixture	pmol/20 μl	2. Polar lipids IS mixture	pmol/20 μl
1,3-14:0, 2-16:1-d_5-TG	150	d17:1 sphingosine	50
1,2,3-17:0 TG	1000	12:0 GluCER	10
1-15:0, 2-$d_7$18:1 DG	80	12:0 LacCER	40
1,3-di15:0 DG	80	12:0 SPM	100
12:0 Cer	80	d3-16:0 AC	50
12:0 DhCer	25	d17.1 sphingosine-1-P	200

Table 2
Reference standard mixtures

1. Nonpolar lipid reference standard mixture			2. Polar lipid reference standard mixture		
Lipid class	**Acyl molecular species**	**Pmol/ul**	**Lipid class**	**Molecular species**	**Pmol/ul**
TG	1,2,3-16:0; 1,2,3-18:1	27.5	SO	d18.1	1
1,2-DG	1,2-di14:0; 1,2-di16:0; 1-16:0, 2-18:0; 1,2-di18:0; 1-16:0, 2-18:1; 1,2-di18:1; 1-18:0, 2-18:2; 1-18:0, 2-20:4;	1.7	GluCER	16:0; 18:0; and 24:1	0.25
			LacCER	18:0 and 24:1	1
			SPM	18:0 and 24:1	6
1,3-DG	1,3-di-18:2	1.7	AC	18:1	4
CER	18:0; 22:0; 24:0; and 24:1	1.25	S1P	d18.1	4
dhCER	16:0; 18:0; 24:0; and 24:1	0.25			

3 Methods

Tissue samples are homogenized, an aliquot taken for protein assay for normalization to protein content, spiked with internal standards, extracted, and reconstituted in appropriate solvent for the LCMS method.

3.1 Sample Preparation

Lipids are recovered from approximately 50 mg muscle samples using liquid-liquid extraction modified from the MTBE extraction method of Matyash [27].

1. Place weighed tissue sample in 13 mm glass tube.
2. Homogenize in 0.8 ml of Millipore-filtered water.
3. Remove of 10 μl for protein assay.
4. Transfer 750 μl of the homogenate to 13 mm screw cap tube.
5. Add 20 μl of the nonpolar lipid internal standard mixture and 20 μl of the polar lipid internal standard mixture (*see* **Note 7**) to each sample using a repeating pipette and 900 μl methanol.
6. Cap tube with PTFE-lined cap and mix using vortexer.
7. Add 3 ml methyl tert-butyl ether (MTBE); vortex; rotate 5 min at room temperature.
8. Centrifuge 10 min at 2500 rpm (1140 × *g*) to separate phases.

9. Transfer upper layer of organic solvent to a 16 mm glass culture tube.
10. Acidify lower methanol-water layer to 0.5% acetic acid. Repeat extraction **steps 7–9** (*see* **Note 8**).
11. Dry combined extracts under nitrogen at 25 °C.
12. Resuspend lipids in 200 μl dichloromethane, methanol (2:1, v/v); transfer to ALS vial with 500 μl glass insert; repeat with another 200 μl; cap and store at −80 until LCMS analysis is available.
13. Dry the sample extracts in ALS vials under nitrogen, at 25 °C.
14. Resuspend residue in hexane, dichloromethane (95:5, v/v) nonpolar method LCMS injection solvent.

3.2 Standard Curve Preparation

1. Warm one vial of each of the reference standard mixtures and internal standard mixtures prepared as in Subheading 2 to room temperature.
2. Add aliquots from the mixtures according to Table 3 to ALS vial inserts.
3. Evaporate solvent under nitrogen at 25 °C, and resuspend lipids in 100 μl hexane, DCM (95:5) injection solvent for LCMS analysis of the nonpolar lipids.

3.3 Setting Up the Periods of MRMs for Samples

Neutral loss and/or precursor ion scans unique to each lipid class can identify the molecular species present [24] in a typical sample. It may be necessary to crudely separate lipid classes of a sample using solid-phase extraction [28] first and then infuse the fractions separately to determine the molecular species present.

1. Infuse a typical sample lipid extract using syringe pump, along with ionization buffer through mixing tee.
2. Scan neutral loss of 35u (H_2O + NH_3) to determine the $MNH4^+$ m/z for DGs. From the molecular adduct ions, determine the C:DB species present. Set up all the possible combinations of MRMs for collision-induced neutral losses of each fatty acid+NH_3 from the MNH_4^+ ion. After running the sample with HPLC introduction, if some of the MRMs are not detectable and are unlikely combinations, these can be removed so that the dwell time of the other MRMs can be increased without increasing the overall cycle time.
3. Scan precursors of m/z 264.2 to determine the MH+ m/z of species of Cer, GluCer, and LacCer present.
4. Scan precursors of m/z 266.2, or scan neutral loss of water (18u) to determine the species of dhCer present (*see* **Note 9**).

Table 3
Standard curve (SC) ranges in pmol of lipid molecular species in each lipid class

SC	ul ref. std. mixtures 1 and 2	ul IS mixtures 1 and 2	TG	DG	Cer	dhCer	GluCer	LacCer	SPM	SO	S1P	AC
1	2	20	55	3.4	2.5	0.5	0.5	2	12	2	8	8
2	5	20	137.5	8.5	6.25	1.25	1.25	5	30	5	20	20
3	10	20	275	17	12.5	2.5	2.5	10	60	10	40	40
4	25	20	687.5	42.5	31.25	6.25	6.25	25	150	25	100	100
5	50	20	1375	85	62.5	12.5	12.5	50	300	50	200	200
6	100	20	2750	170	125	25	25	100	600	100	400	400
7	200	20	5500	340	250	50	50	200	1200	200	800	800

5. Scan precursors of m/z 184 to determine SPM present. As PCs may dominate, it may be necessary to use HPLC introduction instead for determining SPM species.
6. Scan precursors of m/z 85 to determine the MH+ ions of the acylcarnitine species present.
7. Enter MRMs for the species identified in periods of the MS method.

3.4 Nonpolar Lipid Method

This nonaqueous normal-phase method is used to separate the 1,2- and 1,3-DGs, ceramides, and dihydroceramides (see example in Fig. 1).

1. Purge pumps with solvents to remove air bubbles.
2. Set post-column ionization buffer flow to 20 μl/min.
3. Attach HiLiC Kinetex, 2.6 μm, 100 Å, 2.1 × 100 mm, fitted with a Security Guard silica 2 × 4 mm cartridge pre-column.
4. Set flow to 0.22 ml/min, and equilibrate column with 2%B for several minutes until the pressure stabilizes (*see* **Note 10**).
5. Gradient program: 2%B held for 4 min, increased to 64%B over 10 min, held for 3 min before ramping up to 98%B in 1 min, and held for 3 min before re-equilibrating at 2%B for 10 min (*see* **Note 11**).
6. Mass spectrometer parameters: Sequential periods of MRMs for the molecular species of each class of lipid are listed in Table 4 along with their optimal DP and CE voltages, period times, and MRM dwell times (*see* **Note 12**).
7. Run a 10 μl blank injection solvent, followed by analysis of a procedural blank, prior to the samples.

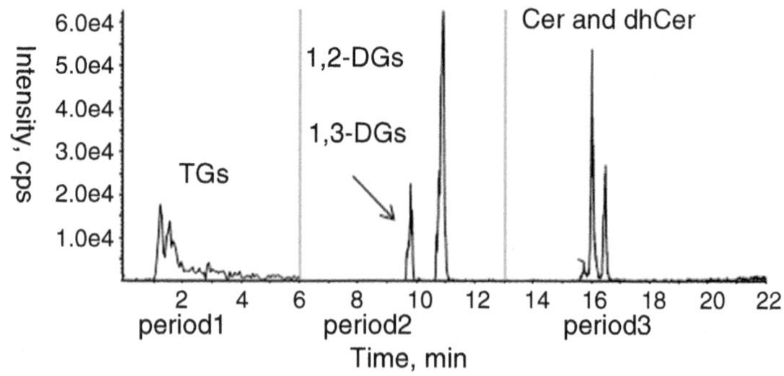

Fig. 1 TIC of MRMs from skeletal muscle lipid extract using nonaqueous normal-phase HPLC/MS/MS "nonpolar" lipid method (Subheading 3.4)

8. Submit sample queue to inject 10 μl of the samples dissolved in the hexane, dichloromethane (95:5, v/v), injection solvent (*see* **Note 13**). Adjust injection volume for samples if necessary so that both the minor lipids and the highly concentrated lipids are within the dynamic range of the detector.
9. For standard curve, inject 10 μl of solutions 1–6 and 2 μl of 7.

3.5 Polar Lipid Method

This aqueous HiLiC method is used to measure the GluCer, LacCer, SPM, and AC molecular species. Sphingosine (*see* **Note 14**) and S-1-P may also be analyzed with this method (*see* **Note 15**).

1. Purge pumps A and B with the polar method mobile phase solvents.
2. Set flow of post-column ionization buffer to 20 μl/min.
3. Attach Luna HiLiC, 3 μm, 200 Å, 3 × 100 mm, fitted with a Security Guard silica 2 × 4 mm cartridge pre-column.
4. Set column flow rate to 0.25 ml/min, and equilibrate the column with 15%B until the pressure is stable.
5. Dry under N2 remaining sample from prior analyses with nonpolar lipid method.
6. Redissolve samples and standard curve solutions in 90 μl of polar lipid method injection solvent, and place in autosampler rack.
7. Gradient program: 15%B held for 4 min, increased over 4 min to 65%B, then to 75%B over 12 min, and then to 90%B in 2 min, and held for 1 min, returned 15%B over 1 min and equilibrated at initial conditions for 10 min.
8. Mass spectrometer parameters: MRMs for the molecular species of each class of lipid are listed in the Table 5, along with

Table 4
Nonpolar lipid method, time periods, MRMs, and parameters

Period	Time	ID	MRM	MS parameters
1. TGs	0–6 min	Tri17:0 TG IS	866.8/579.5	DP 22v; CE 38v; ISV 4000v; Curtain gas 20; dwell time 20 ms plus 2 ms pause
		14:0/16:1/14:0-d_5TG IS	771.7/526.5 MNH4+/MNH4+, for each observed m/z for species of same C:DB	
2. 1,3- and 1,2-DGs	6–13 min	1,3-di15:0 IS	558.5/299.2	DP 7v; CE 25v; ISV 5000; curtain gas 12; dwell time 10 ms plus 2 ms pause
		1,2-15:0,d_7-18:1 IS	605.6/299.2; 605.6/346.3	
		14:0/14:0	530.5/285.2	
		14:0/16:0	558.5/285.2; 558.5/313.2	
		14:0/16:1	556.5/285.2; 556.5/311.2	
		14:0/18:0	586.5/285.2; 586.5/341.3	
		14:0/18:1	584.4/285.2; 584.5/339.2	
		16:0/16:0	586.5/313.2	
		16:0/16:1	584.5/313.2; 584.5/311.2	
		16:1/16:1	582.5/311.2	
		16:0/18:0	614.6/313.2; 614.6/341.3	
		16:0/18:1	612.6/313.2; 612.6/339.2	
		16:0/18:2	610.6/313.2; 610.6/337.2	
		16:1/18:1	610.6/311.2; 610.6/339.2	
		18:0/16:1	612.6/341.3; 612.6/311.2	
		16:0/20:4	634.6/313.2; 634.6/361.2	
		18:0/18:0	642.6/341.3	
		18:0/18:1	642.6/341.3; 640.6/339.2	
		18:0/18:2	638.6/341.3; 638.6/337.2	
		18:0/18:3	636.6/341.3; 636.6/335.2	
		18:1/18:1	638.6/339.2	
		18:1/18:2	636.6/339.2; 636.6/337.2	
		18:2/18:2	634.6/337.2	
		18:0/20:4	662.6/341.3; 662.6/361.2	
		18:0/22:6	686.5/341.3; 686.5/385.2	
3. Cer and dhCer	13–32 min	12:0Cer IS	482.4/264.2	DP 14v; CE 34v; dwell time 10 ms plus 2 ms pause
		14:0Cer	510.5/264.2	
		16:0Cer	538.5/264.2	
		18:0Cer	566.5/264.2	
		18:1Cer	564.5/264.2	
		20:0Cer	594.5/264.2	
		22:0Cer	622.6/264.2	
		23:0Cer	638.6/264.2	
		24:2Cer	646.6/264.2	
		24:1Cer	648.6/264.2	
		24:0Cer	650.6/264.2	
		12:0dhCer IS	484.4/466.4	DP 42v; CE 39v; dwell time 10 ms with 2 ms pause
		16:0dhCer	540.5 / 522.5	
		18:0dhCer	568.5 / 550.5	
		20:0dhCer	596.5 / 578.5	
		22:0dhCer	624.6 / 606.6	
		24:1dhCer	650.6 / 632.6	
		24:0dhCer	652.6 / 634.6	

Table 5
Aqueous polar lipid method, retention times, MRMs, and parameters

Polar lipid class	Retention time, min[a]	ID	MRM	MS parameters
Sphingosines	3.96	d17.1 IS	286.2/268.2	DP 34; CE 22
	3.86	d18.1	300.2/282.2	
	15.85	d17.1-1-P IS	366.2/250.2	DP 49; CE 22
	15.76	d18.1-1-P	380.2/264.2	
GluCer	3.58	12:0 IS	644.6/264.2	DP 25; CE 62
	3.38	16:0	700.6/264.2	
	3.28	18:0	728.6/264.2	
	(3.24)	20:0	756.6/264.2	
	(3.17)	22:0	784.6/264.2	
	(3.13)	23:0	798.7/264.2	
	3.18	24:1	810.7/264.2	
	(3.1)	24:0	812.7/264.2	
LacCer	11.53	12:0	806.6/264.2	DP 25; CE 73
	(10.80)	16:0	862.7/264.2	
	10.45	18:0	890.7/264.2	
	(10.15)	20:0	918.7/264.2	
	(9.77)	22:0	946.7/264.2	
	(9.68)	23:0	960.7/264.2	
	9.79	24:1	972.7/264.2	
	(9.60)	24:0	974.7/264.2	
SPM and PC	13.31	12:0 SPM IS	647.6/184.0	DP 40; CE 40
	(13.19)	14:0SPM	675.6/184.0	
	(13.06)	16:0SPM	703.6/184.0	
	(13.13)	16:1SPM	701.6/184.0	
	12.95	18:0SPM	731.7/184.0	
	(12.37)	$^{13}C_1$-32:2 PC	731.7/184.0	
	(12.37)	32:2PC	730.7/184.0	
	(12.98)	18:1SPM	729.7/184.0	
	(12.88)	20:0SPM	759.7/184.0	
	(12.26)	$^{13}C_1$.34:2 PC	759.7/184.0	
	(12.26)	34:2PC	758.7/184.0	
	(12.92)	20:1SPM	757.7/184.0	
	(12.30)	$^{13}C_1$.34:3 PC	757.7/184.0	
	(12.30)	34:3PC	756.7/184.0	
	(12.77)	22:0SPM	787.7/184.0	
	(12.14)	$^{13}C_1$-36:2 PC	787.7/184.0	
	(12.15)	36:2 PC	786.7/184.0	
	(12.80)	22:1SPM	785.7/184.0	
	(12.18)	$^{13}C_1$-36:3PC	785.7/184.0	
	(12.20)	36:3PC	784.7/184.0	
	(12.67)	24:0SPM	815.7/184.0	
	12.72	24:1SPM	813.7/184.0	
	(12.77)	24:2SPM	811.7/184.0	
	(12.10)	$^{13}C_1$-38:4 PC	811.7/184.0	
	(12.10)	38:4PC	810.8/184.0	
	12.04	di19:0 PC IS	818.8/184.0	

(continued)

Table 5
(continued)

Polar lipid class	Retention time, min[a]	ID	MRM	MS parameters
AC	13.35	d_3-16:0 IS	403.4/85.0	DP 15; CE 56
	13.24	18:1	426.4/85.0	
	(13.35)	18:2	424.4/85.0	
	(13.12)	18:0	428.4/85.0	
	(13.36)	16:0	400.4/85.0	
	(13.43)	16:1	398.4/85.0	
	(13.58)	14:0	372.3/85.0	
	(13.86)	12:0	344.3/85.0	
	(14.20)	10:0	316.3/85.0	

[a]The retention times in parentheses were not confirmed by a reference standard

optimal DP and CE voltages, retention times, and MRMs. A dwell time of 10 ms plus a 2 ms pause time was used for each MRM.

9. Submit sample queue with 10 μl blank injection solvent, followed by analysis of a procedural blank, followed by the samples. Adjust injection volume for samples if necessary so that both the minor lipids and the highly concentrated lipids are within the dynamic range of the detector.
10. For standard curve, inject 10 μl of solutions 1–6 and 2 μl of 7.

3.6 Calibration Curve Assignments

Careful and consistent assignment of calibration reference standards for molecular species with no available reference standard is important for consistent results between experiments.

1. Quantitate components using the standard curves from an identical reference standard if it is available.
2. If a reference standard is not available for a species, use the standard curve from a reference standard with the nearest retention time, according to Table 6.
3. When quantifying DGs, use the sum of the MRMs for loss of each fatty acid from asymmetric DG species. This improves quantitation by increasing the signal to noise of the peak.

4 Notes

1. Although derivatization of the free hydroxyl fixes the isomeric distribution of DGs, it has the disadvantage of modifying the multi-hydroxyl sphingolipid species to an unknown extent that would then require division of the sample and a separate analy-

Table 6
Calibration assignments for lipid molecular species

Lipid class	Molecular species	Calibration reference standard
1.3-DGs	All	1,3-18:2/18:2
1,2-DGs	14:0/14:0; 14:0/16:1; 14:0/16:0; 16:1/16:1	1,2-14:0/14:0
	16:0/16:0; 14:0/18:0; 14:0/18:1; 16:0/16:1; 18:2/18:2	1,2-16:0/16:0
	16:0/18:0	1,2-16:0/18:0
	16:0/18:1; 16:0/20:2; 18:0/16:1; 16:1/18:1	1,2-16:0/18:1
	18:1/18:1; 18:2/18:1; 16:0/20:4; 16:0/20:3	1,2-18:1/18:1
	18:0/18:2; 16:0/20:2; 18:0/18:3	1,2-18:0/18:2
	18:0/18:0; 16:0/20:0; 18:0/18:1	1,2-18:0/18:0
	18:0/20:4; 18:0/22:6[a]	1,2-18:0/20:4
Cer	14:0; 16:0; 18:1; 18:0; 20:0	18:0
	22:0; 22:1	22:0
	23:0; 24:0	24:0
	24:1; 24:2[a]	24:1
dhCer	16:0	16:0
	18:0; 20:0	18:0
	24:0	24:0
	22:0; 24:1	24:1
GluCer	16:0	16:0
	18:0; 20:0	18:0
	22:0; 23:0; 24:0; 24:1	24:1
LacCer	16:0; 18:0; 20:0	18:0
	22:0; 23:0; 24:0; 24:1	24:1
SPM	All species with N-acyl groups less than 22 carbons	18:0
	All species with N-acyl groups of 22 or more carbons	24:1
AC	All	18:1

[a]Reference standard curve chosen because presence of polyunsaturated fatty acid in component

sis. Isomerization of the 1,2-DGs to 1,3-isomer, or vice versa, following isolation is minimal if samples are stored in a freezer in polar nonaqueous solvents [29] and the rate of acyl migration is negligible at room temp over typical time (24–48 h) samples are undergoing analyses [30].

2. TGs are not quantified with this method as they elute too close to solvent front and are beyond the dynamic range of instrument when injecting enough sample to quantify the species of the less abundant lipid classes. A reversed-phase system for TAGs, similar to [26], was found to be adequate. The MRM signals from the highly concentrated TG species eluting near the solvent front along with common polymeric contaminants can be deceptive in the normal-phase method. Ion suppression in the electrospray source often leads to nearly complete loss of

signal followed by a sharp peak at the trailing edge of the eluting TG peak that seems to be a nicely resolved peak, but it is not reliable for quantitation.

3. As both PC and SPM species fragment to the phosphocholine ion, at m/z 184, there may be two closely eluting peaks in the MRM trace for some SPM species due an isobaric $^{13}C_1$-isotope of a PC species. Without reference standards (not all are available), it can be confusing to determine the correct peak to identify as SPM especially since the $^{13}C_1$-PC peaks may be large relative to the isobaric SPM peak. Automatic peak detection parameters may misassign $^{13}C_1$-PCs as SPM. The ^{12}C-PC MRM trace can be used to determine the retention time of the PC species and correct misassignment of a $^{13}C_1$-PC isotope peak as an SPM species.
4. For all mobile phases requiring ammonium acetate, completely dissolve the ammonium acetate in the most polar solvent of the mixture (isopropanol or water), before adding the nonpolar components.
5. For polar mobile phase B, dissolve the NH_4OAc in the water, and then mix in the isopropanol, followed by the hexane and HOAc to ensure no separation of phases develops. The polar method mobile phase B has the maximum percent of water that will dissolve in this ratio of hexane to isopropanol.
6. Although the 1,2-DGs and the 1,3-DGs undergo the same neutral losses of FA + NH3, separate reference and internal standards for DG isomers are required as 1,3-DGs elute earlier than the 1,2-DGs and experience different underlying background in the ion source. The background components affect the extent of ionization in the source differently for the 1,3-DGs than for the later eluting 1,2-DGs. This background from the sample matrix is unlikely to be present (to same extent) in the SC solutions and would lead to inaccuracy if only one isomer of DG is used as internal standard or reference standard for both the 1,2- and 1,3-DGs.
7. Allow internal standard mixture and standard curve vials to reach room temperature and mix before withdrawing aliquots so that all components are uniformly in solution.
8. A neutral extraction followed by an acidic extraction improves overall recovery of d17.1-sphingosine and d17.1-sphingosine-1-phosphate internal standards.
9. Dihydroceramide species may be measured using MRM transitions that monitor the decomposition of MH+ ions to at the sphingoid backbone ion m/z 266 or by MRMs that monitor the loss of water. While the loss of water is a less specific transition, it is a more abundant fragment ion for the dhCer than the ion at m/z 266 on our equipment, enabling detection of these minor components. However, careful attention must be given

to the retention times for both the dhCer species and ceramide species to prevent misidentifying a peak as a dhCER. dhCer that are isobaric with a ceramide species (e.g., 24:1 dhCer and 24:0Cer or the M + 2 isotope peak from the more abundant 24:1 ceramide, $^{13}C_2$-24:1Cer) may be misidentified as all undergo loss of water.

10. HiLiC columns are shipped in 90:10 acetonitrile, water, 10 mM ammonium formate. Any water in the column will drastically reduce the retention of the nonpolar lipids. Before introducing method mobile phase, which is immiscible with the shipping solvent, wash column with several column volumes of series of solvents to remove the water and buffer. Start with 90:10 acetonitrile water with no buffer followed by methanol, isopropanol, and then mobile phase B, and finally equilibrate with gradient starting condition.
11. Adequate equilibration of the column at the starting point of gradient is critical for reproducible retention times. The exact time required varies with the gradient delay volume of the HPLC used. Delay volumes vary significantly between HPLC models.
12. Separate time periods of MRMs for each lipid class were used to maintain short cycle times and maximize the number of scans taken to define each peak (10–15 points recommended) while still having sufficient dwell times (10–20 ms) to detect each MRM transition with good s/n. At least 10–15 points across the peak to define the peak for quantitation and dwell times of 10 ms or more are recommended by Sciex. For example, the total number of DG MRM transitions monitored in the DAG period is about 50. Using a 10 ms dwell time plus 2 ms pause time results in a scan time of 0.60 s, gathering about 15–20 points, baseline to baseline across the peaks, sufficiently defining the peaks, as DG species are about 0.15–0.20 min baseline to baseline (9–12 s).
13. Although this solvent minimizes solubilization of the polar lipid classes, some buildup on column will occur over time requiring flushing out the column regularly with more polar solvent such as IPA to elute them. Blank solvent injections interspersed in the sample queue and monitoring a sphingomyelin MRM will give an indication when column needs flushing out.
14. Sphingosine (SO) is a tricky molecule to analyze as its peak shape and retention time are severely affected by pH of the mobile phase. Non-protonated SO elutes between the glucosyl- and lactosylceramides.
15. S-1-P was generally too low in concentration to be detected on this machine, but d17.1 S-1-P internal standards were readily detected in samples. 4 pmol S-1-P injected on column for the

standard curve was detectable. A more sensitive mass spectrometer or one capable of switching polarity quickly in order to monitor the negative ion may be able to detect this component, if it is indeed present in skeletal muscle.

Acknowledgments

This work was partially supported by the National Institutes of Health General Clinical Research Center (RR-00036) NIDDK (R01DK089170 to BCB).

References

1. Adams JM, Pratipanawatr T, Berria R, Wang E, DeFronzo RA, Sullards MC, Mandarino LJ (2004) Ceramide content is increased in skeletal muscle from obese insulin-resistant humans. Diabetes 53(1):25–31
2. Turinsky J, O'Sullivan DM, Bayly BP (1990) 1,2-Diacylglycerol and ceramide levels in insulin-resistant tissues of the rat in vivo. J Biol Chem 265(28):16880–16885
3. Moro C, Galgani JE, Luu L, Pasarica M, Mairal A, Bajpeyi S, Schmitz G, Langin D, Liebisch G, Smith SR (2009) Influence of gender, obesity, and muscle lipase activity on intramyocellular lipids in sedentary individuals. J Clin Endocrinol Metab 94(9):3440–3447
4. Straczkowski M, Kowalska I, Baranowski M, Nikolajuk A, Otziomek E, Zabielski P, Adamska A, Blachnio A, Gorski J, Gorska M (2007) Increased skeletal muscle ceramide level in men at risk of developing type 2 diabetes. Diabetologia 50(11):2366–2373. https://doi.org/10.1007/s00125-007-0781-2
5. Avignon A, Yamada K, Zhou X, Spencer B, Cardona O, Saba-Siddique S, Galloway L, Standaert ML, Farese RV (1996) Chronic activation of protein kinase C in soleus muscles and other tissues of insulin-resistant type II diabetic Goto-Kakizaki (GK), obese/aged, and obese/Zucker rats. A mechanism for inhibiting glycogen synthesis. Diabetes 45(10):1396–1404
6. Schmitz-Peiffer C, Oakes ND, Browne CL, Kraegen EW, Biden TJ (1997) Reversal of chronic alterations of skeletal muscle protein kinase C from fat-fed rats by BRL-49653. Am J Phys 273(5 Pt 1):E915–E921
7. Schmitz-Peiffer C, Browne CL, Oakes ND, Watkinson A, Chisholm DJ, Kraegen EW, Biden TJ (1997) Alterations in the expression and cellular localization of protein kinase C isozymes epsilon and theta are associated with insulin resistance in skeletal muscle of the high-fat-fed rat. Diabetes 46(2):169–178
8. Griffin ME, Marcucci MJ, Cline GW, Bell K, Barucci N, Lee D, Goodyear LJ, Kraegen EW, White MF, Shulman GI (1999) Free fatty acid-induced insulin resistance is associated with activation of protein kinase C theta and alterations in the insulin signaling cascade. Diabetes 48(6):1270–1274
9. Itani SI, Ruderman NB, Schmieder F, Boden G (2002) Lipid-induced insulin resistance in human muscle is associated with changes in diacylglycerol, protein kinase C, and IkappaB-alpha. Diabetes 51(7):2005–2011
10. Perreault L, Newsom SA, Strauss A, Kerege A, Kahn DE, Harrison KA, Snell-Bergeon JK, Nemkov T, D'Alessandro A, Jackman MR, MacLean PS, Bergman BC (2018) Intracellular localization of diacylglycerols and sphingolipids influences insulin sensitivity and mitochondrial function in human skeletal muscle. JCI Insight 3(3):96805. https://doi.org/10.1172/jci.insight.96805
11. Bergman BC, Hunerdosse DM, Kerege A, Playdon MC, Perreault L (2012) Localisation and composition of skeletal muscle diacylglycerol predicts insulin resistance in humans. Diabetologia 55(4):1140–1150. https://doi.org/10.1007/s00125-011-2419-7
12. van Hees AM, Jans A, Hul GB, Roche HM, Saris WH, Blaak EE (2011) Skeletal muscle fatty acid handling in insulin resistant men. Obesity 19(7):1350–1359. https://doi.org/10.1038/oby.2011.10
13. Amati F, Dube JJ, Alvarez-Carnero E, Edreira MM, Chomentowski P, Coen PM, Switzer GE, Bickel PE, Stefanovic-Racic M, Toledo FG, Goodpaster BH (2011) Skeletal muscle triglycerides, diacylglycerols, and ceramides in insulin resistance: another paradox in endurance-trained athletes? Diabetes 60(10):2588–2597. https://doi.org/10.2337/db10-1221
14. Dube JJ, Amati F, Toledo FG, Stefanovic-Racic M, Rossi A, Coen P, Goodpaster BH (2011)

Effects of weight loss and exercise on insulin resistance, and intramyocellular triacylglycerol, diacylglycerol and ceramide. Diabetologia 54(5):1147–1156. https://doi.org/10.1007/s00125-011-2065-0

15. Holloway GP, Han XX, Jain SS, Bonen A, Chabowski A (2014) Chronic muscle stimulation improves insulin sensitivity while increasing subcellular lipid droplets and reducing selected diacylglycerol and ceramide species in obese Zucker rats. Diabetologia 57(4):832–840. https://doi.org/10.1007/s00125-014-3169-0
16. Bergman BC, Brozinick JT, Strauss A, Bacon S, Kerege A, Bui HH, Sanders P, Siddall P, Wei T, Thomas MK, Kuo MS, Perreault L (2016) Muscle sphingolipids during rest and exercise: a C18:0 signature for insulin resistance in humans. Diabetologia 59(4):785–798. https://doi.org/10.1007/s00125-015-3850-y
17. Turner N, Kowalski GM, Leslie SJ, Risis S, Yang C, Lee-Young RS, Babb JR, Meikle PJ, Lancaster GI, Henstridge DC, White PJ, Kraegen EW, Marette A, Cooney GJ, Febbraio MA, Bruce CR (2013) Distinct patterns of tissue-specific lipid accumulation during the induction of insulin resistance in mice by high-fat feeding. Diabetologia 56(7):1638–1648. https://doi.org/10.1007/s00125-013-2913-1
18. Raichur S, Wang ST, Chan PW, Li Y, Ching J, Chaurasia B, Dogra S, Ohman MK, Takeda K, Sugii S, Pewzner-Jung Y, Futerman AH, Summers SA (2014) CerS2 haploinsufficiency inhibits beta-oxidation and confers susceptibility to diet-induced steatohepatitis and insulin resistance. Cell Metab 20(4):687–695. https://doi.org/10.1016/j.cmet.2014.09.015
19. Tonks KT, Coster AC, Christopher MJ, Chaudhuri R, Xu A, Gagnon-Bartsch J, Chisholm DJ, James DE, Meikle PJ, Greenfield JR, Samocha-Bonet D (2016) Skeletal muscle and plasma lipidomic signatures of insulin resistance and overweight/obesity in humans. Obesity (Silver Spring) 24(4):908–916. https://doi.org/10.1002/oby.21448
20. Turpin SM, Nicholls HT, Willmes DM, Mourier A, Brodesser S, Wunderlich CM, Mauer J, Xu E, Hammerschmidt P, Bronneke HS, Trifunovic A, LoSasso G, Wunderlich FT, Kornfeld JW, Bluher M, Kronke M, Bruning JC (2014) Obesity-induced CerS6-dependent C16:0 ceramide production promotes weight gain and glucose intolerance. Cell Metab 20(4):678–686. https://doi.org/10.1016/j.cmet.2014.08.002
21. Coen PM, Dube JJ, Amati F, Stefanovic-Racic M, Ferrell RE, Toledo FG, Goodpaster BH (2010) Insulin resistance is associated with higher intramyocellular triglycerides in type I but not type II myocytes concomitant with higher ceramide content. Diabetes 59(1):80–88
22. Merrill AH Jr, Sullards MC (2017) Opinion article on lipidomics: inherent challenges of lipidomic analysis of sphingolipids. Biochim Biophys Acta 1862(8):774–776. https://doi.org/10.1016/j.bbalip.2017.01.009
23. Cajka T, Fiehn O (2014) Comprehensive analysis of lipids in biological systems by liquid chromatography-mass spectrometry. Trends Anal Chem 61:192–206. https://doi.org/10.1016/j.trac.2014.04.017
24. Merrill AH Jr, Sullards MC, Allegood JC, Kelly S, Wang E (2005) Sphingolipidomics: high-throughput, structure-specific, and quantitative analysis of sphingolipids by liquid chromatography tandem mass spectrometry. Methods 36(2):207–224. https://doi.org/10.1016/j.ymeth.2005.01.009
25. Leiker TJ, Barkley RM, Murphy RC (2011) Analysis of diacylglycerol molecular species in cellular lipid extracts by normal-phase LC-electrospray mass spectrometry. Int J Mass Spectrom 305(2-3):103–109. https://doi.org/10.1016/j.ijms.2010.09.008
26. Hu C, van Dommelen J, van der Heijden R, Spijksma G, Reijmers TH, Wang M, Slee E, Lu X, Xu G, van der Greef J, Hankemeier T (2008) RPLC-ion-trap-FTMS method for lipid profiling of plasma: method validation and application to p53 mutant mouse model. J Proteome Res 7(11):4982–4991. https://doi.org/10.1021/pr800373m
27. Matyash V, Liebisch G, Kurzchalia TV, Shevchenko A, Schwudke D (2008) Lipid extraction by methyl-tert-butyl ether for high-throughput lipidomics. J Lipid Res 49(5):1137–1146. https://doi.org/10.1194/jlr.D700041-JLR200
28. Kaluzny MA, Duncan LA, Merritt MV, Epps DE (1985) Rapid separation of lipid classes in high yield and purity using bonded phase columns. J Lipid Res 26(1):135–140
29. Li W, Du W, Li Q, Li RW, Liu D (2010) Dependence on the properties of organic solvent: study on acyl migration kinetics of partial glycerides. Bioresour Technol 101(15):5737–5742. https://doi.org/10.1016/j.biortech.2010.03.018
30. Laszlo J, Compton D, Vermillion K (2008) Acyl migration kinetics of vegetable oil 1,2-diacylglycerols. J Am Oil Chem Soc 85:307–312. https://doi.org/10.1007/s11746-008-1202-5

Part III

Metabolomics of Animal and Plant Models

Chapter 10

Quantification of D- and L-2-Hydroxyglutarate in *Drosophila melanogaster* Tissue Samples Using Gas Chromatography-Mass Spectrometry

Hongde Li and Jason M. Tennessen

Abstract

The fruit fly *Drosophila melanogaster* has emerged as an ideal system in which to study 2-hydroxyglutarate (2HG) metabolism. Unlike many mammalian tissues and cell lines, which primarily accumulate D- or L-2HG as the result of genetic mutations or metabolic stress, *Drosophila* larvae accumulate high concentrations of L-2HG during normal larval growth. As a result, flies represent one of the few model systems that allows for studies of endogenous L-2HG metabolism. Moreover, the *Drosophila* genome not only encodes key enzymes involved in the synthesis and degradation of D-2HG, but the fly has also been used as to investigate the in vivo effects of oncogenic isocitrate dehydrogenase 1 and 2 (IDH1/2) mutations. All of these studies, however, rely on mass spectrometry-based methods to distinguish between the D- and L-2HG enantiomers. While such approaches are common among labs studying mammalian cell culture, few *Drosophila* studies have attempted to resolve and measure the individual 2HG enantiomers. Here we describe a highly reproducible gas chromatography-mass spectrometry (GC-MS)-based protocol that allows for quantitative measurements of both 2HG enantiomers in *Drosophila* homogenates.

Key words 2-Hydroxyglutarate, *Drosophila*, Gas chromatography-mass spectrometry, Oncometabolite, Metabolomics

1 Introduction

D- and L-2HG are commonly referred to as oncometabolites and primarily studied in the context of tumor growth and cancer cell proliferation [1–3]. The production of these compounds (Fig. 1), however, is not limited to cancer cells but also is produced by a variety of healthy tissues, such as the brain, testis, activated T-cells, and primary human cell cultures that have been exposed to hypoxia [4–10]. Such observations indicate that both 2HG enantiomers likely serve endogenous roles in normal cellular metabolism and physiology; however, nearly all 2HG studies rely on cell culture

Angelo D'Alessandro (ed.), *High-Throughput Metabolomics: Methods and Protocols*, Methods in Molecular Biology, vol. 1978, https://doi.org/10.1007/978-1-4939-9236-2_10,

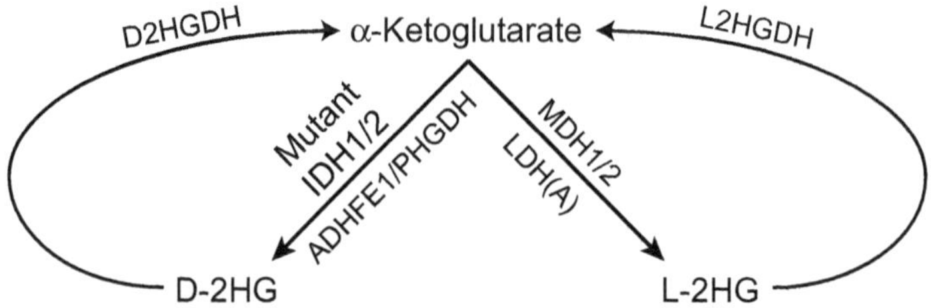

Fig. 1 A schematic diagram illustrating the metabolic reactions that generate 2HG from α-ketoglutarate. L-2HG can be synthesized by either lactate dehydrogenase (LDH(A)) or malate dehydrogenase 1 or 2 (MDH1/2). D-2HG can be synthesized by D-3-phosphoglycerate dehydrogenase (PHGDH), alcohol dehydrogenase iron-containing protein 1 (ADHFE1), or mutant isocitrate dehydrogenase 1 and 2 (IDH1/2). Both D- and L-2HG can be converted back to α-ketoglutarate by D-2HG dehydrogenase (D2HGDH) and L-2HG dehydrogenase (L2HGDH), respectively

models, and relatively little is known about the functions of either D- or L-2HG in vivo. We recently discovered that the concentration of L-2HG exceeds 2 mM in larvae of the fruit fly *Drosophila melanogaster*, which is among the highest 2HG levels observed in healthy animal tissues [11]. Unlike human cells, however, which generate L-2HG in response to hypoxia and disruption of the electron transport chain [4, 5, 12, 13], *Drosophila* larvae generate this compound under standard growth conditions, indicating that fly metabolism is uniquely adapted to synthesize and accumulate L-2HG. Moreover, *Drosophila* also accumulates D-2HG, albeit at significantly low levels when compared with L-2HG [11], and the fly genome encodes the enzyme, D-2-hydroxyglutarate dehydrogenase (D2HGDH; FBgn0023507) [14], that degrades D-2HG. Finally, this system is also amenable to studies of the oncogenic IDH1/2 mutations [14], which result in aberrant D-2HG production. These studies demonstrate that *Drosophila* can be used to study nearly any aspect of 2HG metabolism and suggest that future studies in the fly will be essential for elucidating the mechanisms that control 2HG accumulation in vivo.

In order to facilitate additional 2HG studies in *Drosophila*, here we provide a detailed version of the GC-MS-based protocol that we've previously used to measure both D- and L-2HG [11]. This method, which is adapted from an established protocol used to measure D- and L-2HG in human bodily fluids [15], provides the user with detailed instruction on how to reproducibly extract 2HG from *Drosophila* homogenates and quantitatively measure each molecule. Moreover, the two-step derivatization protocol coupled with GC-MS allows for D- and L-2HG to be easily separated and quantified in a reproducible manner (Fig. 2).

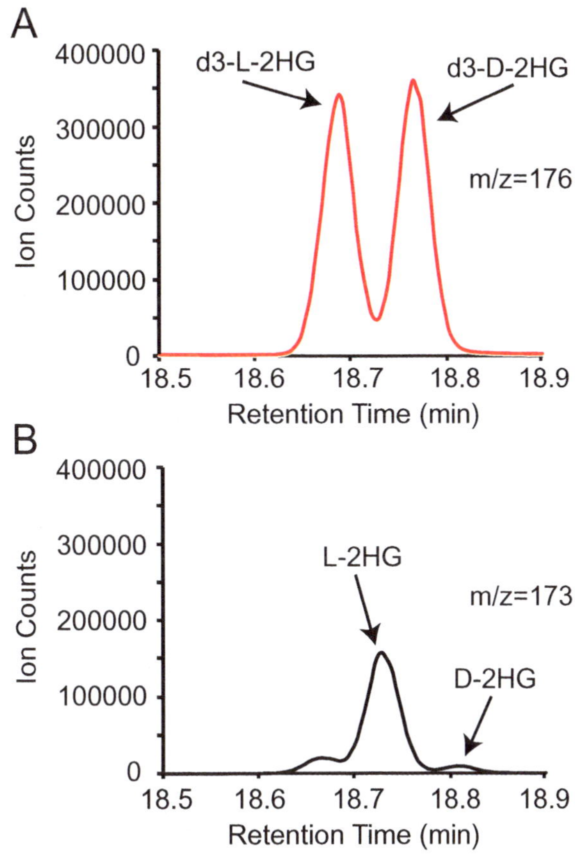

Fig. 2 A representative annotated spectrum generated from a *Drosophila* larval sample. Peaks representing either (**a**) the deuterated D- and L-2HG standard ($m/z = 176$) or (**b**) endogenous D- and L-2HG levels ($m/z = 173$)

2 Materials

2.1 Sample Collection

1. Molasses agar plates (35 × 10 mm): Mix 115 mL molasses, 29 g agar, and 700 mL of H_2O in a 2 L flask. Boil on a hot plate for 15 min while stirring at 350 rpm. Cool to 70 °C. Add 25 mL of acid mix (20 mL of 85% phosphoric acid and 209 mL propionic acid in 1 L of H_2O; store in the dark) and 10 mL of 10% p-hydroxy-benzoic acid methyl ester in 95% ethanol. Pour the molasses agar into the lid and bottom of a 35 × 10 mm plastic tissue culture dish (*see* **Note 1**).
2. Yeast paste: Mix 5 g of baker's yeast with 7 mL of water. The exact ratio will depend of the brand of yeast. The final mixture should have the creamy consistency (*see* **Note 2**).
3. Mating bottles: Use a 22-gauge needle to poke holes in a 6 oz. plastic *Drosophila* stock bottle. The egg-laying caps should fit snuggly in the mouth of the bottle and can be secured in place using laboratory tape.

4. Phosphate buffer saline (pH 7.4): For 1 L of 1× stock, dissolve 8 g NaCl, 0.2 g KCl, 1.44 g Na_2HPO_4, 0.24 g KH_2PO_4 in 800 mL of H_2O. Adjust pH to 7.4 with HCl. Add H_2O to 1 L.
5. 1.5 mL microfuge tubes (*see* **Note 3**).
6. 2 mL screw cap tubes containing 1.4 mm ceramic beads (*see* **Note 4**).
7. Analytical balance capable of measuring 0.01 mg (*see* **Note 5**).
8. Liquid nitrogen.

2.2 Sample Processing

1. Benchtop cooler pre-chilled to −20 °C.
2. 90% methanol: Dilute methanol using H_2O. Store at −20 °C. Only use HPLC-grade reagents.
3. *R*,*S*-[2,3,3-$^2H_3$]-2-hydroxyglutarate standard.
4. Bead mill homogenizer (*see* **Note 6**).
5. Vacuum centrifuge with attached vapor trap.

2.3 Chiral Derivatization

1. *R*-2-butanol.
2. 12 N hydrochloric acid.
3. 1.5 mL microfuge tubes.
4. H_2O (HPLC grade).
5. Hexane (HPLC grade).
6. Anhydrous pyridine (*see* **Note 7**).
7. Acetic anhydride.
8. Autosampler vials and caps: use a 150 μL glass insert.

2.4 GC-MS Analysis

1. GC-MS equipped with a 30 m Phenomenex ZB5-5 MSi column (or equivalent column) and an autosampler.

2.5 Data Analysis

1. A software package capable of analyzing the resulting spectra and integrating the peaks associated with D- and L-2HG. This software is usually provided by the manufacturer of the GC-MS (e.g., MassHunter from Agilent).

3 Methods

L-2HG levels vary dramatically during the course of the *Drosophila* life cycle (i.e., L-2HG levels will drop tenfold during the last 24 h of the third larval instar). Therefore, all samples collected during this analysis must be precisely staged. Only use HPLC-grade reagents when preparing samples for GC-MS analysis. Conduct all derivatization steps in a fume hood. Follow all waste disposal regulations when disposing of organic solvents and contaminated solid waste.

3.1 Sample Collection

1. Separately collect males and virgin females of the desired genotypes. Age these animals for 3–5 days in standard food vials (*see* **Note 8**).
2. Set up crosses of males and virgin females in mating bottles. Collect eggs on a molasses agar plate with ~1.5 g of yeast paste spread on the surface (egg-laying cap). Replace the egg-laying cap at least once a day for 3 days. Discard the egg-laying caps collected during this time (*see* **Note 9**).
3. After Day 3, place a fresh egg-laying cap into the mating cage. After 2 h, replace the egg-laying cap. Discard the first egg-laying cap used for the day (*see* **Note 10**).
4. Remove the second egg-laying cap after a defined period of time (usually 2 or 4 h, depending on the desired level of synchronization) and place inside a larger 60 × 15 mm dish. This step can be repeated multiple times during a single day.
5. Embryos are aged to the desired stage in a 25 °C incubator.
6. If collecting embryos, use a paintbrush to gently transfer 200 embryos onto a piece of PBS-soaked black filter paper that has been placed in the bottom a 35 mm tissue culture plate. Remove any large debris with a paintbrush. Tilt the culture dish and use a 1 mL pipette to slowly rinse the embryos with PBS (pH 7.4). This step will remove any yeast or agar stuck to the outside of the embryos. The embryos should remain in place during the rinse step. Remove the excess PBS from the bottom corner of the plate and discard prior to placing flat on the benchtop. Use the paintbrush to transfer the embryos to a pre-tared 2 mL screw cap tube containing 1.4 mm ceramic beads (*see* **Notes 11** and **12**). Record the mass of the sample and tubes using an analytical balance (*see* **Note 5**). Freeze samples in liquid nitrogen.
7. If collecting larvae, use a paintbrush to transfer synchronized populations of larvae (*see* **Note 13**) into a 1.5 mL microfuge tube (*see* **Note 3**). Individual samples should remain on ice (*see* **Note 14**). Typical larval sample sizes for this assay are 50 mid-first instar larvae (mid-L1), 25 mid-second instar larvae (mid-L2), and 10 mid-third instar larvae (mid-L3). Wash the samples three times with ice-cold PBS. Briefly centrifuge the sample at 2000 × *g* for 1 min, and remove all PBS from the sample tube with a 200 μL pipette. Freeze in liquid nitrogen. Remove the tube from the liquid nitrogen, and sharply pound the lid against the benchtop to dislodge the larval pellet (*see* **Note 3**). Quickly pour the larval pellet into a pre-tared 2 mL screw cap tube. Record the mass and immediately return the tube to liquid nitrogen (*see* **Note 15**).

8. If collecting pupae or adults (*see* **Note 16**), place 15 individuals into a pre-tared 2 mL screw cap tube containing 1.4 mm ceramic beads (*see* **Note 12**). Record the mass of the sample and freeze the sample in liquid nitrogen.
9. Samples can be stored at −80 °C for 3 months (*see* **Note 17**).

3.2 Sample Processing

1. Remove the samples from the −80 °C freezer and transfer into a benchtop cooler that was pre-chilled to −20 °C (*see* **Note 18**).
2. Add 800 μL of 90% methanol (pre-chilled to −20 °C) and 5 μL of 1 mg/mL *R*,*S*-[2,3,3-$^2H_3$]-2-hydroxyglutarate to the 2 mL bead tubes that contain the sample (*see* **Note 19**).
3. Add 800 μL of 90% methanol (pre-chilled to −20 °C) and 5 μL of 1 mg/mL *R*,*S*-[2,3,3-$^2H_3$]-2-hydroxyglutarate to an empty 2 mL bead tube. This sample will serve as a negative control.
4. Homogenize the samples in bead mill homogenizer (*see* **Notes 6** and **20**).
5. Return the homogenized samples to the benchtop cooler and place in a −20 °C freezer for at least 1 h. A white precipitate should form in the homogenized sample.
6. Centrifuge the samples in a 4 °C cold room (or a refrigerated centrifuge) for 5 min at 20,000 × *g*.
7. Carefully remove the samples from centrifuge, and return to the benchtop cooler. Be careful to not disturb the precipitate.
8. Transfer 600 μL of supernatant from the bead tube to a 1.5 mL microfuge tube. Do not disturb the precipitate. If debris from the precipitate is released into the supernatant, repeat **step 6**. If necessary, the transferred supernatant can be centrifuged at maximum speed in the 1.5 mL microfuge tube and transferred again to a new tube (*see* **Note 21**).
9. Dry the samples in a vacuum centrifuge overnight at room temperature. If necessary, the samples can be stored at −80 °C following this step.

3.3 Chiral Derivatization

1. Remove the samples from the vacuum centrifuge and place into a microfuge tube rack. Move the sample tubes to a fume hood (*see* **Note 22**).
2. Add 50 μL of *R*-2-butanol and 5 μL of 12 N HCl (37–38%) to the dried samples.
3. Place the sample tubes into a thermomixer. Incubate the sample tubes at 90 °C for 3 h at 300 rpm (*see* **Note 23**).
4. Remove the samples from the thermomixer and place into a microfuge tube rack. Allow the samples to cool to room temperature.

5. Extract the modified 2HG by adding 200 μL of water and 500 μL of hexane to the sample. Collect the organic phase and place in a new 1.5 mL microfuge tube (the organic phase is the top layer). Add an additional 500 μL of hexane to the sample. Again, collect the organic phase and combine with the organic phase from the first extraction.
6. Dry the sample with a vacuum centrifuge for 30 min at room temperature.
7. Remove the dried samples from the vacuum centrifuge and place into a microfuge tube rack. Move the sample tubes to a fume hood, and add 60 μL of anhydrous pyridine and 60 μL of acetic anhydride (*see* **Note 7**).
8. Place the sample tubes into a thermomixer. Incubate the sample tubes at 80 °C for 1 h at 300 rpm (*see* **Note 24**).
9. Dry the sample with a vacuum centrifuge.
10. Resuspend in 60 μL of hexane.
11. Transfer 50 μL of resuspended sample into an autosampler vial with a 150 μL glass insert.

3.4 GC-MS Analysis

1. Set the flow rate of the helium carrier gas to 1 mL/min.
2. Set the inlet temperature to 250 °C.
3. Set the GC-MS to execute the following temperature gradient:
   (a) Initial temperature of 95 °C with a hold of 1 min.
   (b) Increase to 110 °C at a rate of 40 °C per min. Hold for 2 min.
   (c) Increase to 200 °C at a rate of 5 °C per min.
   (d) Increase to 330 °C at the rate of 40 °C/min with a final hold of 4 min.
4. Set a solvent delay of 5 min (*see* **Note 25**).
5. Randomize the sample injection order.
6. Set the detector to selectively monitor ions with a *m/z* ratio of 173 (2HG) and 176 ($^2H_3$-2HG standard).
7. Inject 1 μL of the derivatized sample into the GC-MS (split ratio of 5:1; *see* **Note 26**).

3.5 Data Analysis

1. Integrate the peak area for D- and L-2HG with $m/z = 173$; the retention time of L-2HG is a little earlier than that of D-2HG. Denote the integration value as A_{target}.
2. Integrate the peak areas for internal standard (*R,S*-[2,3,3-$^2H_3$]-2-hydroxyglutarate) with $m/z = 176$. Add the integration values together for the two internal standard peaks. Denote the sum value as $A_{internal}$.

3. The concentration (μg/mg body mass) of L- or D-2HG in samples:

$$\text{Con}_{(\text{L or D-2HG})} = A_{\text{target}} / A_{\text{internal}} \times 5 / M,$$

where M is the sample pellet mass with unit of mg.

4 Notes

1. Prior to pouring the agar, be sure that both the lid and bottom of the tissue culture plate fit snuggly in the mating bottle. Some brands do fit snuggly and allow flies to escape. We find that both the lids and bottoms of Falcon plates (catalog # 353001) work well for this purpose.
2. Be sure to use the same brand and strain of yeast for all experiments. We use Fleischmann's Active Dry Yeast #2192; however, a variety of brands can be used for this purpose.
3. Frozen larval pellets are difficult to dislodge from some brands of microfuge tubes. Prior to collecting a large number of samples, be sure to determine if larval pellets can be easily removed from the brand of tube used for collection. We recommend 1.5 mL Eppendorf Flex-Tubes due to the ease with which we can dislodge pellets.
4. Some 2 mL screw cap tubes will disintegrate when used in a bead mill homogenizer. Be sure that the screw cap tubes used for sample collection can withstand the homogenization step.
5. Analytical balances that are only capable of measuring 0.1 mg are insufficient for conducting this analysis.
6. Insect cuticles are robust and require a large amount of force to destroy quickly. Moreover, samples must be homogenized while frozen because endogenous metabolic enzymes will reactivate in thawed tissue. Therefore, samples must be homogenized using an instrument capable of completely destroying a frozen sample in less than 1 min. We recommend the Omni Beadruptor 24.
7. Anhydrous pyridine should be stored in a desiccator and the vial flushed with an inert gas (e.g., nitrogen or argon) following every use.
8. L-2HG levels in *L-2-hydroxyglutarate dehydrogenase* (*dL2HGDH*; FBgn0032729) mutant larvae are sensitive to the maternal genotype. F1 progeny from *dL2HGDH/+* heterozygous mothers exhibit lower L-2HG levels than those born from *dL2HGDH* homozygous mutants. As a result, studies based on the *dL2HGDH* mutations should only use homozygous mutant adults for generating offspring.

9. Egg-laying caps can be replaced by inverting the mating bottle and tapping the top against the benchtop. Quickly remove the old egg-laying cap and replace with a new one.
10. This step ensures that females are not holding their eggs. Since L-2HG levels undergo dramatic changes during embryogenesis, the inadvertent collection of asynchronous embryo samples will result in unreliable measurements.
11. Embryos do not need to be dechorionated. The bead mill homogenizer will destroy intact embryos.
12. Be careful that all beads remain in the tube during this transfer.
13. L3 larvae must be carefully synchronized because L-2HG levels drop precipitously during the latter half of this developmental stage. We recommend resynchronizing larvae at the L2–L3 molt. Methods for synchronizing larvae have been previously described [16, 17].
14. Sample should remain on ice for a maximum of 30 min. If necessary, wash and freeze subsets of samples.
15. Do not collect larvae in 2 mL bead tubes. The residual wash buffer will remain attached to the beads, resulting in inaccurate mass measurements and GC-MS artifacts.
16. While white prepupae only exhibit basal levels of D- and L-2HG, the abundance of these molecules during most of pupal development remains undefined. Therefore, any studies of 2HG metabolism must first determine how D- and L-2HG change in precisely staged pupal samples.
17. 2HG appears to degrade in samples stored at −80 °C. Therefore, all samples should be processed within 3 months of collection.
18. This section of the protocol is optimized to quench metabolism and must be performed in an efficient and careful manner. Be sure all solutions are prepared in advance and chilled to −20 °C. Do not allow samples to thaw prior to homogenization in methanol.
19. We recommend using positive-placement pipettes for all steps involving the transfer of organic solvents.
20. Rapid homogenization will heat the sample. We recommend that the bead mill homogenizer is stored and used in a 4 °C controlled temperature room.
21. If sufficient sample material was collected at the beginning of the analysis (>15 mg), individual samples can be split in half immediately following this centrifugation step. One half of the sample is then derivatized and analyzed as described in the remainder of the protocol. Meanwhile, the other half of the sample can then be analyzed using a previously described metabolomic protocol that relies on derivatization by methoxylamine hydrochloride

and *N*-methyl-*N*-trimethylsilyltrifluoracetamide [11–13]. This dual approach provides for quantitative measurements of D- and L-2HG while also allowing for a more extensive metabolomic comparison of the samples. In theory, samples derivatized as described in Subheading 3.3 could also be analyzed for other metabolites; however, there is no standard library currently available for this method.

22. If the samples were stored at −80 °C, place the unopened tubes in a vacuum centrifuge. Dry the outside of the sample tube for 30 min at room temperature. This step will remove and condense water from the outside of the tube.
23. Heating can dislodge the lid of microfuge tubes. Be sure that the tube lid is securely fastened throughout the 3-h incubation. Occasionally check to make sure that the sample tube lid does not become dislodged as the temperature reaches 90 °C.
24. Be sure that the microfuge tube lid is securely fastened throughout the 1 h incubation. Occasionally check to make sure that the sample tube lid does not become dislodged as the temperature reaches 80 °C. Limit the samples' exposure to water vapor.
25. The time can be changed according to the GC-MS system.
26. This protocol is optimized for the sample sizes described herein. 2 μL of sample can be injected if the intensity of the peaks is too low.

Acknowledgment

J.M.T. is supported by a R35 Maximizing Investigators' Research Award (MIRA; 1R35GM119557) from the National Institute of General Medical Sciences of the National Institutes of Health.

References

1. Losman JA, Kaelin WG Jr (2013) What a difference a hydroxyl makes: mutant IDH, (R)-2-hydroxyglutarate, and cancer. Genes Dev 27(8):836–852
2. Ye D, Guan KL, Xiong Y (2018) Metabolism, activity, and targeting of D- and L-2-hydroxyglutarates. Trends Cancer 4(2):151–165
3. Mishra P et al (2018) ADHFE1 is a breast cancer oncogene and induces metabolic reprogramming. J Clin Invest 128(1):323–340
4. Oldham WM et al (2015) Hypoxia-mediated increases in l-2-hydroxyglutarate coordinate the metabolic response to reductive stress. Cell Metab 22(2):291–303
5. Intlekofer AM et al (2015) Hypoxia induces production of L-2-hydroxyglutarate. Cell Metab 22(2):304–311
6. Tyrakis PA et al (2016) S-2-hydroxyglutarate regulates CD8+ T-lymphocyte fate. Nature 540(7632):236–241
7. Fan J et al (2015) Human phosphoglycerate dehydrogenase produces the oncometabolite D-2-hydroxyglutarate. ACS Chem Biol 10(2):510–516

8. Teng X et al (2016) Lactate dehydrogenase C produces S-2-hydroxyglutarate in mouse testis. ACS Chem Biol 11(9):2420–2427
9. Struys EA et al (2005) Kinetic characterization of human hydroxyacid-oxoacid transhydrogenase: relevance to D-2-hydroxyglutaric and gamma-hydroxybutyric acidurias. J Inherit Metab Dis 28(6):921–930
10. Becker-Kettern J et al (2016) *Saccharomyces cerevisiae* Forms D-2-hydroxyglutarate and couples its degradation to D-lactate formation via a cytosolic transhydrogenase. J Biol Chem 291(12):6036–6058
11. Li H et al (2017) Drosophila larvae synthesize the putative oncometabolite L-2-hydroxyglutarate during normal developmental growth. Proc Natl Acad Sci U S A 114(6):1353–1358
12. Reinecke CJ et al (2012) Metabolomics of urinary organic acids in respiratory chain deficiencies in children. Metabolomics 8(2):264–283
13. Nadtochiy SM et al (2016) Acidic pH Is a metabolic switch for 2-hydroxyglutarate generation and signaling. J Biol Chem 291(38):20188–20197
14. Reitman ZJ et al (2015) Genetic dissection of leukemia-associated IDH1 and IDH2 mutants and D-2-hydroxyglutarate in Drosophila. Blood 125(2):336–345
15. Gibson KM et al (1993) Stable-isotope dilution analysis of D- and L-2-hydroxyglutaric acid: application to the detection and prenatal diagnosis of D- and L-2-hydroxyglutaric acidemias. Pediatr Res 34(3):277–280
16. Ashburner M, Bownes M, Abrahamssen N, Gilman C (2005) Drosophila: a laboratory manual, 2nd edn. Cold Spring Harbor Laboratory Press, New York, pp 121–184
17. Li H, Tennessen JM (2017) Methods for studying the metabolic basis of Drosophila development. Wiley Interdiscip Rev Dev Biol 6(5):e280

Chapter 11

Comprehensive LC-MS-Based Metabolite Fingerprinting Approach for Plant and Fungal-Derived Samples

Kirstin Feussner and Ivo Feussner

Abstract

Liquid chromatography-mass spectrometry (LC-MS)-based nontargeted metabolome approaches aim to detect chemotypes as markers for stress, disease, developmental, or genetic perturbation. Herein, we present a metabolite fingerprinting workflow, which is applicable for the analysis of tissues and fluids derived from plants and fungi. This is based on a broad metabolite coverage by a two-phase extraction and the separate analysis of polar, and nonpolar compounds by positive as well as negative electrospray ionization. For analysis of the resulting comprehensive data sets, the interactive and user-friendly data mining software MarVis-Suite is used. It supports statistical analysis, adduct correction, data merging, as well as visualization of multivariate data. Finally, MarVis shapes marker identification to the organism of interest. Therefore, it provides access to the species-specific databases KEGG and BioCyc and to custom databases tailored by the user.

Key words Custom databases, Data mining, Fungus, Metabolome, Nontargeted metabolomics, Metabolite fingerprinting, Two-phase extraction

1 Introduction

Metabolite fingerprinting is a comprehensive and comparative nontargeted metabolomics approach [1]. It aims to describe biological processes on the metabolite level and to identify metabolites or even pathways as biomarker of environmental, developmental, or genetic perturbation [2–4]. This requires the coverage of almost all metabolites of the biological material during metabolite extraction, separation by LC, and detection by high-resolution mass spectrometry (HR-MS) [5, 6] as well as interactive tools for exploratory data analysis [7–11].

The ultra performance liquid chromatography coupled with a time-of-flight mass spectrometer (UPLC ESI-TOF MS)-based metabolite fingerprinting approach, which we describe here, has been established for the comprehensive metabolome analysis of plant and fungal-derived samples (Fig. 1). Beside the common

Angelo D'Alessandro (ed.), *High-Throughput Metabolomics: Methods and Protocols*, Methods in Molecular Biology, vol. 1978,
https://doi.org/10.1007/978-1-4939-9236-2_11,

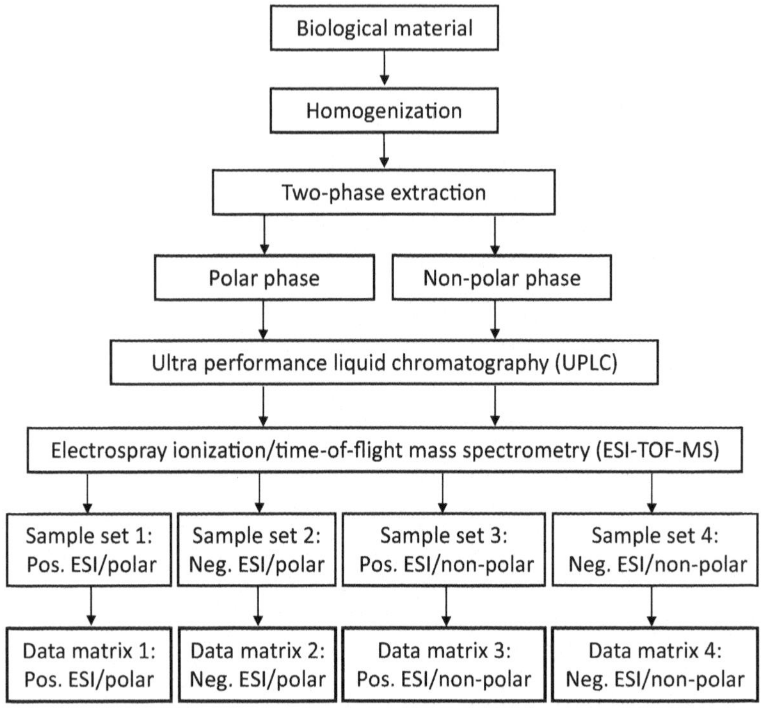

Fig. 1 Workflow of the metabolite fingerprinting approach. The workflow starts with the homogenization of the plant or fungal-derived material and proceeds via extraction and LC-MS analyses up to the generation of data matrixes

primary metabolites, the metabolome of plants and fungi is characterized by a huge variety of so-called secondary or specialized metabolites [12]. These highly diverse and species-specific compound classes enable the respective organism to survive challenging environmental conditions by adaptation processes, like activating defense programs against pathogens, and they play therefore a central role as metabolite markers [1, 13].

Our approach deals with the great variability and species specificity of the metabolome of plants and fungi at the level of metabolite extraction and analysis as well as during data processing and data mining. It has been applied for the analysis of a huge variety of plant species, like dicotyledons [14–17], monocotyledons [18, 19], mosses, and algae [20] and the respective plant organs (leaves, flowers, roots, hypocotyls, seedlings, gametophores, etc.) as well as for the mycelia of different fungi [21–24]. Besides these also plant and fungi-derived fluids, like xylem sap, apoplastic wash fluid, root extracts, and supernatants of fungal cultures were analysis with the described approach [17, 21, 24]. The workflow includes a two-phase extraction, which facilitates the extraction of polar, semipolar, and nonpolar metabolites. This is followed by an UPLC separation of the complex matrixes by using different column materials and solvent gradients for the polar as well as the nonpolar phase, respectively. After this separation, TOF MS analysis is performed in positive and in negative ESI mode to ensure the accurate mass detection of all ionizable metabolites. Peak picking and alignment of the LC-MS analyses lead finally to four data matrixes with several thousand features each (Fig. 1).

The most challenging and time-consuming part of the non-targeted approach is the identification of metabolite feature of interest and their unequivocal assignment to metabolites and their corresponding chemical structures. This task is supported by the MarVis-Suite toolbox (http://marvis.gobics.de/, Fig. 2), which provides an interactive workflow for data analysis, data visualiza-

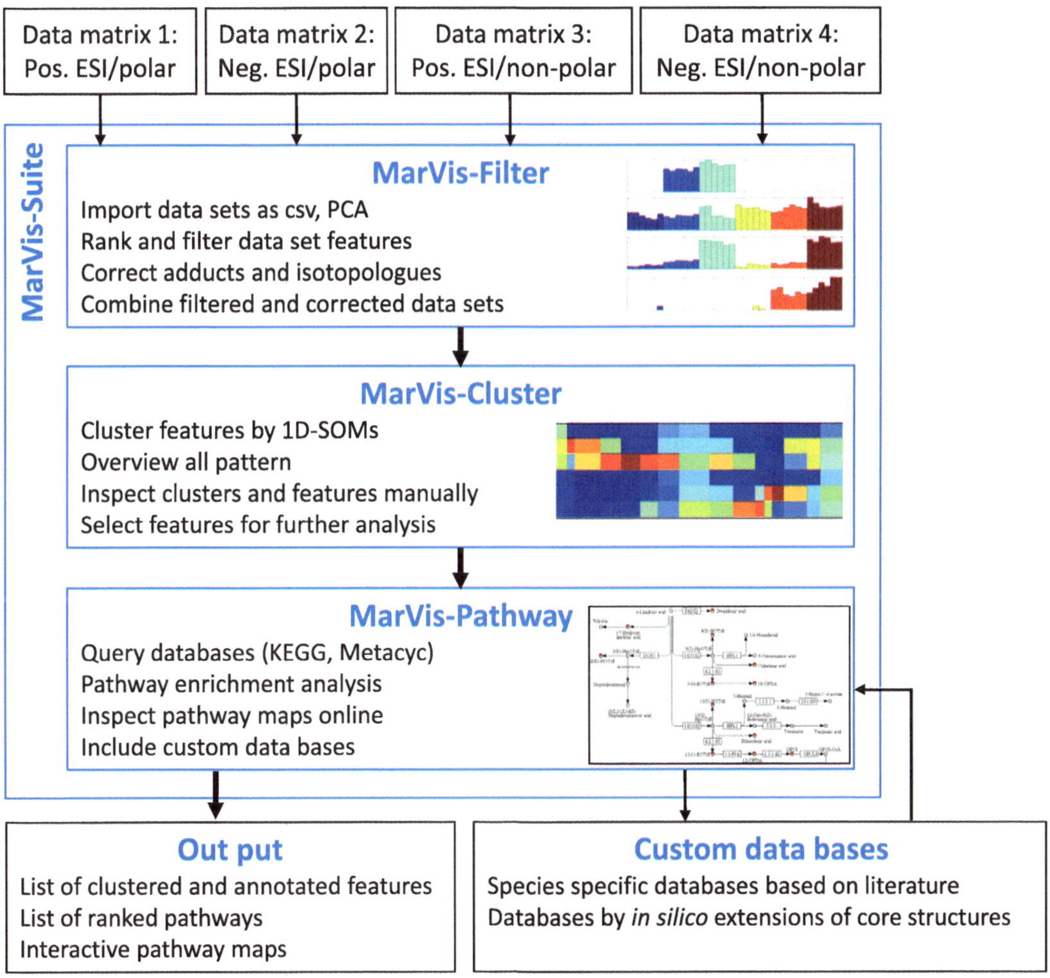

Fig. 2 Interactive workflow for data analysis, data visualization, and data mining by the MarVis-Suite toolbox. MarVis-Filter facilitates statistical analyses of the imported data sets and supports the adduct correction and generation of accurate mass information as well as merging of data sets. The MarVis-Cluster interface supports visualization of large numbers of multivariate feature profiles based on one-dimensional self-organizing maps (1D-SOMs). The accurate mass information of the selected feature is used further for the framework of the metabolite set enrichment analysis in MarVis-Pathway

tion, and data mining [25]. MarVis-Filter, the first part of the tool box, facilitates statistical analyses of the imported data sets and supports the correction of m/z values to accurate mass information and merging of data sets [26]. The MarVis-Cluster interface supports the visualization of large numbers of multivariate feature profiles based on one-dimensional self-organizing maps (1D-SOMs). The heat map representation of all metabolite features of the metabolite fingerprinting experiment by a 1D-SOM allows to overview the complete data set and to select feature of interest for further analysis [27, 28]. Finally, MarVis-Pathway facilitates the database-dependent metabolite annotation by the accurate mass-based search of the databases KEGG and BioCyc in combination with a framework for metabolite set enrichment analysis. A major bottleneck for marker identification, especially in the metabolome of non-model plants and fungi, is the limited number of the highly species-specific secondary metabolites in public available databases. This problem can be addressed in MarVis-Pathway by implementing custom databases into the framework of metabolite set enrichment analysis. The custom databases are created by the scientist for the respective organism and sum up the expert and literature knowledge in respect to the organism-specific metabolome. In addition, the number of putative metabolites of the custom databases can be increased by in silico extension of prominent species-specific core structures with common extensions like hexose, pentose, or amino acid moieties. The implementation of custom databases improves the coverage of the metabolite set enrichment analysis in a highly species-specific way and supports the data mining significantly.

2 Materials

2.1 Samples

1. Plant material (leaves, flowers, hypocotyls, roots, seedlings, fruits) keep at −80 °C until grinding, homogenization and extraction starts.
2. Fungal material (mycelium) keep at −80 °C until the grinding, homogenization and extraction starts.
3. Fluids/aqueous solutions derived from plants (xylem sap, apoplastic wash fluid, root exudates) or from fungi (culture supernatants) keep at −80 °C until the extraction starts.

2.2 Chemicals and Standards

Compounds for quality control (QC) and as references were obtained from Sigma-Aldrich (St. Louis, USA), Phytolab (Vestenbergsgreuth, Germany) and Matreya (Pennsylvania, USA).

2.3 Solvents and Solutions

All solvents used (methanol, acetonitrile) were LC-MS grade quality unless indicated otherwise. Water was always freshly generated by the Arium pro VF TOC ultrapure water system (Sartorius, Goettingen, Germany). Methyl-*tert*-butylether (MTBE, Sigma-Aldrich) used for two-phase extraction was HPLC grade.

1. Solvent system for all U(H)PLC analyses with the ACQUITY HSS T3 column as well as ACQUITY BEH C18 (Waters Corporation, Milford, USA): solvent A (water with 0.1% formic acid), solvent B (acetonitrile with 0.1% formic acid).
2. Calibration solution for LCT Premier: 0.01% phosphoric acid (*v/v*) in acetonitrile/water (1:1, *v/v*).
3. Reference mass solution for accurate mass measurement with LCT Premier: 0.5 mg/l leucine-enkephalin in acetonitrile/water (1:1, *v/v*).
4. Tuning mixture for 6540 UHD Accurate-Mass Q-TOF: low concentration tuning mixture (ESI-L, Agilent Technologies, Santa Clara, USA).
5. Reference mass correction for the 6540 UHD Accurate-Mass Q-TOF: trifluoroacetic acid ($[M-H]^-$ 112.985587) and HP-921 hexakis-($^1H,^1H,^3H$-tetrafluoro-pentoxy) phosphazene ($[M+H_2CO_2-H]^-$ 966.00073).
6. Standard mixtures for QC of the LC-MS performance (retention time stability, mass accuracy and sensitivity): polar QC-standard mixture (Table 1) and non-polar QC-standard mixture (Table 2).

2.4 LC-MS System

1. For metabolite fingerprinting analysis: UPLC (Waters Corporation) coupled to the photodiode array (PDA) detector eλ (Waters Corporation) and the orthogonal TOF MS LCT Premier (Waters Corporation).
2. For structure confirmation and structure elucidation of marker compounds as well as for metabolite fingerprinting analysis: UHPLC 1290 Infinity (Agilent Technologies) coupled to a PDA detector and the 6540 UHD Accurate-Mass Q-TOF LC-MS instrument with Agilent Dual Jet Stream Technology as ESI source (Agilent Technologies).
3. For chromatographic separation of the extracts of the polar extraction phase and the extracts of the aqueous solutions: ACQUITY HSS T3 column (1.0 × 100 mm, 1.8 μm particle size, Waters Corporation) with inlet filter system (Waters Corporation).
4. For chromatographic separation of the extracts of the nonpolar extraction phase: ACQUITY BEH C18 column (1.0 × 100 mm, 1.7 μm particle size, Waters Corporation) with inlet filter system (Waters Corporation).

Table 1
Polar standard mixture for quality control. The concentration of the stock solution is 1 mg/ml. The concentration of each compound in the final solution differs between 1 and 10 μg/ml

Standard compound	Formula	Monoisotopic mass	Detectable as	
Valine	$C_5H_{11}NO_2$	117.0790	$[M+H]^+$ 118.0868	
Phenylalanine	$C_9H_{11}NO_2$	165.0790	$[M+H]^+$ 166.0868	$[M-H]^-$ 164.0712
Esculetin	$C_9H_6O_4$	178.0266	$[M+H]^+$ 179.0344	$[M-H]^-$ 177.0188
Scopoletin	$C_{10}H_8O_4$	192.0423	$[M+H]^+$ 193.0501	$[M-H]^-$ 191.0344
Caffeine	$C_8H_{10}N_4O_2$	194.0804	$[M+H]^+$ 195.0882	
2,5 Diphenyloxazole	$C_{15}H_{11}NO$	221.0841	$[M+H]^+$ 222.0919	$[M-H]^-$ 220.0762
Dinitrophenyl acetate	$C_8H_6N_2O_6$	226.0226		$[M-H]^-$ 225.0148
Aesculin	$C_{15}H_{16}O_9$	340.0794	$[M+H]^+$ 341.0873	$[M-H]^-$ 339.0716
Leucine-enkephalin	$C_{28}H_{37}N_5O_7$	555.2693	$[M+H]^+$ 556.2771	$[M-H]^-$ 554.2615
			$[2M+H]^+$ 1111.5464	$[2M-H]^-$ 1109.5308

Table 2
Nonpolar standard mixture for quality control. The concentration of the stock solution is 1 mg/ml. The concentration of each compound in the final solution differs between 1 and 10 μg/ml

Standard compound	Formula	Monoisotopic mass	Detectable as	
2,5 Diphenyloxazole	$C_{15}H_{11}NO$	221.0841	$[M+H]^+$ 222.0919	$[M-H]^-$ 220.0762
Methyl jasmonate	$C_{13}H_{20}O_3$	224.1412	$[M+H]^+$ 225.1491	
16-Hydroxy-hexadecanoic acid	$C_{16}H_{32}O_3$	272.2351	$[M+H]^+$ 273.2430	$[M-H]^-$ 271.2273
20-Hydroxy-eicosanoic acid	$C_{20}H_{40}O_3$	328.2977	$[M+H]^+$ 329.3056	$[M-H]^-$ 327.2899
22-Hydroxy-docosanoic acid	$C_{22}H_{44}O_3$	356.3290	$[M+H]^+$ 357.3369	$[M-H]^-$ 355.3212
Leucine-enkephalin	$C_{28}H_{37}N_5O_7$	555.2693	$[M+H]^+$ 556.2771	$[M-H]^-$ 554.2615
			$[2M+H]^+$ 1111.5464	$[2M-H]^-$ 1109.5308

2.5 Software

1. Data acquisition on UPLC-TOF MS (LCT Premier): MassLynx 4.1 (Waters Corporation).
2. Peak picking and peak alignment: MarkerLynx XS for MassLynx V.4.1 (Waters Corporation).
3. Data acquisition on 6540 UHD Accurate-Mass Q-TOF LC-MS: Mass Hunter Acquisition B.03.01 (Agilent Technologies).
4. Data analysis of Q-TOF LC-MS data: Mass Hunter Qualitative Analysis (Agilent Technologies).
5. Peak picking and peak alignment: Profinder B.06.00 (Agilent Technologies, Santa Clara, USA).
6. Data processing, statistics, data mining and visualization: MarVis-Suite toolbox (http://marvis.gobics.de/).

2.6 Other Equipment

1. Eppendorf tubes (1.5 ml and 2 ml).
2. Kimble extraction tubes (Kimax-51, 13 × 100 mm) with screw caps (Gerresheimer Glass Inc., New Jersey, USA).
3. Adjustable pipettes and Pasteur pipettes.
4. Tips for water, methanol, and acetonitrile.
5. Chemical-resistant tips for MTBE (Safe Seal Tips Premium, Biozym, Oldendorf, Germany).
6. Chemical-resistant dispenser for MTBE (Dispensette Organic, Brand GmbH, Wertheim, Germany).
7. Micro glass vials for analysis (12 mm, 250 μl, Macherey-Nagel GmbH, Dueren, Germany) fixed by a spring in HPLC glass vials (1.5 ml, N9, Macherey-Nagel GmbH).
8. Eppendorf Microcentrifuge 5415D (Eppendorf, Hamburg, Germany).
9. Eppendorf Centrifuge 5810R (Eppendorf).
10. Nitrogen evaporator.
11. Mixer Ball Mill MM200 with stainless steel grinding jars or PTEE jars (Retsch GmbH, Haan, Germany).
12. Porcelain mortar and pistil.
13. Hood.

3 Methods

3.1 Harvesting Plant or Fungal Material

1. Keep the harvesting procedure of plant or fungal material as short and reproducible as possible. It should not exceed 30 s at all (*see* **Note 1**).
2. Shock freeze the harvested material immediately in liquid nitrogen (*see* **Note 2**).

3.2 Homogenization of Plant or Fungal Material

1. Ground the plant or fungal material by using a porcelain mortar and pistil or homogenize it by using the Mixer Ball Mill MM200 (*see* **Note 3**).
2. Use stainless steel grinding jars (for large sample amounts) or PTFE jars for Eppendorf cups (for small sample amounts) with the corresponding size of stainless steel balls for mill homogenization.
3. Time and repetitions of the homogenization cycles depend on the amount and rigidity of the biological material (*see* **Note 4**).
4. Ensure that the biological material stay always completely frozen under fluid nitrogen.

3.3 Extraction of Plant or Fungal Material

The two-phase extraction method with MTBE, methanol, and water described by Matyash et al. [29] was modified for the extraction of plant and fungal material as follows:

1. Prior to use Kimble glass tubes are kept on ice. All solvents are kept at room temperature.
2. Weigh 80 mg of homogenized and deeply frozen material into Kimble glass tubes. Immediately add 0.75 ml of methanol and vortex. Ensure that the biological material stays frozen during weighing.
3. Add 2.5 ml of MTBE (*see* **Note 5**). Ensure that all material is washed down the side of the tube. Overlay with argon. All following steps are performed at room temperature, unless indicated otherwise.
4. Vortex the sample strongly.
5. Shake for 1 h at 6 °C in the dark.
6. Add 0.6 ml of water and vortex.
7. Incubate the samples for 10 min at room temperature.
8. Centrifuge the samples for 15 min at 800 × *g*.
9. Phase separation occurs. The upper phase is the nonpolar MTBE phase, which contains mainly the nonpolar metabolites. The lower phase contains mainly the polar metabolites.

Upper extraction phase (nonpolar phase):

10. Collect the upper phase with a Pasteur glass pipette, and transfer it into a new Kimble glass tube.
11. For re-extraction add 0.7 ml of methanol/water (3:2.5 (*v*/*v*)) to the lower phase and vortex.
12. Add 1.3 ml of MTBE to the lower phase and vortex.
13. Incubate for 10 min and centrifuge for 15 min at 800 × *g*.
14. Collect the upper phase with a Pasteur glass pipette and combine both upper phases.

15. Dry the combined upper extraction phases under a stream of nitrogen.
16. Resolve the samples in 500 μl of methanol and vortex. Ensure that all material has been removed from the glass border.
17. Transfer the mixture into a 1.5 ml Eppendorf tube, centrifuge (2 min, 16,000 × *g*), and dry carefully under a stream of nitrogen.
18. Resolve the sample in 100 μl of acetonitrile/water (2:3 (*v/v*), (*see* **Note 6**)), shake for 15 min, and centrifuge (10 min, 16,000 × *g*).
19. Transfer 80 μl of the sample into a glass micro vial.
20. Cover the samples of the nonpolar extraction phase with argon and use immediately for metabolite fingerprinting analysis.

Lower extraction phase (polar phase):

21. Transfer the lower phase with a glass pipette to a 2 ml Eppendorf tube and dry it under a stream of nitrogen (*see* **Note 7**).
22. Resolve the samples in 100 μl of methanol, vortex, and shake for 5 min.
23. Centrifuge (5 min, 16,000 × *g*) and dry carefully under a stream of nitrogen.
24. Resolve the sample in 15 μl of methanol and shake for 10 min.
25. Add 15 μl of acetonitrile and shake for 10 min.
26. Add 100 μl of water, shake for 10 min, and centrifuge (10 min, 16,000 × *g*).
27. Transfer 100 μl of the sample into a glass micro vial.
28. Cover the sample with argon and use immediately for metabolite fingerprinting analysis.

To avoid autoxidation immediately cover the samples with argon after each extraction step, particularly after the last solvation step.

3.4 Extraction of Metabolites from Aqueous Solutions of Plant or Fungal Origin

The two-phase extraction method with MTBE, methanol, and water described by Matyash et al. [29] was modified for the extraction of fluids of plant and fungal origin as follows:

1. Thaw the deeply frozen fluids on ice. Vortex the samples after complete thawing.
2. Give 150 μl of methanol to a glass vial, add 200 μl of the sample, and vortex. All following steps are performed at room temperature, unless indicated otherwise.
3. Add 500 μl MTBE (*see* **Note 5**) and vortex.
4. Shake for 1 h in the dark.
5. Add 120 μl of water (to induce phase separation) and vortex.

6. Incubate the samples for 10 min.
7. Centrifuge the samples for 15 min at 800 × *g*.
8. Phase separation occurs. Take the upper phase and the lower phase separately (avoid transferring the interphase and if it exists, the pellet) and combine both phases in a new glass vial.
9. Dry the combined extraction phases carefully under a stream of nitrogen.
10. Resolve the sample in 60 μl of methanol/acetonitrile/water (1:1:12, *v/v/v*, *see* **Note 8**), shake for 15 min, and centrifuge (10 min, 16,000 × *g*).
11. Transfer 50 μl of the sample into a glass micro vial.
12. Cover the sample with argon and use immediately for metabolite fingerprinting analysis.

3.5 UHPLC Parameter

The performance of the U(H)PLC has to be controlled regularly by inspecting, e.g., the back pressure of the system and the retention time stability with the respective QC standard mixtures (Tables 1 and 2).

1. Set the temperature of the sample manager to 10 °C and the column oven temperature to 40 °C.
2. Set the flow rate to 0.2 ml/min and the injection volume to values between 0.5 and 2 μl/min.
3. Set the PDA detector to 190–800 nm with a resolution of 1.2 nm.
4. Use acetonitrile as strong and acetonitrile/water (1:9, *v/v*) as weak wash solution.
5. Use the gradients shown in Tables 3 and 4 as mobile phases for the column ACQUITY HSS T3 and the column BEH C18, respectively.

3.6 HR-MS Parameter

The performance of the LCT Premier has to be controlled regularly. Inspect mass accuracy and sensitivity by using the QC standard mixtures (*see* Tables 1 and 2) before running sample sets.

1. Operate the LCT Premier in W optics to enable mass resolution >10,000 in positive and negative ESI mode.
2. Acquire the data in centroid format within a mass range of *m/z* 50–1200 (negative ESI) and *m/z* 85–1200 (positive ESI, *see* **Note 9**).
3. Set scan duration to 0.5 s and an interscan delay to 0.1 s.
4. Set capillary and cone voltage to 2700 V and 30 V, respectively.

Table 3
Solvent gradient for U(H)PLC separation with the ACQUITY HSS T3 column

Time (min)	Solvent A (%)	Solvent B (%)
0	99	1
0.5	99	1
3	80	20
8	0	100
10	0	100
10.1	99	1
14	99	1

Table 4
Solvent gradient for U(H)PLC separation with the ACQUITY BEH C18 column

Time (min)	Solvent A (%)	Solvent B (%)
0	54	46
0.5	54	46
5.5	0	100
10	0	100
10.1	54	46
13	54	46

5. Set desolvation and source temperature to 350 °C and 80 °C, respectively.
6. Set the nitrogen flow of the cone to 30 l/h and of the desolvation gas to 800 l/h.
7. Use the dynamic range enhancement mode for data acquisition.
8. Use leucine-enkephalin as lock spray reference: ($[M-H]^-$ 554.2615 or $[M+H]^+$ 556.2771 as lock mass and the ^{13}C isotopologue $[M-H]^-$ 555.2615 and the double ^{13}C isotopologue $[M+H]^+$ 558.2836 as attenuated lock mass) to ensure highest mass accuracy.

The performance of the 6540 UHD Accurate-Mass Q-TOF LC-MS instrument has to be controlled regularly. Calibrate the Q-TOF MS in positive and negative ESI mode by using the calibration mixture ESI-L. Inspect mass accuracy and sensitivity by using the QC standard mixtures (*see* Tables 1 and 2) before running sample sets.

1. Operate the Q-TOF MS instrument with a detection frequency of 4 GHz in the targeted MS/MS mode.
2. Acquire the data in a mass range of *m/z* 50–1700 for negative and positive ESI.
3. Set capillary voltage to 3500 V, nozzle voltage to 100 V, and fragmentor voltage to 175 V, respectively.
4. Set drying gas flow to 8 l/min and sheath gas flow to 11 l/min, respectively.
5. Set nebulizer pressure to 35 psi, sheath gas temperature to 350 °C, and gas temperature to 300 °C.
6. Use reference mass correction with the reference mass solution.

For MS/MS fragmentation experiments:

1. Use the targeted MS/MS mode and vary collision energies in dependence of the requirements of the analyte ion.
2. Set the isolation window for precursor ion in Q1 to narrow width of *m/z* 1.3.

3.7 Peak Picking and Peak Alignment

The mass data of all LC-MS analyses of one sample set (e.g., polar sample fraction analyzed by pos. ESI, *see* Fig. 1) has to be combined to a sample matrix. Therefore, all peaks/mass signals have to be picked in each of the respective samples and aligned with the mass signals of all other samples of the sample set.

Generate a method for peak picking and peak alignment in MarkerLynx with the following parameters:

1. For MS peak detection set function to channel 1 and analysis type to peak detection.
2. Set retention time range from 0.3 to 10 min (*see* **Note 10**).
3. Set the mass range from 50 to 1200 Da (*see* **Note 10**).
4. Set the extracted ion current (EIC) window to 0.03 Da.
5. Do not use relative retention time correction.
6. Set check to the apex track peak parameter for peak width at 5% height and for peak-to-peak baseline noise to allow automatic calculation of both parameters.
7. Do not apply smoothing.
8. For peak alignment use the following collection parameters: marker intensity threshold, 50 counts (*see* **Note 11**); mass window, 0.03 Da; retention time window, 0.2 min; noise elimination level, 2.
9. Allow to deisotope the raw data (*see* **Note 12**).

Process the selected samples of one sample set and create the respective data matrix by running MarkerLynx algorithm (*see* **Note 13**). The resulting data matrixes 1–4 are stored as mds-file and exported as csv-file (Fig. 1).

3.8 Data Processing and Data Mining by MarVis

For further data processing, like ranking and filtering, adduct and isotopologue correction, merging of data sets, as well as data mining and visualization like clustering, database search, and metabolite set enrichment analysis, the MarVis-Suite [25] (http://marvis.gobics.de/) is used (Fig. 2).

Use MarVis-Filter [25, 26] to upload the data matrixes, rank and filter the data, correct for adducts and isotopologues, and merge the filtered and corrected data.

1. Open MarVis-Filter.
2. Upload the csv-file of data matrix 1.
3. Rank the data by ANOVA and use Benjamini-Hochberg for multiple testing to obtain the false discovery rates (FDR).
4. Cut the ranked data set by a suitable FDR.
5. Correct adducts based on the rules for positive as well as negative ESI (*see* **Note 14**). Use a window of 10 mDa for mass and of 0.05 min for retention time tolerance. Allow isotopologue correction, if necessary.
6. Store the filtered and corrected data in the MarVis-Filter clipboard repository.
7. Upload, rank, filter, and correct data sets 2–4 (Figs. 1 and 2).
8. Combine all four data sets (*see* **Note 15**).
9. Select all features of the combined data set and continue with MarVis-Cluster.

Use MarVis-Cluster [25, 27] for clustering the intensity profiles of the selected features by means of 1D-SOMs. MarVis-Cluster provides an interactive surface to overview the complete data set and investigate intensity patterns for a large number of feature profiles.

10. For sample aggregation and normalization, use average by mean for replicate intensities per condition and marker scaling by 2-norm (Euclidean distance).
11. Select an appropriate number of clusters for the representation of the clustering results (*see* **Note 16**).
12. The main window displays the prototype plot, the intensity profiles of features in the selected prototype, detailed information about the features, and the retention time vs. mass scatter plot of features.

13. Use the interactive display to overview the data set and select feature or even complete prototypes of interest for further analysis in MarVis-Pathway.

Continue data analysis and data mining by using MarVis-Pathway [25, 30]. MarVis-Pathway implements databases from the KEGG [31] (http://www.kegg.jp) and the BioCyc collection [32] (http://biocyc.org) as well as custom databases, respectively (Fig. 2). MarVis-Pathway supports mapping of the selected (or even all) features of MarVis-Cluster to metabolite entries of the chosen databases by using the extensive framework for metabolite set enrichment analysis.

14. Select from BioCyc and KEGG the reference pathway databases or the respective species-specific ones.
15. Load additional custom databases as csv-files (*see* **Note 17**).
16. Select mass tolerance of 5 mDa for mass matching (*see* **Note 18**).
17. The main window displays the pathway list, which lists all matched pathways in a ranked order. The profile map shows the heat map representation of all feature profiles, which were mapped to the indicated pathway. The assignments of features to entries in the selected pathway are shown in the entry assignment list box.
18. Inspect pathways and metabolite assignments interactively via the online resources of the queried databases by using the Map and Entry button.
19. Export annotated metabolite markers and/or pathway sets.

3.9 Verification of the Chemical Structure of Metabolite Markers

The tentative chemical identity of metabolite markers based on accurate mass information has to be confirmed unequivocally. MS/MS fragmentation experiments provide important information to confirm the identity of a metabolite marker or to get hints about the chemical structure of so far not described compounds.

1. For MS/MS experiments the UHPLC system LC 1290 Infinity coupled with the 6540 UHD Accurate-Mass Q-TOF LC-MS instrument with Agilent Dual Jet Stream Technology is used.
2. For LC separation, columns, solvent systems, and gradients are used as described above.
3. For MS/MS fragmentation source parameters are used as described above.
4. For fragmentation use the targeted MS/MS mode and vary the collision energies.
5. For the isolation of a precursor ion, use the narrow isolation width of m/z 1.3.

Verification of compound identity is classified to certain identification criteria, which provides a particular level of confidence [33]. The highest level of confidence requires matching of at least two feature properties (e.g., accurate mass, MS/MS spectrum, retention time, isotopic pattern) of the authentic standard with the marker metabolite of interest, while both have to be analyzed by the same analytical method. Based on this definition and in relation to Strehmel 2014 [34], we use the following criteria to describe the level of verification confidence:

(a) Accurate mass, retention time, isotopic pattern, and MS/MS spectrum match with the criteria, which were obtained by authentic standard.

(b) Accurate mass, isotopic pattern, and MS/MS spectrum match with the criteria, which were obtained from literature or MS/MS database (Metlin (https://metlin.scripps.edu), MassBank [35] (http://www.massbank.jp)).

(c) Accurate mass and MS/MS spectrum of the compound core structure (subtraction of the masses for, e.g., sugar, amino acid, malonyl moieties) match with the criteria, which were obtained for the core structure from literature, MS/MS database, or authentic standards.

(d) Accurate mass, isotopic pattern, and UV/Vis spectrum match with the criteria, which were obtained from literature.

3.10 Confirmation of the Intensity Pattern of Metabolite Markers

Beside the chemical identity of a marker metabolite, also its intensity pattern should be confirmed (and if applicable quantified) by an independent targeted approach. The targeted method should be optimized to the requirements of the marker compound or compound class in respect to sample extraction and analysis.

4 Notes

1. Avoid using Miracloth (MilliporeSigma, Schwalbach, Germany) to separate fungal mycelium from liquid cultures. The contact of the later samples (mycelium as well as culture medium) with Miracloth results in large background signals with detergent-like signal characteristics, which elute in the semipolar to nonpolar range of the LC gradient.
2. Wear glasses and cold protective gloves when handling liquid nitrogen.
3. Cool down all equipment (mortar, pistil, mill jars, cups, tubes, spatula) with liquid nitrogen to avoid thawing and clogging of the sample on the equipment.

4. Arabidopsis rosettes are usually completely homogenized to fine powder by using the Mixer Ball Mill MM200 for 1 min at 30 1/s.
5. For working with MTBE, use pipette tips or dispensers, which are chemical resistant to organic solvents.
6. The described solvation procedure is suitable for resolving semipolar and slight nonpolar compounds out of the residue. Use more lipophilic solvent mixtures, like tetrahydrofuran/methanol/water (4:4:1 (*v/v/v*)) to resolve high nonpolar compounds from the residue.
7. Evaporation of the polar extraction phase will last much longer as for the nonpolar phase.
8. Resolve the sample in a solvent mixture with a higher amount of organic solvent, when nonpolar metabolites are expected in the fluids (like sterigmatocystin in culture supernatants of *Aspergillus nidulans* or lipids in plant root exudates).
9. A high intense background signal is created in the positive ESI mode by an acetonitrile cluster (*m/z* 83.0609). Therefore, the mass detection starts at *m/z* 85 for pos. ESI mode.
10. The retention time and mass range used for data deconvolution should be adapted to the parameters of the data acquisition method.
11. The number of features, which are generated by data deconvolution, is highly dependent on the selected intensity threshold.
12. In certain cases, it is useful not to remove the isotopologue from the data and to keep the information, which are enclosed in the isotopologue distribution for further data analyses. Not to deisotope the data results in a two- to fourfold higher amount of features.
13. This data pre-processing step (data deconvolution) results in a data matrix with up to several thousand features. A feature in that respect is characterized by *m/z*, retention time, and the corresponding signal intensities over all samples of a sample set.
14. For the described approach, the following rules were usually used for adduct correction: $[M+H]^+$, $[M+Na]^+$, and $[M+NH_4]^+$ for data derived by positive ESI and $[M-H]^-$, $[M+\text{formic acid}-H]^-$, and $[M+Na+\text{formic acid}-H]^-$ for data derived by negative ESI [26].
15. All features (independent of the ESI mode, from which they derived) are combined in one data set.
16. MarVis-Cluster supports the interactive analysis of putative clusters, even though the true number of meaningful groups of

features is not known [27]. The appropriate number of clusters could be dependent on the number of features, conditions, and the variations in the intensity profiles. Select all features and use the functionality Recluster to adapt the number of clusters.

17. Custom metabolite databases for the organism of interest are created as csv-files. They are important tools to (1) sum up the latest expert and literature knowledge in respect to an organism-specific metabolome and/or (2) to improve the database coverage by in silico extensions of prominent species-specific core structures. For higher plants hexosyl, pentosyl, malonyl, or amino acid moieties are common residues, which were often found to be connected to the core structures of secondary metabolites.
18. Estimate the mass accuracy of your HR device by using the data of the QC analysis, and use this value for adjusting the mass tolerance.

Acknowledgments

This work was supported by the German Federal Ministry of Education and Research (BioFung 0315595A) and the German Research Foundation (ZUK 45/2010). We are very grateful to Alexander Kaever and Manuel Landesfeind for collaborating with us on the development of the MarVis-Suite and improving the workflow of the metabolite fingerprinting approach. We thank Sabine Freitag and Pia Meyer for excellent assistance.

References

1. Scholz M, Gatzek S, Sterling A, Fiehn O, Selbig J (2004) Metabolite fingerprinting: detecting biological features by independent component analysis. Bioinformatics 20(15):2447–2454
2. Kumar R, Bohra A, Pandey AK, Pandey MK, Kumar A (2017) Metabolomics for plant improvement: status and prospects. Front Plant Sci 8(1302). https://doi.org/10.3389/fpls.2017.01302
3. Tenenboim H, Brotman Y (2016) Omic relief for the biotically stressed: metabolomics of plant biotic interactions. Trends Plant Sci 21(9):781–791. https://doi.org/10.1016/j.tplants.2016.04.009
4. Fernandez O, Urrutia M, Bernillon S, Giauffret C, Tardieu F, Le Gouis J, Langlade N, Charcosset A, Moing A, Gibon Y (2016) Fortune telling: metabolic markers of plant performance. Metabolomics 12(10):1–14. https://doi.org/10.1007/s11306-016-1099-1
5. Haggarty J, Burgess KEV (2017) Recent advances in liquid and gas chromatography methodology for extending coverage of the metabolome. Curr Opin Biotechnol 43:77–85. https://doi.org/10.1016/j.copbio.2016.09.006
6. Doerr A (2017) Global metabolomics. Nat Methods 14(1):32–32. https://doi.org/10.1038/nmeth.4112
7. Spicer R, Salek RM, Moreno P, Cañueto D, Steinbeck C (2017) Navigating freely-available software tools for metabolomics analysis. Metabolomics 13(9):106. https://doi.org/10.1007/s11306-017-1242-7

8. Huan T, Forsberg EM, Rinehart D, Johnson CH, Ivanisevic J, Benton HP, Fang M, Aisporna A, Hilmers B, Poole FL, Thorgersen MP, Adams MWW, Krantz G, Fields MW, Robbins PD, Niedernhofer LJ, Ideker T, Majumder EL, Wall JD, Rattray NJW, Goodacre R, Lairson LL, Siuzdak G (2017) Systems biology guided by XCMS online metabolomics. Nat Methods 14(5):461–462. https://doi.org/10.1038/nmeth.4260
9. Baran R (2017) Untargeted metabolomics suffers from incomplete raw data processing. Metabolomics 13(9):107. https://doi.org/10.1007/s11306-017-1246-3
10. Xia J, Wishart DS (2016) Using MetaboAnalyst 3.0 for comprehensive metabolomics data analysis. In: Current protocols in bioinformatics. John Wiley & Sons, Inc. https://doi.org/10.1002/cpbi.11
11. Mahieu NG, Genenbacher JL, Patti GJ (2016) A roadmap for the XCMS family of software solutions in metabolomics. Curr Opin Chem Biol 30:87–93. https://doi.org/10.1016/j.cbpa.2015.11.009
12. Sedio BE (2017) Recent breakthroughs in metabolomics promise to reveal the cryptic chemical traits that mediate plant community composition, character evolution and lineage diversification. New Phytol 214(3):952–958. https://doi.org/10.1111/nph.14438
13. Feussner I, Polle A (2015) What the transcriptome does not tell—proteomics and metabolomics are closer to the plants' patho-phenotype. Curr Opin Plant Biol 26(0):26–31. https://doi.org/10.1016/j.pbi.2015.05.023
14. Bruckhoff V, Haroth S, Feussner K, König S, Brodhun F, Feussner I (2016) Functional characterization of CYP94-genes and identification of a novel jasmonate catabolite in flowers. PLoS One 11(7):e0159875. https://doi.org/10.1371/journal.pone.0159875
15. König S, Feussner K, Kaever A, Landesfeind M, Thurow C, Karlovsky P, Gatz C, Polle A, Feussner I (2014) Soluble phenylpropanoids are involved in the defense response of Arabidopsis against *Verticillium longisporum*. New Phytol 202(3):823–837. https://doi.org/10.1111/nph.12709
16. König S, Feussner K, Schwarz M, Kaever A, Iven T, Landesfeind M, Ternes P, Karlovsky P, Lipka V, Feussner I (2012) Arabidopsis mutants of sphingolipid fatty acid α-hydroxylases accumulate ceramides and salicylates. New Phytol 196(4):1086–1097. https://doi.org/10.1111/j.1469-8137.2012.04351.x
17. Floerl S, Majcherczyk A, Possienke M, Feussner K, Tappe H, Gatz C, Feussner I, Kües U, Polle A (2012) *Verticillium longisporum* infection affects the leaf apoplastic proteome, metabolome, and cell wall properties in *Arabidopsis thaliana*. PLoS One 7(2):e31435. https://doi.org/10.1371/journal.pone.0031435
18. Tanaka S, Brefort T, Neidig N, Djamei A, Kahnt J, Vermerris W, Koenig S, Feussner K, Feussner I, Kahmann R (2014) A secreted *Ustilago maydis* effector promotes virulence by targeting anthocyanin biosynthesis in maize. eLife 3:e01355. https://doi.org/10.7554/eLife.01355
19. Djamei A, Schipper K, Rabe F, Ghosh A, Vincon V, Kahnt J, Osorio S, Tohge T, Fernie AR, Feussner I, Feussner K, Meinicke P, Stierhof Y-D, Schwarz H, Macek B, Mann M, Kahmann R (2011) Metabolic priming by a secreted fungal effector. Nature 478(7369):395–398. https://doi.org/10.1038/nature10454
20. Popko J, Herrfurth C, Feussner K, Ischebeck T, Iven T, Haslam R, Hamilton M, Sayanova O, Napier J, Khozin-Goldberg I, Feussner I (2016) Metabolome analysis reveals betaine lipids as major source for triglyceride formation, and the accumulation of sedoheptulose during nitrogen-starvation of *Phaeodactylum tricornutum*. PLoS One 11(10):e0164673. https://doi.org/10.1371/journal.pone.0164673
21. Bayram Ö, Feussner K, Dumkow M, Herrfurth C, Feussner I, Braus GH (2016) Changes of global gene expression and secondary metabolite accumulation during light-dependent *Aspergillus nidulans* development. Fungal Genet Biol 87:30–53. https://doi.org/10.1016/j.fgb.2016.01.004
22. Sarikaya-Bayram Ö, Bayram Ö, Feussner K, Kim J-H, Kim H-S, Kaever A, Feussner I, Chae K-S, Han D-M, Han K-H, Braus GH (2014) Membrane-bound methyltransferase complex VapA-VipC-VapB guides epigenetic control of fungal development. Dev Cell 29(4):406–420. https://doi.org/10.1016/j.devcel.2014.03.020
23. Gerke J, Bayram Ö, Feussner K, Landesfeind M, Shelest E, Feussner I, Braus GH (2012) Breaking the silence: protein stabilization uncovers silenced biosynthetic gene clusters in the fungus Aspergillus nidulans. Appl Environ Microbiol 78(23):8234–8244. https://doi.org/10.1128/aem.01808-12
24. Stringlis IA, Yu K, Feussner K, de Jonge R, Van Bentum S, Van Verk MC, Berendsen RL, Bakker PAHM, Feussner I, Pieterse CMJ (2018) MYB72-dependent coumarin exudation shapes root microbiome assembly to promote plant health. Proc Natl Acad Sci U S A 115(22):E5213–E5222. https://doi.org/10.1073/pnas.1722335115

25. Kaever A, Landesfeind M, Feussner K, Mosblech A, Heilmann I, Morgenstern B, Feussner I, Meinicke P (2015) MarVis-Pathway: integrative and exploratory pathway analysis of non-targeted metabolomics data. Metabolomics 11(3):764–777. https://doi.org/10.1007/s11306-014-0734-y
26. Kaever A, Landesfeind M, Possienke M, Feussner K, Feussner I, Meinicke P (2012) MarVis-Filter: ranking, filtering, adduct and isotope correction of mass spectrometry data. J Biomed Biotechnol 2012:263910. https://doi.org/10.1155/2012/263910
27. Kaever A, Lingner T, Feussner K, Göbel C, Feussner I, Meinicke P (2009) MarVis: a tool for clustering and visualization of metabolic biomarkers. BMC Bioinformatics 10(1):92. https://doi.org/10.1186/1471-2105-10-92
28. Meinicke P, Lingner T, Kaever A, Feussner K, Göbel C, Feussner I, Karlovsky P, Morgenstern B (2008) Metabolite-based clustering and visualization of mass spectrometry data using one-dimensional self-organizing maps. Algorithms Mol Biol 3(1):9. https://doi.org/10.1186/1748-7188-3-9
29. Matyash V, Liebisch G, Kurzchalia TV, Shevchenko A, Schwudke D (2008) Lipid extraction by methyl-tert-butyl ether for high-throughput lipidomics. J Lipid Res 49(5):1137–1146. https://doi.org/10.1194/jlr.D700041-JLR200
30. Kaever A, Landesfeind M, Feussner K, Morgenstern B, Feussner I, Meinicke P (2014) Meta-analysis of pathway enrichment: combining independent and dependent omics data sets. PLoS One 9(2):e89297. https://doi.org/10.1371/journal.pone.0089297
31. Tanabe M, Kanehisa M (2012) Using the KEGG database resource. Curr Protoc Bioinformatics 38(1):1.12.11–11.12.43. https://doi.org/10.1002/0471250953.bi0112s38
32. Caspi R, Altman T, Dreher K, Fulcher CA, Subhraveti P, Keseler IM, Kothari A, Krummenacker M, Latendresse M, Mueller LA, Ong Q, Paley S, Pujar A, Shearer AG, Travers M, Weerasinghe D, Zhang P, Karp PD (2012) The MetaCyc database of metabolic pathways and enzymes and the BioCyc collection of pathway/genome databases. Nucleic Acids Res 40(Database issue):D742–D753. https://doi.org/10.1093/nar/gkr1014
33. Salek RM, Steinbeck C, Viant MR, Goodacre R, Dunn WB (2013) The role of reporting standards for metabolite annotation and identification in metabolomic studies. GigaScience 2(1):1–3. https://doi.org/10.1186/2047-217X-2-13
34. Strehmel N, Böttcher C, Schmidt S, Scheel D (2014) Profiling of secondary metabolites in root exudates of *Arabidopsis thaliana*. Phytochemistry 108:35–46. https://doi.org/10.1016/j.phytochem.2014.10.003
35. Horai H, Arita M, Kanaya S, Nihei Y, Ikeda T, Suwa K, Ojima Y, Tanaka K, Tanaka S, Aoshima K, Oda Y, Kakazu Y, Kusano M, Tohge T, Matsuda F, Sawada Y, Hirai MY, Nakanishi H, Ikeda K, Akimoto N, Maoka T, Takahashi H, Ara T, Sakurai N, Suzuki H, Shibata D, Neumann S, Iida T, Tanaka K, Funatsu K, Matsuura F, Soga T, Taguchi R, Saito K, Nishioka T (2010) MassBank: a public repository for sharing mass spectral data for life sciences. J Mass Spectrom 45(7):703–714. https://doi.org/10.1002/jms.1777

Chapter 12

Untargeted Metabolomics of Plant Leaf Tissues

Federica Gevi, Giuseppina Fanelli, Lello Zolla, and Sara Rinalducci

Abstract

Untargeted metabolomics is a useful approach for the simultaneous analysis of a vast array of compounds from a single extract. Metabolomic profiling is the relative multi-parallel quantification of a mixture of low molecular weight compounds, or classes of compounds, and it is most often performed by using ultra performance liquid chromatography (UPLC) coupled with mass spectrometry (MS). Being an extension of the classical targeted methods, this approach allows a broader view of the main biochemical events within a particular sample. This chapter exemplifies and provides experimental details on the basic steps to perform a non-targeted metabolomic analysis on plant leaf tissues: sample collection and homogenization, extraction of metabolites, raw data acquisition, and processing into formats for data mining and informatics. In particular, the approach was applied to two spring wheat varieties with different level of drought tolerance (Kavir, drought-resistant; Bahar, drought-sensitive) developed by the CIMMYT (International Center for the Improvement of Corn and Wheat).

Key words Metabolomics, Mass spectrometry, Wheat, Plant metabolite extraction, UPLC

1 Introduction

1.1 Plant Metabolomics

Metabolites are biological molecules within a cell and represent the phenotypic status of a biological system. Therefore, their analysis provides information on the metabolic pathways participating in different biological processes and on the state of cell perturbations. Metabolomics is being widely applied to studies on plant responses to abiotic stresses, with the aim of investigating the plant system at the molecular level to provide a characterization of the total metabolite pool of a plant tissue in response to its environment [1, 2]. Plant metabolic studies generate enormous amounts of information about chemical species with different physical properties, from ionic inorganic compounds to hydrophilic, biochemical, organic, and amino hydrophilic carbohydrates [3]. Generally, these low

Federica Gevi and Giuseppina Fanelli contributed equally to this work and share the first authorship.

Angelo D'Alessandro (ed.), *High-Throughput Metabolomics: Methods and Protocols*, Methods in Molecular Biology, vol. 1978, https://doi.org/10.1007/978-1-4939-9236-2_12,

molecular weight compounds are classified into primary and secondary metabolites. The former are indispensable for the growth and development of a plant, while the latter, although not essential, are crucial for plant survival under stress conditions. Using Bahar and Kavir wheat leaves as a model, we generated 200 mg sample metabolomics data sets [4].

1.2 Workflow for Untargeted Metabolomics

The protocol here described may provide a practical guide to the preparation of plant leaf samples for metabolomic analyses. Since sample preparation is a critical step for reproducibility and reliability of the data obtained, all practical considerations on the process should be taken into consideration from the beginning. To provide accurate and meaningful results, it is crucial to minimize the experimental variation that occurs from all stages of sample generation and processing before statistical analysis. The typical workflow of an untargeted metabolomics study involves five phases: (1) collection, (2) drying, (3) extraction of plant material, (4) MS analysis, and (5) data processing, statistical treatment, and interpretation (Fig. 1). All sample preparation steps must be planned in order to identify potential sources of experimental variation and errors. Sample collection is performed following standardized procedures, taking into account the age of the plant or the phase of phenotypic growth, sampling, and processing time [5]. Furthermore, it must be carried out as quickly as possible to prevent any metabolic changes. The second phase takes place immediately after the sample collection and provides for the storage of the tissue at extremely low temperature with liquid nitrogen and subsequent freezing at

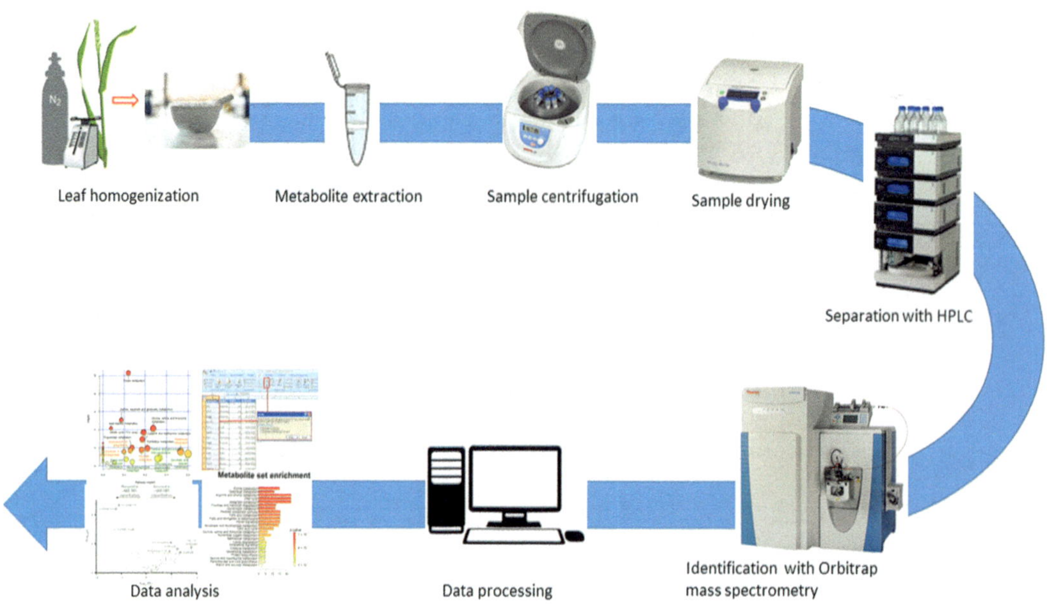

Fig. 1 Workflow showing the general strategy and experimental steps of a LC-MS experiment

−80 °C. This phase occurs before the extraction of metabolites and guarantees the inactivation of metabolic processes, thus ensuring the correct measurement of metabolic variation.

1.3 Extraction Methods

Different approaches can be used to extract metabolites from plant material. Extractions can be either "selective" or "total." The most appropriate solvent to use is based on its selectivity for the substances to be extracted. In a selective extraction, the plant material is extracted using a solvent of an appropriate polarity following the principle of "like dissolves like." Thus, nonpolar solvents are used to solubilize mostly lipophilic compounds (e.g., alkanes, fatty acids). Medium-polarity solvents are used to extract compounds of intermediate polarity (e.g., some alkaloids, flavonoids), while more polar ones are used for more polar compounds (e.g., flavonoid glycosides, tannins, some alkaloids). In this protocol a total extraction by using polar organic solvent (e.g., ethanol, methanol, or an aqueous alcoholic mixture) is employed with the attempt to extract as many compounds as possible. Extraction procedures include maceration, Soxhlet extraction, pressurized solvent extraction, assisted ultrasonic solvent extraction, reflux extraction, and steam distillation [6]. In our study we performed conventional maceration through the use of pestle mortar and liquid nitrogen.

1.4 Metabolite Identification and Data Interpretation

After metabolic extractions, samples are most often analyzed by mass spectrometry coupled with GC or LC. However, LC-MS is the most frequently applied analytical technique in plant metabolomics studies due to its sensitivity and the wide coverage of biological information relevant to metabolism. Several approaches can be adopted for the LC-MS analysis; one of these concerns the redissolution of the sample in a solvent with the same composition of the LC mobile phase at the beginning of the gradient [7]. By using an UPLC-MS method, we successfully annotated more than 300 compounds covering the main classes involved in central metabolic pathways of Bahar and Kavir wheat plants. Metabolic profiling was performed for primary metabolites such as amino acids, organic acids, and sugars. Biological and technical triplicates are suggested to add statistical power to the analysis. Moreover, for the untargeted approach, the use of the web-based free MetaboAnalyst software was associated to the LC-MS analysis, thus ensuring not only the identification of the metabolites but also their quantification and statistically significant grouping in metabolic pathways.

2 Materials

2.1 Equipment

1. Vortexer.
2. Pestle.
3. Mortar.

4. 2 mL original Eppendorf tubes (*see* **Note 1**).
5. Rotary vacuum evaporator.
6. Bench centrifuge.
7. Glass vials for LC analysis.
8. Ultrapure water purification system.
9. Reprosil C18 column (2.0 mm × 150 mm; 2.5 μm, Dr. Maisch, Germany).
10. Ultimate 3000 Rapid Resolution UPLC system (DIONEX, Sunnyvale, CA, USA).
11. Q-Exactive Hybrid Quadrupole-Orbitrap Mass Spectrometer (Thermo Scientific™).

2.2 Reagent Setup

1. Liquid nitrogen (*see* **Note 2**).
2. High-purity (MS grade) acetonitrile, formic acid, and HPLC-grade water for LC mobile phases (*see* **Note 3**).
3. High-purity (MS grade) methanol at −20 °C.
4. High-purity (MS grade) chloroform at −20 ° C.
5. Ultrapure water (18 MΩ-cm at 25 °C).
6. Positive and negative ion calibration mixes (Thermo Scientific™).

3 Methods

3.1 Harvesting of Plant Leaves

1. Harvest leaf tissue (200 mg) from wheat plants grown under the desired experimental condition. Either use randomized positions of leaves to be harvested or select leaves at the same position (*see* **Note 4**).
2. Homogenize leaves in liquid nitrogen into a fine powder using a pestle and mortar (*see* **Note 5**).
3. Collect the obtained fine frozen powder in a pre-cooled Eppendorf tube.

3.2 Extraction of Metabolites

1. Add a volume of 0.15 mL of ice-cold ultrapure water (18 mΩ-cm) to the Eppendorf tube containing the powdered tissue (*see* **Note 6**).
2. Extract metabolites from leaf samples in three replicates.
3. Transfer the tubes into a water bath at 37 °C for 0.5 min.
4. Sonicate 3 times for 1 s each time.
5. Add 400 μL of −20 °C methanol and vortex vigorously (*see* **Note 7**).
6. Add 500 μL of −20 °C chloroform and vortex well.

7. Add 100 μL of ice-cold ultrapure water to each tube.
8. Centrifuge at 14,000 × *g* for several minutes. After centrifugation, the different phases are now separated, and each one can be sampled for further analyses with multiple techniques.
9. Transfer the tubes to a −20 °C freezer for 2–8 h.
10. Dry in a rotational vacuum concentrator at 50 °C for at least 3 h (*see* **Note 8**).
11. Resuspend in 200 μL of water with 5% formic acid.
12. Transfer the solution to glass auto-sampler vials for LC/MS analysis with the help of a pipette.
13. For control and quality assessment of analytical precision, add the leucine enkephalin standard to all samples at a concentration of 0.5 μg/mL and use it as lock mass ($[M+H]^+$ = 556.276575 Da; $[M-H]^-$ = 554.260925 Da) [8] (*see* **Note 9**).

3.3 LC-MS Analysis

1. Calibrate the instrument in both positive and negative ion mode by using the appropriate calibrator mix (Thermo Scientific™) (*see* **Notes 10–15**).
2. Metabolites are separated via reversed phase liquid chromatography (RP-LC) using a 2.0 mm × 150 mm Reprosil C18 column (2.5 μm—Dr Maisch, Germany).
3. LC parameters: injection volume, 20 μL; column temperature, 30 °C; flow rate, 0.2 mL/min. LC solvent gradient and timetable are identical during the whole period of the analyses. For positive ion mode, a 0–65% linear gradient of solvent A (ddH_2O, 0.1% formic acid) to B (acetonitrile, 0.1% formic acid) is employed over 17 min, returning to 100% A in 2 min and a 4-min post-time solvent A hold (*see* **Notes 16** and **17**). For negative ion mode, a 0–100% linear gradient of solvent A (ddH_2O, 2.5 mM ammonium acetate) to B (acetonitrile, 2.5 mM ammonium acetate) is employed over 20 min, returning to 100% A in 2 min and a 6-min post-time solvent A hold (*see* **Note 18**).
4. Ion detection is achieved using an Orbitrap mass spectrometer (Q-Exactive, Thermo Scientific™). Select the appropriate electrospray ionization (ESI) mode (positive/negative) to operate. Scanning is in full MS mode (2 μscans) at 70,000 resolutions in the 67 to 1000 *m/z* range.
5. Source ionization parameters: spray voltage, 3.8 kV; capillary temperature, 300 °C; sheath gas, 40; auxiliary gas, 25; S-Lens level, 45.
6. Set up the sample table (*see* **Note 19**).

7. Prepare pooled quality control (QC) samples by mixing together a small aliquot of each biological sample to be analyzed in the study set. QC sample is injected at regular intervals (every 7–9 sample batch). QCs are applied to assess and ensure that the analytical process is performed appropriately and meets the predefined criteria (*see* **Note 20**).
8. Register the TIC (total ion chromatogram; Fig. 2).

3.4 Data Analysis

1. Process LC-MS raw data by using Xcalibur™ Software (Thermo Fisher Scientific).
2. Files from each run should be exported as mzXML with MASS MATRIX ms convert.
3. For data deconvolution, use the freely Maven software (http://genomics-pubs.princeton.edu/mzroll/index.php).
4. Data can be plotted with GraphPad Prism 5.0 (GraphPad Software Inc. san Diego, CA, USA).
5. More than one thousand peaks per sample will be obtained referring to the KEGG database. Among them, a list of selected metabolites is analyzed more precisely and identified.
6. Perform univariate (two-sample *t*-test; volcano plot) and multivariate (PCA, PLS-DA) statistics analyses by using the MetaboAnalyst 3.0 software (http://metpa.metabolomics.ca/) (Fig. 3; *see* **Note 21**).
7. Normalize the data set in order to reduce any systematic error and improve overall data consistency for meaningful biological comparisons (*see* **Note 22**).
8. Use the web-based tools MSEA (Metabolite Set Enrichment Analysis) and MetPA (Metabolomics Pathway Analysis) to perform metabolite enrichment and pathway analyses, respectively (*see* **Notes 23** and **24**).

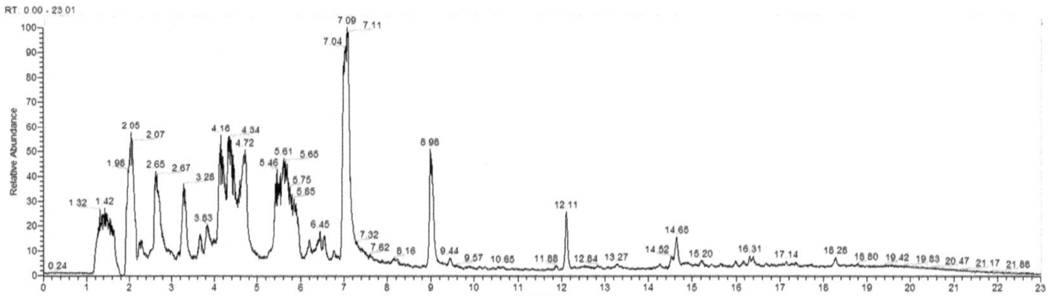

Fig. 2 Total ion chromatogram (TIC) profile obtained by positive ion mode electrospray ionization mass spectrometry (ESI/MS) for Kavir plants

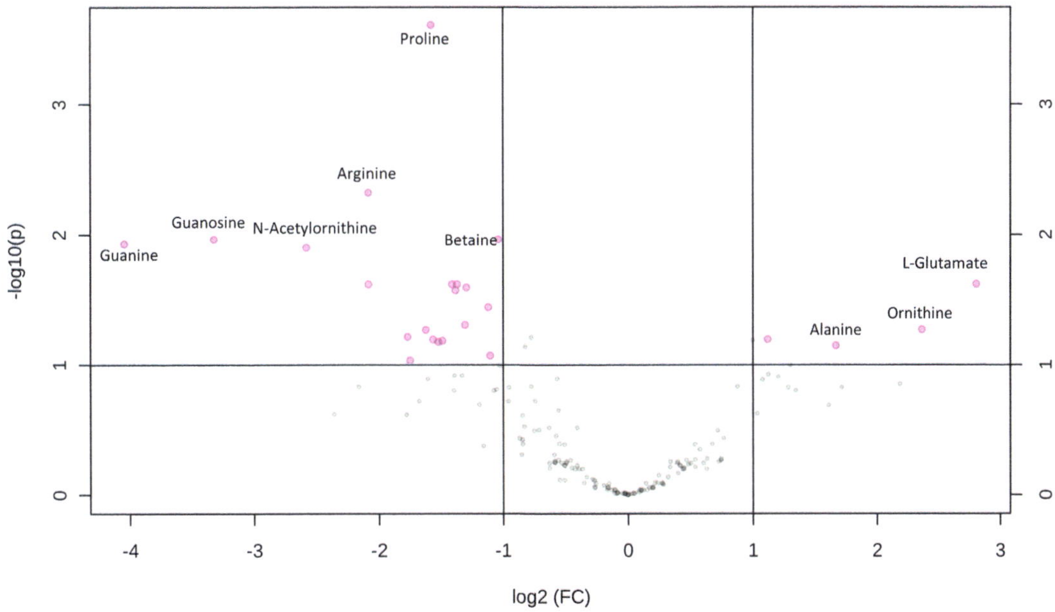

Fig. 3 Univariate analysis (volcano plot) of metabolites measured in Kavir leaves using MetaboAnalyst 3.0. Metabolites with absolute fold change >2 and adjusted *p*-value (FDR < 0.05) are indicated in pink

4 Notes

1. Only original Eppendorf tubes should be used, to avoid contaminants from polymers released by other plasticizers.
2. Use liquid nitrogen with caution. Wear eye protection and avoid direct skin contact.
3. Prepare all solutions using ultrapure water (prepared by purifying deionized water to reach a resistivity of 18 MΩ-cm at 25 °C) and analytical grade reagents.
4. Always wear gloves when handling tissue samples and for all sample preparation and extraction steps.
5. Take care that tissues stay well frozen during homogenization; discard any samples that start to thaw. If needed, carefully let the nitrogen evaporate and repeat homogenization several times.
6. In general, all sample preparation and processing steps should be as fast as possible and performed at low temperature, in order to exclude any metabolite degradation due to residual enzymatic contamination of the samples. Carry out all procedures at room temperature unless otherwise specified.
7. Methanol and chloroform are toxic; therefore, handle them with care.

8. When samples are dried, raise temperature on the rotary vacuum evaporator to room temperature for 30 min before removing the vials to avoid condensation.
9. Internal standards are ideally compounds not present in the biological extract and are included prior to or during metabolite extraction.
10. Sonicate phases A and B to avoid bubbles in the system after mixing them.
11. Perform a purge before starting the LC analysis to eliminate the bubbles in the tubes.
12. Clean the syringe before starting the analysis batch.
13. Clean the ion source before starting an analysis batch.
14. Be careful for high temperature when sweep cone is removed for cleaning.
15. To perform external calibration, use Hamilton gastight syringe #31750RN 500 μL.
16. It is recommended to prepare solutions fresh each time.
17. For positive ion mode, formic acid (FA) is the optimal coupling agent, as it works best in enhancing MS signals.
18. Ammonium acetate is used as mobile phase additive in negative ion mode to improve detection of anionic compounds.
19. Carry out at least two blank runs prior to the first analysis (this helps regularizing pressures) and one blank every five runs to avoid contaminations that could invalidate the interpretations.
20. Preparation of the QCs should follow the same sample procedure adopted to study biological samples.
21. Pay attention to the format (.csv UTF8) used to load the data in the software.
22. MetaboAnalyst offers three types of normalization: (1) sample normalization; (2) data transformation; (3) data scaling.
23. MSEA directly investigates if a group of functionally related metabolites are significantly enriched, eliminating the need to preselect compounds based on some arbitrary cut-off threshold. MetPA refers to quantitative enrichment analysis directly based on the compound concentration values and has the potential to discover "subtle but consistent" changes among compounds within the same biological pathway.
24. MSEA and MetPA tools analyze several pathways at the same time; therefore, perform multiple testing corrections of the raw *p*-values via false discovery rate (FDR) estimation and/or Holm-Bonferroni method.

Acknowledgments

Leaves from the spring wheat cultivars (Bahar, drought-susceptible; Kavir, drought-tolerant) used in the experiment were kindly provided by Prof. Mahmoud Toorchi research group at the Department of Biotechnology and Plant Breeding, University of Tabriz, Tabriz, Iran.

References

1. Arbona V, Manzi M, de Ollas C, Gómez-Cadenas A (2013) Metabolomics as a tool to investigate abiotic stress tolerance in plants. Int J Mol Sci 14:4885–4911
2. Fiehn O (2002) Metabolomics – the link between genotypes and phenotypes. Plant Mol Biol 48:155–171
3. Jorge TF, Rodrigues JA, Caldana C, Schmidt R, van Dongen JT, Thomas-Oates J, António C (2016) Mass spectrometry-based plant metabolomics: metabolite responses to abiotic stress. Mass Spectrom Rev 35:620–649
4. Michaletti A, Reza Naghavi M, Toorchi M, Zolla L, Rinalducci S (2018) Metabolomics and proteomics reveal drought-stress responses of leaf tissues from spring-wheat. Sci Rep 8:5710
5. William Allwood J, Winder C, Dunn W, Goodacre R (2013) Considerations in sample preparation, collection, and extraction approaches applied in microbial, plant, and mammalian metabolic profiling. Methodologies for Metabolomics: Experimental Strategies and Techniques Cambridge: Cambridge University Press. pp 79–118
6. Seidel V (2012) Initial and bulk extraction natural products isolation. Methods Mol Biol 864:27–41
7. Kim HK, Choi YH, Verpoorte R (2010) NMR-based metabolomic analysis of plants. Nat Protoc 5:536–549
8. Sztáray J, Memboeuf A, Drahos L, Vékey K (2011) Leucine enkephalin-a mass spectrometry standard. Mass Spectrom Rev 30:298–320

Part IV

Tracing Experiments and Metabolic Flux Analysis

Chapter 13

Analysis of Arginine Metabolism Using LC-MS and Isotopic Labeling

Gretchen L. Seim, Emily C. Britt, and Jing Fan

Abstract

Arginine metabolism is linked to several important metabolic processes, and reprogramming of arginine metabolism occurs in various physiological and pathological conditions. Here we describe a method, using a LC-MS-based metabolomics and $^{15}N_4$-arginine tracing approach, to quantitatively analyze arginine metabolism. This method can reliably quantify the abundance of important intermediates and fluxes of major metabolic reactions in arginine metabolism in a variety of cultured mammalian cell models.

Key words Arginine, Metabolic flux analysis, Isotopic labeling, LC-MS

1 Introduction

Arginine metabolism is connected to several important metabolic pathways including nitric oxide (NO) production, the urea cycle, polyamine synthesis, and creatine synthesis (Fig. 1). It plays a major role in the functions of immune cells [1–3], and alterations in arginine metabolism have been linked to various diseases, such as cancer [4–7], vascular dysfunction [8, 9], and asthma [10]. In myeloid cells, arginase (ARG) and nitric oxide synthase (NOS) are differentially regulated by type 1 and type 2 cytokines, and the reprogramming of arginine metabolism is critical for inflammation regulation [1, 2]. CD8+ T cells have increased survival capacity and antitumor activity when L-arginine levels are increased, further exemplifying the connection between immune cells and arginine [3]. Additionally, a number of cancers including melanoma, hepatocellular carcinoma, prostate cancer, and renal cell carcinoma are deficient in argininosuccinate synthase (ASS) and auxotrophic for arginine [4]. Therefore, lowering arginine availability has become an effective strategy to treat these cancers [5–7]. The relevance of arginine metabolism goes beyond immune cell function and cancer. For instance, changes in arginine metabolism can affect the production of nitric oxide, which is important in vasodilation of

Angelo D'Alessandro (ed.), *High-Throughput Metabolomics: Methods and Protocols*, Methods in Molecular Biology, vol. 1978, https://doi.org/10.1007/978-1-4939-9236-2_13,

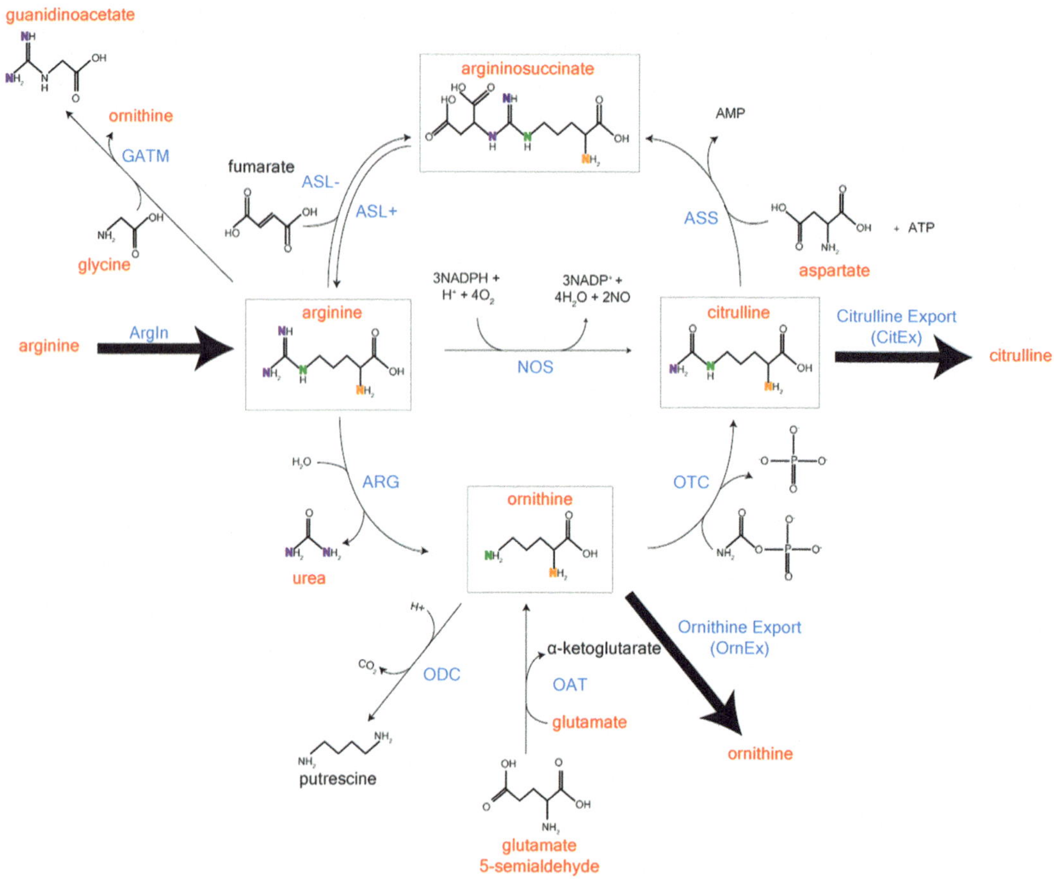

Fig. 1 Murine arginine metabolism pathway. Metabolites detected using described method are indicated in red. Those metabolic reactions whose fluxes are determined using described method are indicated in blue: *NOS* nitric oxide synthase, *ARG* arginase, *ASS* argininosuccinate synthase, *ASL* argininosuccinate lyase, *OTC* ornithine transcarbamylase, *OAT* ornithine aminotransferase, *ODC* ornithine decarboxylase, *GATM* glycine amidinotransferase, *CitEx* citrulline export, *OrnEx* ornithine export, and *ArgIn* Arginine import. Arginine nitrogens are color coded to track their transfer to downstream metabolites

coronary arterioles and regulation of hyperreactive airways in asthma [8–10]. Analysis of metabolites and metabolic fluxes in arginine metabolism enables a quantitative understanding of how this pathway is reprogrammed in different physiological and pathological conditions and thus can bring insight into the specific roles of arginine metabolism in cell function and disease pathophysiology.

Here we describe a protocol to quantitatively measure the levels of arginine metabolism intermediates using a liquid chromatography-mass spectrometry (LC-MS)-based method and to determine the fluxes of arginine metabolism reactions using ^{15}N tracer-based metabolic flux analysis [11]. Specifically, cells are cultured in the presence of $^{15}N_4$-arginine until pseudo-steady state is reached; at this point, the incorporation of labeled nitrogen into downstream metabolites quantitatively reflects the flux distribution of different metabolic pathways [11–13]. The abundance and labeling patterns of relevant intracellular metabolites, such as arginine, ornithine, citrulline, and argininosuccinate, are measured using LC-MS. Relative flux ratios are calculated based on the principle that at steady state; the observed labeling pattern of a metabolite is a weighted average of all the labeling patterns from each possible precursor, where the corresponding weights indicate the fractional flux contributions. The exchange fluxes, including arginine uptake, NO production, urea production, and excretion rate of citrulline and ornithine, can be directly measured by monitoring the concentration change of these compounds in the media. Lastly, the absolute intracellular fluxes can be determined by integrating the exchange fluxes and relative flux ratios. We demonstrate this method can reliably quantify the abundance of important intermediates (Fig. 1, labeled in red) and fluxes of major metabolic reactions (Fig.1, labeled in blue) in arginine metabolism in cultured mammalian cells. The following procedure is outlined using RAW 264.7 murine macrophages. However, this method can be applied to various cell types of interest.

2 Materials

All media or solutions that will be in contact with the cells must be sterile. Make solutions inside biological safety cabinet or sterile filter solutions before use. All solutions that will go into the LC-MS should be LC-MS grade.

2.1 Cell Culture and Isotopic Labeling

1. Cells: Any cell type of interest can be used. In this example, we use Raw 264.7 cells.
2. Unlabeled media: RPMI-1640 medium, supplemented with 10% dialyzed fetal bovine serum (dFBS) and 100 U/mL penicillin-streptomycin (*see* **Notes 1** and **2**).
3. Arginine-labeled media: Arginine, leucine, and lysine-free RPMI-1640 medium, supplemented with 0.05 g/L L-lysine, 0.05 g/L L-leucine, 1.6 g/L L-arginine-$^{15}N_4$ hydrochloride, 10% dFBS, and 100 U/mL penicillin-streptomycin (*see* **Note 3**).

2.2 Metabolite Extraction

1. Dulbecco's phosphate buffered saline (dPBS).
2. Extraction solvent: 80:20 methanol:H_2O (v/v).
3. Methanol.

2.3 Metabolite Analysis

1. Column: SeQuant® PEEK HPLC Column with ZIC-pHILIC (5 m) Polymeric Beads Sorbent, 150 × 2.1 mm (1.50460.0001, Millipore Sigma).
2. Mobile Phase A: 95:5 Acetonitrile:H_2O (v/v).
3. Mobile Phase B: 10 mM ammonium acetate buffer, pH 5.5.

4. Metabolite standards: urea, citrulline, ornithine. Prepare standard stock solutions (10 mM) in water and store at −80 °C.
5. Sample loading solvent: 80:20 Acetonitrile:H_2O (v/v).

2.4 Assay Kits

1. Griess Reagent System.
2. Urea Assay Kit.

3 Methods

3.1 Cell Culture and Isotopic Labeling

1. Plate cells in 6-well plates in unlabeled media. Plate enough replicates ($n \geq 3$) for metabolite analysis and additional wells for normalization at the same seeding density (*see* **Note 4**). Let cells attach and maintain cells at 37 °C.
2. If the goal of the experiment is to compare arginine metabolism under different conditions, treat the cells with perturbation of interest. In this example, we leave the macrophages in resting state (M0) or activate them to M1 state with LPS and interferon-γ stimulation (*see* **Note 5**).
3. Change media to chemically identical L-$^{15}N_4$-arginine-labeled media, and incubate until isotopic steady state is reached (*see* **Notes 6** and 7).

3.2 Metabolite Extraction from Cells (See Note 8)

1. Before collection, pre-chill extraction solvent in −80 °C.
2. Collect media from each well in addition to aliquots of fresh media. Centrifuge at 1000 × *g* for 2 min for removal of suspended cells. Transfer supernatant to new tube and freeze at −20 °C until media metabolite extraction (*see* Subheading 3.3).
3. Count cells or collect protein and perform protein quantification from extra normalization wells.
4. Aspirate media from remaining cells.
5. Wash each well three times with dPBS.
6. Immediately add 1000 μL cold extraction solvent to each well.
7. Immediately place plate on dry ice or in −80 °C (plate can be left here for 15–60 min before proceeding with protocol).
8. With plate on dry ice, scrape bottom of well to dislodge cellular material and transfer to an Eppendorf tube.
9. Centrifuge at 20,000 × *g* for 5 min at 4 °C.
10. Transfer all of supernatant to a new Eppendorf tube.
11. Add 0.5 mL of cold extraction solvent to the remaining pellet for re-extraction. Pipette to resuspend. Centrifuge at 20,000 × *g* for 5 min at 4 °C.
12. Transfer the supernatant and combine with the previous extract supernatant.
13. Dry down the extracted samples under a nitrogen stream.

14. Add 50 μL of 80:20 acetonitrile:H_2O (v/v) to the dried metabolite pellet and resuspend.
15. Centrifuge at 20,000 × *g* for 5 min at 4 °C.
16. Transfer supernatant to LC-MS vial for analysis.

3.3 Metabolite Extraction from Media

1. Thaw media if frozen, vortex to mix.
2. Combine media with methanol in a 1:5 ratio (100 μL media + 400 μL of methanol), mix well by vortexing.
3. Centrifuge at 20,000 × *g* for 5 min at 4 °C.
4. Transfer 20 μL of supernatant and combine with 180 μL 80:20 ACN:H_2O (v/v) (1:10 dilution), mix well.
5. Place solution (now diluted 1:50) into LC-MS vial for analysis.

3.4 [Optional] Prepare Samples with Metabolite Standards

Metabolite standards allow for quantification of absolute intracellular and extracellular metabolite quantities. Absolute metabolite quantity in media is required to determine some of the exchange fluxes (Subheading 3.8), which are ultimately required to calculate absolute intracellular fluxes (Subheading 3.9). If only relative abundance of intracellular metabolites and relative flux ratios are desired (Subheading 3.7), this step is not required. Quantification of absolute metabolite levels can be either done with internal standards or with an external calibration curve. The internal standard method [14] accounts for the impact of the sample matrix on the ion peak size. The external calibration curve is advantageous as it can be used continually across experiments and does not require sacrifice of a portion of the sample extract. Either are viable options, and both are described below and in Subheading 3.6, **steps** 7 and **8**.

1. Internal standards: Use the protocol described in [14] to quantify absolute concentration of metabolites in samples. Prepare a mixture of metabolite standards. Spike mixture into a separate aliquot of each labeled sample. Analyze these samples with spiked internal standards using LC-MS alongside parallel unspiked samples as described in Subheading 3.5. *See* Subheading 3.6, **step** 7 for details on data analysis (*see* **Note 9**).
2. Calibration curve: Prepare a mixture of metabolite standards. Dilute mixture to a range of concentrations (0.05 μM–200 μM) in the sample loading solvent (80:20 Acetonitrile:H_2O (v/v)). Analyze using LC-MS alongside samples as described in Subheading 3.5. *See* Subheading 3.6, **step 8** for details on data analysis.

3.5 Metabolite Analysis by LC-MS

Samples are analyzed using a Thermo Q-Exactive mass spectrometer coupled to a Vanquish Horizon UHPLC. Metabolites are separated on a 150 × 2.1 mm SeQuant PEEK HPLC column with ZIC-pHILIC (5 μm) polymeric beads with a gradient of solvent A

(95% ACN, 5% water) and solvent B (10 mM NH_4Ac, pH 5.5) at a 0.2 mL/min flow rate and 30 °C column temperature. The gradient is 0 min, 5% B; 2 min, 5% B; 18 min, 60% B; 19 min, 90% B; 24 min, 90% B; and 25.5 min, 5% B. Data are collected on a full scan positive mode at a resolution of 70 K with a maximum injection time of 100 ms and automatic gain control of 3E6. Settings for the ion source are 12 aux gas flow rate, 40 sheath gas flow rate, 1 sweep gas flow rate, 3.5 kV spray voltage, 340 °C capillary temperature, and 250 °C heater temperature. Multiple ($n \geq 3$) injections of sample loading solvent (80:20 acetonitrile/H_2O (v/v)) are analyzed alongside samples as blanks (*see* **Notes 10** and **11**).

3.6 Data Analysis

1. Quantify compounds of interest (including all isotopic forms) using software such as Xcalibur or Maven [15]. Compounds are identified based on exact mass determined from formula and retention time (Table 1).
2. Subtract blank signal from the signal of each of the measured isotopes.
3. Average signals from replicates.
4. Normalize total signal (all isotopic forms) to protein content or cell number to obtain relative intracellular metabolite abundances (Fig. 2).

Table 1
Observed retention times for arginine metabolite intermediates using indicated LC-MS method (Subheading 3.5). Formulas listed are neutral formulas

Compound	Formula	Retention time (min)
Arginine	$C_6H_{14}N_4O_2$	20.58
Argininosuccinate	$C_{10}H_{18}N_4O_6$	13.55
Aspartate	$C_4H_7NO_4$	12.38
Citrulline	$C_6H_{13}N_3O_3$	12.17
Glutamate	$C_5H_9NO_4$	12.12
Glutamate-5-semialdehyde	$C_5H_9NO_3$	10.83
Glycine	$C_2H_5NO_2$	11.95
Guanidinoacetate	$C_3H_7N_3O_2$	11.69
Ornithine	$C_5H_{12}N_2O_2$	19.98
Urea	CH_4N_2O	6.50

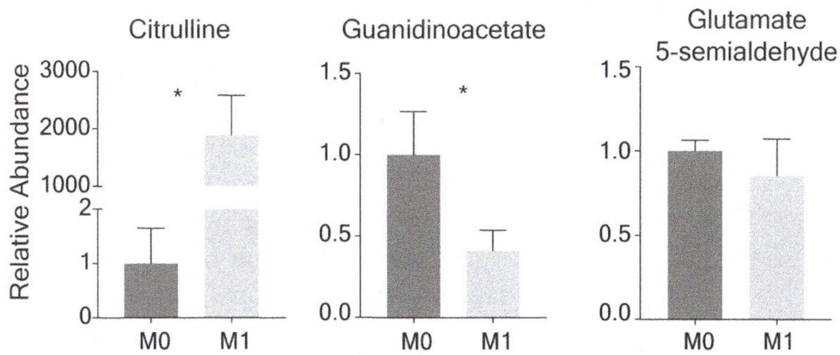

Fig. 2 Relative metabolic abundance of arginine metabolism intermediates in unstimulated (M0) and LPS and interferon-γ stimulated (M1) RAW 264.7 cells. * indicates p-value <0.05

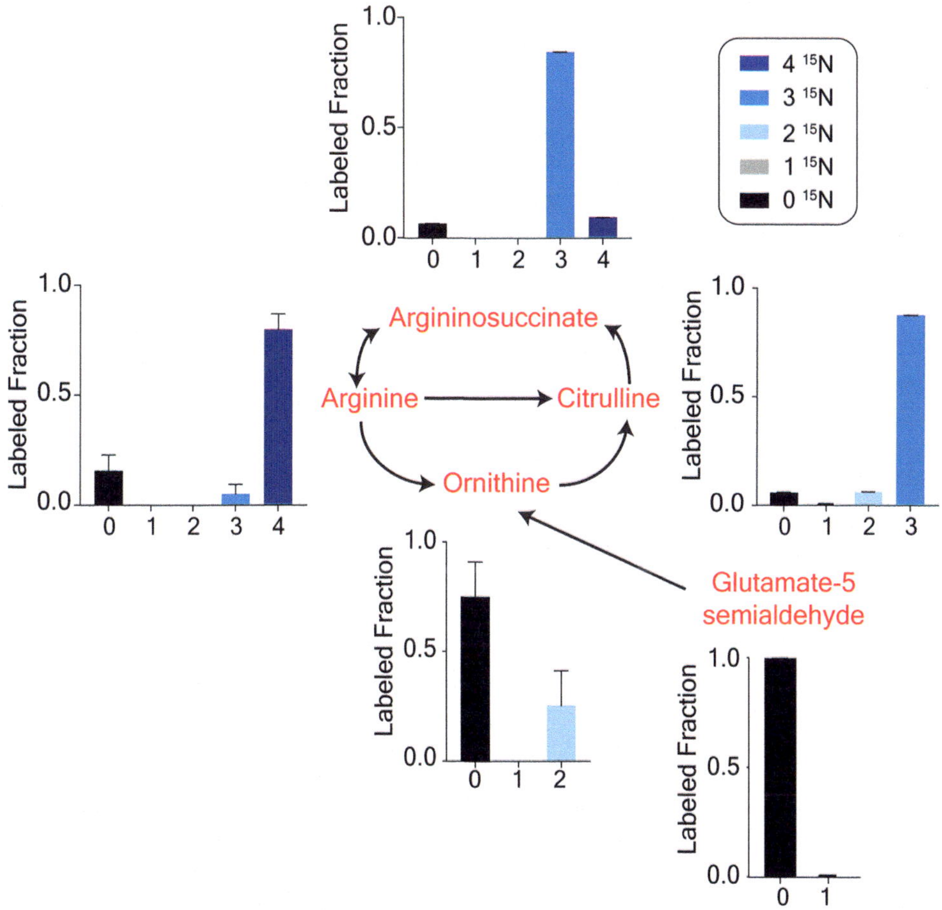

Fig. 3 Labeling pattern of arginine metabolism intermediates in LPS and interferon-γ stimulated (M1) RAW 264.7 cells

5. Generate the labeling pattern of each compound of interest: Calculate the fraction of i-labeled form (M_i) by normalizing the signal of the i-^{15}N-labeled isotope to the total signal of all label forms of this compound (Fig. 3).

$$M_i = \frac{\text{Signal of } i\text{-labeled form}}{\text{Total signal of the compound}} \quad (1)$$

6. Adjust for natural abundance of ^{15}N. The natural abundance of ^{15}N is 0.4%. To correct for this, use the following equations to adjust the measured labeling pattern to obtain the adjusted values [16]. Here M_i is the measured fraction of compound with i-labeled nitrogen atoms, and X_i is the corrected fraction of i-^{15}N-labeled form. For a compound with total of n nitrogen atoms (*see* **Note 12**):

$$X_0 = \frac{M_0}{.996^n} \quad (2)$$

$$X_1 = \frac{M_1 - \left(\binom{n}{1}.996^{n-1}.004\right)X_0}{.996^{n-1}} \quad (3)$$

$$X_2 = \frac{M_2 - \left(\left(\binom{n}{2}.996^{n-2}.004^2\right)X_0 + \left(\binom{n-1}{1}.996^{n-2}.004\right)X_1\right)}{.996^{n-2}} \quad (4)$$

$$X_3 = \frac{M_3 - \left(\left(\binom{n}{3}.996^{n-3} \times .004^3\right)X_0 + \left(\binom{n-1}{2}.996^{n-3} \times .004^2\right) X_1 + \binom{n-2}{1}.996^{n-3} \times .004)X_2\right)}{.996^{n-3}} \quad (5)$$

$$X_4 = \frac{M_4 - \left(\left(\binom{n}{4}.996^{n-4} \times .004^4\right)X_0 + \left(\binom{n-1}{3}.996^{n-4} \times .004^3\right) X_1 + \left(\binom{n-2}{2}.996^{n-4} \times .004^2\right)X_2 + \left(\binom{n-3}{1}.996^{n-4} \times .004\right)X_3\right)}{.996^{n-4}} \quad (6)$$

7. [Optional] Quantify absolute metabolite levels using internal standards.

   In a $^{15}N_4$ arginine labeling experiment, the compounds in arginine metabolism pathway get labeled to various degrees. This labeled fraction is measured in metabolite samples without internal standard spiked-in. When the same samples are spiked with unlabeled internal standards, the labeled fractions shift toward more unlabeled forms. Absolute abundance of metabolites in samples can be determined by comparing the labeled fractions in samples with or without unlabeled internal standard spike-in [14].

For each compound of interest, calculate the following parameters:

In the spiked samples:

$$P = \frac{\text{Sum of fraction of all labeled isotopes } \sum_{i=1}^{n} X_i}{\text{Fraction of unlabeled isotope } X_0} \quad (7)$$

In the un-spiked samples:

$$L = \text{Sum of fraction of all labeled isotopes} \left(\sum_{n}^{i=1} X_i \right) \quad (8)$$

where n is the total number of nitrogens in the compound and X_i is the corrected fraction of i-^{15}N-labeled form (*see* Subheading 3.6, **step 6**). Use the following equation to calculate the absolute quantity of each metabolite in the sample (C):

$$C = \frac{P}{L - P(1-L)} \times S \times D \quad (9)$$

where S is the amount of internal standard spiked in the sample, D is the dilution factor if relevant.

8. [Optional] Quantify absolute abundance of metabolites using external calibration curves.

   Fit the relationship of signal with injected standard quantity by linear regression for each compound of interest. Determine the absolute quantity of compounds in samples based on these calibration curves (*see* **Note 13**).

3.7 Calculate Relative Flux Ratios

The labeling pattern of each compound is dictated by the labeling pattern of its precursors and the relative fluxes at which those precursors are feeding into that compound. Under metabolic and isotopic steady state, we can create a series of equations that balance the influx and outflux of each isotope of a compound. This will allow for determination of relative flux ratios (R) feeding into this compound. In this protocol, we determine relative fluxes using the measured labeling pattern of the following compounds:

Intracellular Ornithine: $X_{Orn\text{-}0}$, $X_{Orn\text{-}2}$.

Intracellular Citrulline: $X_{Cit\text{-}0}$, $X_{Cit\text{-}2}$, $X_{Cit\text{-}3}$.

Intracellular Arginine: $X_{Arg\text{-}0}$, $X_{Arg\text{-}2}$, $X_{Arg\text{-}3}$, $X_{Arg\text{-}4}$.

Intracellular Argininosuccinate: $X_{Argsucc\text{-}0}$, $X_{Argsucc\text{-}2}$, $X_{Argsucc\text{-}3}$, $X_{Argsucc\text{-}4}$.

Extracellular Arginine: $X_{ECArg\text{-}0}$, $X_{ECArg\text{-}2}$, $X_{ECArg\text{-}3}$, $X_{ECArg\text{-}4}$ (*see* **Note 14**).

Here the variable $X_{compound\text{-}i}$ represents the measured fraction of n-labeled form of a compound obtained from calculations described in Subheading 3.6. For example, $X_{\text{Arg-2}}$ represents the fraction of measured intracellular arginine that is labeled with two heavy nitrogen atoms.

A number of isotopomers of compounds relevant to this pathway can be excluded based on the chemistry of the pathway and thus allows for simplification of the equations. When intracellular arginine is labeled with two or more nitrogen atoms, the α and ε-N are dominantly labeled (orange and green in Fig. 1) which restricts the possible resulting labeling pattern of ornithine and citrulline. In addition, in most investigated cases, the nitrogen atoms in aspartate, glutamate, carbamoyl phosphate, and glutamate-5-semialdehyde do not get significant labeling from $^{15}N_4$-Arg, which can be confirmed by experimental data. In addition, the singly labeled forms of arginine, ornithine, citrulline, and argininosuccinate can be ignored, which can also be confirmed by experimental data.

In the procedure below, we determine the relative flux ratios of reactions producing each of the following compounds: citrulline, ornithine, arginine, and argininosuccinate. Here F_{enzyme} represents the flux of reaction catalyzed by indicated enzyme.

3.7.1 Intracellular Citrulline

The labeling pattern of citrulline is dictated by the fluxes from ornithine (F_{OTC}) and arginine (F_{NOS}) (Fig. 1). Under metabolic steady state, the total flux out of citrulline equals the flux into citrulline:

$$F_{\text{ASS}} + F_{\text{CitEx}} = F_{\text{OTC}} + F_{\text{NOS}} \tag{10}$$

The following series of equations balance the influx and outflux of each labeled form of citrulline under isotopic steady state (*see* **Note 15**):

$$X_{\text{Cit-0}}(F_{\text{ASS}} + F_{\text{CitEx}}) = X_{\text{Orn-0}}F_{\text{OTC}} + X_{Arg-0}F_{\text{NOS}} \tag{11}$$

$$X_{\text{Cit-2}}(F_{\text{ASS}} + F_{\text{CitEx}}) = X_{\text{Orn-2}}F_{\text{OTC}} + (X_{Arg-2} + .5X_{Arg-3})F_{\text{NOS}} \tag{12}$$

$$X_{\text{Cit-3}}(F_{\text{ASS}} + F_{\text{CitEx}}) = (.5X_{Arg-3} + X_{Arg-4})F_{\text{NOS}} \tag{13}$$

To simplify the process of solving these equations, we create a matrix form equivalent to the series of equations above. Define R_{OTC} as the relative flux from ornithine into citrulline and R_{NOS} as the relative flux from arginine into citrulline, from Eq. 10:

$$R_{\text{OTC}} = \frac{F_{\text{OTC}}}{F_{\text{OTC}} + F_{\text{NOS}}} = \frac{F_{\text{OTC}}}{F_{\text{ASS}} + F_{\text{CitEx}}} \tag{14}$$

$$R_{\text{NOS}} = \frac{F_{\text{NOS}}}{F_{\text{OTC}} + F_{\text{NOS}}} = \frac{F_{\text{NOS}}}{F_{\text{ASS}} + F_{\text{CitEx}}} \tag{15}$$

Thus, equation set 11–13 is equivalent to:

$$X_{\text{Cit-0}} = X_{\text{Orn-0}} R_{\text{OTC}} + X_{Arg-0} R_{\text{NOS}} \tag{16}$$

$$X_{\text{Cit-2}} = X_{\text{Orn-2}} R_{\text{OTC}} + (X_{Arg-2} + .5X_{Arg-3}) R_{\text{NOS}} \tag{17}$$

$$X_{\text{Cit-3}} = (.5X_{Arg-3} + X_{Arg-4}) R_{\text{NOS}} \tag{18}$$

Or expressed in matrix form:

$$\begin{bmatrix} X_{\text{Cit-0}} \\ X_{\text{Cit-2}} \\ X_{\text{Cit-3}} \end{bmatrix} = \begin{bmatrix} X_{\text{Orn-0}} & X_{Arg-0} \\ X_{\text{Orn-2}} & X_{Arg-2} + .5X_{Arg-3} \\ 0 & .5X_{Arg-3} + X_{Arg-4} \end{bmatrix} \times \begin{bmatrix} R_{\text{OTC}} \\ R_{\text{NOS}} \end{bmatrix} \tag{19}$$

R_{OTC} and R_{NOS} can be solved via matrix division.

3.7.2 Intracellular Argininosuccinate

Following the same workflow as above with citrulline, the following equations describe the mass balance of total argininosuccinate and each measured labeled form of argininosuccinate (Fig. 1).

$$F_{\text{ASL+}} = F_{\text{ASS}} + F_{\text{ASL-}} \tag{20}$$

$$X_{\text{Argsucc-0}} F_{\text{ASL+}} = F_{\text{ASL-}} X_{Arg-0} + F_{\text{ASS}} X_{\text{Cit-0}} \tag{21}$$

$$X_{\text{Argsucc-2}} F_{\text{ASL+}} = F_{\text{ASL-}} X_{Arg-2} + F_{\text{ASS}} X_{\text{Cit-2}} \tag{22}$$

$$X_{\text{Argsucc-3}} F_{\text{ASL+}} = F_{\text{ASL-}} X_{Arg-3} + F_{\text{ASS}} X_{\text{Cit-3}} \tag{23}$$

$$X_{\text{Argsucc-4}} F_{\text{ASL+}} = F_{\text{ASL-}} X_{Arg-4} \tag{24}$$

In the equivalent matrix form:

$$\begin{bmatrix} X_{\text{Argsucc-0}} \\ X_{\text{Argsucc-2}} \\ X_{\text{Argsucc-3}} \\ X_{\text{Argsucc-4}} \end{bmatrix} = \begin{bmatrix} X_{Arg-0} & X_{\text{Cit-0}} \\ X_{Arg-2} & X_{\text{Cit-2}} \\ X_{Arg-3} & X_{\text{Cit-3}} \\ X_{Arg-4} & 0 \end{bmatrix} \times \begin{bmatrix} R_{\text{ASL-}} \\ R_{\text{ASS}} \end{bmatrix} \tag{25}$$

where:

$$R_{\text{ASL-}} = \frac{F_{\text{ASL-}}}{F_{\text{ASL+}}} \tag{26}$$

$$R_{\text{ASS}} = \frac{F_{\text{ASS}}}{F_{\text{ASL+}}} \tag{27}$$

The relative flux ratios into argininosuccinate from arginine ($R_{\text{ASL-}}$) and citrulline (R_{ASS}) can be solved via matrix division from measured citrulline, intracellular arginine, and argininosuccinate labeling patterns.

3.7.3 Intracellular Arginine

To balance the influx and outflux of total intracellular arginine (Fig. 1):

$$F_{\mathrm{ARG}} + F_{\mathrm{GATM}} + F_{\mathrm{NOS}} + F_{\mathrm{ASL-}} = F_{\mathrm{ArgIn}} + F_{\mathrm{ASL+}} \tag{28}$$

To balance the influx and outflux of each labeled form of intracellular arginine:

$$(F_{\mathrm{ARG}} + F_{\mathrm{GATM}} + F_{\mathrm{NOS}} + F_{\mathrm{ASL-}})X_{Arg-0} = F_{\mathrm{ArgIn}} X_{\mathrm{ECArg-0}} + F_{\mathrm{ASL+}} X_{\mathrm{Argsucc-0}} \tag{29}$$

$$(F_{\mathrm{ARG}} + F_{\mathrm{GATM}} + F_{\mathrm{NOS}} + F_{\mathrm{ASL-}})X_{Arg-2} = F_{\mathrm{ArgIn}} X_{\mathrm{ECArg-2}} + F_{\mathrm{ASL+}} X_{\mathrm{Argsucc-2}} \tag{30}$$

$$(F_{\mathrm{ARG}} + F_{\mathrm{GATM}} + F_{\mathrm{NOS}} + F_{\mathrm{ASL-}})X_{Arg-3} = F_{\mathrm{ArgIn}} X_{\mathrm{ECArg-3}} + F_{\mathrm{ASL+}} X_{\mathrm{Argsucc-3}} \tag{31}$$

$$(F_{\mathrm{ARG}} + F_{\mathrm{GATM}} + F_{\mathrm{NOS}} + F_{\mathrm{ASL-}})X_{Arg-4} = F_{\mathrm{ArgIn}} X_{\mathrm{ECArg-4}} + F_{\mathrm{ASL+}} X_{\mathrm{Argsucc-4}} \tag{32}$$

In the equivalent matrix form:

$$\begin{bmatrix} X_{Arg-0} \\ X_{Arg-2} \\ X_{Arg-3} \\ X_{Arg-4} \end{bmatrix} = \begin{bmatrix} X_{\mathrm{ECArg-0}} & X_{\mathrm{Argsucc-0}} \\ X_{\mathrm{ECArg-2}} & X_{\mathrm{Argsucc-2}} \\ X_{\mathrm{ECArg-3}} & X_{\mathrm{Argsucc-3}} \\ X_{\mathrm{ECArg-4}} & X_{\mathrm{Argsucc-4}} \end{bmatrix} \times \begin{bmatrix} R_{\mathrm{ArgIn}} \\ R_{\mathrm{ASL+}} \end{bmatrix} \tag{33}$$

where flux ratios:

$$R_{\mathrm{ArgIn}} = \frac{F_{\mathrm{ArgIn}}}{F_{\mathrm{ARG}} + F_{\mathrm{GATM}} + F_{\mathrm{NOS}} + F_{\mathrm{ASL-}}} \tag{34}$$

$$R_{\mathrm{ASL+}} = \frac{F_{\mathrm{ASL+}}}{F_{\mathrm{ARG}} + F_{\mathrm{GATM}} + F_{\mathrm{NOS}} + F_{\mathrm{ASL-}}} \tag{35}$$

The relative flux ratios into intracellular arginine from extracellular arginine (R_{ArgIn}) and argininosuccinate ($R_{\mathrm{ASL+}}$) can be solved using the measured labeling patterns of these compounds and matrix division of Eq. 33 (*see* **Note 16**).

3.7.4 Intracellular Ornithine (See Note 17)

To balance the influx and outflux of total ornithine:

$$F_{\mathrm{OTC}} + F_{\mathrm{ODC}} + F_{\mathrm{OrnEx}} = F_{\mathrm{ARG}} + F_{\mathrm{GATM}} + F_{\mathrm{OAT}} \tag{36}$$

To balance the influx and outflux of each labeled form of ornithine (*see* **Note 18**):

$$(F_{\mathrm{OTC}} + F_{\mathrm{ODC}} + F_{\mathrm{OrnEx}})X_{\mathrm{Orn-0}} = (F_{\mathrm{ARG}} + F_{\mathrm{GATM}})X_{Arg-0} + F_{\mathrm{OAT}} \times 1 \tag{37}$$

$$(F_{\mathrm{OTC}} + F_{\mathrm{ODC}} + F_{\mathrm{OrnEx}})X_{\mathrm{Orn-2}} = (F_{\mathrm{ARG}} + F_{\mathrm{GATM}})(X_{Arg-2} + X_{Arg-3} + X_{Arg-4}) \tag{38}$$

In the equivalent matrix form:

$$\begin{bmatrix} X_{\mathrm{Orn-0}} \\ X_{\mathrm{Orn-2}} \end{bmatrix} = \begin{bmatrix} X_{\mathrm{arg-0}} & 1 \\ X_{Arg-2} + X_{Arg-3} + X_{Arg-4} & 0 \end{bmatrix} \times \begin{bmatrix} R_{\mathrm{ARG+GATM}} \\ R_{\mathrm{OAT}} \end{bmatrix} \tag{39}$$

where:

$$R_{\text{ARG+GATM}} = \frac{F_{\text{ARG}} + F_{\text{GATM}}}{F_{\text{OTC}} + F_{\text{ODC}} + F_{\text{OrnEx}}} \tag{40}$$

$$R_{\text{OAT}} = \frac{F_{\text{OAT}}}{F_{\text{OTC}} + F_{\text{ODC}} + F_{\text{OrnEx}}} \tag{41}$$

The relative flux ratios into ornithine from arginine ($R_{\text{ARG + GATM}}$) and glutamate-5-semialdehyde and glutamate (R_{OAT}) can be solved via matrix division.

3.8 [Optional] Calculate Exchange Fluxes

3.8.1 Determining F_{ArgIn}

F_{ArgIn} corresponds to the uptake of arginine from the media. Measure this flux via HPLC-MS analysis of arginine in fresh and spent media. Because the concentration of arginine in the fresh media is known from the standard formulation, we can compare the total ion count (all labeled forms) of fresh and spent media to calculate the amount taken up by the cells. Normalize to the duration the collected media were incubated with the cells and to either protein content (μg) or the number of cells per well.

$$F_{\text{ArgIn}}\left(\frac{\mu\text{mol}}{\text{Hour} \times \mu g \text{ protein}}\right) = \frac{\dfrac{\text{Fresh media total arginine ion count} - \text{spent media total arginine ion count}}{\text{Fresh media total arginine ion count}} \times \mu M \text{ arginine in fresh media} \times \text{liter of media per well}}{\text{Hours of incubation} \times \mu g \text{ protein per well}} \tag{42}$$

3.8.2 Determining F_{NOS}

F_{NOS} can be determined by the production of nitric oxide by the cells using a Griess assay on the spent media (*see* **Note 19**). Follow manufacturer's protocol to determine concentration of nitric oxide in sample and use the following equation to calculate F_{NOS}. Normalize to protein content or cell number.

$$F_{\text{NOS}}\left(\frac{\mu\text{mol}}{\text{Hour} \times \mu g \text{ protein}}\right) = \frac{\mu M \text{ nitric oxide} \times \text{liters of media in each well}}{\text{Hours of incubation} \times \mu g \text{ protein per well}} \tag{43}$$

3.8.3 Determining F_{ARG}

F_{ARG} flux corresponds to urea production and excretion from the cell. We can calculate this flux either via HPLC-MS analysis of urea content in the media (Subheadings 3.4 and 3.6, **step 7/8**) or, if urea production is sizable enough, via a urea assay kit following manufacturer's protocol.

$$F_{\text{ARG}}\left(\frac{\mu\text{mol}}{\text{Hour} \times \mu g \text{ protein}}\right) = \frac{\mu M \text{ urea in media} \times \text{Liters of media in each well}}{\text{Hours of incubation} \times \mu g \text{ protein per well}} \tag{44}$$

3.8.4 Determining F_{CitEx} and F_{OrnEx}

Secretion rate of citrulline (F_{CitEx}) and ornithine (F_{OrnEx}) can be measured by quantifying the concentration of citrulline and ornithine in spent media using LC-MS analysis with internal or external standards (Subheadings 3.4 and 3.6, **step 7/8**).

$$F_{\text{CitEx or OrnEx}}\left(\frac{\mu\text{mol}}{\text{Hour}\times\mu g\text{ protein}}\right)=\frac{\mu M\text{ citrulline or ornithine in spent media }\times\text{Liters of media per well}}{\text{Hours of incubation}\times\mu g\text{ protein per well}} \quad (45)$$

3.9 [Optional] Determining Absolute Intracellular Flux

Absolute intracellular fluxes can be determined using the calculated relative flux ratios (obtained in Subheading 3.7) and the exchange fluxes (obtained in Subheading 3.8). Below is an example of this process (*see* **Note 20**).

Because we have measured F_{NOS} and F_{CitEx} (Eqs. 43 and 45, respectively) and determined R_{NOS} and R_{OTC} from fitting the labeling of citrulline (Subheading 3.7, Eq. 19) from Eqs. 14 and 15, we can calculate F_{OTC} and F_{ASS}:

$$F_{OTC}=\frac{R_{OTC}}{R_{NOS}}F_{NOS} \quad (46)$$

$$F_{ASS}=\frac{F_{NOS}}{R_{NOS}}-F_{CitEx} \quad (47)$$

Next, because we have calculated F_{ASS} (Eq. 47), R_{ASS} and R_{ASL-} (Eq. 25), from Eqs. 26 and 27, we can determine F_{ASL+} and F_{ASL-}

$$F_{ASL+}=\frac{F_{ASS}}{R_{ASS}} \quad (48)$$

$$F_{ASL-}=F_{ASS}\times\frac{R_{ASL-}}{R_{ASS}} \quad (49)$$

Having now calculated F_{ASL-} and F_{ASL+} and having previously determined F_{NOS} (Eq. 43) and R_{ASL+} (Eq. 33), from Eq. 35, we can calculate (F_{ARG} + F_{GATM})

$$F_{ARG}+F_{GATM}=\frac{F_{ASL+}}{R_{ASL+}}-\left(F_{NOS}+F_{ASL-}\right) \quad (50)$$

Alternatively, if F_{ArgIn} (Eq. 42) and R_{ArgIn} (Eq. 33) can be determined more accurately, we can determine F_{ARG} + F_{GATM} from Eq. 34 using previously calculated F_{NOS} (Eq. 43) and F_{ASL-} (Eq. 49):

$$F_{ARG}+F_{GATM}=\frac{F_{ArgIn}}{R_{ArgIn}}-\left(F_{NOS}+F_{ASL-}\right) \quad (51)$$

Because F_{ARG} can be measured by urea production (Eq. 44), F_{GATM} can be calculated using calculated (F_{Arg} + F_{GATM}) (Eq. 50 or 51) as such:

$$F_{GATM} = \left(F_{ARG} + F_{GATM}\right) - F_{ARG} \tag{52}$$

As F_{OrnEx} (Eq. 45), F_{OTC} (Eq. 46), F_{ARG} + F_{GATM} (Eq. 50 or 51), and $R_{ARG+GATM}$ (Eq. 39) have been determined F_{ODC} can be determined from Eq. 40 as:

$$F_{ODC} = \frac{F_{ARG} + F_{GATM}}{R_{ARG+GATM}} - \left(F_{OTC} + F_{OrnEx}\right) \tag{53}$$

Lastly, with calculated R_{OAT} (Eq. 39), $R_{ARG+GATM}$ (Eq. 39) and F_{ARG} + F_{GATM} (Eqs. 50 and 51), F_{OAT} can be determined from Eqs. 40 and 41:

$$F_{OAT} = \left(F_{ARG} + F_{GATM}\right)\frac{R_{OAT}}{R_{ARG+GATM}} \tag{54}$$

4 Notes

1. Any growth media, appropriate for the cell type can be used as long as an arginine-free equivalent can be obtained.
2. In order to obtain consistent quantitation of metabolites and achieve better enrichment of isotopic labeling, it is important to use dialyzed serum, as undialyzed serum contains various metabolites, including arginine.
3. There exist numerous labeled compounds for utilization in a metabolic flux experiment, and the specific research question will dictate the best choice. In general, a labeled compound that would allow best differentiation among various branches of the metabolic pathway in question is a good choice. In this illustrated example, we use $^{15}N_4$-L-arginine. *See* Fig. 1 for illustration of how this compound allows for differentiation between branches of arginine metabolism.
4. In this example, cells were plated at a density of 3×10^5 cells per well in 6-well plates. Make sure that cells are not over-confluent.
5. In order to activate Raw 264.7 macrophages to classical activation state (M1), cells were stimulated with 50 ng/mL LPS and 10 ng/μL IFN-γ for 24 h.
6. We aim to incubate cells with labeled tracer for a duration that allows for achievement of isotopic steady state by the time of collection (labeling pattern of compound of interest is stable). We also aim to perform experiments when cells are grown in

metabolic pseudo-steady state. Pseudo-steady state implies that changes in metabolite levels and fluxes are negligible during the time they are being measured. Both of these will vary depending on the system being studied and may require additional experiments to ensure these assumptions are correct. In our system, we stimulate macrophages for 24 h and label for 6 h because previous data have suggested that total metabolic pools were relatively stable during the 6-h period and that 6 h was sufficient to reach steady isotopic labeling.

7. Choose a volume of media that is sufficient to adequately cover the cells while also small enough to reveal differences in arginine uptake and urea, ornithine, and citrulline secretion.
8. Move as quickly as possible during intracellular metabolite extraction. This will help minimize alteration or degradation of metabolites in the sample.
9. For most accurate determination of absolute metabolite concentration, internal standards should be spiked in at a quantity close to the endogenous pool. A couple of trials may be necessary to achieve this.
10. Fifteen microliter was injected for both intracellular and extracellular samples. Adjust this as needed to ensure best quantitation. If calibration curve is performed (Subheading 3.4, **step 2**), refer to the curves and aim to get signal from sample within linear range of detection.
11. Multiple LC-MS methods are available using a variety of systems. Depending on the available system, some parameters may have to be adjusted to ensure high-quality data [3, 17–19].
12. The equations shown (Eqs. 2–6) are for those compounds which contain up to 4 nitrogen atoms. For those with fewer nitrogen atoms, use only those that apply to that compound (e.g., for ornithine only use 2–4). This adjustment can also be done using a variety of software packages adjusting for natural abundance such as IsoCor [16], Corrector [20], and Isotope Correction Toolbox [21].
13. Note that standards can fall below detection at low concentration or can become saturated at high concentration; adjust standard curve to include only the linear range and aim for this range when analyzing samples.
14. Extracellular arginine is not fully labeled due to impurities in the $^{15}N_4$ isotopic tracer and potential contamination due to incomplete aspiration of unlabeled media prior to application of labeled media.
15. The two guanidine amino nitrogen atoms (purple in Fig. 1) are equivalent. Therefore, there are two equivalent isotopomers that contribute to the pool of 3-labeled arginine.

In the reaction catalyzed by NOS, there is a 50% chance to lose the labeled nitrogen as NO from 3-labeled arginine. To account for this, in the flux equations we adjust the $X_{\text{Arg-3}}$ factor by 0.5.

16. If ASL reaction is highly reversible, labeling of intracellular arginine and argininosuccinate would be very similar, and the gross flux of ASL at each direction cannot be accurately determined. In this case, intracellular arginine and argininosuccinate can be treated as one node for the fitting, and net flux of ASL is equal to the flux of ASS.
17. If OAT reaction is reversible, glutamate-5-semialdehyde can be labeled from ^{15}N-labeled arginine, and ornithine can be singly labeled. However, as in most standard culturing conditions, the contribution of reverse OAT flux to the total production of glutamate is minimal; glutamate labeling from arginine is usually negligible, which can be easily verified by experimental data. If experimental measurement reveals glutamate-5-semialdehyde (X_{G5S}) labeling is significant, Eqs. 37 and 38 should be adjusted to:

$$(F_{\text{OTC}} + F_{\text{ODC}} + F_{\text{OrnEx}} + F_{\text{OAT-}})X_{\text{Orn-0}} = (F_{\text{ARG}} + F_{\text{GATM}})X_{Arg-0} + F_{\text{OAT}}X_{\text{G5S-0}} \tag{55}$$

$$(F_{\text{OTC}} + F_{\text{ODC}} + F_{\text{OrnEx}} + F_{\text{OAT-}})X_{\text{Orn-1}} = (F_{\text{ARG}} + F_{\text{GATM}})X_{Arg-1} + F_{\text{OAT}}X_{\text{G5S-1}} \tag{56}$$

$$(F_{\text{OTC}} + F_{\text{ODC}} + F_{\text{OrnEx}} + F_{\text{OAT-}})X_{\text{Orn-2}} = (F_{\text{ARG}} + F_{\text{GATM}})(X_{Arg-2} + X_{Arg-3} + X_{Arg-4}) \tag{57}$$

Here $X_{\text{G5S-0}}$ and $X_{\text{G5S-1}}$ represent fraction of glutamate-5-semialdehyde with 0- or 1-labeled nitrogens. And F_{OAT} represent gross OAT flux from glutamate-5-semialdehyde to ornithine; $F_{\text{OAT-}}$ represent gross reverse OAT flux. Equation 36 also needs to be modified to include reverse OAT flux. In this case equations balancing the labeling of arginine, citrulline, and arginosuccinate must be modified accordingly to include 1-labeled forms labeled on α-nitrogen (orange in Fig. 1). Labeling of aspartate and carbamoyl phosphate from arginine is also highly unlikely and can be experimentally verified.

18. As F_{ARG} and F_{GATM} both contribute to ornithine production with the same nitrogen transfer from arginine to ornithine (Fig. 1), they are combined in the relative flux equations.
19. F_{NOS} can be determined by the production of nitric oxide by the cells only if NOS is the only or dominant reaction that produces nitric oxide. Verify this before utilizing this test.
20. Here we show an example process for determining absolute intracellular fluxes using previously determined relative flux ratios (R) and exchange fluxes. In order to determine absolute intracellular fluxes in Fig. 1, not all of the exchange fluxes we

have outlined in Subheading 3.8 are required. Additionally, depending on the quality of data and extent of labeling, the relative flux ratios obtained in Subheading 3.7 have different accuracies. In general, fluxes with a very small relative contribution cannot be determined very accurately. When calculating intracellular fluxes, try to use exchange fluxes and flux ratios with the highest confidence as much as possible.

References

1. Kelly B, O'Neill LA (2015) Metabolic reprogramming in macrophages and dendritic cells in innate immunity. Cell Res 25(7):771–784. https://doi.org/10.1038/cr.2015.68
2. Hesse M, Modolell M, La Flamme AC, Schito M, Fuentes JM, Cheever AW, Pearce EJ, Wynn TA (2001) Differential regulation of nitric oxide synthase-2 and arginase-1 by type 1/type 2 cytokines in vivo: granulomatous pathology is shaped by the pattern of L-arginine metabolism. J Immunol 167(11):6533–6544
3. Geiger R, Rieckmann JC, Wolf T, Basso C, Feng Y, Fuhrer T, Kogadeeva M, Picotti P, Meissner F, Mann M, Zamboni N, Sallusto F, Lanzavecchia A (2016) L-arginine modulates T cell metabolism and enhances survival and antitumor activity. Cell 167(3):829–842.e813. https://doi.org/10.1016/j.cell.2016.09.031
4. Dillon BJ, Prieto VG, Curley SA, Ensor CM, Holtsberg FW, Bomalaski JS, Clark MA (2004) Incidence and distribution of argininosuccinate synthetase deficiency in human cancers: a method for identifying cancers sensitive to arginine deprivation. Cancer 100(4):826–833. https://doi.org/10.1002/cncr
5. McAlpine JA, Lu HT, Wu KC, Knowles SK, Thomson JA (2014) Down-regulation of argininosuccinate synthetase is associated with cisplatin resistance in hepatocellular carcinoma cell lines: implications for PEGylated arginine deiminase combination therapy. BMC Cancer 14:621. https://doi.org/10.1186/1471-2407-14-621
6. Ensor CM, Holtsberg FW, Bomalaski JS, Clark MA (2002) Pegylated arginine deiminase (ADI-SS PEG20,000 mw) inhibits human melanomas and hepatocellular carcinomas in vitro and in vivo. Cancer Res 62(19):5443–5450
7. Shen LJ, Lin WC, Beloussow K, Shen WC (2003) Resistance to the anti-proliferative activity of recombinant arginine deiminase in cell culture correlates with the endogenous enzyme, argininosuccinate synthetase. Cancer Lett 191(2):165–170
8. Moncada S, Higgs A (1993) The L-arginine-nitric oxide pathway. N Engl J Med 329(27):2002–2012. https://doi.org/10.1056/nejm199312303292706
9. Zhang C, Hein TW, Wang W, Miller MW, Fossum TW, McDonald MM, Humphrey JD, Kuo L (2004) Upregulation of vascular arginase in hypertension decreases nitric oxide-mediated dilation of coronary arterioles. Hypertension 44(6):935–943. https://doi.org/10.1161/01.HYP.0000146907.82869.f2
10. Morris CR, Poljakovic M, Lavrisha L, Machado L, Kuypers FA, Morris SM Jr (2004) Decreased arginine bioavailability and increased serum arginase activity in asthma. Am J Respir Crit Care Med 170(2):148–153. https://doi.org/10.1164/rccm.200309-1304OC
11. Antoniewicz MR (2015) Methods and advances in metabolic flux analysis: a mini-review. J Ind Microbiol Biotechnol 42(3):317–325. https://doi.org/10.1007/s10295-015-1585-x
12. Zamboni N (2011) 13C metabolic flux analysis in complex systems. Curr Opin Biotechnol 22(1):103–108. https://doi.org/10.1016/j.copbio.2010.08.009
13. Wittmann C, Heinzle E (1999) Mass spectrometry for metabolic flux analysis. Biotechnol Bioeng 62(6):739–750
14. Bennett BD, Yuan J, Kimball EH, Rabinowitz JD (2008) Absolute quantitation of intracellular metabolite concentrations by an isotope ratio-based approach. Nat Protoc 3(8):1299–1311. https://doi.org/10.1038/nprot.2008.107
15. Clasquin MF, Melamud E, Rabinowitz JD (2012) LC-MS data processing with MAVEN: a metabolomic analysis and visualization engine. Curr Protoc Bioinformatics 14:Unit14.11. https://doi.org/10.1002/0471250953.bi1411s37
16. Millard P, Letisse F, Sokol S, Portais JC (2012) IsoCor: correcting MS data in isotope labeling experiments. Bioinformatics 28(9):1294–1296. https://doi.org/10.1093/bioinformatics/bts12

17. Monticelli LA, Buck MD, Flamar AL, Saenz SA, Tait Wojno ED, Yudanin NA, Osborne LC, Hepworth MR, Tran SV, Rodewald HR, Shah H, Cross JR, Diamond JM, Cantu E, Christie JD, Pearce EL, Artis D (2016) Arginase 1 is an innate lymphoid-cell-intrinsic metabolic checkpoint controlling type 2 inflammation. Nat Immunol 17(6):656–665. https://doi.org/10.1038/ni.3421
18. Shin S, Fung SM, Mohan S, Fung HL (2011) Simultaneous bioanalysis of L-arginine, L-citrulline, and dimethylarginines by LC-MS/MS. J Chromatogr B Analyt Technol Biomed Life Sci 879(7–8):467–474. https://doi.org/10.1016/j.jchromb.2011.01.006
19. Luo X, Gu X, Li L (2017) Development of a simple and efficient method of harvesting and lysing adherent mammalian cells for chemical isotope labeling LC-MS-based cellular metabolomics. Analytica Chimica Acta. Millard P, Letisse F, Sokol S, Portais JC (2012) IsoCor: correcting MS data in isotope labeling experiments. Bioinformatics 28(9):1294–1296. https://doi.org/10.1093/bioinformatics/bts12
20. Huege J, Goetze J, Dethloff F, Junker B, Kopka J (2014) Quantification of stable isotope label in metabolites via mass spectrometry. Methods Mol Biol 1056:213–223. https://doi.org/10.1007/978-1-62703-592-7_20
21. Jungreuthmayer C, Neubauer S, Mairinger T, Zanghellini J, Hann S (2016) ICT: isotope correction toolbox. Bioinformatics 32(1):154–156. https://doi.org/10.1093/bioinformatics/btv514

Chapter 14

Quantifying Intermediary Metabolism and Lipogenesis in Cultured Mammalian Cells Using Stable Isotope Tracing and Mass Spectrometry

Thekla Cordes and Christian M. Metallo

Abstract

Metabolism plays a central role in virtually all diseases, including diabetes, cancer, and neurodegeneration. Detailed analysis is required to identify the specific metabolic pathways dysregulated in the context of a given disease or biological perturbation. Measurement of metabolite concentrations can provide some insights into altered pathway activity or enzyme function, but since most biochemicals are metabolized by various enzymes in distinct pathways within cells and tissues, these approaches are somewhat limited. By applying metabolic tracers to a biological system, one can visualize pathway-specific information depending on the tracer used and analytes measured. To this end, stable isotope tracers and mass spectrometry are emerging as important tools for the examination of metabolic pathways and fluxes in cultured mammalian cells and other systems. Here, we describe a detailed workflow for quantifying metabolic processes in mammalian cell cultures using stable isotopes and gas chromatography coupled to mass spectrometry (GC-MS). As a case study, we apply ^{13}C isotopic labeled glucose and glutamine to a cancer cell line to quantify substrate utilization for TCA metabolism and lipogenesis. Guidelines are also provided for interpretation of data and considerations for application to other cell systems. Ultimately, this approach provides a robust and precise method for quantifying stable isotope labeling in metabolite pools that can be applied to diverse biological systems.

Key words Stable isotope tracer, Metabolite extraction, Metabolic flux, Isotopologue distribution, Metabolism, Gas chromatography mass spectrometry (GC-MS), Isotopomer spectral analysis (ISA), Fragment, TBDMS, MTBSTFA

1 Introduction

Living cells require metabolic activity to generate energy, synthesize biomass, and execute specific biochemical functions. Alterations in nutrient availability, gene expression/function, exposure to toxins, or other stresses contribute to cellular dysfunction and disease, which often results in metabolic reprogramming. Many pharmaceuticals target enzymes that play direct or indirect roles in metabolic pathways; therefore, identifying quantitative differences in pathway activity or flux is critically important for elucidating

Angelo D'Alessandro (ed.), *High-Throughput Metabolomics: Methods and Protocols*, Methods in Molecular Biology, vol. 1978, https://doi.org/10.1007/978-1-4939-9236-2_14,

mechanisms that can be exploited therapeutically. As such, interest in metabolic flux analysis (MFA) has increased significantly within the biomedical research community in the past several decades. This approach and related methods require a command of cell biology, analytical chemistry, and biochemical engineering principles to resolve fluxes in a given system [1].

The dynamic nature of metabolism can be quantitatively assessed with stable, nonradioactive isotope tracers in conjunction with a wide variety of analytical platforms [2, 3]. For example, mass spectrometry is typically combined with gas chromatography (GC-MS), liquid chromatography (LC-MS), or capillary electrophoresis (CE-MS) to quantify abundance and isotope enrichment in specific metabolites [4, 5]. Exposure of living cell cultures to stable isotopic tracer leads to an incorporation of isotopes into molecules within downstream biochemical pathways, providing quantitative information on the origin and turnover of a given metabolite. Tracer choice (e.g., substrates labeled with ^{13}C, ^{15}N, and/or ^{2}H) directly impacts how and if a given metabolite will be labeled and can also impact flux resolution. Since mammalian cells consume diverse nutrients and various tracers are often available, the specific tracer(s) used will focus the information provided and hypothesis to be tested [6–8]. Detailed consideration prior to executing studies is recommended.

Here we provide a detailed workflow for quantifying substrate utilization associated with TCA metabolism and de novo lipogenesis in mammalian cell cultures. As a case study, we apply uniformly labeled [U-$^{13}C_6$]glucose and [U-$^{13}C_5$]glutamine to the A549 non-small cell lung carcinoma cell line to quantify carbon utilization for TCA cycle metabolism and lipogenesis. We include protocols for media preparation, metabolite extraction, and GC-MS-based quantitation of isotope enrichment of intracellular metabolites and those present in culture media. The latter allows for quantification of cellular uptake and secretion fluxes. Finally, we provide an overview of tools available for data analysis and visualization. We have applied this workflow to various cell systems, including cancer cell lines [9–12], primary neurons [13], immune cells [14], pluripotent stem cells [15], primary skeletal myotubes [16], and adipocytes [17], highlighting the versatility of this method. This high content data provides insights into metabolic fluxes important for elucidating the regulation and function of metabolic pathways and understanding human physiology [1, 18, 19].

2 Materials

2.1 List of Chemicals

1. 2,2,2-trifluoro-*N*-methyl-*N*-trimethylsilyl-acetamide (MSTFA) (Macherey-Nagel, 24589-78-4).
2. Chloroform Chromasolv® for HPLC, equilibrated to −20 °C.

3. Dialyzed fetal bovine serum (dFBS).
4. DMEM without glucose and glutamine (Sigma-Aldrich, D5030) (*see* **Note 1**).
5. Glucose (unlabeled or naturally labeled).
6. Glutamine (unlabeled or naturally labeled).
7. Hexane Chromasolv® for HPLC.
8. Internal standard (e.g., norvaline (Sigma-Aldrich, N7502), [U-$^{2}H_{31}$]palmitate (D31) (Cambridge Isotope Laboratories, CLM-215-1)).
9. Methanol Chromasolv® for HPLC, equilibrated to −80 °C.
10. Methoxyamine hydrochloride (MOX) (Supelco, 33045-U).
11. Milli-Q-purified water or equivalently pure water.
12. *N*-tertbutyldimethylsilyl-*N*-methyltrifluoroacetamide (MTBSTFA) with 1% tert-butyldimethylchlorosilane (Regis Technologies, 1-270143-200).
13. Phosphate-buffered saline (PBS).
14. Pyridine.
15. Sodium bicarbonate ($NaHCO_3$).
16. Sodium chloride (NaCl).
17. Sulfuric acid (H_2SO_4).
18. [U-$^{13}C_5$]glutamine (Cambridge Isotope Laboratories, CML-1822).
19. [U-$^{13}C_6$]glucose (Cambridge Isotope Laboratories, CLM-1396).

2.2 List of Instruments and Materials

1. 1.5 ml microcentrifuge tubes (e.g., Eppendorf interlocking caps (Cat. No. 22363204)).
2. Analytical glass sample vials (e.g., Agilent 5182-7616) with glass inserts (Agilent 5183-2088) for fatty acid methyl esters (FAMEs).
3. Analytical platform (e.g., gas chromatograph coupled to mass spectrometry).
4. Analytical sample vials with inserts (e.g., Thermo Scientific, C4000-11) for media samples and metabolites.
5. Freeze dryer or rotary vacuum evaporator, e.g., CentriVap Concentrator (Labconco), pre-cooled to 4 °C.
6. Heating block for microcentrifuge tubes.
7. Microcentrifuge tube shaker, pre-cooled to 4 °C.
8. Nitrogen/air flow evaporator for microcentrifuge tubes.
9. Refrigerated table centrifuge for microcentrifuge tubes, pre-cooled to 4 °C.

10. Sterile filter unit to sterilize 500 ml medium.
11. Syringe filter with sterile syringe for small volumes of medium.

2.3 Reagents and Solutions

Use Milli-Q-purified or equivalently pure water and analytical grade reagents in all recipes and protocol steps.

1. 2% (v/v) sulfuric acid (H_2SO_4) in methanol: Mix 98 ml methanol and 2 ml anhydrous sulfuric acid in a glass bottle. Store at 4 °C for up to four weeks.
2. Chloroform with internal standard [U-$^2H_{31}$]palmitate (D31) for organic metabolite fraction: Mix 40 μl palmitate D31 (stock 10 mM in methanol) with 20 ml chloroform to prepare a 20 μM solution. Store at −20 °C.
3. Extraction fluid for media metabolites: Prepare a mixture of 8 parts methanol and 1 part water. Store at −20 °C.
4. Methoxyamine hydrochloride (MOX) reagent 2% (w/v): Prepare a 2% MOX solution in pyridine by dissolving 20 mg MOX in 1 ml pyridine.
5. NaCl 0.9% (w/v): Dissolve 4.5 g NaCl in 500 ml deionized water. Store at room temperature.
6. Norvaline (stock 1 mg/ml): Dissolve 10 mg norvaline powder in 10 ml water for a 1 mg/ml stock solution. Store at −20 °C.
7. Saturated sodium chloride (NaCl) solution: Dissolve 15 mg NaCl with 20 ml water and vortex for 2 min. If salt does not dissolve after mixing, the solution has become saturated. Store at room temperature.
8. Water with internal standard norvaline for aqueous metabolite fraction: Mix 200 μl norvaline (stock 1 mg/ml) with 40 ml water to prepare a 5 μg/ml norvaline solution. Store at 4 °C (*see* **Note 2**).

3 Methods

A detailed workflow for stable isotope tracing in cultured mammalian cells (e.g., A549 non-small cell lung carcinoma cell line (ATCC, CCL-185)) is shown in Fig. 1. We provide protocols for media preparation (Subheading 3.1), cell preparation and medium change (Subheading 3.2), metabolite extraction (Subheading 3.3), derivatization (Subheading 3.4), GC-MS analysis (Subheading 3.5), as well as metabolite identification and data analysis (Subheading 3.6). Protocols and calculations refer to adherent mammalian cell cultures grown on a 12-well tissue culture plate, but volumes may be adjusted accordingly if using different culture formats (*see* **Note 3**).

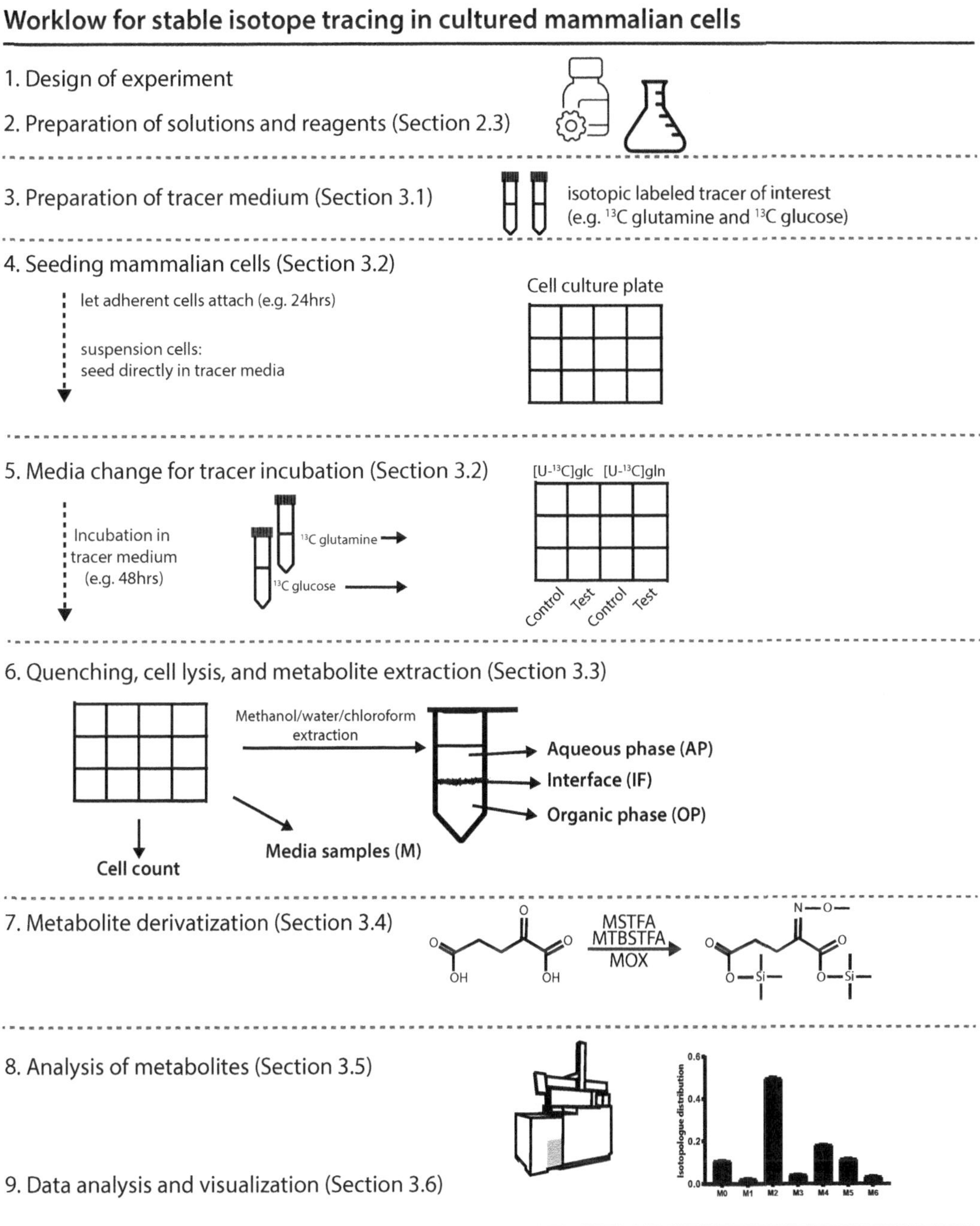

Fig. 1 Workflow for stable isotope tracing in cultured mammalian cells. The scheme summarizes the main procedures described in this chapter, including media preparation, metabolite extraction, and GC-MS-based quantitation intracellular and medium metabolite abundances/enrichments. Metabolite extraction using methanol/water/chloroform will form an upper aqueous phase (AP) (containing hydrophilic metabolites), an interface (IF), and a lower organic phase (OP) (containing more hydrophobic metabolites)

3.1 Media Preparation

This protocol describes the preparation of [U-$^{13}C_6$]glucose and [U-$^{13}C_5$]glutamine Dulbecco's modified Eagle medium (DMEM) containing 25 mM glucose and 4 mM glutamine. In stable isotope tracer experiments, specialized culture medium is used lacking glucose, glutamine, and/or amino acids. Here, we add back labeled ^{13}C substrates to commercially available glucose and glutamine-free DMEM medium (Sigma-Aldrich, D5030) supplemented with 10% dialyzed fetal bovine serum (dFBS) (*see* **Note 4**). Figure 2 depicts a detailed map of carbon atom transitions from [U-$^{13}C_6$] glucose and [U-$^{13}C_5$]glutamine.

3.1.1 Preparation of Glucose and Glutamine-Free DMEM Stock Medium

1. Dissolve DMEM5030 medium powder and 3.7 g $NaHCO_3$ in 1 l Milli-Q water.
2. Adjust pH to 7.4.
3. Sterile filter medium and store at 4 °C until use.

3.1.2 Preparation of [U-$^{13}C_6$]Glucose-Labeled Medium (10 ml)

4. Mix 9 ml DMEM5030 medium (from **step 3**) with 1 ml dFBS to a final concentration of 10% (v/v).
5. Add 25 mM (46.5 mg) [U-$^{13}C_6$]glucose (MW 186 g/mol).
6. Add 4 mM (5.84 mg) unlabeled ^{12}C glutamine (MW 146 g/mol).
7. Adjust pH to 7.4 and sterile filter (*see* **Note 5**).

3.1.3 Preparation of [U-$^{13}C_5$]Glutamine-Labeled Medium (10 ml)

8. Repeat **steps 4–7** but add 25 mM (45 mg) unlabeled glucose (MW 180 g/mol) and 4 mM (6.04 mg) [U-$^{13}C_5$]glutamine (MW 151 g/mol).

3.1.4 Preparation of Unlabeled Medium for Counting Wells (10 ml)

9. Repeat **steps 4–7** but add 25 mM (45 mg) unlabeled glucose (MW 180 g/mol) and 4 mM (5.84 mg) unlabeled ^{12}C glutamine (MW 146 g/mol).

3.2 Cell Preparation and Medium Change to Isotopic Labeled Medium

1. Seed cells into 12-well tissue culture plates in 1 ml growth medium (e.g., DMEM medium containing 25 mM glucose, 4 mM glutamine, and 10% FBS). Six wells will be used for [U-$^{13}C_6$]glucose and six wells for [U-$^{13}C_5$]glutamine trace (Fig. 1) (*see* **Note 6**).
2. Incubate the cells at 37 °C and 5% CO_2 in a humidified cell culture incubator to allow cells to attach to the culture plate (e.g., overnight).
3. Remove the growth medium by aspiration and wash cells once with 1 ml PBS.
4. Add 1 ml prewarmed tracer medium ([U-$^{13}C_6$]glucose and [U-$^{13}C_5$]glutamine-labeled DMEM medium, Subheading 3.1) per well. For counting wells, change medium to 1 ml unlabeled medium (*see* **Note 7**).

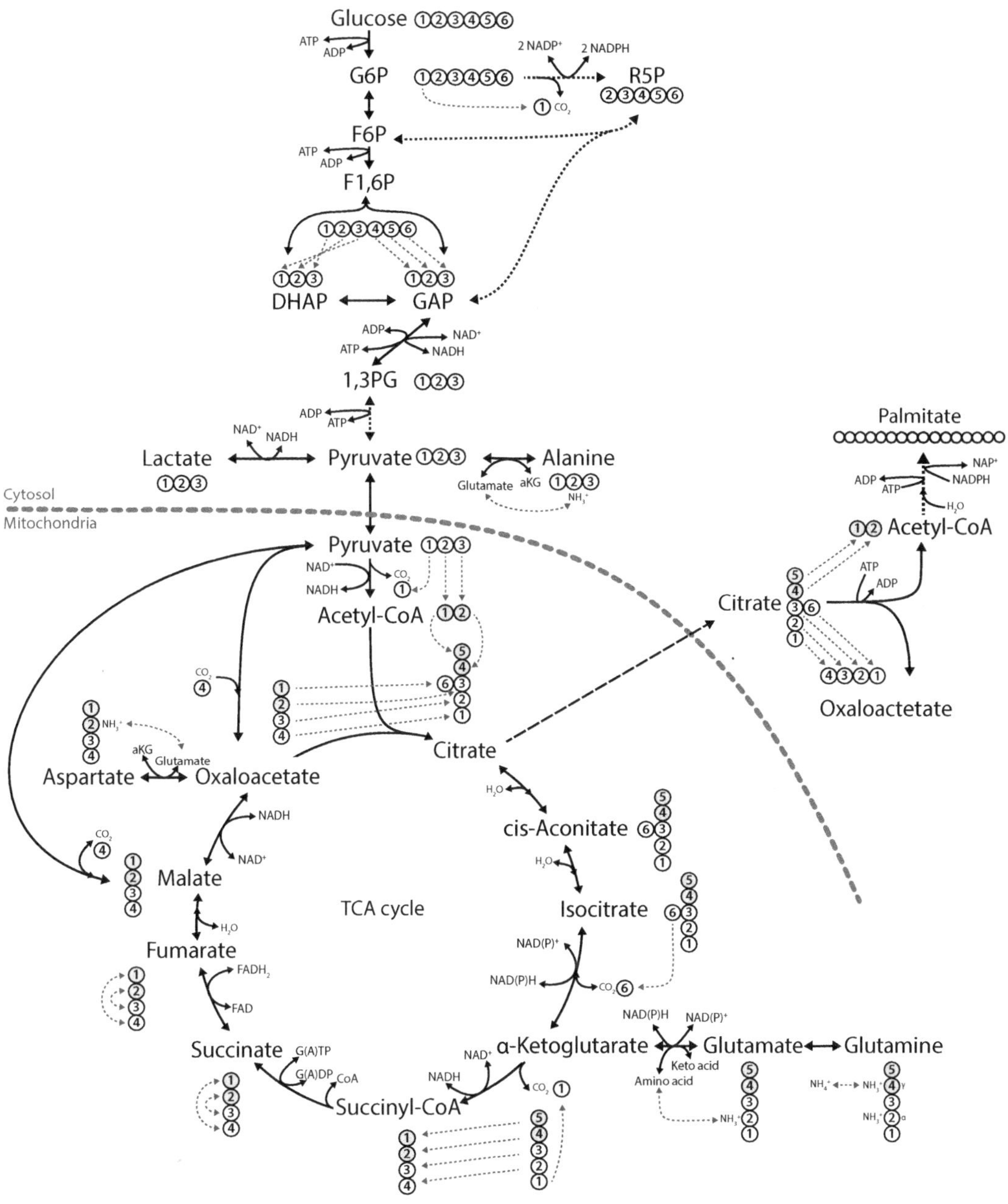

Fig. 2 Metabolic map depicting atom transitions in central carbon metabolism. Labeling patterns arising from symmetry are considered for succinate and fumarate. Carbon atoms entering TCA cycle from acetyl-CoA are depicted in gray. Some compartment-specific reactions are omitted for brevity. *G6P* glucose-6-phosphate, *F6P* fructose-6-phosphate, *F1,6P* fructose-1,6-bisphosphate, *DHAP* dihydroxyacetone phosphate, *GAP* glyceraldehyde-3-phosphate, *1,3PG* 1,3-phosphoglycerate

5. Add 1 ml isotopic labeled medium into one 12-well without cells as control condition for analysis of media secretion and uptake (*see* **Note 8**).
6. Incubate cells in tracer medium at 37 °C and 5% CO_2 in a humidified cell culture incubator for the desired length of time. In the present case study, we incubated cells for 48 h, at which time isotope labeling in glycolysis and the TCA cycle is at steady state (or more likely pseudo-steady state), while lipid labeling is not (*see* **Note 9**).

3.3 Quenching, Cell Lysis, and Metabolite Extraction

This protocol describes a robust method for quenching metabolism, extracting, and separating metabolites from culture media and adherent cells into aqueous and organic/hydrophobic fractions. We employ a modified Bligh and Dyer method using methanol/water/chloroform [20], though various approaches may be effective depending on the metabolite(s) to be studied. Rapid quenching is critical to halt metabolism and minimize residual enzymatic activities that can influence metabolite pool sizes and labeling. To account for potential non-cell-derived metabolites (such as lipid contamination from microcentrifuge tubes), blank extractions from wells without cells may be processed in parallel (*see* **Note 10**).

1. Remove the cell plate from the incubator and collect spent media samples to a new microcentrifuge tube (labeled with medium (M)) and continue immediately with **step 2**.
2. Quickly wash each well once with 1 ml 0.9% NaCl using a 10 ml serological pipette and aspirator (*see* **Note 11**).
3. Place culture plate on ice.
4. Quickly add 250 μl −80 °C methanol to each well to quench metabolism.
5. Add 100 μl ice-cold water containing internal standard norvaline (Subheading 2.3).
6. Scrape cells off the plate.
7. Quickly transfer cell extract from wells into pre-labeled and −20 °C pre-cooled microcentrifuge tubes containing 250 μl chloroform (*see* **Note 12**).
8. Cap each tube and invert three times.
9. Vortex microcentrifuge tube for 5–10 min at 4 °C (*see* **Note 13**).
10. Centrifuge samples for 5–10 min at 21,000 × *g* at 4 °C.
11. Two distinct phases will form upon centrifugation (Fig. 1). The upper aqueous phase (AP) contains hydrophilic metabolites (e.g., organic acids and amino acids), while the lower organic phase (OP) contains more hydrophobic metabolites

such as lipids (with some exceptions). The interface (IF) is a thin layer of proteins, nucleic acids, and other macromolecules that may be utilized if desired to provide additional information for flux calculations [21–23].

3.3.1 Upper Aqueous Phase (AP)

12. Collect 200 μl of upper aqueous phase into an analytical sample vial with insert.
13. Evaporate each sample in a freeze dryer or a rotary vacuum evaporator at 4 °C (*see* **Note 14**).
14. Continue with metabolite derivatization protocols (Subheading 3.4). Alternatively, capped samples can be stored at −80 °C until further use.

3.3.2 Lower Organic Phase (OP)

15. Collect 200 μl of lower organic phase in a new pre-labeled microcentrifuge tube.
16. Evaporate each sample under air flow at room temperature in a fume hood (*see* **Note 15**).
17. Cap samples tightly and continue with metabolite derivatization protocols (Subheading 3.4). Alternatively, capped samples can be stored at −80 °C until further use.

3.3.3 Interface (IF)

18. Add 500 μl methanol to microcentrifuge tube containing interface.
19. Centrifuge at 21,000 × *g* for 5 min at 4 °C.
20. When centrifugation is complete, remove supernatant. The pellet contains the interface that can be stored at −80 °C until further use.

3.3.4 Media Samples (M)

21. Centrifuge samples for 5 min at 300 × *g* at 4 °C and collect the supernatant to a new microcentrifuge tube. Medium can be stored at −20 °C until further use (*see* **Note 16**).
22. Add 90 μl −20 °C extraction fluid for medium metabolites (Subheading 2.3) to a new, pre-cooled microcentrifuge tube on ice.
23. Add 10 μl medium from **step 21**.
24. Vortex for 15 s.
25. Centrifuge for 5 min at 21,000 × *g* at 4 °C.
26. Transfer 60 μl of upper phase into an analytical sample vial with insert.
27. Evaporate samples under vacuum at 4 °C using a freeze dryer or a rotary vacuum evaporator.
28. Store samples at −80 °C or continue with metabolite derivatization (Subheading 3.4) (*see* **Note 17**).

3.4 Metabolite Derivatization

While various MS-based platforms can be effectively used for metabolite quantitation and isotopologue analysis, we predominantly use GC-MS systems. Although derivatization is required (as outlined below), GC-MS offers a cost-effective and robust platform for measurement of amino acids, organic acids, sugars, and fatty acids. Commonly used derivatization reagents for polar metabolites include 2,2,2-trifluoro-*N*-methyl-*N*-trimethylsilyl-acetamide (MSTFA), *N*-tertbutyldimethylsilyl-*N*-methyltrifluoroacetamide (MTBSTFA), and/or methoxyamine hydrochloride (MOX). MSTFA derivatives are preferable for analysis of sugars and non-targeted studies, as more extensive libraries are available. MTBSTFA derivatives are ideally used for targeted analyses of organic and amino acids, as the derivatives are highly stable. Here, we apply MTBSTFA-MOX derivatization for polar metabolites and sulfuric acid/methanol/hexane-based transesterification of fatty acids.

3.4.1 MTBSTFA-MOX Derivatization of Aqueous Intracellular (AP) and Media Metabolites (M)

Metabolite derivatization can be performed by hand using a heating block. Alternatively, reaction steps can be performed by using an automated sample robot (*see* **Note 18**).

1. Add 15 μl MOX reagent (Subheading 2.3) to each dried sample tube and vortex for 5 s (*see* **Note 19**).
2. Incubate at 45 °C for 60 min with shaking.
3. Add 15 μl MTBSTFA to each sample tube.
4. Incubate at 45 °C for 30 min with shaking.
5. Measure samples on analytical platform, e.g., GC-MS (*see* **Note 20**).

3.4.2 Transesterification of Fatty Acids

Preparation of fatty acid methyl esters (FAMEs) is ideal for analysis of free fatty acids.

6. Add 500 μl derivatization reagent (2% (v/v) H_2SO_4 in methanol) (Subheading 2.3) to microcentrifuge tube containing dried, organic phase from metabolite extraction (OP).
7. Incubate at 50 °C for 2 h on a heating block.
8. Add 100 μl saturated NaCl solution (Subheading 2.3).
9. Add 500 μl hexane to each microcentrifuge tube.
10. Vortex samples at room temperature for 2 min. About 1 min after mixing, the phases will be separated.
11. Carefully collect the upper hexane phase (approximately 400 μl) in a new microcentrifuge tube (*see* **Note 21**).
12. Repeat **steps 9–11**, collecting the second extract in the same microcentrifuge tube as **step 11**.
13. Evaporate the combined hexane extracts (from **steps 6** and **7**) at room temperature under nitrogen or airflow until dry (*see* **Note 22**).

14. Add 75 μl of hexane to the microcentrifuge tube and vortex for 10 s to resuspend the dried samples.
15. Transfer hexane solution from microcentrifuge tube to an analytical glass vial with glass insert and cap immediately to avoid evaporation as hexane is highly volatile. Measure FAMEs on analytical platform, e.g., GC-MS.

3.5 Mass Spectrometry-Based Platform for Metabolite Analysis

Mass spectrometry-based analytical platforms are ideal for quantifying small molecules in biological systems. In this case study, we provide protocols to measure a broad range of polar and nonpolar metabolites using a GC-MS-based analytical platform.

1. Instrumental setup: Derivatized samples are measured using a DB-35MS UI (Agilent J&W Scientific, 122-3832 UI) installed in an Agilent 7890A gas chromatograph (GC) interfaced with an Agilent 5977 mass spectrometer (MS) (*see* **Note 23**).
2. For MTBSTFA-derivatized samples (aqueous phase (AP) and media samples (M)), the GC oven is held at 100 °C for 1 min, increased to 255 °C at 3.5 °C/min, increased to 320 °C at 15 °C/min, and held at 320 °C for 3 min. The total run time for one sample is 54.6 min (*see* **Note 24**).
3. For measurements of FAMEs (from organic phase (OP)), e.g., palmitate detection, the GC oven is held at 100 °C for 3 min, increased to 205 °C at 25 °C/min, increased to 230 °C at 5 °C/min, increased to 300 °C at 25 °C/min, and held at 300 °C for 2 min.
4. The MS operates under electron impact ionization mode at 70 eV. The source is held at 230 °C and the quadrupole at 150 °C. 1 μl of derivatized sample is injected, and helium is used as carrier gas at a flow rate of 1 ml/min.
5. When measurement is complete, export data and continue with metabolite identification and data analysis (Subheading 3.6).

3.6 Metabolite Identification and Data Analysis

A wide variety of excellent software is commercial available to determine metabolite abundances and to calculate isotopologue distributions [24]. In this present case study, we identified metabolites and calculate abundances as well as isotopic labeling patterns by integrating metabolite ion fragments (summarized in Table 1) using in-house algorithms and correcting for natural isotope abundance using the method described by Fernandez et al. [25]. Figures 3 and 4 depict substrate utilization for TCA cycle metabolism and de novo palmitate synthesis in mammalian A549 cells.

1. *Metabolite identification*: Metabolites measured by GC-MS are identified by their characteristic fragmentation pattern and retention index. A wide variety of mass spectral and retention

Table 1
Metabolite fragment ions used for GC-MS analysis. Metabolites No. 1–46 are derivatized with MTBSTFA and MOX, while metabolites No. 47–52 represent fatty acids methyl esters

No.	Metabolite	Abbreviation	Retention time (min)	Monitored ions (*m/z*)	Carbons	Formula
1	Pyruvate	Pyr	5.91	174–180	1–2–3	$C_6H_{12}O_3NSi$
2	2-Oxo-isovalerate	KIV	7.46	202–210	1–2–3–4–5	$C_8H_{16}O_3NSi$
3	2-Oxo-3-methyl-valerate	KMV	9.19	216–224	1–2–3–4–5–6	$C_9H_{18}O_3NSi$
4	2-Oxo-isocaproate	KIC	9.54	216–224	1–2–3–4–5–6	$C_9H_{18}O_3NSi$
5	Lactate	Lac	10.03	233–240	2–3	$C_{10}H_{25}O_2Si_2$
				261–269	1–2–3	$C_{11}H_{25}O_3Si_2$
6	Alanine	Ala	11.15	232–239	2–3	$C_{10}H_{26}ONSi_2$
				260–268	1–2–3	$C_{11}H_{26}O_2NSi_2$
7	Glycine	Gly	12.42	218–223	2	$C_9H_{24}ONSi_2$
				246–252	1–2	$C_{10}H_{24}O_2NSi_2$
8	3-Hydroxybutyrate	3Hb	12.7	275–279	1–2–3–4	$C_{12}H_{27}O_3Si_2$
9	Oxalate	Oxa	13.04	261–266	1–2	$C_{10}H_{21}O_4Si_2$
10	Valine	Val	14.29	260–268	2–3–4–5	$C_{12}H_{30}ONSi_2$
				288–296	1–2–3–4–5	$C_{13}H_{30}O_2NSi_2$
11	Norvaline (internal standard)	Nor	14.71	288–294	1–2–3–4–5	$C_{13}H_{30}O_2NSi_2$
12	Leucine	Leu	15.27	200–208	2–3–4–5–6	$C_{11}H_{26}NSi$
				274–282	2–3–4–5–6	$C_{13}H_{32}ONSi_2$
13	Methylmalonate	MMA	15.75	289–295	1–2–3–4	$C_{12}H_{25}O_4Si_2$

14	Isoleucine	Ile	16.35	200–208	2–3–4–5–6	$C_{11}H_{26}NSi$
				274–282	2–3–4–5–6	$C_{13}H_{32}ONSi_2$
15	Urea	Urea	17.14	231–236	1	$C_9H_{23}ON_2Si_2$
16	y-Aminobutyrate	GABA	18.35	274–280	1–2–3–4	$C_{12}H_{28}O_2NSi_2$
17	Proline	Pro	18.52	258–266	2–3–4–5	$C_{12}H_{28}ONSi_2$
18	Succinate	Suc	19.21	289–294	1–2–3–4	$C_{12}H_{25}O_4Si_2$
19	Uracil	Ura	19.29	283–289	1–2–3–4	$C_{12}H_{23}O_2N_2Si_2$
20	Fumarate	Fum	19.67	287–292	1–2–3–4	$C_{12}H_{23}O_4Si_2$
21	Itaconate	Ita	20.33	301–310	1–2–3–4–5	$C_{13}H_{25}O_4Si_2$
22	Serine	Ser	23.13	288–294	2–3	$C_{14}H_{34}ONSi_2$
				302–308	1–2	$C_{14}H_{32}O_2NSi_2$
				362–368	2–3	$C_{16}H_{40}O_2NSi_3$
				390–398	1–2–3	$C_{17}H_{40}O_3NSi_3$
23	Threonine	Thr	23.67	376–384	2–3–4	$C_{17}H_{42}O_2NSi_3$
				404–412	1–2–3–4	$C_{18}H_{42}O_3NSi_3$
24	Methionine	Met	24.74	218–226	2–3–4–5	$C_{10}H_{24}NSiS$
				320–328	1–2–3–4–5	$C_{13}H_{30}O_2NSi_2S$
25	Cytosine	Cyt	25.81	282–286	1–2–3–4	$C_{12}H_{24}ON_3Si_2$
26	Aminolevulinate	Ala	26.51	331–339	1–2–3–4–5	$C_{14}H_{31}O_3N_2Si_2$
27	a-Ketoglutarate	Akg	26.94	346–355	1–2–3–4–5	$C_{14}H_{28}O_5NSi_2$

(continued)

Table 1
(continued)

No.	Metabolite	Abbreviation	Retention time (min)	Monitored ions (*m/z*)	Carbons	Formula
28	Malate	Mal	27.38	419–428	1–2–3–4	$C_{18}H_{39}O_5Si_3$
29	Phenylalanine	Phe	28.03	302–307	1–2	$C_{14}H_{32}O_2NSi_2$
				308–318	2–3–4–5–6–7–8–9	$C_{16}H_{30}ONSi_2$
				336–346	1–2–3–4–5–6–7–8–9	$C_{17}H_{30}O_2NSi_2$
30	Aspartate	Asp	28.43	302–310	1–2	$C_{14}H_{32}O_2NSi_2$
				390–400	2–3–4	$C_{17}H_{40}O_3NSi_3$
				418–428	1–2–3–4	$C_{18}H_{40}O_4NSi_3$
31	Proline	Pro	28.58	258–266	2–3–4–5	$C_{12}H_{28}ONSi_2$
				330–342	1–2–3–4–5	$C_{16}H_{36}O_2NSi_2$
32	Cysteine	Cys	29.86	406–418	1–2–3	$C_{17}H_{40}O_2NSi_3S$
33	2-Hydroxyglutarate	2HG	30.1	433–443	1–2–3–4–5	$C_{19}H_{41}O_5Si_3$
34	Ornithine	Orn	30.78	286–294	1–2–3–4–5	$C_{13}H_{30}ON_2Si_2$
				474–482	1–2–3–4–5	$C_{23}H_{54}O_2N_2Si_3$
35	Glutamate	Glu	31.5	330–340	2–3–4–5	$C_{16}H_{36}O_2NSi_2$
				432–442	1–2–3–4–5	$C_{19}H_{42}O_4NSi_3$
36	Phosphoenolpyruvate	PEP	31.82	453–459	1–2–3	$C_{17}H_{38}O_6Si_3P$
37	Asparagine	Asn	33.03	417–427	1–2–3–4	$C_{21}H_{47}O_3N_2Si_3$
38	Lysine	Lys	33.26	329–338	2–3–4–5–6	$C_{17}H_{41}N_2Si_2$
				431–443	1–2–3–4–5–6	$C_{20}H_{47}O_2N_2Si_3$

39	Glyceraldehyde-3-Phosphate	GAP	34.62	484–492	1–2–3	$C_{18}H_{43}O_6NSi_3P$
40	Glutamine	Gln	35.97	431–441	1–2–3–4–5	$C_{19}H_{43}O_3N_2Si_3$
41	Glycerol-3-phosphate	Glyc3P	36.56	571–580	1–2–3	$C_{23}H_{56}O_6Si_4P$
42	Citrate	Cit	38.68	459–469	1–2–3–4–5–6	$C_{20}H_{39}O_6Si_3$
				591–601	1–2–3–4–5–6	$C_{26}H_{55}O_7Si_4$
43	3-Phosphoglycerate	3PG	39.45	585–589	1–2–3	$C_{23}H_{54}O_7Si_4P$
44	Tyrosine	Tyr	40.12	302–308	1–2	$C_{14}H_{32}O_2NSi_2$
				466–480	1–2–3–4–5–6–7–8–9	$C_{23}H_{44}O_3NSi_3$
45	Histidine	His	40.65	338–350	2–3–4–5–6	$C_{17}H_{36}N_3Si_2$
				440–452	1–2–3–4–5–6	$C_{20}H_{42}O_2N_3Si_3$
46	Tryptophan	Trp	44.68	375–388	1–11	$C_{19}H_{31}O_2N_2Si_2$
47	Myristate	C14:0	15.59	242–258	1–14	$C_{15}H_{30}O_2$
48	Pentadecanoate	C15:0	17.24	256–272	1–15	$C_{16}H_{32}O_2$
49	Palmitate D31 (internal standard)	C16:0 D31	18.43	301		$C_{17}H_{35}O_2$
50	Palmitate	C16:0	19.23	270–284	1–16	$C_{17}H_{34}O_2$
51	Heptadecanoate	C17:0	21.56	284–294	1–17	$C_{18}H_{37}O_2$
52	Stearate	C18:0	24.22	298–316	1–18	$C_{19}H_{38}O_2$

index libraries are commercially available, e.g., FiehnLib [26]. Table 1 depicts metabolite ion fragments, retention times, and fragment formula for various metabolites using MTBSTFA derivatization. A representative total ion chromatogram (TIC) of intracellular metabolites from aqueous phase is depicted in Fig. 3a.

2. *Metabolite abundance*: To account for variabilities throughout metabolite extraction procedure, metabolite abundance may be normalized to internal standard of interest and cell number.

3. *Metabolite uptake and secretion*: To quantify metabolite uptake and secretion, calculate the difference of metabolite abundance in spent media and control, fresh media (incubated without cells). Positive values depict metabolite secretion and negative values uptake from the medium. These differences should be divided by the integral viable cell number obtained from cell counts.

4. *Isotopologue distribution*: Isotopologue distribution is the fractional abundance of isotopologues determined by integrating metabolite ion fragments. Isotopologue distributions can be corrected for natural isotope abundances [25]. As an example, Fig. 3c depicts isotopologue distribution for intracellular lactate, as well as TCA cycle intermediates and related amino acids.

5. *Quantification of ^{13}C enrichment*: Enrichment from stable isotope tracers can be expressed by calculating mole percent enrichment (MPE) from isotopologue distributions corrected for natural abundances [17, 27]. MPE is defined as the percent of all atoms within the metabolite pool that are labeled and can be calculated using the following equation:

$$\mathrm{MPE} = \sum_{i=1}^{n} \frac{M_i \cdot i}{n}$$

where E is the percent ^{13}C enrichment, i iterates the number of possible ^{13}C-labeled carbons on a metabolite fragment, n is the number of carbon atoms in the metabolite, and M_i is the relative abundance of the ith isotopologue. Figure 3b depicts MPE for lactate as well as TCA cycle intermediates and related amino acids.

6. *Isotopomer spectral analysis (ISA) approach to quantify de novo lipogenesis* [28]: The relative contribution to the de novo lipogenesis (e.g., palmitate) can be calculated from [U-$^{13}C_6$]glucose and [U-$^{13}C_5$]glutamine as both substrates contribute to the precursor pool of lipogenic acetyl-CoA (Fig. 2). First, calculate the fractional contribution of glucose and glutamine ^{13}C-derived carbons into the precursor pool acetyl-CoA for de

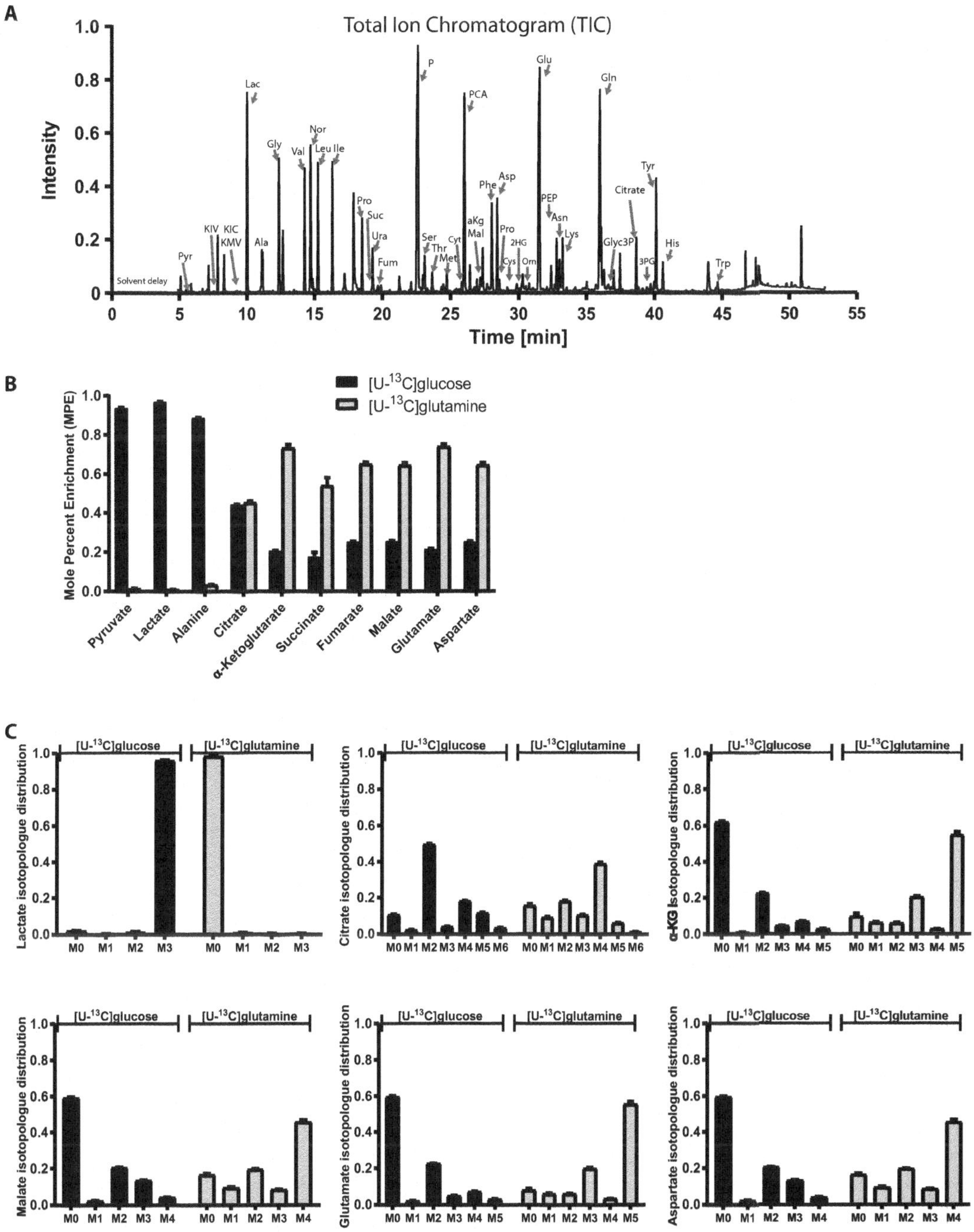

Fig. 3 Quantifying substrate utilization for TCA cycle metabolism and related amino acids. A549 cells were offered [U-$^{13}C_6$]glucose (black) and [U-$^{13}C_5$]glutamine (gray) for 48 h prior metabolite extraction. Metabolites were analyzed using GC-MS upon MTBSTFA/MOX derivatization. (**a**) Total ion chromatogram (TIC) of intracellular metabolites from aqueous phase. Identified metabolites are highlighted and summarized with retention time in Table 1. *P* phosphate, *PCA* pyroglutamate. (**b**) Isotope incorporation (^{13}C) from glucose and glutamine into pyruvate, lactate, TCA cycle intermediates, and related amino acids (mean ± SEM). (**c**) Isotopologue distribution of lactate, citrate, a-ketoglutarate, malate, glutamate, and aspartate (mean ± SEM)

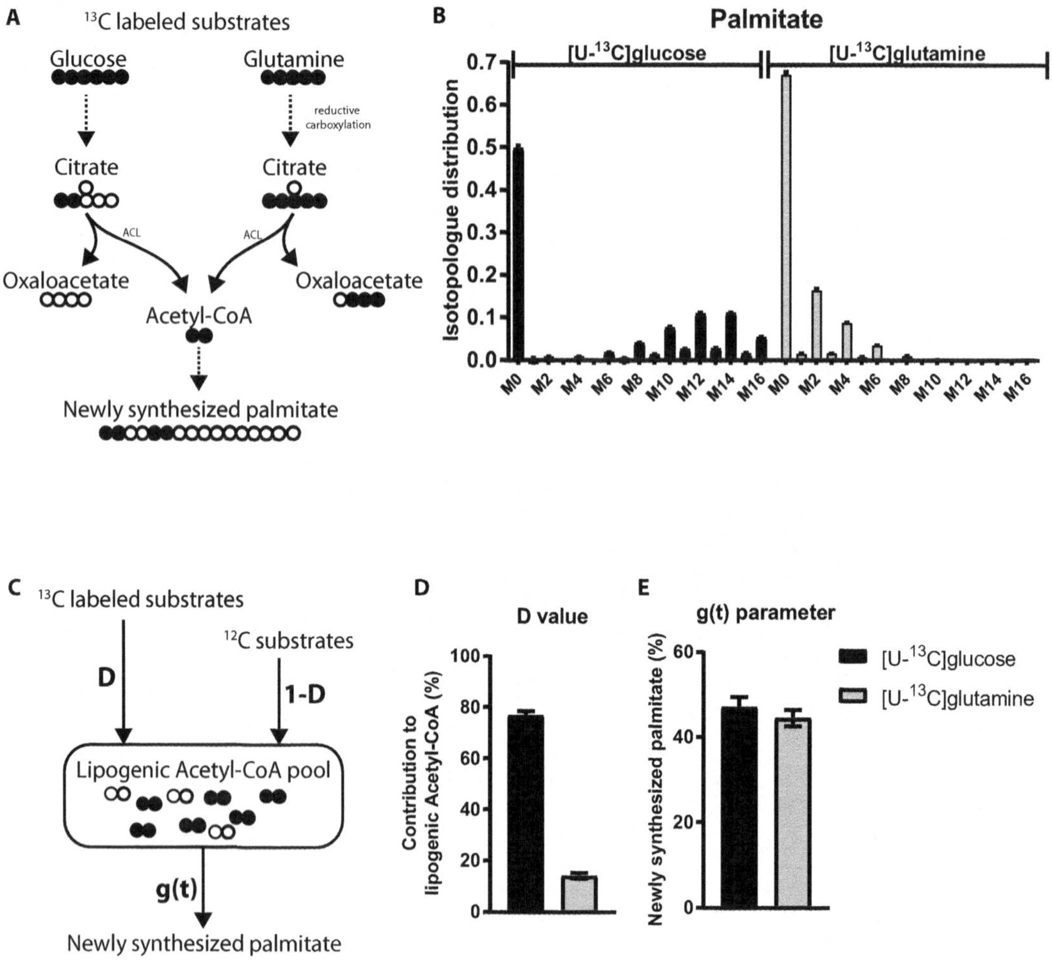

Fig. 4 Quantifying substrate utilization for de novo lipogenesis. A549 cells were cultured in the presence of [U-$^{13}C_6$]glucose or [U-$^{13}C_5$]glutamine for 48 h prior to GC-MS analysis as described. (**a**) Tracer map depicting ^{13}C incorporation from [U-$^{13}C_6$]glucose and [U-$^{13}C_5$]glutamine into palmitate, with open circles depicting ^{12}C and closed circles depicting ^{13}C isotopes. (**b**) Isotopologue distribution of palmitate (mean ± SEM). (**c**) Isotopomer spectral analysis (ISA) approach to quantify de novo palmitate synthesis. (**d**) Fractional contribution of ^{13}C-derived carbons into the precursor pool acetyl-CoA for de novo palmitate synthesis, defined as the dimensionless parameter *D* (mean ± 95% CI). (**e**) Fraction of newly synthesized palmitate from isotopic labeled acetyl-CoA after a period of time *t*, defined as the dimensionless parameter *g*(*t*) (mean ± 95% CI)

novo palmitate synthesis which is defined as the dimensionless parameter *D*. Second, determine the dimensionless parameter *g*(*t*) which is the fraction of newly synthesized product (e.g., palmitate) from isotopic labeled acetyl-CoA; Fig. 4 depicts isotopologue distribution of palmitate as well as *D* and *g*(*t*) values from [U-$^{13}C_6$]glucose and [U-$^{13}C_5$]glutamine.

4 Notes

1. Culture media formulations will have a significant impact on metabolism, and different cell types often require specific media types or additives. Care should be taken when choosing media, and various factors should be considered, including biological impact, custom media availability, and the relevant physiological system to be modeled. For example, one recent study highlighted how medium that mimics plasma metabolite concentrations can influence cell phenotype and metabolic activity [29]. The most effective media will be application-specific. Here we use DMEM since it is widely used with a long track record of application in research using cancer cell lines.
2. Norvaline is not endogenously present in mammalian cells, making it a suitable internal standard for normalization of abundances of MTBSTFA-derivatized metabolites across samples. Choice and concentration of internal standard may be adjusted depending on the biological system, derivatization, and analytical platform used. Isotopically labeled internal standards may also be used if more precise quantitation of a particular analyte is desired (e.g., [U-$^{13}C_3$]serine), as we do for fatty acids.
3. For non-adherent cell cultures, such as suspension cells or spheroids, incubate cells in isotopic labeled medium as described in Subheading 3.1. For metabolite extraction, cells should be pelleted by centrifugation (5 min, 300 $\times g$, 4 °C), washed once quickly with 1 ml 4 °C 0.9% NaCl, and pelleted again (5 min, 300 $\times g$, 4 °C) before removing fluid. Metabolites from the cell pellet can then be extracted by adding 250 μl −80 °C methanol to each tube using protocols as described in Subheading 3.3.
4. Fetal bovine serum (FBS) contains unspecified molecules that may dilute isotopic labeling of metabolites. While this dilution can be accounted for in subsequent data analysis, one can supplement tracer media with dialyzed FBS (dFBS) containing reduced concentrations of small molecules, such as amino acids. However, such media may lack other important small molecules such as vitamins, cofactors [30], and signaling lipids [31], so biological phenotypes should be validated when changing sera.
5. Medium can be stored at 4 °C, but freshly prepared medium is preferred for quantitative metabolic studies as some components may degrade over time, i.e., glutamine. Adjusting pH may cause an unwanted change in the osmolality that may affect cell metabolism. Minimizing salt addition can be particularly difficult if adding low-pH amino acid concentrates to media.

6. Plating equal amounts of cells per well is critical to minimize variability across replicates and experimental arms. If cells are exposed to treatments that affect cell growth, additional wells may be seeded to obtain cell counts for metabolite abundance normalization. Cell number may be adjusted depending on cell type and duration of tracing. At time of extraction, cells should exhibit consistent confluency across experimental arms. Since metabolites all have specific limits of detection and quantitation, cell number may be adjusted to facilitate measurements of specific metabolites.
7. Medium should be pre-equilibrated in a humidified cell incubator at 37 °C and 5% CO_2.
8. Prolonged incubation even in a humidified cell culture incubator results in medium evaporation and metabolite degradation over time. This step is critical to more accurately quantify changes in metabolite levels.
9. Depending on study goals, kinetic or steady-state labeling data may be preferred. In the latter case, isotopic labeling patterns do not change appreciably with time. Since dynamics of isotope incorporation varies across pathways, the time required to reach isotopic steady state may be specific for certain pathways (e.g., glycolysis vs TCA cycle). Analysis of kinetic labeling may be required for linear pathways or other applications. However, data analysis and interpretation are more intensive, as pool size information must be accounted for in models and flux comparisons.
10. We recommend working as fast as possible and performing all steps on ice/cold conditions. Extract only one plate at a time as performing more extractions concurrently will extend waiting times and increase variability across replicates. Sample volumes refer to a metabolite extraction from a 12-well tissue culture plate, but volumes may be adjusted accordingly if using different culture formats.
11. Avoid PBS as large phosphate contents in the sample might interfere with metabolite derivatization and/or MS analysis.
12. Fill 250 μl chloroform to microcentrifuge tubes before metabolite extraction and freeze sample tubes at −20 °C. Do not add chloroform into culture dishes as chloroform will dissolve culture plates.
13. If multiple plates will be extracted, this is an ideal time to start extraction of the second plate or to count cells.
14. Samples must be completely dry as water will hinder the derivatization reaction. If water condensate inside vials while transferring from cold (e.g., vacuum evaporator or freezer), samples may be re-evaporated at room temperature.

15. Different methods are available for evaporation, depending on the timing and application. For most applications, airflow at room temperature is acceptable. In some instances, nitrogen flow is preferable to avoid oxidation of products.
16. Centrifugation at low speed will pellet cells. Centrifugation at higher speeds may result in leaking cells affecting metabolite concentration in media. However, an additional centrifuge step (5 min at 21,000 × *g* at 4 °C) may be needed to remove any cell debris.
17. For quantification of lipids from the medium, use 100 μl medium from **step 21**, and proceed with a methanol-/water-/chloroform-based extraction as described above.
18. Derivatization reagents are very hygroscopic, and they are best stored at 4 °C with desiccant.
19. When reaction of intracellular keto acids, such as pyruvate and alpha-ketoglutarate, is desired, methoxyamine hydrochloride derivatization should be performed to protect keto groups. If only amino acids are to be analyzed, this reactant can be replaced with pyridine.
20. Derivatization by hand may form precipitations that may be removed by centrifugation prior analytical analysis. If derivatization is performed using an automated sample robot, washing of the syringe with acetone in between prevents carry over. However, an additional washing step with pyridine directly prior to pipetting derivatization reagents such as MOX is critical as acetone might interfere with derivatization.
21. The lower phase contains sulfuric acid which may interact and damage GC columns. Therefore, collect the upper hexane phase only without disturbing the lower phase.
22. Different methods are available for evaporation, depending on the application. For most targeted applications, airflow at room temperature is acceptable. In some instances, nitrogen flow is preferable to avoid oxidation of products.
23. Column length and polarity may affect metabolite separation. More specific columns, such as Select FAME (Agilent J&W Scientific, CP7420), may increase metabolite separation, especially trans/cis separation of monounsaturated FAMEs.
24. To reduce run time for one sample from 55 to 26 min, the GC oven temperature may be adjusted. The GC oven is held at 100 °C for 2 min, increased to 300 °C at 10 °C/min, and held at 325 °C for 3 min. This method is suitable for most metabolites, including TCA cycle intermediates and amino acids, and a good alternative for media measurements. However, shorter elution times lead to decreased metabolite separation that may result in co-elution of specific metabolites of interest.

Acknowledgments

We thank members of the Metallo Lab for helpful discussions. Figure 2 was adapted with permission from unpublished work by Jamey D. Young (Vanderbilt University). This work was supported, in part, by NIH grant R01CA188652, a Searle Scholar Award, a NSF CAREER Award (1454425), a Camille Dreyfus Teacher Scholar Award (all to C.M.M.), and Deutsche Forschungsgesellschaft (German Research Foundation) (CO1488/1-1 to T.C.).

References

1. Cordes T, Metallo CM (2016) Tracing insights into human metabolism using chemical engineering approaches. Curr Opin Chem Eng 14:72–81. https://doi.org/10.1016/j.coche.2016.08.019
2. Wiechert W (2001) 13C metabolic flux analysis. Metab Eng 3:195–206. https://doi.org/10.1006/mben.2001.0187
3. Sauer U (2006) Metabolic networks in motion: 13C-based flux analysis. Mol Syst Biol 2:62. https://doi.org/10.1038/msb4100109
4. Zamboni N, Saghatelian A, Patti GJ (2015) Defining the metabolome: size, flux, and regulation. Mol Cell 58:699–706. https://doi.org/10.1016/j.molcel.2015.04.021
5. Wishart DS (2016) Emerging applications of metabolomics in drug discovery and precision medicine. Nat Rev Drug Discov 15:473–484. https://doi.org/10.1038/nrd.2016.32
6. Metallo CM, Walther JL, Stephanopoulos G (2009) Evaluation of 13C isotopic tracers for metabolic flux analysis in mammalian cells. J Biotechnol 144:167–174. https://doi.org/10.1016/j.jbiotec.2009.07.010
7. Walther JL, Metallo CM, Zhang J, Stephanopoulos G (2012) Optimization of 13C isotopic tracers for metabolic flux analysis in mammalian cells. Metab Eng 14:162–171. https://doi.org/10.1016/j.ymben.2011.12.004
8. Crown SB, Long CP, Antoniewicz MR (2016) Optimal tracers for parallel labeling experiments and 13C metabolic flux analysis: a new precision and synergy scoring system. Metab Eng 38:10–18. https://doi.org/10.1016/j.ymben.2016.06.001
9. Linares JF, Cordes T, Duran A et al (2017) ATF4-induced metabolic reprograming is a synthetic vulnerability of the p62-deficient tumor stroma. Cell Metab 26:817–829.e6. https://doi.org/10.1016/j.cmet.2017.09.001
10. Metallo CM, Gameiro PA, Bell EL et al (2012) Reductive glutamine metabolism by IDH1 mediates lipogenesis under hypoxia. Nature 481:380–384. https://doi.org/10.1038/nature10602
11. Parker SJ, Svensson RU, Divakaruni AS et al (2017) LKB1 promotes metabolic flexibility in response to energy stress. Metab Eng 43:208–217. https://doi.org/10.1016/j.ymben.2016.12.010
12. Zhao D, Badur MG, Luebeck J et al (2018) Combinatorial CRISPR-Cas9 metabolic screens reveal critical redox control points dependent on the KEAP1-NRF2 regulatory axis. Mol Cell 69:699–708.e7. https://doi.org/10.1016/j.molcel.2018.01.017
13. Divakaruni AS, Wallace M, Buren C et al (2017) Inhibition of the mitochondrial pyruvate carrier protects from excitotoxic neuronal death. J Cell Biol 216:1091–1105. https://doi.org/10.1083/jcb.201612067
14. Cordes T, Wallace M, Michelucci A et al (2016) Immunoresponsive gene 1 and itaconate inhibit succinate dehydrogenase to modulate intracellular succinate levels. J Biol Chem 291:14274–14284. https://doi.org/10.1074/jbc.M115.685792
15. Zhang H, Badur MG, Divakaruni AS et al (2016) Distinct metabolic states can support self-renewal and lipogenesis in human pluripotent stem cells under different culture conditions. Cell Rep 16:1536–1547. https://doi.org/10.1016/j.celrep.2016.06.102
16. Vacanti NM, Divakaruni AS, Green CR et al (2014) Regulation of substrate utilization by the mitochondrial pyruvate carrier. Mol Cell 56:425–435. https://doi.org/10.1016/j.molcel.2014.09.024

17. Green CR, Wallace M, Divakaruni AS et al (2016) Branched-chain amino acid catabolism fuels adipocyte differentiation and lipogenesis. Nat Chem Biol 12:15–21. https://doi.org/10.1038/nchembio.1961
18. Metallo CM, Vander Heiden MG (2013) Understanding metabolic regulation and its influence on cell physiology. Mol Cell 49:388–398. https://doi.org/10.1016/j.molcel.2013.01.018
19. Badur MG, Metallo CM (2018) Reverse engineering the cancer metabolic network using flux analysis to understand drivers of human disease. Metab Eng 45:95–108. https://doi.org/10.1016/j.ymben.2017.11.013
20. Bligh EG, Dyer WJ (1959) A rapid method of total lipid extraction and purification. Can J Biochem Physiol 37:911–917. https://doi.org/10.1139/y59-099
21. Badur MG, Zhang H, Metallo CM (2015) Enzymatic passaging of human embryonic stem cells alters central carbon metabolism and glycan abundance. Biotechnol J 10:1600–1611. https://doi.org/10.1002/biot.201400749
22. Miranda-Santos I, Gramacho S, Pineiro M et al (2015) Mass isotopomer analysis of nucleosides isolated from RNA and DNA using GC/MS. Anal Chem 87:617–623. https://doi.org/10.1021/ac503305w
23. Long CP, Au J, Gonzalez JE, Antoniewicz MR (2016) 13C metabolic flux analysis of microbial and mammalian systems is enhanced with GC-MS measurements of glycogen and RNA labeling. Metab Eng 38:65–72. https://doi.org/10.1016/j.ymben.2016.06.007
24. Dandekar T, Fieselmann A, Majeed S, Ahmed Z (2014) Software applications toward quantitative metabolic flux analysis and modeling. Brief Bioinform 15:91–107. https://doi.org/10.1093/bib/bbs065
25. Fernandez CA, Des Rosiers C, Previs SF et al (1996) Correction of 13C mass isotopomer distributions for natural stable isotope abundance. J Mass Spectrom 31:255–262. https://doi.org/10.1002/(SICI)1096-9888(199603)31:3<255::AID-JMS290>3.0.CO;2-3
26. Kind T, Wohlgemuth G, Lee DY et al (2009) FiehnLib: mass spectral and retention index libraries for metabolomics based on quadrupole and time-of-flight gas chromatography/mass spectrometry. Anal Chem 81:10038–10048. https://doi.org/10.1021/ac9019522
27. Vincent G, Bouchard B, Khairallah M, Des Rosiers C (2004) Differential modulation of citrate synthesis and release by fatty acids in perfused working rat hearts. Am J Physiol Circ Physiol 286:H257–H266. https://doi.org/10.1152/ajpheart.00717.2003
28. Young JD (2014) INCA: a computational platform for isotopically non-stationary metabolic flux analysis. Bioinformatics 30:1333–1335. https://doi.org/10.1093/bioinformatics/btu015
29. Cantor JR, Abu-Remaileh M, Kanarek N et al (2017) Physiologic medium rewires cellular metabolism and reveals uric acid as an endogenous inhibitor of UMP synthase. Cell 169:258–272.e17. https://doi.org/10.1016/j.cell.2017.03.023
30. Kuri-Harcuch W, Wise LS, Green H (1978) Interruption of the adipose conversion of 3T3 cells by biotin deficiency: differentiation without triglyceride accumulation. Cell 14:53–59
31. Aoki J, Taira A, Takanezawa Y et al (2002) Serum lysophosphatidic acid is produced through diverse phospholipase pathways. J Biol Chem 277:48737–48744. https://doi.org/10.1074/jbc.M206812200

Chapter 15

Insights into Dynamic Network States Using Metabolomic Data

Reihaneh Mostolizadeh, Andreas Dräger, and Neema Jamshidi

Abstract

Metabolomic data is the youngest of the high-throughput data types; however, it is potentially one of the most informative, as it provides a direct, quantitative biochemical phenotype. There are a number of ways in which metabolomic data can be analyzed in systems biology; however, the thermodynamic and kinetic relevance of these data cannot be overstated. Genome-scale metabolic network reconstructions provide a natural context to incorporate metabolomic data in order to provide insight into the condition-specific kinetic characteristics of metabolic networks. Herein we discuss how metabolomic data can be incorporated into constraint-based models in a flexible framework that enables scaling from small pathways to cell-scale models, while being able to accommodate coarse-grained to more detailed, allosteric interactions, all using the well-known principle of mass action.

Key words Systems biology, Dynamic network states, Metabolomics

1 Introduction

The importance of kinetics in biochemical systems has long been recognized, and historically the characterization of metabolic networks has been built upon kinetic and dynamic analysis [19, 47, 53]. The subsequent development of constraint-based modeling and flux balance analysis has resulted in cellular and ordinal scale metabolic characterization to be implemented [3, 16, 56]. These developments were largely enabled by sequencing technologies and subsequent genomic, procurement, and metabolism mixing technologies enabled biologically meaningful relationships to be defined between the components with resulting construction of genome-scale metabolic models (GeMMs) [51]. The growth and development of this field is too large and broad to summarize in a few statements, but suffice it to say there are examples in the utility of these models ranging from environmental engineering applications to human disease [15, 26, 29, 30, 45, 54]. Metabolite uptake and secretion rates can be used to constrain flux bounds for flux

Angelo D'Alessandro (ed.), *High-Throughput Metabolomics: Methods and Protocols*, Methods in Molecular Biology, vol. 1978, https://doi.org/10.1007/978-1-4939-9236-2_15, © Springer Science+Business Media, LLC, part of Springer Nature 2019

balance analysis (FBA) [1, 38], however basic FBA does not inherently account for absolute, quantified metabolomic data; these data map between the row and column spaces of the network and provide complementary information to the null spaces [22], informing how the system responds dynamically to alterations in the external environment as well as genetic changes.

In more complex scenarios, particularly multi-cellular eukaryotes have disease phenotypes that are more subtle and often only reflected in dynamic states. In such cases, it is necessary to extend the mathematical description to also cover temporal changes, long-term development, or effects of perturbations. With increased appreciation of this fact and the continued maturity of high-throughput data, particularly metabolomic data, in recent years there has been renewed interest and resurgence in the development of kinetic models [6, 12, 18, 22, 39, 55, 57]. For example, analysis of network dynamics enables the assessment of system responses to metabolic perturbations. Studying such effects requires applying different methods for model building, incorporating in vivo data sets, simulation, and analysis.

This chapter describes a scalable approach, the mass action stoichiometric simulation (MASS) procedure [22, 25, 43]. This is approach to dynamic modeling and analysis of network states is flexible as it can account for varying degrees of granularity without any change in the formulation. Furthermore, it is amenable to direct use of metabolomic data, without necessitating in vitro or ex situ enzymatic measurements.

Herein we provide a step-wise description of all necessary steps to implement the MASS approach in form of a protocol for building models. Basic challenges will be discussed using the well-known urea cycle in yeast as a realistic example for metabolic modeling. Literature data will be incorporated to demonstrate the feasibility of the approach.

2 Materials

The proposed MASS approach for building models can provide insight into dynamic states of networks, however it requires preexisting knowledge (i.e., metabolic network reconstructions) of the network, in conjunction with additional data (either available in the literature or by direct experimental measurement). The type(s) of data that has been measured (e.g. metabolomic, proteomic, etc) determines the granularity or level of mechanistic detail accounted for in the resulting model. The objective of the approach is to define and parameterize the following first order, non-linear differential equation system,

$$\frac{d\mathbf{x}}{dt} = \mathbf{S} \cdot \mathbf{v}(\mathbf{x}, \mathbf{k}) \tag{1}$$

which represents the dynamic mass balance for a metabolic network with metabolite vector, **x**, flux vector, **v**, and stoichiometric matrix, **S**. Of note $\mathbf{v} = f(\mathbf{x}, \mathbf{k})$, with kinetic parameters **k** that are functions of metabolite concentrations, time, and other factors depending on the underlying assumptions of mechanism (enzyme concentration, turnover rate, etc.). There are a number of different approaches for formulating kinetic parameters [55]; we favor the approach based on mass action kinetics, because it is, in principle, consistent and flexible at multiple levels of mechanistic granularity and due to the stoichiometrically based formulation, enables a biologically meaningful interpretation of the row and column spaces, complementing the left and right null spaces [22, 24].

2.1 The GeMM Scaffold

The field of metabolic network reconstructions has now been well established for over a decade [42]. A model that has been constructed with the appropriate quality control and quality assurance protocols (Table 1) forms the basis of any large or cell-scale dynamic network and can be found at various sites [17, 32]. While there are varying degrees of annotation with respect to gene–protein–reaction relationships, the critical component for these networks is to have an elementally (and charge) balanced network; this provides the basis for a number of downstream applications and analyses, including construction of kinetic models [42, 43].

2.2 Estimates of Equilibrium Constants, K_{eq}

Flux directionality is a critical constraint in FBA and while smaller networks often permit physiologic insight to determine reversibility for key reactions, as the complexity and size of networks increases, there is a greater potential for thermodynamically futile [52] cycles or even unfeasible flux states. The development of the group contribution approach for estimating equilibrium constants

Table 1
An abbreviated summary of the key basic quality control measures for the steps outlined in Fig. 1

	Quality control/quality assurance measure
Network reconstruction	Elemental mass balance for reactants and products
Network reconstruction	Charge balance for products and reactants
Calculated flux points	Thermodynamic consistency for reaction directionality (K_{eq})
Calculated flux points	Thermodynamic consistency for metabolomic data
Kinetic model	Physically plausible rate constants
Kinetic model	Dynamically stable behavior

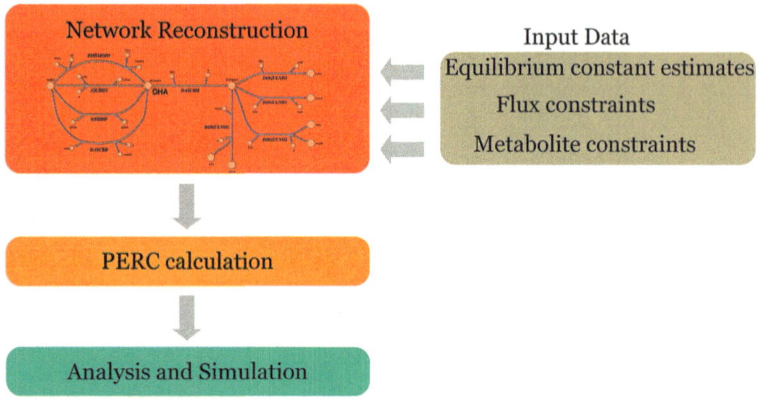

Fig. 1 Workflow for constructing kinetic models using metabolomic data in conjunction with genome-scale (or cell-scale) metabolic network models

[40] with the subsequent improvements in the approach [13, 27] has led to great improvements and estimations of the equilibrium constant in biologically relevant conditions (i.e., accounting for temperature, ionic strength, and pH). Estimates of the K_{eq} for biochemical transformations of many of the small metabolites involved in prokaryotes as well as multi-cellular eukaryotes can be estimated through the advancements made in Gibbs free energy estimations based upon the group contribution method. These estimates enable explicit description of mass action kinetics for these networks.

2.3 Flux Constraints

Although the identification of unique internal flux states remains a challenging problem, measurements of extracellular metabolites over time allow estimates of cellular uptake and secretion rates, providing valuable information for characterizing metabolic network states (through reduction of the size of the feasible steady state solution space). Since the scaffold of these kinetic models is built upon FBA based models, the greater the reduction of the null space of the model results in decreased uncertainty of the resulting kinetic models.

2.4 Literature Estimates of Metabolite Concentrations

Current, practical limitations in experimental capabilities as well as mass spectrometry instrumentation preclude the ability to comprehensively assess the entire metabolome. However, in many cases the approximate order of magnitude (or at least a range of values) for a class of metabolites is often known; this information can help constrain directionality of fluxes as well as the range of feasible kinetic parameters.

2.5 Software and Tools for Data Analysis and Integration

Stoichiometric matrices of metabolic networks are more than just "data matrices", but rather represent knowledge some of which is described quantitatively and numerically and others as relationships. Thus, the computation and simulation environments need to support data structures that account for different physicochemical properties as well as biological relationships. Efforts for model representation and standardization, most notably of which is perhaps the systems biology markup language (SBML) [8, 21] have been an enabling format for sharing models and improving reproducibility of results. However, there are still areas that provide ambiguities [11], particularly for novice users, thus we expect continued development in this area [7].

Nevertheless, suitable software environments require the ability to manipulate biological data structures (stoichiometric matrix, associated metabolites, and reactions), with their associated physical and chemical properties (formulas, charge, numerical coefficients, flux bounds, etc.) as well as an optimization engine, differential equation solver, and algebraic (symbolic) equation solver. Some environments enable all of these together and many freeware alternatives have provided APIs to enable function calls to the different programs to solve different steps of the problem.

Academic and commercial software packages have been instrumental in the advancement and development of systems and computational biology and their utility in the development of the constraint-based modeling field has been critical and will arguably be even more instrumental in use of metabolomic data, particularly for kinetic models. The MASS Toolbox [49] has been developed specifically for implementation and use of constraint-based models for construction of dynamic models, but other tool packages can be used with open source scripting languages, open source computational packages (SAGE (www.sagemath.org), Octave (www.gnu.org/software/octave), Scilab (www.scilab.org)) as well as commercial programs (Mathematica, MATLAB) and solvers. Different types of numerical solver packages will be required for constructing these models, including linear programming solvers, non-linear optimization programs, and algebraic solvers (GAMS (www.gams.com), Gurobi (www.gurobi.com), CPLEX (www.ibm.com/software/), MOSEK (www.mosek.com), SNOPT (https://ccom.ucsd.edu), TOMLAB (https://tomopt.com/tomlab/)).

Currently, Python (www.python.org) has been a very popular scripting language with an ever growing presence in systems biology and computational biology, thus accordingly there are a growing number of libraries, tools, and development environments [35, 36] that enable flexible workflows and APIs for integration with third-party solvers.

3 Methods

3.1 Basic Protocol Outline

In recent years developments in applying thermodynamic constraints have enabled different approaches for constructing kinetic models, notably the simultaneous application of data constraints to reduce network size and calculate kinetic parameters versus the sequential incorporation of layers of data. We will follow the latter approach in order to be systematic in the description, however we will also comment on the steps in the protocol that can be combined, given adequate coverage of the metabolome and/or sufficient knowledge of the flux state of the system.

There are multiple ways of formulating the equations needed to construct mass action models. As noted above, the stoichiometrically based (i.e., mass action) formulation enables identification of complementary relationships between the concentrations and fluxes, motivated by the underlying mathematical relationships between the rank of the stoichiometric matrix and its row and column spaces [22]. Fundamentally, molecular interactions involve two-component interactions to form an intermediate,

$$x_i + x_j \rightleftharpoons x_i x_j \tag{2}$$

in which the general case x may be a small metabolite or an enzyme to which the metabolite can bind; in these cases then Eq. 2 can be used to construct all of the different varieties of metabolic reaction types (from the most coarse-grained in which only substrates and products are explicitly described to the more mechanistic descriptions accounting for enzymatic intermediates with allosteric effects [25, 53]). Figure 1 provides a general overview of the steps needed to construct a kinetic model. This approach can be implemented as a single complex optimization problem, or a set of simpler consequential optimizations and calculations. In this description we defer to the latter, as it is more transparent and easier to follow.

3.2 Calculate Candidate Flux States

Although our interest is in assessing the dynamic states of networks, Eq. 1 is solved at the steady state,

$$\mathbf{S} \cdot \mathbf{v} = \mathbf{0} \tag{3}$$

in order to identify points in the right null space needed to calculate kinetic parameters.

Although advancements in high-throughput capabilities have enabled the measurement of a significantly larger number of metabolites, faster and with greater accuracy, there is still a discrepancy between the size of the datasets that can reasonably be achieved and the desired number of measured metabolites. Since even the largest reported metabolomic data sets have on the order of 100 metabolites (with absolute quantification), but GeMMs have grown to the size of 2000+ metabolites, it is necessary to significantly reduce the content in the models in order to produce cell-scale kinetic models. Furthermore, as noted in Subheading 3, in general there is not enough information to uniquely specify a flux state, thus the steady state solution space is underdetermined for which sampling the solution space is commonly used to evaluate different feasible states [37, 48, 50].

3.3 Incorporate K_{eq} and Metabolomic Data

The challenge of data-incompleteness particularly with respect to metabolomic data results in different approaches for incorporating thermodynamic constraints using equilibrium constants and metabolomic data. The difference in the chemical potentials of interacting metabolites determines the direction of the catalyzed reactions. The Gibbs free energy of formation can be described on a reaction by reaction basis, for which the free energy of formation at reference conditions is equal to the equilibrium constant ($|_0$ indicates the equilibrium state),

$$\alpha x_i + \beta x_j \rightleftharpoons \delta x_k \gamma x_l \tag{4}$$

$$\Delta G_0 = RT \ln(K_{\mathrm{eq}}) = RT \ln\left(\frac{x_k^{\delta} x_l^{\gamma}}{x_i^{\alpha} x_j^{\beta}}\right)\Big|_0 \tag{5}$$

from which it follows that at non-equilibrium conditions (i.e., when there is a net forward or reverse flux for Eq. 5),

$$\Delta G = \Delta G_0 + RT \ln\left(\frac{x_k^{\delta} x_l^{\gamma}}{x_i^{\alpha} x_j^{\beta}}\right) \tag{6}$$

for ideal gas constant, R, and temperature, T, and stoichiometric coefficient, $\{\alpha, \beta, \gamma, \delta\} \in \mathbb{N}$. Historically, thermodynamic constraints in flux balance analysis were incorporated as reaction reversibility (i.e., reversible for $0.1 \leq K_{\mathrm{eq}} \leq 10$, otherwise irreversible). However, the development of energy balance analysis (EBA) [2] and non-equilibrium thermodynamic considerations [33] led to more systematic assessment of feasible flux states [18, 20, 27, 52], which can in turn be used in the kinetic model construction procedure.

Ultimately, these approaches are still based upon the necessity to use K_{eq} estimates in conjunction with condition-specific metabolomic measurements to determine feasible flux states. Additionally, as described in the next section, the concentration data can be further employed to calculate kinetic parameters.

Since FBA optimizes network fluxes, the free energy constraints need to be linked to the flux directionality; this can be done after the fact (i.e., sample candidate flux states and then remove fluxes that violate thermodynamic considerations) or, more elegantly, the constraints can be enforced by introduction of a binary variable linking the flux upper and lower bounds to the free energy changes, as described by Hamilton et al. [18],

$$(1-\delta) v_{i,\min} \leq v_i \leq \delta v_{i,\max} \tag{7}$$

$$-\Omega\delta + \varepsilon \le \Delta G \le \Omega(1-\delta) - \varepsilon \tag{8}$$

in which Ω is an upper and lower bound on ΔG and ϵ is a small numerical factor on the order of the expected precision of calculations.

The formulation of the thermodynamic constraints and whether this is done sequentially (by calculating a flux state and then reducing the number of acceptable feasible points in the solution space versus simultaneously calculating states) is dependent on the size of the model, the amount of data (i.e., how large the coverage of metabolomic data is), as well as the desired formulation (addition of coupling constraints using a linear programming framework versus a mixed-integer linear programming formulation with a trade-off in calculation time and performance).

3.4 Constrain and Calculate Rate Constants

Consider a simple (coarse-grained) mass action equation involving two substrates and two products (e.g., a cofactor redox reaction),

$$x_1 + x_2 \rightleftharpoons x_3 + x_4 \tag{9}$$

with corresponding reaction flux equation given by

$$v = v^+ - v^- = k^+ x_1 x_2 - k^- x_3 x_4 \tag{10}$$

Given a non-equilibrium steady state for non-zero $v = \alpha$, with a known estimation of the equilibrium constant, $K_{eq} = \frac{k^+}{k^-}$, then

$$\alpha = k^+ \left(x_1 x_2 - \frac{x_3 x_4}{K_{eq}} \right) \tag{11}$$

If the measured metabolome for the model is fully defined (i.e., measured), then this is a trivial algebraic equation to solve. However, if not all metabolites have been measured, then for a network with metabolite connectivities >1 (which is typically the case for the majority of metabolites in GeMMs), then complex non-linear dependencies will appear. Algebraic equation solvers (*see* Subheading 2.5) can provide great utility here and the calculated value of the rate constants serves as a quality control measure (e.g., if the constants are negative, then there is a problem with constraints or assumptions in the earlier model definition). Since the rate constants are calculated indirectly, based on fluxes that have an associated error (most notably from experimental measurements are typically near a steady state, as opposed to at a defined steady state), these have been termed pseudo-elementary rate constants (PERCs) [12, 22]. Interestingly these have shown to have biologically meaningful interpretations [25] and to potentially better reflect phenotypes than genotypic alterations [4].

3.5 Assessment of System Dynamics

Basic dynamic systems analysis revolves around linearization of the dynamic flux balance equation, whose eigenvalues reflect the characteristic time scales that each of the corresponding eigenvectors (or biochemical reaction modes) take place on. Time scale decomposition involves linearization of Eq. 1 around a particular steady state and calculating the corresponding eigenvalues and eigenvectors. Transformation into the so-called modal matrix provides an assessment of the metabolic interactions in terms of dynamically independent reaction vectors [23, 28, 44].

A simple yet effective means of assessing the stability (as well as physical and biological relevance) of dynamic models is interpretation of the eigenvalues. Any eigenvalue with a real component is not physically meaningful and likely identifies a problem with either the data or assumptions that were made in the absence of data; in either case, this would require further evaluation of the model construction steps. The negative reciprocal of the eigenvalue corresponds to the relaxation time constant of the metabolic modes (eigenvectors of the Jacobian matrix), thus negative real eigenvalues define the hierarchical relaxation spectrum of the network from a perturbation to a specified steady state. Complex conjugate pairs with real and imaginary values correspond to complementary oscillatory states; when complementary modes are combined into "joint modes" (by taking the square root of product of exponentials), then the oscillatory component is removed and the time constant is simply α, if $\alpha +/- i\beta$ are the eigenvalues. When $\alpha = 0$, the complex conjugates reflect the presence of a Hopf bifurcation. The calculation of a positive PERC requires an iterative cycle to revise the model. Since the right null space is underdetermined, multiple flux states will be calculated, resulting in an ensemble of kinetic models [4], thus one can simply remove any candidate models that fail the quality control tests.

4 Pathway Example: Urea Cycle in Yeast

Baker's yeast, *Saccharomyces cerevisiae*, was the first eukaryotic organism for which a metabolic network reconstruction was constructed and has since continued to grow and develop through multiple iterations [14, 34, 41]. We consider the case of the urea cycle *S. cerevisiae*, extracting the pathway from *iND*750 [10] for the interpretation of metabolomic profiles from yeast grown under batch culture or chemostat conditions [5]. The growth rate with the chemostat was 0.1/h compared to 0.4/h for the batch culture. Estimates for the unmeasured metabolites were obtained from the Yeast Metabolomic Database [46].

The stoichiometric matrix for the cycle (Fig. 2) is provided in Table 2, which is clearly 14×14, with rank 13, indicating a right null space of 1 reflecting by the forward loop in the cycle. The left null space is in turn also of dimension 1 and is composed of the metabolites that are not transported in or out of the system

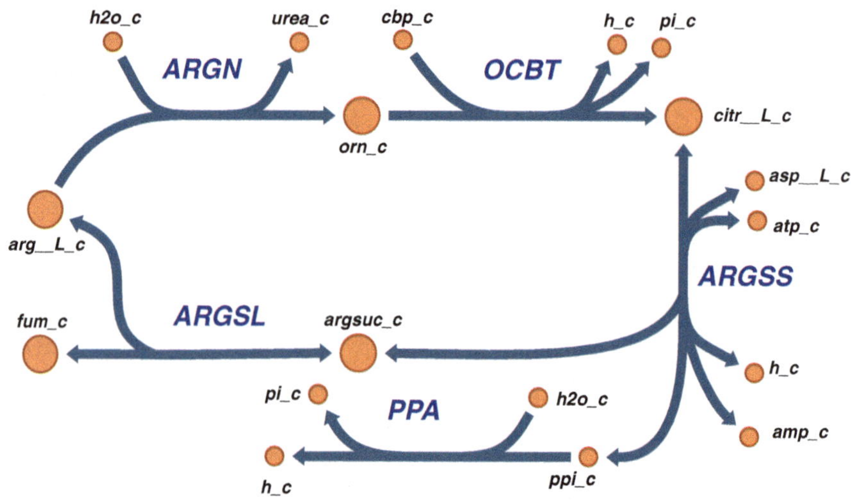

Fig. 2 Schematic representation of the urea cycle in *S. cerevisiae*. "_c" reflects the cytoplasm compartment

(arg_L_c + argsuc_c + citr_L_c + orn_c) (Tables 3 and 4). The size and relative simplicity of the pathway considered allow for transparency and lucidity in assessing the results. The gradient matrices [22] are provided in Tables 4 and 5 for the chemostat and batch culture conditions, respectively. Further the modal matrices [44] are summarized in Tables 6 and 7. For simplicity and to avoid potential noise introduced by the variables that did not have any experimental measurements, only the metabolites that had reported measurements by Canelas et al. [5] were analyzed. For a small network or pathway, one can carry out such an analysis without "contaminating" the other modes. In larger networks (with a larger set of measured metabolites), the noise may need to be filtered through the use of ensemble methods [4, 12, 57] (sampling across different states and then selecting only those features that are statistically significant). The cycle is described by three dynamically independent modes. In both cases, one mode occurs on the time scale of ~10 min for fluctuations in fumarate and another on approximately 2 h for the metabolism of ornithine. The other mode, which corresponds to the dynamics of OCBT and ARGSS, is more interesting in the context of these data. Under chemostat growth conditions there is an exchange between ATP and L-citrulline balanced by L-ornithine (with smaller fractions of L-aspartate and AMP) with a relaxation time scale of about 1.3 h. In batch culture conditions, however, this mode is more than four times slower at nearly 4 h, with a shift in directionality suggesting a change (or at least decrease) in the driving force of the cycle.

There is a noted difference in the growth rates of yeast under the two conditions and the accumulation of biomass is necessarily linked with protein metabolism. Since the urea cycle provides a means to dispense of accumulating nitrogen products, it makes

Table 2
Stoichiometric matrix for the urea cycle (corresponding to the map in Fig. 1) using Escher maps [31]; each row corresponds to a different metabolite and each column corresponds to a different enzymatic reaction

Metabolite	ARGN	ARGSL	ARGSS	OCBT	PPA	EXATP	EXAMP	EXFUM	EXUREA	EXCBP	EXPI	EXASP	EXH2O	EXH
h2o_c	-1	0	0	0	-1	0	0	0	0	0	0	0	1	0
h_c	0	0	1	1	1	0	0	0	0	0	0	0	0	-1
atp_c	0	0	-1	0	0	1	0	0	0	0	0	0	0	0
ppi_c	0	0	1	0	-1	0	0	0	0	0	0	0	0	0
amp_c	0	0	1	0	0	0	-1	0	0	0	0	0	0	0
fum_c	0	1	0	0	0	0	0	-1	0	0	0	0	0	0
asp__L_c	0	0	-1	0	0	0	0	0	0	0	0	1	0	0
pi_c	0	0	0	1	2	0	0	0	0	0	-1	0	0	0
urea_c	1	0	0	0	0	0	0	0	-1	0	0	0	0	0
orn_c	1	0	0	-1	0	0	0	0	0	0	0	0	0	0
arg__L_c	-1	1	0	0	0	0	0	0	0	0	0	0	0	0
argsuc_c	0	-1	1	0	0	0	0	0	0	0	0	0	0	0
citr__L_c	0	0	-1	1	0	0	0	0	0	0	0	0	0	0
cbp_c	0	0	0	-1	0	0	0	0	0	1	0	0	0	0

Reaction names with "EX" prefix correspond to source or sink reactions required to balance the inputs and outputs of the pathway

Table 3
Equilibrium constants for the urea cycle pathway [13]

Enzyme	K_{eq}
ARGN	100,000
ARGSL	0.0066
ARGSS	0.497
OCBT	120,000
PPA	574

Table 4
Gradient matrix for the chemostat growth condition

	atp_c	amp_c	fum_c	asp__L_c	orn_c	citr__L_c
ARGN	0.	0.	0.	0.	0.	0.
ARGSL	0.	0.	-5.97	0.	0.	0.
ARGSS	0.288	-0.039	0.	0.128	0.	0.288
OCBT	0.	0.	0.	0.	0.513	0.
PPA	0.	0.	0.	0.	0.	0.

Table 5
Gradient matrix for the batch growth condition

	atp_c	amp_c	fum_c	asp__L_c	orn_c	citr__L_c
ARGN	0.	0.	0.	0.	-0.001	0.
ARGSL	0.	0.	-14.815	0.	0.	0.
ARGSS	2.187	-1.146	0.	0.692	0.	3.462
OCBT	0.	0.	0.	0.	0.533	0.
PPA	0.	0.	0.	0.	0.	0.

Table 6
Modal matrix for the growth in the chemostat (with time scales along left-hand side)

TS (h)	atp_c	amp_c	fum_c	asp__L_c	orn_c	citr__L_c
0.168	0.	0.	1.	0.	0.	0.
1.345	0.21	-0.03	0.	0.09	-0.46	0.21
1.949	0.	0.	0.	0.	1.	0.

Table 7
Modal matrix for batch fed culture conditions (with time scales along left-hand side)

TS (h)	atp_c	amp_c	fum_c	asp__L_c	orn_c	citr__L_c
0.168	0.	0.	1.	0.	0.	0.
1.949	0.	0.	0.	0.	-1.	0.
3.891	-0.15	0.08	0.	-0.05	-0.48	-0.24

sense that changes in the growth rate will manifest, at least in part, as changes in the nitrogen disposition of the organism, and in this case, change the relative ratios of the nitrogen storing metabolites.

The energy charge [47] for the chemostat and batch culture conditions were 0.86 and 0.82, respectively [5]. While this is a numerically small difference, it may still be biologically meaningful. Unfortunately, not all of the intermediates of the urea cycle were measured, so calculation of the mass action ratios for OCBT and ARGSS cannot be performed precisely, however assuming that carbamoyl phosphate, argininosuccinate, and inorganic phosphate concentrations do not change significantly between the two conditions, there is a 10-fold decrease in the mass action ratio of OCBT between chemostat and batch culture conditions. Since these intermediate metabolites may change in their values, the actual change in the thermodynamic driving force may not be as dramatic, but nevertheless it is consistent with the time scale decomposition analysis. Further consistent with the differences in the mass action ratios and the time scales, the left null space pools (e.g., cyclic intermediates of the urea cycle) increase from 26 to 32.5 mM in the chemostat versus batch growth conditions. Thus, there is an increased substrate urea cycle storage pool in the batch growth conditions. The example considered herein, although a simple pathway (single dimension in the right and left null spaces, respectively), provides some interesting observations about the difference in pooling of metabolite intermediates in the pathway that is in turn also reflected in the time scale of the metabolite interactions.

5 Conclusion

The development of multiple high-throughput data types in the past two decades, with genomics, proteomics, and more recently and most notably, metabolomics, has enabled the construction of more detailed and larger scope metabolic models. In particular, the improved ability to predict free energy of formation in conjunction with large, quantitative metabolomic data has provided means to construct kinetic models of metabolism using in vivo data.

Quality Control/Quality Assurance measures can be used to ensure thermodynamic consistency and assessment of the dynamic network properties can ensure stability of the resulting models. The principle of mass action is simple yet also applicable to coarse-grained as well as more resolved descriptions of enzymatic biochemical transformations. Thus, the determination of the degree of mechanistic detail in a model is driven by the available data and not a potential bias from the investigator. Further, recent studies have supported the judicious, selective targeting of enzymes to characterize with more detail [9]. Future developments will likely focus on means to achieve more mechanistic details within different pathways that are represented in these models, thus improving the specificity of the predictions.

References

1. Ahn SY, Jamshidi N, Mo ML, Wu W, Eraly SA, Dnyanmote A, Bush KT, Gallegos TF, Sweet DH, Palsson BØ, Nigam SK (2011) Linkage of organic anion transporter-1 to metabolic pathways through integrated "omics"-driven network and functional analysis. J Biol Chem 286(36):31522–31531
2. Beard DA, Liang S-d, Qian H (2002) Energy balance for analysis of complex metabolic networks. Biophys J 83(1):79–86
3. Bordbar A, Monk JM, King ZA, Palsson BØ (2014) Constraint-based models predict metabolic and associated cellular functions. Nat Rev Genet 15:107–120
4. Bordbar A, McCloskey D, Zielinski DC, Sonnenschein N, Jamshidi N, Palsson BØ (2015) Personalized whole-cell kinetic models of metabolism for discovery in genomics and pharmacodynamics. Cell Syst 1(4):283–292
5. Canelas AB, ten Pierick A, Ras C, Seifar RM, van Dam JC, van Gulik WM, Heijnen JJ (2009) Quantitative evaluation of intracellular metabolite extraction techniques for yeast metabolomics. Anal Chem 81:7379–7389
6. Chakrabarti A, Miskovic L, Soh KC, Hatzimanikatis V (2013) Towards kinetic modeling of genome-scale metabolic networks without sacrificing stoichiometric, thermodynamic and physiological constraints. Biotechnol J 8(9):1043–1057
7. Dräger A, Palsson BØ (2014) Improving collaboration by standardization efforts in systems biology. Front Bioeng 2(61). https://doi.org/10.3389/fbioe.2014.00061
8. Dräger A, Zielinski DC, Keller R, Rall M, Eichner J, Palsson BO, Zell A (2015) SBMLsqueezer 2: context-sensitive creation of kinetic equations in biochemical networks. BMC Syst Biol 9(1):1–17
9. Du B, Zielinski D, Dräger A, Tan J, Zhang Z, Ruggiero K, Arzumanyan G, Palsson BO (2016) Evaluation of rate law approximations in bottom-up kinetic models of metabolism. BMC Syst Biol 10(1):1–15
10. Duarte NC, Herrgard MJ, Palsson BØ (2004) Constraint-based models predict metabolic and associated cellular functions. Genome Res 14(7):1298–1309
11. Ebrahim A, Almaas E, Bauer E, Bordbar A, Burgard AP, Chang RL, Dräger A, Famili I, Feist AM, Fleming RMT, Fong SS, Hatzimanikatis V, Herrgård MJ, Holder A, Hucka M, Hyduke D, Jamshidi N, Lee SY, Le Novère N, Lerman JA, Lewis NE, Ma D, Mahadevan R, Maranas C, Nagarajan H, Navid A, Nielsen J, Nielsen LK, Nogales J, Noronha A, Pal C, Palsson BO, Papin JA, Patil KR, Price ND, Reed JL, Saunders M, Senger RS, Sonnenschein N, Sun Y, Thiele I (2015) Do genome-scale models need exact solvers or clearer standards? Mol Syst Biol 11(10):831
12. Famili I, Mahadevan R, Palsson BO (2005) k-cone analysis: determining all candidate values for kinetic parameters on a network scale. Biophys J 88(3):1616–1625
13. Flamholz A, Noor E, Bar-Even A, Milo R (2011) Equilibrator–the biochemical thermodynamics calculator. Nucleic Acids Res 40(D1):D770–D775
14. Förster J, Famili I, Fu P, Palsson BØ, Nielsen J (2003) Genome-scale reconstruction of the Saccharomyces cerevisiae metabolic network. Genome Res 13(2):244–253

15. Frezza C, Zheng L, Folger O, Rajagopalan KN, MacKenzie ED, Jerby L, Micaroni M, Chaneton B, Adam J, Hedley A, Kalna G, Tomlinson IP, Pollard PJ, Watson DG, Deberardinis RJ, Shlomi T, Ruppin E, Gottlieb E (2011) Haem oxygenase is synthetically lethal with the tumour suppressor fumarate hydratase. Nature 477(7363):225–228
16. Gianchandani EP, Chavali AK, Papin JA (2010) The application of flux balance analysis in systems biology. Wiley Interdiscip Rev Syst Biol Med 2(3):372–382
17. Glont M, Nguyen TVN, Graesslin M, Hälke R, Ali R, Schramm J, Wimalaratne SM, Kothamachu VB, Rodriguez N, Swat MJ, Eils J, Eils R, Laibe C, Malik-Sheriff RS, Chelliah V, Le Novère N, Hermjakob H (2018) BioModels: expanding horizons to include more modelling approaches and formats. Nucleic Acids Res 46:D1248–D1253
18. Hamilton JJ, Dwivedi V, Reed JL (2013) Quantitative assessment of thermodynamic constraints on the solution space of genome-scale metabolic models. Biophys J 105(2):512–522
19. Heinrich R, Rapoport SM, Rapoport TA (1978) Metabolic regulation and mathematical models. Prog Biophys Mol Biol 32:1–82
20. Henry CS, Broadbelt LJ, Hatzimanikatis V (2007) Thermodynamics-based metabolic flux analysis. Biophys J 92(5):1792–1805
21. Hucka M, Finney A, Sauro HM, Bolouri H, Doyle JC, Kitano H, Arkin AP, Bornstein BJ, Bray D, Cornish-Bowden A, Cuellar AA, Dronov S, Gilles ED, Ginkel M, Gor V, Goryanin II, Hedley WJ, Hodgman TC, Hofmeyr J-H, Hunter PJ, Juty NS, Kasberger JL, Kremling A, Kummer U, Le Novère N, Loew LM, Lucio D, Mendes P, Minch E, Mjolsness ED, Nakayama Y, Nelson MR, Nielsen PF, Sakurada T, Schaff JC, Shapiro BE, Shimizu TS, Spence HD, Stelling J, Takahashi K, Tomita M, Wagner J, Wang J (2003) The systems biology markup language (SBML): a medium for representation and exchange of biochemical network models. Bioinformatics 19(4):524–531
22. Jamshidi N, Palsson BØ (2008) Formulating genome-scale kinetic models in the post-genome era. Mol Syst Biol 4(1):171
23. Jamshidi N, Palsson BØ (2008) Top-down analysis of temporal hierarchy in biochemical reaction networks. PLoS Comput Biol 4(9): e1000177
24. Jamshidi N, Palsson BØ (2009) Flux-concentration duality in dynamic nonequilibrium biological networks. Biophys J 97(5): L11–L13
25. Jamshidi N, Palsson BØ (2010) Mass action stoichiometric simulation models: incorporating kinetics and regulation into stoichiometric models. Biophys J 98(2):175–185
26. Jamshidi N, Miller FJ, Mandel J, Evans T, Kuo MD (2011) Individualized therapy of HHT driven by network analysis of metabolomic profiles. BMC Syst Biol 5:200
27. Jankowski MD, Henry CS, Broadbelt LJ, Hatzimanikatis V (2008) Group contribution method for thermodynamic analysis of complex metabolic networks. Biophys J 95(3): 1487–1499
28. Kauffman KJ, Pajerowski JD, Jamshidi N, Palsson BØ, Edwards JS (2002) Description and analysis of metabolic connectivity and dynamics in the human red blood cell. Biophys J 83(2):2646–2662
29. Kim TY, Sohn SB, Kim YB, Kim WJ, Lee SY (2012) Recent advances in reconstruction and applications of genome-scale metabolic models. Biotechnol Adv 23(4):617–623
30. Kim B, Kim WJ, Kim DI, Lee SY (2015) Applications of genome-scale metabolic network model in metabolic engineering. J Ind Microbiol Biotechnol 42(3):339–348
31. King ZA, Dräger A, Ebrahim A, Sonnenschein N, Lewis NE, Palsson BO (2015) Escher: a web application for building, sharing, and embedding data-rich visualizations of biological pathways. PLoS Comput Biol 11(8):e1004321
32. King ZA, Lu JS, Dräger A, Miller PC, Federowicz S, Lerman JA, Ebrahim A, Palsson BO, Lewis NE (2016) BiGG Models: a platform for integrating, standardizing, and sharing genome-scale models. Nucleic Acids Res 44(D1):D515–D522
33. Kümmel A, Panke S, Heinemann M (2006) Putative regulatory sites unraveled by network-embedded thermodynamic analysis of metabolome data. Mol Syst Biol 2:2006.0034
34. Lopes H, Rocha I (2017) Genome-scale modeling of yeast: chronology, applications and critical perspectives. FEMS Yeast Res 17(5). https://doi.org/10.1093/femsyr/fox050
35. Lopez CF, Muhlich JL, Bachman JA, Sorger PK (2013) Programming biological models in python using PySB. Mol Syst Biol 9:646
36. Medley K, König M, Galdzicki M, Choi K, Sauro H, Stocking K, Gu S, Smith LP, Asifullah S, Somogyi A (2014–2018) Tellurium. https://github.com/sys-bio/tellurium
37. Megchelenbrink W, Huynen M, Marchiori E (2014) optGpSampler: an improved tool for uniformly sampling the solution-space of genome-scale metabolic networks. PLoS One 9(2):e86587

38. Mo M, Palsson BØ, Herrgård MJ (2009) Connecting extracellular metabolomic measurements to intracellular flux states in yeast. BMC Syst Biol 3:37
39. Nielsen LK, Saa PA (2016) Construction of feasible and accurate kinetic models of metabolism: a Bayesian approach. Sci Rep 6:29635
40. Okino MS, Mavrovouniotis ML (1998) Simplification of mathematical models of chemical reaction systems. Chem Rev 98(2):391–408
41. Osterlund T, Nookaew I, Nielsen J (2012) Fifteen years of large scale metabolic modeling of yeast: developments and impacts. Biotechnol Adv 30(5):979–988
42. Palsson BØ (2006) Systems biology: determining the capabilities of reconstructed networks. Cambridge University Press, Cambridge
43. Palsson BØ (2011) Systems biology: simulation of dynamic network states, 1st edn. Cambridge University Press, Cambridge
44. Palsson BØ, Joshi A, Ozturk SS (1987) Reducing complexity in metabolic networks: making metabolic meshes manageable. Fed Proc 46(8):2485–2489
45. Pan S, Reed JL (2017) Advances in gap-filling genome-scale metabolic models and model-driven experiments lead to novel metabolic discoveries. Curr Opin Biotechnol 51:103–108
46. Ramirez-Guana M, Marcu A, Pon A, Guo AC, Sajed T, Wishart NA, Karu N, Djoumbou Y, Arndt D, Wishart DS (2017) Ymdb 2.0: a significantly expanded version of the yeast metabolome database. Nucleic Acids Res 45(D1): D440–D445
47. Reich JG, Sel'kov EE (1981) Energy metabolism of the cell a theoretical treatise. Academic, London
48. Saa PA, Nielsen LK (2016) ll-ACHRB: a scalable algorithm for sampling the feasible solution space of metabolic networks. Bioinformatics 32(15):2330–2337
49. Sastry A, Sonnenschein N (2013–2018) Mass-toolbox. https://github.com/opencobra/MASS-Toolbox
50. Schellenberger J, Palsson BØ (2009) Use of randomized sampling for analysis of metabolic networks. J Biol Chem 284(9):5457–5461
51. Schellenberger J, Park JO, Conrad TM, Palsson BØ (2010) BiGG: a biochemical genetic and genomic knowledgebase of large scale metabolic reconstructions. BMC Bioinf 11(1):213
52. Schellenberger J, Lewis NE, Palsson BØ (2011) Elimination of thermodynamically infeasible loops in steady-state metabolic models. Biophys J 100(3):544–553
53. Segel IH (1975) Enzyme kinetics: behavior and analysis of rapid equilibrium and steady-state enzyme systems. Wiley-Interscience, New York
54. Shoaie S, Karlsson F, Mardinoglu A, Nookaew I, Bordel S, Nielsen J (2013) Understanding the interactions between bacteria in the human gut through metabolic modeling. Sci Rep 3:2532
55. Srinivasan S, Cluett W, Mahadevan R (2015) Constructing kinetic models of metabolism at genome-scales: a review. Biotechnol J 10(9): 1345–1359
56. Terzer M, Maynard ND, Covert MW, Stelling J (2009) Genome-scale metabolic networks. Wiley Interdiscip Rev Syst Biol Med 1(3): 285–297
57. Tran LM, Rizk ML, Liao JC (2008) Ensemble modeling of metabolic networks. Biophys J 95(12):5606–5617

Chapter 16

Analysis of Endothelial Fatty Acid Metabolism Using Tracer Metabolomics

Joanna Kalucka, Bart Ghesquière, Sarah-Maria Fendt, and Peter Carmeliet

Abstract

Blood vessels are lined by a streamlined monolayer of quiescent endothelial cells (ECs). Although these cells can remain quiescent for years, different stimuli (ischemia, inflammation) and growth factors can activate them and drive a process of new vessel formation (angiogenesis). Emerging evidence reveals that cellular metabolism is a key determinant of the EC subtype specification. The use of stable isotope tracing and mass spectrometry analysis has been essential for the discovery that fatty acid metabolism contributes to EC proliferation and lymphatic EC differentiation. This chapter describes the methodology for setting up palmitate-based tracer metabolomics and the subsequent liquid chromatography-mass spectrometry (LC-MS)-based analysis. As such, tracer metabolomics can be used: (1) to identify the different metabolic pathways relying on carbons provided by fatty acid oxidation and (2) to quantify the relative contributions of palmitate-derived carbons. We begin by providing a background and general principles regarding the use of stable isotopes to study fatty acid metabolism. We then proceed with detailed procedures for the labeling conditions, sample preparation, and subsequent LC-MS analysis.

Key words EC metabolism, Angiogenesis, Stable isotopes, Mass spectrometry, Fatty acids

1 Introduction

Endothelial cells (ECs) are very plastic cell types. They can rapidly switch from periods of quiescence, into intense proliferation during angiogenesis [1]. Until recently, primarily growth factors, receptors, signaling molecules, and related molecules were recognized to drive these processes. However, emerging evidence points out that blood vascular and lymphatic vessel sprouting mechanisms rely on the metabolism of glucose, fatty acids (FAs), and amino acids [2–11]. Over the past years, our group demonstrated that despite close proximity of ECs to oxygen in the blood, ECs mainly

Joanna Kalucka and Bart Ghesquière contributed equally to this work.

Angelo D'Alessandro (ed.), *High-Throughput Metabolomics: Methods and Protocols*, Methods in Molecular Biology, vol. 1978, https://doi.org/10.1007/978-1-4939-9236-2_16, © Springer Science+Business Media, LLC, part of Springer Nature 2019

rely on glycolysis to produce ATP [2]. Activated ECs are hyperglycolytic and depend largely on glycolysis for angiogenic growth. Hence, anti-glycolytic strategies impair pathological angiogenesis [7, 12]. In addition to glycolysis, other metabolic pathways, including FA metabolism, contribute to the EC differentiation, motility, and growth. FAs play an important role in cells as structural elements of membranes, energy, storage, and signaling molecules. Long-chain FAs (e.g., palmitate) are the products of fatty acid synthesis, which takes place in the cytoplasm, and can be precursors for longer fatty acids and/or substrates for fatty acid ß-oxidation (FAO). FAO progresses in mitochondria in four-step cycles, each time cleaving two carbons that form acetyl-CoA at the end of every cycle. Successively, these acetyl-CoA units can be used for further oxidation in the tricarboxylic acid (TCA) cycle.

We recently reported that FAO is critical for vessel sprouting in vivo [3]. Endothelial-selective deletion of carnitine palmitoyltransferase 1A (CPT1A), a mitochondrial FA importer and rate-controlling enzyme of FAO, causes vessel sprouting defects in vitro and in vivo by impairing EC proliferation, without affecting migration [3]. Moreover, FAO and its key enzyme CPT1A play an important function in the lymphatic development of murine embryos [4]. FAO in lymphatic ECs promotes lymphatic vessel growth and stimulates lymphatic EC proliferation by sustaining the TCA cycle, in conjunction with an anaplerotic substrate, for dNTP synthesis during DNA replication. Recent evidence indicates that FAO in quiescent ECs ensures redox homeostasis by promoting NADPH production, presumably to cope with the high oxygen, and therefore oxidative stress-prone, microenvironment that these cells are exposed to [9]. ECs are also capable of synthesizing FAs, but fatty acid synthetase (FASN), the enzyme synthesizing FAs, also regulates the levels of malonyl-CoA, the substrate of this enzyme. Notably, FASN inhibition elevates cytosolic malonyl-CoA levels, which leads to posttranslational malonylation of mTOR, thereby reducing mTOR's activity, which impairs vessel sprouting [10]. Given the important role of FA metabolism in both normal cellular functioning and disease, a variety of different techniques were developed that allow us to study this process.

One of the earliest approaches used radiolabeled substrates such as ^{14}C-palmitate to measure the movement of FAs across the myocardial wall from its surface into intracellular sites and their distribution between subcellular fractions [13]. In addition, labeling with ^{14}C-acetate (that can be metabolized to acetyl-coA) was established to measure isotopic enrichment in the lipid fraction [14]. The downside of this approach is that it is not specific to particular fatty acid types and therefore does not provide information on isotope enrichment per molecule. Therefore, it cannot be used to determine the contribution of individual fatty acid metabolic reaction.

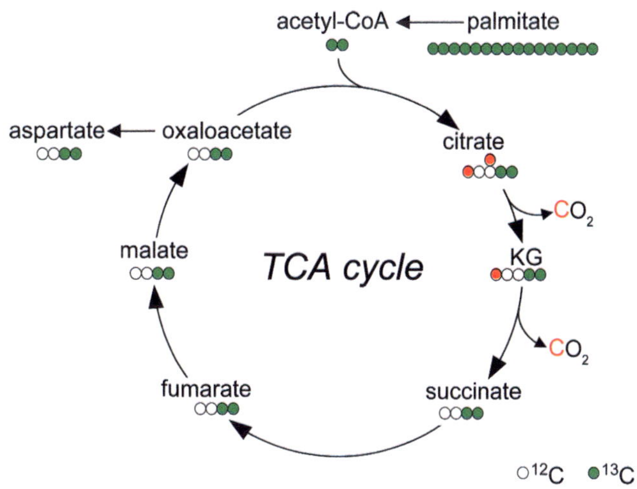

Fig. 1 A schematic representation of the utilization of ^{13}C palmitate carbons into TCA cycle intermediates. Circles symbolize the carbon backbone of the molecules. Green circles mark ^{13}C palmitate-derived carbons. White circles mark ^{12}C carbons. Carbons released as CO_2 are indicated in red

Besides radioactive tracers, nonradioactive isotopically labeled substrates (carrying, e.g., ^{13}C-palmitate, ^{13}C-glucose) can be used to generate information on the metabolic network and relative activity of metabolic pathways linked to the labeled substrate. ^{13}C-palmitate carbons are metabolized to acetyl-CoA and can be further incorporated into TCA cycle intermediates (*see* Fig. 1). Consequently, these carbons eventually stretch out into a multitude of metabolic paths. Using ^{13}C-palmitate, we documented that proliferating ECs incorporate palmitate-derived carbons into TCA intermediates, as well as in glutamate and aspartate. Remarkably, the amount of carbons of citrate that were derived from palmitate were nearly equal to those from glucose and glutamine in ECs [3]. Furthermore, knockdown of CPT1A depleted the levels of glutamate and aspartate, precursors of critical nucleotides, and proteins, further supporting evidence of the importance of FA-derived carbons in EC function [3]. Below, we provide a detailed description on how to perform ^{13}C-palmitate experiments using liquid chromatography-based mass spectrometry (LC-MS) in order to study the central aspects of fatty acid metabolism in ECs.

2 Materials

2.1 General

1. Regular endothelial M199 (ThermoFischer Scientific) cell culture medium containing 5.5 mM glucose, 2 mM glutamine, 20% FBS, supplemented with endothelial cell growth factors and heparin.

2. Labeling and control of endothelial cell culture medium: complete M199 medium containing 20% fatty acid-free bovine serum albumin (e.g., charcoal-stripped FBS) and BSA-conjugated with sodium palmitate ([U-^{13}C]-palmitate or control ^{12}C-sodium palmitate). A final concentration of 100 μM palmitate in the medium is used.
3. 5 M NaCl stock solution.

2.2 BSA-Conjugated ([U-^{13}C]-) Palmitate Solution

1. Sodium palmitate (Sigma) and sodium [U-^{13}C]-palmitate (Cambridge Isotopes).
2. Ultra Fatty Acid-Free BSA (Roche Applied Science).
3. 150 ml filter unit, 0.22 μm.
4. 150 mM NaCl solution, e.g., prepare approximately 300 ml by adding 9 ml of 5 M NaCl stock to 291 ml mqH_2O.

2.3 Extraction Buffers

1. Prepare ice cold 150 mM NaCl washing solution, e.g., by adding 9 ml of the 5 M NaCl stock to 291 ml mqH_2O. The solution can be prepared in advance and kept at 4 °C.
2. Cellular extraction buffer: 50% methanol/30% acetonitrile/20% water (containing 10 mM ammonium acetate pH 9.3). The extraction buffer also contains a final concentration of 2 μM deuterated (d27) myristic acid. All of the organic buffers are LC-MS grade. The solution can be prepared in advance and kept at 4 °C.
3. Medium extraction buffer: 80% methanol containing a final concentration of 2 μM deuterated (d27) myristic acid. Methanol is LC-MS grade. The solution can be prepared in advance and stored at −80 °C.

2.4 MS Analysis

1. iHILIC-fusion(P) column (200A 150 × 2.1 mm ID × 5 μm polymer-based) (Achrom).
2. Ultimate 3000 UPLC (ultra performance liquid chromatography) in line connected to a Q-Exactive OrbiTRAP mass spectrometer (Thermo Scientific).
3. Solvent A (100% acetonitrile, LC-MS grade), solvent B (10 mM ammonium acetate pH 9.3).

3 Methods

3.1 Preparation of Palmitate-BSA Labeling (Protocol Adapted from Seahorse Bioscience)

1. To prepare the sodium palmitate-BSA solution: prepare 1 mM sodium palmitate (either labeled [U-^{13}C]- or unlabeled sodium palmitate) and 0.17 mM BSA solution (molar ratio should be kept 6:1; palmitate: BSA) (*see* **Note 1**).
2. In order to prepare 50 ml of BSA solution, weigh out 1.1335 g Ultra Fatty Acid-Free BSA. Dissolve BSA in 50 ml of 150 mM

NaCl solution and transfer to the pre-warmed 250 ml beaker. Mix the BSA solution using magnetic stirrer with temperature control and cover well the beaker to avoid liquid evaporation (e.g., with parafilm). Adjust the temperature as needed to maintain approximately 37 °C but not higher than 40 °C (use thermometer to control liquid temperature) (*see* **Note 2**).

3. Once BSA is completely dissolved, filter the solution with a vacuum in a laminar flow hood.
4. To prepare palmitate solution, weight out 15.3 mg of sodium palmitate ([U-^{13}C]-palmitate or unlabeled) and add it to 22 ml of 150 mM NaCl solution. Pre-warm the solution to 70 °C (*see* **Note 3**).
5. To conjugate palmitate and BSA, transfer pre-warmed palmitate solution to 25 ml of BSA solution and keep on stirring at 37 °C (*see* **Note 4**). Stir for approximately 1 h at 37 °C, monitoring the temperature to keep it between 35 °C and 40 °C. Adjust the final volume to 100 ml with 150 mM NaCl to create 1 mM palmitate-conjugate working solution. If needed, adapt the pH to 7.4 with 1 N NaOH. Aliquot palmitate-BSA solution and freeze at −20 °C (*see* **Note 5**).

3.2 Cultivation of ECs and Labeling Procedure

Carry out all procedures under sterile conditions unless otherwise specified. Plan labeling experiment carefully and make sure that all control conditions are included (*see* Fig. 2). The following protocol describes the [U-^{13}C]-palmitate labeling procedure for proliferating ECs. The labeling of ECs with palmitate is performed throughout 72 h to ensure steady-state metabolite concentrations.

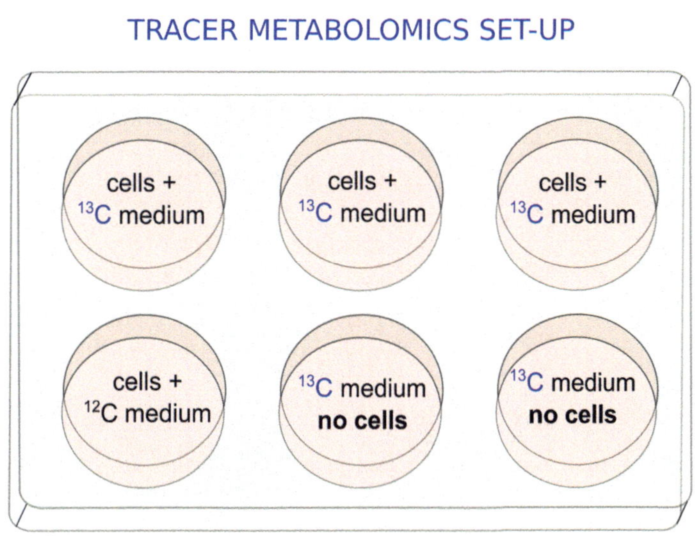

Fig. 2 A scheme illustrating a typical approach of setting up a tracer metabolomic experiment. Apart from experimental conditions (upper row), the following control conditions are included: cells cultivated with ^{12}C medium and ^{13}C medium with no cells

1. Day 1: the day before the experiment, seed 300,000 ECs per well in a gelatin coated 6-well plate (*see* **Notes 6** and **7**).
2. Day 2: wash cells twice with 2 ml of PBS to remove traces of unlabeled palmitate and other fatty acids. Add 1.5 ml of medium containing labeled [U-^{13}C]-palmitate. Perform the same procedure for the control conditions that will be cultivated with ^{12}C-palmitate medium (*see* **Notes 8** and **9**).
3. Day 3: refresh the medium by gently aspirating old medium and replacing it with new labeled (for experimental conditions) or unlabeled (for control conditions) medium.
4. Day 4: remove the medium (in case of medium analysis, *see* Subheading 3.2.2) and proceed with extraction protocol.

3.2.1 Cellular Metabolite Extraction

Cell metabolism is a very dynamic process, and consequently, the metabolic activity can be correlated with the cell cycle phases. Hence, one should carefully control cell density while performing these steps and apply a normalization control, if mandatory (*see* **Note 10**).

1. Wash cells with ice cold (4 °C) 150 mM NaCl. Ensure washing solution is removed completely and transfer plate on ice for the following steps (*see* **Note 11**).
2. Add 300 μl of cellular extraction buffer (stored at −20 °C) and incubate 2–3 min on ice and scrape cells using cell scraper.
3. Transfer extraction mix to labeled Eppendorf tubes and centrifuge extract at 20,000 × *g* in table top centrifuge for 15 min at 4 °C. Next, transfer supernatant to glass vials equipped with conical inserts.
4. If normalization of the results to protein content is required, keep the obtained from this step protein pellet and proceed with, e.g., BCA analysis [15] (*see* **Notes 10** and **12**).

3.2.2 Metabolite Extraction from Medium

1. To 10 μl of the medium, add 990 μl of medium extraction buffer (stored at −80 °C) and store this extraction mixture overnight at −80 °C. Next, centrifuge the extract at 20,000 × *g* in a cooled (4 °C) table top centrifuge for 15 min.
2. Transfer of 250 μl of the supernatant to glass vials equipped with conical inserts suited for mass spectrometry.

3.3 LC-MS Analysis

Preconditioning of the iHILIC-fusion(P) column (200A 150 × 2.1 mm ID × 5 μm polymer based) (Achrom) is done by running 30 random samples (cellular, medium, or tissue extracts). Typical backpressure profiles start from 40 to 70 bar at t0 (*see* below) and reach up to maximum of 150–180 bar. Quality control (QC) samples can be generated by taking aliquots from each sample and pooling these into a QC sample. This QC is run prior,

Table 1
Linear LC-MS gradient for the HILIC analysis of metabolites, solvent A is LC-MS grade acetonitrile, and B is 10 mM ammonium acetate pH 9.3

Time (min)	% Solvent A	% Solvent B
0	90	10
2	90	10
20	20	80
23	20	80
25	40	60
27	90	10
35	90	10

intermediate, and post each batch of samples (*see* **Note 13**). Variation in intensity and retention time should not differ more than 10% and 0.5 min, respectively (*see* **Note 13**).

1. Inject either 10 μl (medium samples) or 20 μl (cellular samples) of the metabolite extracts onto the Ultimate 3000 UPLC.
2. A linear gradient is carried out (*see* Table 1) at a constant flow rate of 200 μl/min. The temperature of the column compartment is kept constant at 25 °C.
3. Metabolites eluting between 0.5 and 28 min are analyzed onto the Q-Exactive OrbiTRAP. The mass spectrometer operates in negative ion mode; the settings of the HESI probe are as follows: sheath gas flow rate at 35 and auxiliary gas flow rate at 10 (at a temperature of 260 °C). Spray voltage was set at 4.8 kV, the temperature of the capillary at 300 °C, and the S-lens RF level at 50. A full scan (resolution of 140,000 and scan range of m/z 70–1050) was applied. For this, the ion fill time was set at 512 ms and the AGC target at 3E6 ions.
4. Data analysis of the obtained raw files is performed using vendor-specific or in-house software platforms (*see* **Note 14**).

4 Notes

1. The usage of palmitate in metabolomics is challenging due to palmitate's low solubility in aqueous solution. Therefore, it is recommended to conjugate palmitate to BSA in order to ensure good solubility and further uptake.
2. In order to achieve high quality of conjugation, maintaining specified temperatures is critical.

3. Palmitate solution may become hazy between 50 and 60 °C. It becomes clear as it reaches 70 °C.
4. Palmitate will precipitate when left to cool down and can be avoided by stepwise pipetting.
5. BSA-palmitate solution is stable for 2 months when stored properly at −20 °C.
6. Cell metabolism is a very dynamic process, and metabolic activity of ECs has been shown to be correlated with the cell cycle phases [2, 16]. Therefore, it is of utmost importance that labeling experiments are performed at the same conditions to ensure reproducibility of obtained results.
7. It is not advisable to scale down to smaller well formats or use less cells than 300,000. Only using sufficient number of cells can guarantee reproducible results.
8. Make sure that all medium ([U-^{13}C]-palmitate and control medium) are adequately pre-warmed and no precipitates of palmitate tracer are visible. Precipitated palmitate can be toxic for cells.
9. Make sure that the cells are properly washed with PBS before adding labeling medium. This step reassures that all unlabeled fatty acids will be removed from the culture conditions and will increase reproducibility of the obtained results.
10. Although the analysis of [U-^{13}C]-palmitate labeling patterns and more specifically fractional contribution do not require a normalization step, one could include it if the assessment of metabolite concentration (abundance) is necessary. Normalization to the total protein content, cell number, and DNA content are considered as the golden standards in metabolomics. In order to normalize results to the protein content (e.g., using BCA analysis), protein pellets should be kept after step described in Subheading 3.2.1, **step 4**.
11. When having multiple experimental conditions, always extract samples stepwise. Avoid extracting samples from multiple plates simultaneously. To ensure good quality of MS analysis, work as precisely and rapidly as possible.
12. To dissolve protein pellets, add up to 200 μl of 0.2 N NaOH and boil the samples 3–5 min at 95 °C. Vortex the samples and check if pellet is dissolved. In case if the pellet is not fully re-solubilized, perform the boiling step again.
13. When running cellular and medium samples in the same batch, a separate QC for both the cellular and medium samples should be generated. Variations in intensity and retention time are monitored by extracting the same target compounds.

14. Data extraction of isotopic profiles occurs via vendor-specific software or *in-house* developed software platforms (such as BIOMEX™) [17]. In general, the workflow starts with *in-house* libraries containing information on the exact mass of the target molecule and its retention time during the chromatographic separation to identify the metabolite of interest. Next, the abundances of the isotopes of different metabolites of interest are quantified. Following a correction for natural abundances [18, 19] to facilitate proper interpretation of the labeling data, the fractional contribution is calculated [19].

Acknowledgments

J.K. is supported by the Research Foundation Flanders (FWO); S.M.F. is supported by Marie Curie CIG, grants from the Research Foundation Flanders (FWO-Vlaanderen) Belgium; P.C. is supported by the VIB TechWatch program, a Federal Government Belgium grant (IUAP7/03), long-term structural Methusalem funding by the Flemish Government, grants from the Research Foundation Flanders (FWO-Vlaanderen), Foundation against Cancer (2012-175 and 2016-078), and ERC Advanced Research grant (EU-ERC743074).

References

1. Potente M, Gerhardt H, Carmeliet P (2011) Basic and therapeutic aspects of angiogenesis. Cell 146(6):873–887
2. De Bock K, Georgiadou M, Schoors S, Kuchnio A, Wong BW, Cantelmo AR, Quaegebeur A, Ghesquiere B, Cauwenberghs S, Eelen G, Phng LK, Betz I, Tembuyser B, Brepoels K, Welti J, Geudens I, Segura I, Cruys B, Bifari F, Decimo I, Blanco R, Wyns S, Vangindertael J, Rocha S, Collins RT, Munck S, Daelemans D, Imamura H, Devlieger R, Rider M, Van Veldhoven PP, Schuit F, Bartrons R, Hofkens J, Fraisl P, Telang S, Deberardinis RJ, Schoonjans L, Vinckier S, Chesney J, Gerhardt H, Dewerchin M, Carmeliet P (2013) Role of PFKFB3-driven glycolysis in vessel sprouting. Cell 154(3):651–663
3. Schoors S, Bruning U, Missiaen R, Queiroz KC, Borgers G, Elia I, Zecchin A, Cantelmo AR, Christen S, Goveia J, Heggermont W, Godde L, Vinckier S, Van Veldhoven PP, Eelen G, Schoonjans L, Gerhardt H, Dewerchin M, Baes M, De Bock K, Ghesquiere B, Lunt SY, Fendt SM, Carmeliet P (2015) Fatty acid carbon is essential for dNTP synthesis in endothelial cells. Nature 520(7546):192–197
4. Wong BW, Wang X, Zecchin A, Thienpont B, Cornelissen I, Kalucka J, Garcia-Caballero M, Missiaen R, Huang H, Bruning U, Blacher S, Vinckier S, Goveia J, Knobloch M, Zhao H, Dierkes C, Shi C, Hagerling R, Moral-Darde V, Wyns S, Lippens M, Jessberger S, Fendt SM, Luttun A, Noel A, Kiefer F, Ghesquiere B, Moons L, Schoonjans L, Dewerchin M, Eelen G, Lambrechts D, Carmeliet P (2017) The role of fatty acid beta-oxidation in lymphangiogenesis. Nature 542(7639):49–54
5. Huang H, Vandekeere S, Kalucka J, Bierhansl L, Zecchin A, Bruning U, Visnagri A, Yuldasheva N, Goveia J, Cruys B, Brepoels K, Wyns S, Rayport S, Ghesquiere B, Vinckier S, Schoonjans L, Cubbon R, Dewerchin M, Eelen G, Carmeliet P (2017) Role of glutamine and interlinked asparagine metabolism in vessel formation. EMBO J 36(16):2334–2352
6. Kim B, Li J, Jang C, Arany Z (2017) Glutamine fuels proliferation but not migration of endothelial cells. EMBO J 36(16):2321–2333

7. Cantelmo AR, Conradi LC, Brajic A, Goveia J, Kalucka J, Pircher A, Chaturvedi P, Hol J, Thienpont B, Teuwen LA, Schoors S, Boeckx B, Vriens J, Kuchnio A, Veys K, Cruys B, Finotto L, Treps L, Stav-Noraas TE, Bifari F, Stapor P, Decimo I, Kampen K, De Bock K, Haraldsen G, Schoonjans L, Rabelink T, Eelen G, Ghesquiere B, Rehman J, Lambrechts D, Malik AB, Dewerchin M, Carmeliet P (2016) Inhibition of the glycolytic activator PFKFB3 in endothelium induces tumor vessel normalization, impairs metastasis, and improves chemotherapy. Cancer Cell 30(6):968–985
8. Vandekeere S, Dubois C, Kalucka J, Sullivan MR, Garcia-Caballero M, Goveia J, Chen R, Diehl FF, Bar-Lev L, Souffreau J, Pircher A, Kumar S, Vinckier S, Hirabayashi Y, Furuya S, Schoonjans L, Eelen G, Ghesquiere B, Keshet E, Li X, Vander Heiden MG, Dewerchin M, Carmeliet P (2018) Serine synthesis via PHGDH is essential for heme production in endothelial cells. Cell Metab 28(4):573–587.e13. https://doi.org/10.1016/j.cmet.2018.06.009
9. Kalucka J, Bierhansl L, Conchinha NV, Missiaen R, Elia I, Bruning U, Scheinok S, Treps L, Cantelmo AR, Dubois C, de Zeeuw P, Goveia J, Zecchin A, Taverna F, Morales-Rodriguez F, Brajic A, Conradi LC, Schoors S, Harjes U, Vriens K, Pilz GA, Chen R, Cubbon R, Thienpont B, Cruys B, Wong BW, Ghesquiere B, Dewerchin M, De Bock K, Sagaert X, Jessberger S, Jones EAV, Gallez B, Lambrechts D, Mazzone M, Eelen G, Li X, Fendt SM, Carmeliet P (2018) Quiescent endothelial cells upregulate fatty acid beta-oxidation for vasculoprotection via redox homeostasis. Cell Metab 28(6):881–894.e13. https://doi.org/10.1016/j.cmet.2018.07.016
10. Bruning U, Morales-Rodriguez F, Kalucka J, Goveia J, Taverna F, Queiroz KCS, Dubois C, Cantelmo AR, Chen R, Loroch S, Timmerman E, Caixeta V, Bloch K, Conradi LC, Treps L, Staes A, Gevaert K, Tee A, Dewerchin M, Semenkovich CF, Impens F, Schilling B, Verdin E, Swinnen JV, Meier JL, Kulkarni RA, Sickmann A, Ghesquiere B, Schoonjans L, Li X, Mazzone M, Carmeliet P (2018) Impairment of angiogenesis by fatty acid synthase inhibition involves mTOR malonylation. Cell Metab 28(6):866–880.e15. https://doi.org/10.1016/j.cmet.2018.07.019
11. Eelen G, Dubois C, Cantelmo AR, Goveia J, Bruning U, DeRan M, Jarugumilli G, van Rijssel J, Saladino G, Comitani F, Zecchin A, Rocha S, Chen R, Huang H, Vandekeere S, Kalucka J, Lange C, Morales-Rodriguez F, Cruys B, Treps L, Ramer L, Vinckier S, Brepoels K, Wyns S, Souffreau J, Schoonjans L, Lamers WH, Wu Y, Haustraete J, Hofkens J, Liekens S, Cubbon R, Ghesquiere B, Dewerchin M, Gervasio FL, Li X, van Buul JD, Wu X, Carmeliet P (2018) Role of glutamine synthetase in angiogenesis beyond glutamine synthesis. Nature 561(7721):63–69. https://doi.org/10.1038/s41586-018-0466-7
12. Schoors S, De Bock K, Cantelmo AR, Georgiadou M, Ghesquiere B, Cauwenberghs S, Kuchnio A, Wong BW, Quaegebeur A, Goveia J, Bifari F, Wang X, Blanco R, Tembuyser B, Cornelissen I, Bouche A, Vinckier S, Diaz-Moralli S, Gerhardt H, Telang S, Cascante M, Chesney J, Dewerchin M, Carmeliet P (2014) Partial and transient reduction of glycolysis by PFKFB3 blockade reduces pathological angiogenesis. Cell Metab 19(1):37–48
13. Fields M, Gloster J, Harris P (1975) Transfer of [1(−14)C] palmitate between myocardial cell fractions. J Mol Cell Cardiol 7(12):907–916
14. Daniels VW, Smans K, Royaux I, Chypre M, Swinnen JV, Zaidi N (2014) Cancer cells differentially activate and thrive on de novo lipid synthesis pathways in a low-lipid environment. PLoS One 9(9):e106913
15. Walker JM (1994) The bicinchoninic acid (BCA) assay for protein quantitation. Methods Mol Biol 32:5–8
16. Kalucka J, Missiaen R, Georgiadou M, Schoors S, Lange C, De Bock K, Dewerchin M, Carmeliet P (2015) Metabolic control of the cell cycle. Cell Cycle 14(21):3379–3388
17. Sugimoto M, Kawakami M, Robert M, Soga T, Tomita M (2012) Bioinformatics tools for mass spectroscopy-based metabolomic data processing and analysis. Curr Bioinforma 7(1):96–108
18. Fernandez CA, Des Rosiers C, Previs SF, David F, Brunengraber H (1996) Correction of 13C mass isotopomer distributions for natural stable isotope abundance. J Mass Spectrom 31(3):255–262
19. Buescher JM, Antoniewicz MR, Boros LG, Burgess SC, Brunengraber H, Clish CB, DeBerardinis RJ, Feron O, Frezza C, Ghesquiere B, Gottlieb E, Hiller K, Jones RG, Kamphorst JJ, Kibbey RG, Kimmelman AC, Locasale JW, Lunt SY, Maddocks OD, Malloy C, Metallo CM, Meuillet EJ, Munger J, Noh K, Rabinowitz JD, Ralser M, Sauer U, Stephanopoulos G, St-Pierre J, Tennant DA, Wittmann C, Vander Heiden MG, Vazquez A, Vousden K, Young JD, Zamboni N, Fendt SM (2015) A roadmap for interpreting (13)C metabolite labeling patterns from cells. Curr Opin Biotechnol 34:189–201

Chapter 17

Stable Isotope Tracers for Metabolic Pathway Analysis

Sara Violante, Mirela Berisa, Tiffany H. Thomas, and Justin R. Cross

Abstract

Stable isotope tracing allows a metabolic substrate to be followed through downstream biochemical reactions, thereby providing unparalleled insights into the metabolic wiring of cells. This approach stops short of modeling absolute fluxes but is relatively straightforward and has become increasingly accessible due to the widespread adoption of high-resolution mass spectrometers. Analysis of both dynamic and steady-state labeling patterns in downstream metabolites provides valuable qualitative information as to their origin and relative rates of production. Stable isotope tracing is, therefore, a powerful way to understand the impact of genetic alterations and defined perturbations on metabolism. In this chapter, we describe a liquid chromatography-mass spectrometry (LC-MS) protocol for stable isotope tracing using ^{13}C-L-arginine in a macrophage cell line. A similar approach can be used to follow other stable isotope tracers, and notes are provided with advice on how this protocol can be generalized for use in other settings.

Key words Stable isotope, Metabolomics, Tracing experiments, Fluxomics

1 Introduction

Interpretation of metabolomics data is often complicated by a lack of dynamic information; this is because significant changes in flux through a pathway can occur without corresponding changes in metabolite abundances (pool sizes). Stable isotope tracing, also referred to as qualitative flux analysis, can provide clarity in these situations by allowing visualization of the flow of a metabolic tracer through pathways of interest. Typically, the tracer contains ^{13}C, ^{15}N, or ^{2}H atoms (less commonly ^{18}O and ^{34}S have also been used). These heavy atoms can either be uniformly enriched throughout the tracer molecule or enriched at specific positions. The choice of tracer depends on the pathway being investigated, and examples of common isotopic tracers and their uses have recently been comprehensively reviewed [1].

In these experiments, a stable isotope tracer is introduced into the biological system as a metabolic substrate. Uptake and incorporation

Angelo D'Alessandro (ed.), *High-Throughput Metabolomics: Methods and Protocols*, Methods in Molecular Biology, vol. 1978, https://doi.org/10.1007/978-1-4939-9236-2_17,

of the tracer results in altered isotopic patterns (mass isotopologue distributions) in downstream metabolites. This approach is particularly useful when studying cell culture models which are inherently amenable to the introduction of stable isotope tracers through exchange of the culture medium. However, stable isotope tracing has also been successfully implemented in vivo, for example, by supplementing drinking water with D_2O in animal models [2], and in human studies using preoperative tracer infusions followed by analysis of labeling patterns in resected tumor tissue [3, 4]. Suitable analytical platforms for these experiments include gas chromatography (GC) or liquid chromatography (LC) coupled to mass spectrometry (MS) and nuclear magnetic resonance (NMR) spectroscopy has also been used extensively [5, 6]. While NMR and MS/MS fragmentation can unambiguously determine the position of label incorporations, LC-MS analysis provides the highest throughput approach, particularly when considering a limited number of tracers and time points. When LC-MS is used, the chromatography should be robust and well-annotated, and a high-resolution detector is recommended to distinguish potentially overlapping mass isotopologue envelopes. To illustrate these and other key considerations involved in performing successful stable isotope tracing experiments we describe an in vitro tracing experiment using a macrophage cell line.

Macrophages are frontline sentinels of the innate immune system responsible for the early detection of pathogens and coordinating the adaptive immune response. Macrophage polarization refers to their activation status and has been grouped into two major phenotypic states, termed M1 and M2 activation. M1 macrophages are associated with immunity to intracellular pathogens and bacteria and arise in inflammatory settings in response to Toll-like (TLR) and interferon signaling. M2 macrophages are associated with anti-inflammatory responses and with immunity to helminth infection, asthma, and allergy [7]. Elevated arginase I expression has been considered a marker of M2 differentiation and leads to a concomitant redirection of L-arginine metabolism [8]. Thus, isotope tracing with ^{13}C-L-arginine provides a useful approach to investigate the metabolic rewiring of M2 polarization.

In this chapter, we provide a protocol to study the metabolism of activated macrophages by following the incorporation of ^{13}C-L-arginine using LC-MS. Detailed notes are provided throughout to explain how this protocol can be generalized for use in other settings and with other tracers.

2 Materials

1. Reagents and mobile phases should be prepared using ultrapure water and glassware reserved for mass spectrometry applications. Solvents should be high-performance liquid chromatography (HPLC) grade or higher, and fresh mobile

phases should be prepared at least weekly, especially in the case of aqueous solutions.

2. Cell culture media (SILAC DMEM lacking arginine, lysine, glutamine), phosphate-buffered saline (PBS), L-glutamine, and antibiotics are from Gibco. Low endotoxin fetal bovine serum (FBS) is from Thermo Fisher Scientific and dialyzed FBS (dFBS) is from Gemini Scientific.
3. Chemicals required are 8-Bromoadenosine 3′,5′-cyclic monophosphate (8-Br-cAMP) and L-lysine from Millipore Sigma and [U-^{13}C]-L-arginine ($^{13}C_6$-L-arginine) from Cambridge Isotope Laboratories (>99% atom percent enrichment).

2.1 Cell Culture Media and Conditions

1. The cell model used is a murine macrophage cell line, RAW 264.7 (ATCC).
2. Cells are defrosted, maintained in 10 cm dishes, and grown in Dulbecco's Modified Eagle Medium (DMEM) high glucose (4.5 g/L) supplemented with 10% low endotoxin FBS (FBS), 2 mM L-glutamine, and 1% Pen/Strep.
3. For macrophage subculture aspirate the media, wash with 5 mL PBS, and add 5 mL of new media. Gently scrape the cells making sure that they detach from the dish without trypsinization. Transfer the detached cells to a 15 mL tube, wash the plate with 5 mL of media, and combine with the scraped cells. Ensure a single-cell suspension by pipetting up and down at least five times.
4. Dilute the cells 1:6 into a new dish. Change to new media every 2 days and subculture when cells reach 80–90% confluency (typically within 3–4 days).

2.2 LC-MS

1. Liquid chromatography is performed on an Agilent 1290 Infinity II Binary UHPLC pump, thermostatted column compartment, and multisampler. A 1260 Isocratic pump with a 100:1 flow splitter or equivalent is used to introduce the reference mass solution containing purine and HP-0921 (hexakis(1H, 1H, 3H, 3H-tetrafluoropropoxy)phosphazene) for in-scan mass axis calibration according to the manufacturer's instructions. The system is fitted with a 0.2 μm inline filter after the autosampler.
2. Chromatographic separation is performed by hydrophilic liquid interaction chromatography (HILIC) using an amide stationary phase (Waters Acquity UPLC BEH Amide, 2.1 × 100 mm, 1.7 μm). Optional: a guard column can be added (Waters Acquity UPLC BEH Amide VanGuard Precolumn 2.1 × 5 mm, 1.7 μm).
3. The samples are acquired in positive and negative ion mode using low pH and high pH methods, respectively. A separate column was dedicated to each buffer system.

4. Mobile phase composition for the low pH method: water, 200 mM ammonium acetate (10 mM final concentration), acetonitrile, and acetic acid at a ratio of 85:5:10:0.2 for mobile phase A and 5:5:90:0.2 for mobile phase B (*see* **Note 1**).
5. Mobile phase composition for the high pH method: water, 200 mM ammonium acetate pH 9.8 (10 mM final concentration), and acetonitrile at a ratio of 85:5:10 for mobile phase A and 5:5:90 for mobile phase B (*see* **Note 2**).
6. The LC is interfaced to a quadrupole time-of-flight (Q-TOF) mass spectrometer (Agilent 6545 Q-TOF) with a dual Agilent Jet Stream electrospray ionization source operating in 2 GHz extended dynamic range mode.

2.3 Other Reagents and Materials

A system suitability test (SST) sample is used in every batch to ensure adequate instrument performance and track the longitudinal performance of the analytical system. These samples are HEK293 cells (1 × 10^6) previously extracted with 80% methanol containing a mixture of isotope-labeled internal standards (labeled Metabolomics Amino Acid Mix Standard, Cambridge Isotope Laboratories). The SST aliquots are dried and stored at −80 °C. Before analysis, the SST is resuspended in 60 μL of 60% acetonitrile in water and prepared as indicated in Subheading 3.4. Area responses and retention times of key metabolites and initial and peak column pressures are tracked, and system maintenance is performed if these deviate from historical norms.

3 Methods

3.1 Cell Culture

1. To set up the tracing experiment, prepare a master flask with cells at a density 240,000 cell/mL and add 2.5 mL per well to 6-well plates, for a final density of 600,000 cells/well. When multiple plates are being seeded, swirl the master flask regularly to prevent cells from settling and ensure even plating across all plates.
2. Prepare at least duplicate and preferably triplicate wells for each labeling time point and biological condition. In this example, the time points are 4, 8, 12, and 24 h of labeling, with and without 8-Br-cAMP treatment (*see* **Note 3**).
3. Include samples that will receive unlabeled media (+/− treatments). These are useful controls, and since metabolites will have natural isotope distributions, they can be used to help identify compounds and search spectral libraries. These samples are also useful to confirm that natural isotope abundance correction is working correctly (described in Subheading 3.6, **step 3**, *see* **Note 4**).
4. Optional: if metabolite pool sizes are to be compared across conditions (in addition to percent labeling), then additional

wells should be prepared for final cell counts to allow subsequent data normalization.

5. Optional: we recommend plating some additional wells for each biological condition for use as system test samples.

3.2 Stable Isotope Labeling

Choosing an appropriate experimental design is critical to the success of stable isotope tracing experiments and depends on the experimental question being addressed, the tracer, and the biological system. Ideally, the system will be at metabolic steady state throughout the stable isotope tracing experiment. For proliferating cells in exponential growth phase, a pseudo-metabolic steady state can often be assumed, but care should be taken to make sure they remain in the same growth state throughout the experiment.

If the comparison is between cells of different genetic backgrounds or another pre-existing difference, then a design similar to that shown in Fig. 1a may be appropriate. This staggered start approach ensures all samples have approximately equal biomass

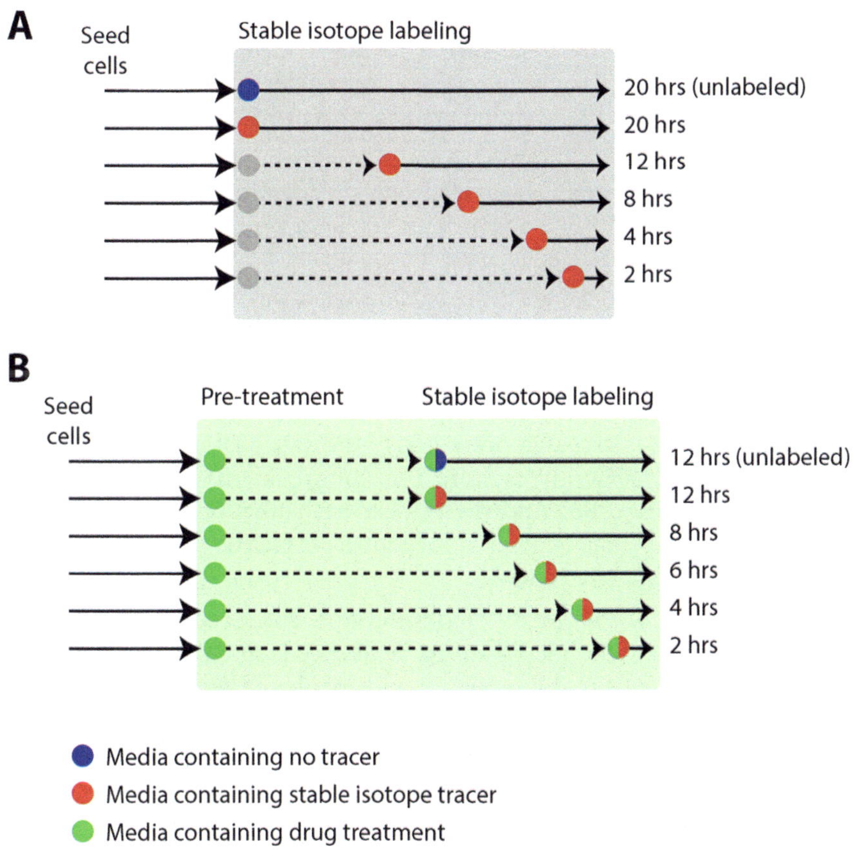

Fig. 1 Choosing the appropriate experimental design for a stable isotope tracing experiment depends on the experimental system, the choice of stable isotope tracer, and whether dynamic or steady-state labeling of downstream metabolites is required. Examples of possible experimental designs are shown: (**a**) to compare cells with pre-existing differences, (**b**) when pre-treatment is required prior to adding the stable isotope tracer

when harvested, an important consideration for consistent LC-MS performance. If labeling windows are very short, they may be further offset by a few minutes to permit ease of handling and reduce the number of samples to be harvested simultaneously.

Alternatively, if a drug treatment or activation phenotype is being studied, it may be preferable to pre-treat the cells and allow the phenotype to develop, prior to adding the stable isotope tracer, and then maintain the treatment throughout the labeling period as shown in Fig. 1b.

Finally, if the phenotype being studied is dynamic, it may be possible to use a labeling window with fixed duration placed at intervals across the time course of the biological change developing. This yields averaged snapshots of the metabolic state within each labeling window.

1. For stable isotope tracing experiments, the base media is Dulbecco's Modified Eagle Medium (DMEM) high glucose formulation without L-arginine, L-lysine, and L-glutamine. This is supplemented with 0.8 mM L-lysine, 4 mM L-glutamine, dialyzed FBS, and 1% Pen/Strep.
2. The base tracing media is further supplemented with either 0.4 mM [U-^{13}C]-L-arginine or 0.4 mM unenriched L-arginine (^{12}C-arginine media, *see* **Note 5**).
3. Media are split and supplemented with 0.5 mM 8-Br-cAMP to establish an M2-like polarization state, M(cAMP), or with PBS as a vehicle control (*see* **Note 6**).
4. The four resulting media are filtered through 50 mL SteriFlip 0.2 μm filters (Millipore Sigma). For each media, an additional volume (20%) is prepared to account for losses during filtration.
5. When the cells reach ~60% confluence, begin the media exchange according to the time course (Subheading 3.1, **step 2**). Gently aspirate the culture media and wash the cells with PBS. Add the respective incubation media with and without the activation reagent 8-Br-cAMP and with or without tracer (*see* **Note 7**).
6. Optional: if consumption and production of metabolites in the media are to be assessed, spare media-alone wells should also be prepared and incubated alongside the cells (method blanks). These wells will give the baseline media concentrations of metabolites, corrected for evaporation and thermal degradation.

3.3 Metabolite Extraction

1. Optional: for metabolite extraction from the media, take an aliquot of the media from each well (1 mL) and add to a 5 mL tube (Eppendorf) pre-filled with 4 mL of cold extraction solvent (100% aqueous methanol). Mix well by inverting.

2. Gently aspirate the media from each well. To ensure that all media is removed, position the plate at an angle and aspirate the remaining media with a small pipette tip (*see* **Note 8**).
3. For metabolite extraction from the cells, add 1 mL of extraction solvent (80% ice-cold aqueous methanol) to each well (*see* **Note 9**).
4. Incubate the plates and media samples at −80 °C for 1 h or overnight to aid protein precipitation.
5. Remove the plates from the −80 °C and keep them on dry ice. Scrape the cells ensuring that cells are detached from the bottom of the well.
6. Also apply **steps 1–5** to the method blank wells.
7. Transfer the extracts to microcentrifuge tubes, keeping them on wet ice.
8. Centrifuge the cell extracts and the collected media at 20,000 × *g* for 20 min at 4 °C.
9. Transfer the supernatant into clean tubes and evaporate to dryness in a vacuum concentrator (Genevac) or under nitrogen gas flow.
10. Store the dried extracts at −80 °C until LC-MS analysis.

3.4 Preparation of Samples for LC-MS Analysis

1. Re-suspend the dried extracts in 60 μL of 60% acetonitrile in water and incubate on ice for 20 min, vortexing every 5 min.
2. Centrifuge the re-suspended extracts at 20,000 × *g* for 20 min at 4 °C or filter using 0.2 μm filter plates (PALL Corporation) to remove any residual particulate material.
3. Transfer 45 μL to a reduced-volume autosampler vial and remove 10 μL from every sample to assemble a pooled QC vial (*see* **Note 10**).

3.5 LC-MS Analysis

1. The samples are analyzed using two separate methods, one at low pH and one at high pH. Data is acquired in positive ion mode using the low pH method and in negative ion mode using the high pH method.
2. Prepare mobile phase A and B. Purge each of the solvents for at least 5 min at high flow (5 mL/min).
3. Set the column compartment temperature to 40 °C. Equilibrate the column with 5% mobile phase A and 95% mobile phase B (initial gradient conditions) for at least 30 min at a flow rate of 0.4 mL/min before the start of the analysis and monitor for the backpressure to become stable.
4. Set the injection volume to 5 μL. Use a multi-step needle wash program ending with a wash solvent closely matched to initial conditions (*see* **Note 11**).
5. Program the elution gradient as described in Table 1 (*see* **Note 12**).

Table 1
LC elution gradient

Time (min)	% A	% B	Flow (mL/min)
0	5	95	0.400
9	30	70	0.400
12	60	40	0.400
13	70	30	0.400
14	70	30	0.400
14.5	5	95	0.400
15	5	95	0.600
20	5	95	0.600

Table 2
MS settings

Parameter	Low pH LC method	High pH LC method
Ion mode	Positive	Negative
Reference masses	121.050873 and 922.009798	119.03632 and 980.016375
Gas temperature	325 °C	300 °C
Drying gas flow	10 L/min	10 L/min
Nebulizer	35 psig	35 psig
Sheath gas temperature	400 °C	350 °C
Sheath gas flow	12 L/min	12 L/min
Capillary voltage	4000 V	3000 V
Nozzle voltage	0 V	0 V
Fragmentor	125 V	125 V
Skimmer	45 V	45 V
Data acquisition	1 spectra/s	1 spectra/s

6. High-resolution mass spectrometers require periodic source cleaning and mass axis calibration. Clean all source parts by sonication in 50% methanol followed by 100% methanol. For Agilent Q-TOF instruments, ensure both reference masses are being delivered and present within the recommended abundance range when the main LC stream is directed to the source. Calibrate the instrument in both polarities when the mass accuracy is not acceptable.

7. Set the MS parameters as described in Table 2.
8. Start by re-suspending and analyzing the SST sample and additional experiment-specific test samples (Subheading 3.1, **step 5**). These can be used to evaluate metabolite coverage, retention time stability, suitable re-suspension and injection volumes, and analytical system performance prior to re-suspending the main sample set.
9. The main injection sequence should start with solvent blank injections, extraction blanks, and five to ten technical replicates of the pooled QC sample to condition the column for the matrix. The pooled QC sample should then be injected every five to ten samples, depending on the size of the worklist and is used to monitor system performance (*see* **Note 13**).

3.6 Data Analysis

There are many high-quality software packages for the analysis of metabolomics data available from instrument vendors and open-source, vendor-neutral packages. In this protocol, Agilent MassHunter and VistaFlux software is used for data processing. Popular open-source alternatives particularly suited to the analysis of stable isotope tracing data are *El-MAVEN* [9, 10] and *XCMS Online* with the *X(13)CMS* extension, a cloud-based platform for processing stable isotope tracing data [11, 12].

1. A mass and retention time list of the compounds of interest is required by most software packages for targeted isotopologue extraction. Using Agilent MassHunter Pathways to PCDL software, an Agilent Personal Compound Database and Library file (.cdb) was created containing this information (Table 3).
2. This PCDL file is then used as a target list for batch isotopologue extraction in MassHunter Profinder software (*see* **Note 14**).

Table 3
Target list for batch isotopologue extraction. These compounds had previously been validated as well-detected in the cell extracts, and retention times were verified by comparison to authentic standards

Compound name	Formula	Retention time (min)
Arginine	$C_6H_{14}N_2O_2$	7.67
Ornithine	$C_5H_{12}N_2O_2$	7.90
Proline	$C_5H_9NO_2$	7.85
Citrulline	$C_6H_{13}N_3O_2$	6.55
Putrescine	$C_4H_{12}N_2$	7.51
Spermidine	$C_7H_{19}N_3$	9.90

3. The Profinder batch extraction results, corrected for ^{13}C natural abundance, are exported as a Profinder archive file (.pfa), which includes sample group information, compound identifiers, and isotopologue abundances. Alternative software packages exist for natural isotope abundance correction [13, 14] (*see* **Note 15**).

3.7 Data Visualization and Interpretation

Choosing an appropriate visualization strategy is an important step in exploring the data produced by stable isotope tracing experiments. For initial experiments, it may be helpful to survey multiple metabolites in a pathway context in order to understand where rate-limiting reactions exist.

1. The Agilent Profinder .pfa file from Subheading 3.6, **step 3** was imported into Omix Premium (Agilent Technologies) and displayed as quilt plots representing the isotopologue distributions over time and bar plots to display total pool sizes, segmented by the unlabeled signal and amount of the metabolite derived from the tracer (Fig. 2).
2. Detailed .csv output files from Profinder or *El-MAVEN* can also be visualized using statistical programs such as *Prism 7*

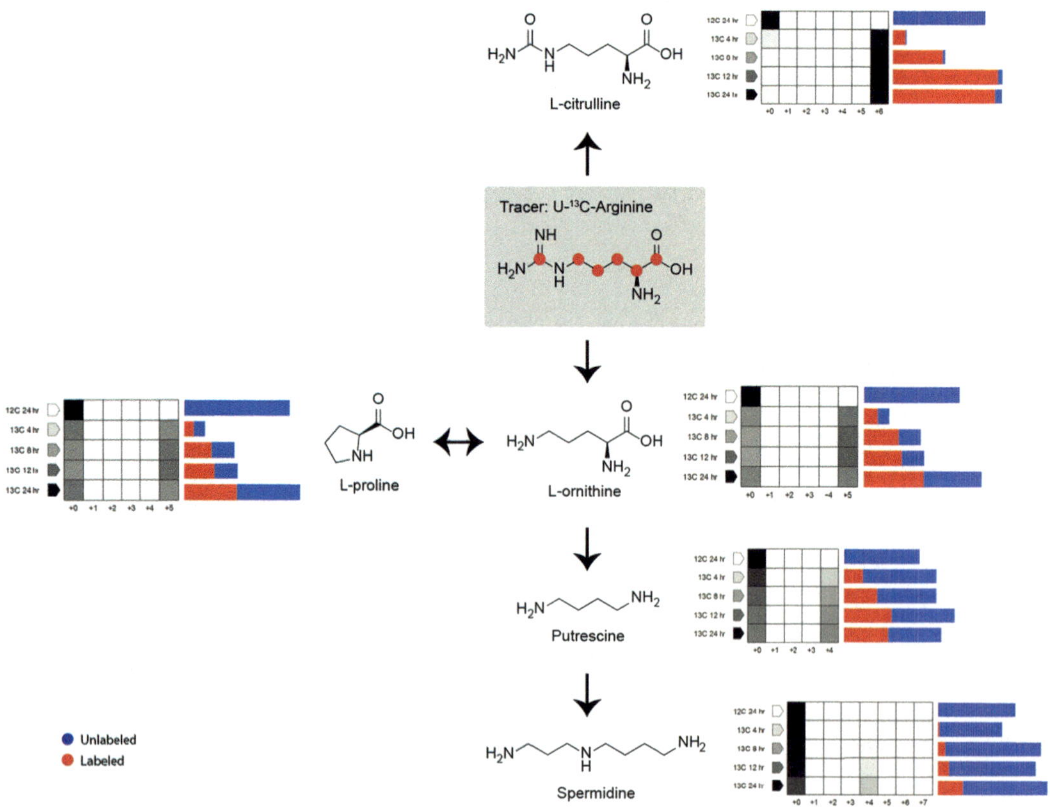

Fig. 2 Stable isotope tracing of [U-^{13}C]-L-arginine in RAW 264.7 macrophages. Quilt plots show individual isotopologues. Bar graphs show metabolite pool sizes segmented by the unlabeled signal (blue) and the sum of the labeled isotopologues (red). For simplicity only one biological condition is shown (non-cAMP-treated cells)

(GraphPad). Alternatively, isotopologue abundances can be extracted using most quantitative analysis packages and, following natural isotope abundance correction [13, 14], labelling patterns visualized by manual plotting in *Prism 7* (GraphPad).

Interpretation of isotope tracing data often requires consideration of total pool sizes, the rate of labeling, and the percent labeled at isotopic steady state. Some metabolites are entirely derived from the tracer (eg. citrulline), while other metabolites reach isotopic steady state but are not completely labeled (eg. putrescine), indicating they are also produced by other pathways (Fig. 2). A source of unlabeled ornithine (and putrescine) could be glutamine, since glutamine contributes to the glutamate, proline and ornithine pools [15]. This could be investigated further by glutamine tracing, which results in labeling of the TCA cycle [16]. For further guidance on interpreting labeling patterns, we refer the reader to excellent recent reviews of this topic [17, 18].

4 Notes

1. Prepare the stock ammonium acetate buffer for the low pH method at 200 mM and add to the mobile phase in the respective concentration. No further pH adjustment is necessary. The stock solution should be stored at 4 °C and can be kept for at least 3 months.
2. Prepare the stock ammonium acetate buffer for the high pH method at 200 mM and adjust the pH to 9.8 using ammonium hydroxide and confirm using pH paper. Add to the mobile phase in the respective concentration without further pH adjustment. The stock solution should be stored at 4 °C and can be kept for at least 3 months.
3. Pilot experiments are usually necessary to establish how quickly downstream metabolites become labeled and to determine whether the experiment will follow dynamic labeling patterns and when the isotopologue distribution for key metabolites reaches steady state, i.e., the labeling pattern becomes constant. We recommend performing a time-course experiment initially, even if subsequent experiments are performed at a single time point. In some situations, it can also be useful to monitor labeling kinetics appearing in end products (protein, lipids, and nucleic acids) or in metabolites secreted into the media. These pools will likely not saturate or reach isotopic steady state but can still provide useful dynamic information.
4. We typically recommend that the unlabeled samples are manipulated in exactly the same way as the most information-rich labeled samples. This may be the longest labeling time point or a time when key compounds have reached isotopic

steady-state labeling. In this example, the 24 h labeled sample is matched by a media change to unlabeled media 24 h prior to harvest.

5. Labeled and unlabeled media should be as similar as possible; this is best achieved by using the same base media stock and supplementing nutrients and drug treatments at the last step. Dialyzed serum is used to avoid a contribution of unlabeled nutrient present in FBS.
6. It has been previously demonstrated that the compound 8-Br-cAMP, a non-hydrolysable analog of cyclic adenosine 5′-monophosphate (cAMP), together with the cytokine interleukin 4 (IL-4) promotes Arginase I expression and polarization of macrophages to an M2 phenotype [19]. In this protocol, we describe an experimental plan to investigating the fate of L-arginine and metabolic rewiring in response to 8-Br-cAMP, denoted M(cAMP).
7. Ideally, the unlabeled nutrient is completely replaced with a labeled nutrient, without any other perturbation of the system. In reality, changing cell culture media and washing cells with PBS depletes some rapidly exchanging metabolite pools, refreshes cytokine stimulation, and resets metabolite gradients that may have been established between intra- and extracellular compartments in the prior incubation period. Fractional labeling data may be relatively forgiving of fluctuations in pool size, but nevertheless, these changes can occur (Fig. 2, compare fractional labeling in the quilt plots to absolute pool sizes displayed in the bar charts). An alternative design that can be considered in these situations is a spike-in approach, where the tracer is added to the pre-existing nutrient pool. In this case, an additional sample should be taken to determine the final tracer enrichment achieved. This design can be useful in situations where a dropout base media lacking the tracer nutrient is not available, for example, when working with complex microbial or embryonic stem cell media formulations with supplements. This is also necessary for in vivo experiments, such as D_2O tracing, where an assessment of total body water enrichment is required.
8. For this qualitative flux experiment, cells were only washed with PBS prior to addition of the stable isotope containing media. At harvest, the extraction solvent was added directly after aspirating the media without additional washing in order to best preserve levels of intracellular metabolites and the associated labeling patterns. However, if such a procedure is followed care must be taken when interpreting the metabolites also present at high levels in the culture media as there may be residual media contamination in the cell extracts.
9. In this example, the extraction solvent is 80% aqueous methanol for cells grown in culture. Optimal extraction con-

ditions may need to be determined for other sample matrices and/or metabolites. For example, 4:4:2 methanol/acetonitrile/water has worked well for microbial and tissue extracts and 10% TCA for B vitamins and CoA molecules. The inclusion of internal standards in the extraction solvent can also be useful to track instrument performance and correct responses; however, care should be taken to choose compounds that will not be labeled in the experiment. $^{13}C_5$,^{15}N-glutamate is an example of a compound that would not interfere with data analysis in a typical ^{13}C-only tracing experiment.

10. A pooled QC is used as a quality control sample that represents all samples in the batch. This sample can be used to condition the column and is injected five times at the start of the analytical run. Furthermore, this pooled QC should be injected regularly throughout the analysis (after every five to ten samples depending on the batch size) to serve as an indication of chromatographic stability over time. The sample is generated by pooling a small aliquot of all samples or just the unlabeled samples (Subheading 3.4, **step 3**).
11. If using an autosampler with more than one wash solvent, this should be programmed to start with a strong wash, followed by a weak wash closely matching the initial chromatographic conditions. In HILIC chromatography the strong solvent is water, and therefore the strong wash is usually composed of 90% water and 10% acetonitrile which helps remove polar compounds from the needle. The weak wash is either 100% acetonitrile or MPB. Failure to end the wash sequence with mobile phase B (MPB), or a closely matched solvent, can be a major contributor to bad peak shapes in HILIC chromatography, particularly for early eluting compounds. For this reason, samples should also be re-suspended in 20–40% water (with 60–80% acetonitrile), and the injection volumes are usually between 2 and 5 μL.
12. Preparing mobile phases with 10% organic in the aqueous mobile phase protects against microbial growth; inclusion of 10% aqueous in the organic mobile phase protects against the possibility of salt precipitating in the pump. We observe most compounds elute on HILIC separations between 20 and 35% MPA, so this does not adversely impact retention and separation.
13. Care should be taken when including solvent-only or pure standards injections during the main injection sequence, i.e., between matrix samples. HILIC chromatography is particularly sensitive to the concentration of salt in the samples, so switching to matrix samples after solvent-only injections sometimes requires column re-equilibration with several matrix injections in order to avoid retention time shifts and drifts in

response. As such, for any metabolomics study, we recommend solvent blanks are part of pilot evaluation experiments to confirm an absence of carry-over but are then subsequently only used at the beginning or end of an injection sequence and not interspersed during main analytical batches. Furthermore, the injections of experimental samples should be randomized whenever possible to prevent chromatography or analytical biases and run order effects.

14. Manual inspection of extracted ion chromatograms (EICs) is always recommended, and mass tolerances should be carefully evaluated since low abundance isotopologues will have worse ion statistics and may, therefore, exhibit higher mass errors. In experiments where only percent labeling information is required, it can sometimes be helpful to integrate just the core of the chromatographic peak where signals are strongest (a concept termed Core Peak Area in Agilent MassHunter and AreaTop in *El-MAVEN*). A useful confirmation that the correct group of mass isotopologues has been identified is the chromatographic coelution of all isotopologues belonging to the same compound, and some software packages use this coelution parameter as part of their scoring. An exception is deuterium labeling where increasing deuterium incorporation leads to earlier chromatographic elution, particularly when samples are analyzed by GC-MS. For lower-resolution instruments (e.g., GC MSD) and when studying larger compounds (e.g., fatty acid methyl ester analysis), it is also important to inspect unlabeled samples to confirm the entire isotopic labeling envelope is clear of any interfering masses, and when using electron impact (EI) ionization, the choice of fragment ion may need to be adjusted to avoid other interfering EI fragments. When working with EI fragment ions, it is also important to consider whether the ion contains all the relevant atoms from the parent metabolite (i.e., has only lost part of the derivatization groups); otherwise there will be ambiguity in the labeling pattern since some mass isotopologues cannot be assessed. There are a few cases where fragmentation can also be useful to obtain positional information by mass spectrometry [20], but in most cases, this is best achieved using NMR or by using tracers labeled at specific atom positions that can be used to distinguish pathway utilization [6].

15. Natural abundance correction can depend on the resolution of the instrumentation being used. Some instruments, such as Fourier transform mass spectrometers, have the resolution to separate fine isotopic structure (composed of ^{13}C, ^{15}N, ^{2}H natural abundance, in addition to the tracer signal), but for most platforms, the ability to distinguish fine isotopic structure varies by mass and the extent of label incorporation. It is

therefore important to understand if the signal being measured is the sum of all contributing elements or unique to a single element and correct the data appropriately. This is particularly relevant for ^{15}N tracing experiments where at low levels of tracer enrichment the natural abundance of ^{13}C can also contribute substantially to the m + 1 labeling pattern. For GC-MS data, natural abundance correction should also take into account the contribution of any derivatization regents used since silicone, in particular, contributes substantially to the observed signal in GC-MS isotope tracing experiments.

References

1. Jang C, Chen L, Rabinowitz JD (2018) Metabolomics and isotope tracing. Cell 173(4):822–837
2. Previs SF et al (2011) Quantifying cholesterol synthesis in vivo using (2)H(2)O: enabling back-to-back studies in the same subject. J Lipid Res 52(7):1420–1428
3. Faubert B et al (2017) Lactate metabolism in human lung tumors. Cell 171(2):358–371.e9
4. Hui S et al (2017) Glucose feeds the TCA cycle via circulating lactate. Nature 551(7678): 115–118
5. DeBerardinis RJ et al (2007) Beyond aerobic glycolysis: transformed cells can engage in glutamine metabolism that exceeds the requirement for protein and nucleotide synthesis. Proc Natl Acad Sci U S A 104(49):19345–19350
6. Lane AN et al (2017) Probing the metabolic phenotype of breast cancer cells by multiple tracer stable isotope resolved metabolomics. Metab Eng 43(Pt B):125–136
7. Murray PJ (2017) Macrophage polarization. Annu Rev Physiol 79:541–566
8. Jha AK et al (2015) Network integration of parallel metabolic and transcriptional data reveals metabolic modules that regulate macrophage polarization. Immunity 42(3):419–430
9. Clasquin MF, Melamud E, Rabinowitz JD (2012) LC-MS data processing with MAVEN: a metabolomic analysis and visualization engine. Curr Protoc Bioinformatics. Chapter 14: Unit14.11
10. Melamud E, Vastag L, Rabinowitz JD (2010) Metabolomic analysis and visualization engine for LC-MS data. Anal Chem 82(23):9818–9826
11. Huang X et al (2014) X13CMS: global tracking of isotopic labels in untargeted metabolomics. Anal Chem 86(3):1632–1639
12. Gowda H et al (2014) Interactive XCMS online: simplifying advanced metabolomic data processing and subsequent statistical analyses. Anal Chem 86(14):6931–6939
13. Millard P et al (2012) IsoCor: correcting MS data in isotope labeling experiments. Bioinformatics 28(9):1294–1296
14. Trefely S, Ashwell P, Snyder NW (2016) FluxFix: automatic isotopologue normalization for metabolic tracer analysis. BMC Bioinformatics 17(1):485
15. Wang R et al (2011) The transcription factor Myc controls metabolic reprogramming upon T lymphocyte activation. Immunity 35(6): 871–882
16. Salamanca-Cardona L et al (2017) In vivo imaging of glutamine metabolism to the oncometabolite 2-hydroxyglutarate in IDH1/2 mutant tumors. Cell Metab 26(6):830–841.e3
17. Buescher JM et al (2015) A roadmap for interpreting (13)C metabolite labeling patterns from cells. Curr Opin Biotechnol 34:189–201
18. Zamboni N, Saghatelian A, Patti GJ (2015) Defining the metabolome: size, flux, and regulation. Mol Cell 58(4):699–706
19. Sheldon KE et al (2013) Shaping the murine macrophage phenotype: IL-4 and cyclic AMP synergistically activate the arginase I promoter. J Immunol 191(5):2290–2298
20. Alves TC et al (2015) Integrated, step-wise, mass-isotopomeric flux analysis of the TCA cycle. Cell Metab 22(5):936–947

Part V

Data Processing in Metabolomics

Chapter 18

Data Processing for GC-MS- and LC-MS-Based Untargeted Metabolomics

Linxing Yao, Amy M. Sheflin, Corey D. Broeckling, and Jessica E. Prenni

Abstract

Gas chromatography and liquid chromatography coupled to mass spectrometry are used extensively in untargeted metabolomics, which involves the profiling of small metabolites in biological samples. The complex raw dataset produced from untargeted metabolomics requires proper processing before it can be statistically analyzed and interpreted. This chapter describes a high-throughput data processing workflow routinely used in our laboratory, including feature detection and alignment, data reduction, and spectral-matching-based annotation. This semiautomated workflow uses vendor neutral data file formats and freely available data processing tools and therefore can be readily implemented on datasets acquired from instruments of different vendors.

Key words GC-MS, LC-MS, Untargeted metabolomics, RAMClustR, RAMSearch, XCMS

1 Introduction

Metabolomics involves broadscale profiling of small molecule metabolites (also termed "the metabolome"). The discovery of significant metabolic changes within a biological system in response to diseases, metabolic pathway alterations, and genetic and environmental changes provides insight for fundamental understanding of biological mechanisms. Applications of metabolomics are described for the fields of plant biology, biomedical research and pharmacology, nutritional studies, and others [1–4]. Among the various modern analytical techniques, gas chromatography and liquid chromatography coupled to mass spectrometry (GC-MS and LC-MS, respectively) are the most commonly used for acquiring metabolomics data. Prior to data acquisition, metabolites are extracted using aqueous and/or organic solvents from biological samples such as cells, blood, urine, tissues, and other biofluids. Chemical derivatization is usually required to increase the volatility of compounds for separation and detection by GC-MS. GC-MS and LC-MS are complementary techniques, which together can yield

Angelo D'Alessandro (ed.), *High-Throughput Metabolomics: Methods and Protocols*, Methods in Molecular Biology, vol. 1978, https://doi.org/10.1007/978-1-4939-9236-2_18,

broad coverage of metabolites and generate a large amount of highly dimensional complex data. This data complexity is intensified by additional acquisition options sometimes specific to one vendor's technology offerings. For example, instrumentation produced by Waters Corporation provides the option of acquiring MS^E-containing LC-MS data, which records data for every detectable compound and its fragment ions, i.e., all precursor ions and product ions are collected without preselection or discrimination [5].

Proper data processing is a key step in the analysis of such comprehensive datasets and requires knowledge of both bioinformatics and analytical chemistry in addition to the field of interest. Although a wide selection of data processing tools are available, either freely or commercially, a generalized data processing workflow is needed for better utilization of the data by a broader field of scientists. Here we describe a data processing workflow routinely used in our laboratory for GC-MS- and LC-MS-based untargeted global metabolite profiling studies. This workflow utilizes three freely available tools, XCMS [6], RAMClustR [7], and RAMSearch [8], to achieve feature detection and alignment, data reduction, and annotation, respectively. An overview of the data processing workflow is shown in Fig. 1. We have included applications of our workflow to MS^E-containing LC-MS data as an example of how to accommodate more complex data from optional acquisition modes. The benefits of this workflow include more effective utilization of instrumentation through efficient use of all available MS data, reduced feature redundancy for downstream statistical analysis, reduction of analytical variance, and a more automated and high throughput workflow for metabolite annotation based on spectral searching.

2 Materials

Data processing tools used in this workflow are freely available.

1. A workstation with sufficient storage space for data files and software tools installed.
2. RAMSearch software [8], https://osf.io/x8bw5/.
3. R or R studio [9], https://www.r-project.org/.
4. R packages:

   xcms [6], https://bioconductor.org/packages/release/bioc/html/xcms.html.

   RAMClustR [7], https://github.com/cbroeckl/RAMClustR.

   pcaMethods [10], https://github.com/hredestig/pcamethods.

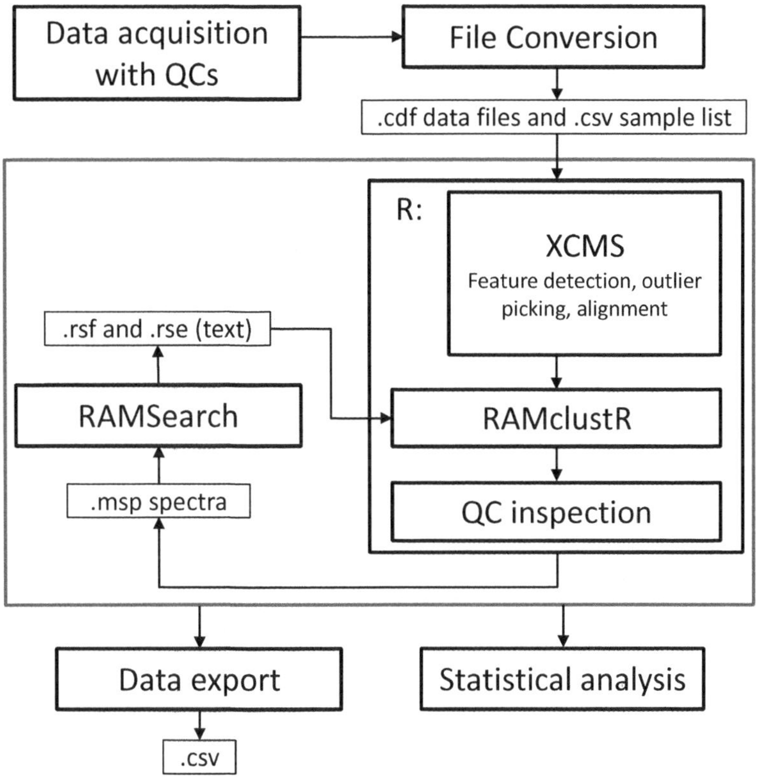

Fig. 1 Summary of data processing workflow. Adapted from ref. [7]

3 Methods

3.1 Data Acquisition

Data acquisition of large-scale untargeted metabolomics studies can take long periods of time, ranging from several hours to years. Over the course of data acquisition, variation can be introduced from multiple sources including instability or fluctuations in ionization, variation in LC mobile phase conditions, autosampler injection variation, source and/or optics becoming "dirty," etc., all of which can influence the downstream quantitative data analysis [11]. Therefore, it is important to monitor instrument performance and minimize analytical variations during data collection. Interspersed injections of quality control (QC) samples, i.e., pooled QC approach, can be used as a means of quality control for monitoring any significant change of instrument conditions and estimating systematic variability [12]. Practices such as scheduled cleaning of ion sources and conditioning the instruments with nonexperimental injections should be included to reduce analytical variations.

1. Prepare pooled QC by mixing equal small aliquots of each biological samples and aliquot into multiple vials for subsequent sample extraction (*see* **Note 1**).
2. Acquire data on a clean instrument (GC-MS or LC-MS) using the corresponding vendor software (*see* **Note 2**).
3. Start the acquisition sequence with "nonexperimental" injections—injections of real samples, which are not to be analyzed as part of the final dataset—followed by actual samples that are arranged in a randomized order (*see* **Note 3**).
4. Inject pooled QC samples after every six to eight actual samples (*see* **Note 4**).
5. Perform visual inspection on the data in the operating software of the instrument (*see* **Note 5**).

3.2 Export the Raw Data

Since all subsequent data processing is performed in the R environment, data files will first be converted to an R acceptable format.

1. For each sample, convert raw data files to Computable Document Format (.cdf) files using the instrument operating software (*see* **Note 6**).
2. Export the sequence list using the instrument software and save it as comma-separated-text (.csv) file such as seq.csv in the same folder as the .cdf files. The content of seq.csv should be modified (if needed) so that it consists of two columns, with the first column containing file names and the second column containing corresponding sample names (*see* **Note 7**).
3. Organize sample names so that the experimental design is encoded in the sample names, using a hyphen as the delimiter to separate factor values for each sample (*see* **Note 8**).

3.3 XCMS Processing for Feature Detection and Alignment

Both GC-MS and LC-MS produce a complex data matrix of molecular features that are defined by retention time, mass (m/z), and intensity. The XCMS software in the R environment [6] is utilized for peak picking, outlier detection, and feature alignment with retention time correction. Functions to be used in the following data processing workflow are summarized in Table 1.

1. Set R working directory to the folder containing the .cdf files and seq.csv and load the required R packages.
2. Perform peak picking using the **`xcmsSet`** function in XCMS R package. The **`matchedFilter`** algorithm should be utilized for low-resolution (quadrupole) GC-MS data, and the **`centWave`** algorithm should be utilized for high-resolution centroided LC-MS data.
3. Remove outliers using total XCMS signal (**`outlTIC`** function), principal component analysis (PCA) on mass intensity

Table 1
R functions used in the data processing workflow[a]

R package	Description	Function	Methods to select
XCMS	Peak peaking	**xcmsSet**	Matched filter (GC-MS), centWave (LC-MS)
	Outlier removal	**outlTIC**[b]	
		outlPCA[b]	
		minlTIC[b]	
	Feature alignment (peak grouping)	**group**	Nearest
	Retention time correction	**retcor**	Loess, linear, obiwarp
	Peak filling	**fillPeaks.chrom**	
RAMClustR	QC evaluation	**RCQC**	
	Incorporate annotation results into the dataset	**impRamSearch**	
	Export data into .csv file	**exportDataset**	
	Evaluate spectral matching	**degolm**[c]	
stat	Statistical analysis	**anova**	
		lm	
		p.adjust	

[a]Methods that can be used in the data processing workflow are not limited to the ones listed in the table
[b]These functions are used in the XCMS processing, but not from XCMS R package. R scripts of these functions can be found in the file "supplemental_R_functions.R" at https://osf.io/x8bw5/. Accessed 1 Feb 2018
[c]The degolm function is used after annotation results are imported to RAMClustR object. Its R scripts can be found in the file "supplemental_R_functions.R" at https://osf.io/x8bw5/. Accessed 1 Feb 2018

(**`outlPCA`**), and/or minimum total signal cutoff (**`minlTIC`**) (*see* **Note 9**).

4. Perform feature alignment with retention time correction using the **`group`** function in the XCMS R package. This step starts with a coarse feature matching to separate peaks into groups using overlapping m/z bins and a wide retention time window, followed by retention time correction using the **`retcor`** function. A second pass of feature alignment is subsequently performed with a narrow retention window (*see* **Note 10**).
5. Fill in missing peaks. Peaks can be missed during peak detection and filtration or are absent or below the limit of detection in a sample. Missing values can reduce the statistical power of the dataset. Thus, we recommend using the **`fillPeaks.chrom`** function in the XCMS R package to replace missing values with intensity values representing baseline noise.

3.4 Perform Feature Grouping Using RAMClustR

RAMClustR is implemented in the R environment and uses the XCMS object directly as input. Either XCMS objects that contain only MS features or objects containing both MS and MS^E data can be used [7]. Features that both co-elute and covary across the dataset are clustered together into a feature group which represents a single compound. The output of RAMClustR includes an R object containing processed data and an exported .mspLib file containing spectra (i.e., feature groups) for each detected compound. RAMClustR was constructed to process both LC-MS and GC-MS data.

1. Normalize data to the total XCMS extracted ion signal to correct for the signal drifts and batch variation (*see* **Note 11**).
2. Clustering of GC-MS data into spectra can be accomplished with these default option settings: hmax = 0.9, minModuleSize = 2, linkage = "average," mslev = 1, and normalize = "TIC." These settings may be adjusted to best fit your GC-MS setup and specific experimental design.
3. Clustering of XCMS object derived from MS^E-containing LC-MS data can be accomplished with these default option settings: hmax = 0.5, minModuleSize = 2, linkage = "average," normalize = "TIC," MStag = "01.cdf," idMSMStag = "02.cdf" (for Waters Databridge derived MS^E cdf datafiles), and taglocation = "filepaths." These settings may be adjusted to best fit your LC-MS setup and specific experimental design.
4. Clustering of XCMS object derived from MS only LC-MS data can be accomplished with these default option settings: hmax = 0.5, minModuleSize = 2, linkage = "average," and normalize = "TIC." These settings may be adjusted to best fit your LC-MS setup and specific experimental design.

3.5 Evaluate Data Quality Using Pooled QC Samples

At this step, the pooled QC samples are separated from the whole dataset in RAMClustR package and then subjected to PCA analysis using `RCQC` function. Diagnostic plots are generated to visualize analytical variation. As an example, Fig. 2a illustrates the observed variation of pooled QC samples relative to a whole dataset. In this example the PC1 score standard deviation of the QCs is 8% of the PC1 score standard deviation of the full sample set, indicating that the analytical variation is small relative to biological (sample-to-sample) variation. Figure 2b presents the coefficient of variation (CV) for each compound in QC samples. The distribution of CV values is depicted as a histogram, demonstrating that 80% of the compounds have CV values of 26.6% or less in the given example (*see* **Note 12**).

3.6 Annotation by Spectra Matching Using RAMSearch

Metabolite annotation is performed based on batch spectral searching using the RAMSearch program [8]. RAMSearch utilizes both retention time and spectral similarity for spectral similarity

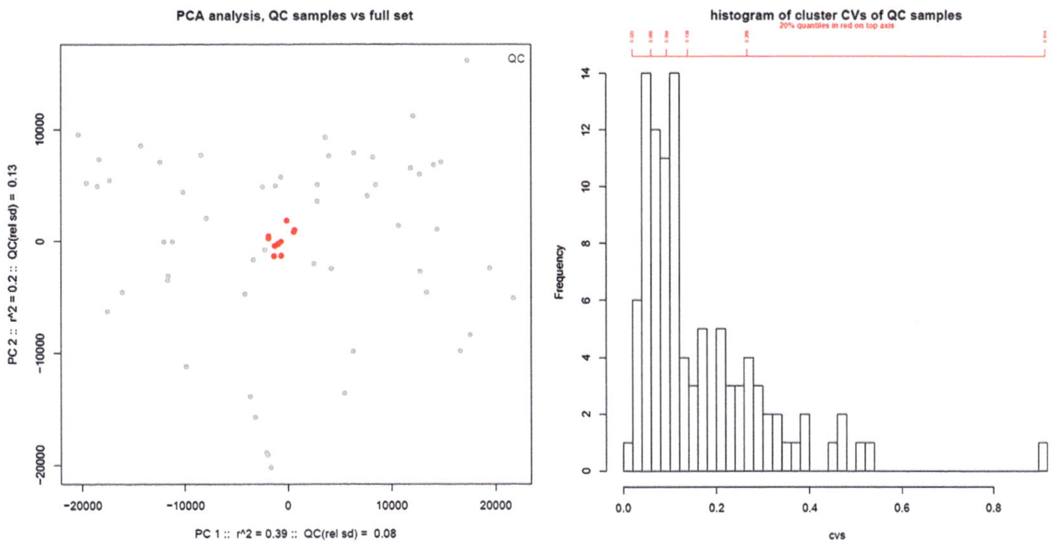

Fig. 2 QCs diagnostic plots: (**a**) variation of QCs relative to whole dataset and (**b**) histogram of cluster CVs of QCs. QCs are colored in red

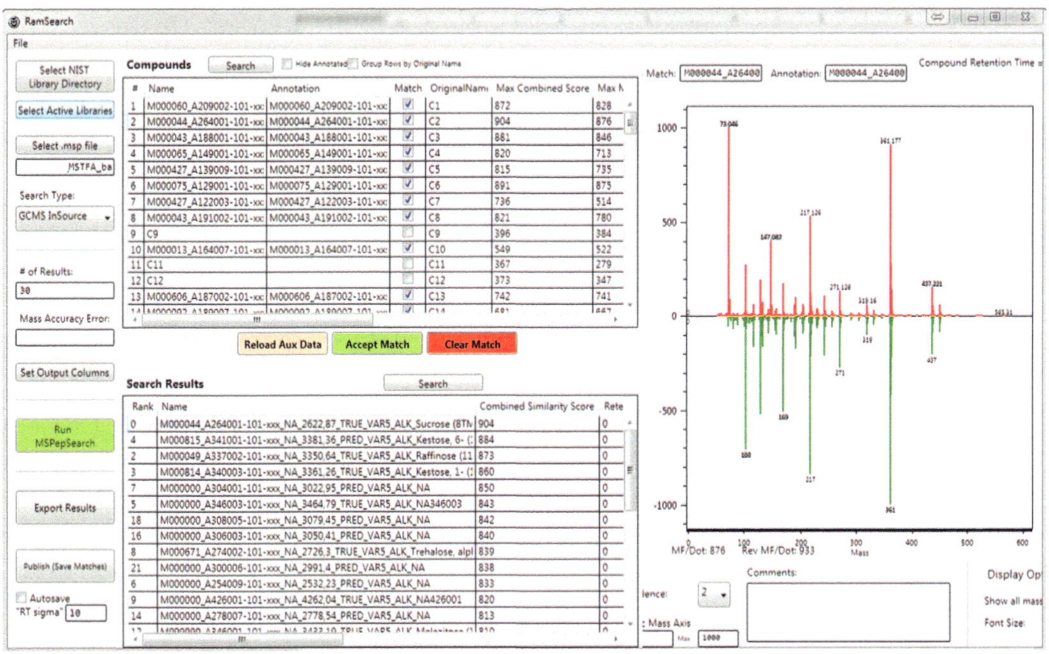

Fig. 3 RAMSearch program user interface

matching, and the results can be displayed and evaluated using a graphical user interface as shown in Fig. 3.

1. Open RAMSearch and select libraries for spectral matching (*see* **Note 13**).

2. Select the experimental .mspLib file (*see* **Note 14**).
3. Choose the type of search you want to perform among the options of InSource or Precursor MS/MS (*see* **Note 15**).
4. Start the search program, evaluate, and accept the spectral matches based on the combined similarity score. Of note, spectral similarity scores can sometimes be misleading and should be manually validated for high confidence annotation (*see* **Note 16**).
5. Save the annotation result as a RAMSearch project file (.rsf) that contains all of the experimental spectral clusters and library spectra for all of the annotated compounds.
6. Export the results to a text format (.rse) file that can be re-integrated with the RAMClustR workflow in R for down-stream statistical analysis.

3.7 Export Final Annotated Dataset

The fully processed and annotated data is exported to an easily shared format, such as .csv file that can be read by most data viewing and analysis software.

1. Update the R object created by RAMClustR to include the annotations performed with RAMSearch using the **`impRamSearch`** function.
2. Use the **`exportDataset`** function to export the dataset to .csv format with samples as rows and columns containing peak intensities for spectral clusters.

3.8 Statistical Analysis

R offers a wealth of statistical analysis and visualization options. The most appropriate statistical procedure will depend on the experimental design. We perform PCA and Analysis of Variance (ANOVA) analyses on nearly every dataset. PCA finds the multivariate pattern and similarity of a dataset in unsupervised way and can be performed on mean-centered and pareto variance-scaled data using the pcaMethods package in R [10]. ANOVA can be conducted on each compound using the **`anova`** function after fitting a model using **`lm`** in R [9], and *p*-values are adjusted for false positives using the Bonferroni-Hochberg method in the **`p.adjust`** function in R [13] (*see* **Note 17**).

4 Notes

1. Generating a pooled QC directly from biological samples is mostly appropriate for liquid samples such as plasma and cell culture. When the biological samples are solid, such as stools or plant tissues, or when the sample volumes are not enough to generate pooled QC, a pooled solvent extract is recommended to be used as QC.

2. The ion source and inlet liner of GC-MS should be cleaned or replaced after a maximum of 100 injections. The ion source of LC-MS can be cleaned as necessary, with complex samples and larger injection volumes requiring more frequent cleaning. About 100–200 injections should be feasible under most circumstances.
3. Prior to data acquisition of the actual sample set, a sequence of samples including one blank (i.e., derivatization blank for GC-MS and solvent blank for LC-MS) and three or four non-experimental samples can be injected to condition the instruments. The nonexperimental samples can be an extra QC or actual sample, which are not analyzed as part of the final dataset, but only serve to condition the system for the most stable operation. An example of a sequence of 18 samples is presented in Fig. 4a.
4. Each injection of the pooled QC should be from a different vial to prevent variation due to evaporation or contamination.
5. A brief inspection sometimes reveals bad injections/samples due to incomplete derivatizations, air bubbles present in the samples, clogged syringes, or other temporary instrument failure. These bad injections can be deleted from the dataset or overwritten by a reinjection (if applicable).

#	Chromatogram	Name
1	None	blank
2	None	non-experimental
3	None	non-experimental
4	None	non-experimental
5	None	trash
6	None	QC-1
7	None	sample-1
8	None	sample-2
9	None	sample-3
10	None	sample-4
11	None	sample-5
12	None	sample-6
13	None	QC-2
14	None	sample-7
15	None	sample-8
16	None	sample-9
17	None	sample-10
18	None	sample-11
19	None	sample-12
20	None	QC-3
21	None	sample-13
22	None	sample-14
23	None	sample-15
24	None	sample-16
25	None	sample-17
26	None	sample-18
27	None	QC-4

(a) Injection sequence in the vendor software

	A	B
1	File name	Sample ID
2	2017_12_25_QC-1_001	QC-1
3	2017_12_25_sample-1_002	sample-1
4	2017_12_25_sample-2_003	sample-2
5	2017_12_25_sample-3_004	sample-3
6	2017_12_25_sample-4_005	sample-4
7	2017_12_25_sample-5_006	sample-5
8	2017_12_25_sample-6_007	sample-6
9	2017_12_25_QC-2_008	QC-2
10	2017_12_25_sample-7_009	sample-7
11	2017_12_25_sample-8_010	sample-8
12	2017_12_25_sample-9_011	sample-9
13	2017_12_25_sample-10_012	sample-10
14	2017_12_25_sample-11_013	sample-11
15	2017_12_25_sample-12_014	sample-12
16	2017_12_25_QC-3_015	QC-3
17	2017_12_25_sample-13_016	sample-13
18	2017_12_25_sample-14_017	sample-14
19	2017_12_25_sample-15_018	sample-15
20	2017_12_25_sample-16_019	sample-16
21	2017_12_25_sample-17_020	sample-17
22	2017_12_25_sample-18_021	sample-18
23	2017_12_25_QC-4_022	QC-4

(b) seq.csv

Fig. 4 An example of (**a**) an injection sequence of 18 samples and (**b**) its seq.csv after data are exported

6. If your data has been collected in profile mode, ensure that centroiding occurs during conversion, as XCMS requires centroided data. Alternative formats which are compatible with XCMS in the R environment include .mzML and .mzXML. The options available will vary by vendors. If vendor software does not support export to a suitable format, ProteoWizard is an alternative for data conversion [14].
7. An example of seq.csv in Fig. 4b shows the format needed for subsequent data processing in R. Blank and nonexperimental samples should be deleted from the seq.csv after data exporting.
8. The factors should be the interest of statistical analysis. Each sample must have the same number of factor values. Replicate injections should have exactly the same sample name, but biological replicates should not. For example, if the design contains biological replicates with factors of treatment, genotype, and date, the sample could be named as Drought-Supersweet-July-rep1.
9. We use a combination of three outlier removal steps. First, an outlier test is applied to the distribution of summed XCMS signal for each injection (excluding MS^E) to remove the abnormally strong or weak chromatograms. Samples can differ dramatically while falling within the normal range of total signal intensity. For example, if a sample is dominated by polyethylene glycol, but lipid replete, it may not have a dramatic shift in total ion signal. However, if we look at the distribution of masses detected, we will be able to distinguish these outlier samples. To do so, the mass values for each sample (excluding MS^E data) are unit mass binned, and PCA is performed on this "average mass spectrum" dataset. Finally, there are situations where the above are not the most appropriate. For these rare situations, we also use on occasion a simple minimum total signal cutoff. This total signal cutoff removes the injection if a total signal value is, for example, tenfold lower than the median. The parameter specifying the cutoff level can be optimized to the dataset.
10. Available regression fitting methods in XCMS for retention time corrections include loess, linear, and obiwarp (Table 1), which are based on different processing algorithms [6].
11. Sample concentration is normalized during solvent extraction by adjusting the solvent-to-mass ratio for solid samples, the solvent-to-volume ratio for liquid samples, or the solvent-to-cell counts ratio for cell culture. Normalization in this step prepares the dataset for downstream statistical analysis and minimizes the impact of instrument dynamic range on normalization efficacy. Various data-driven or method-driven normalization methods have been reported with or without the use of QC samples and/or internal standards [15]. Normalization by total ion sig-

nals (TIC) or quantile method [16] are the available options within the RAMClustR function.

12. Figure 2a, b is generally used together to evaluate the analytical variation of a dataset. For example, when the experimental treatment difference is large, analytical variation of QCs would be relatively small, and thus a tight clustering of QCs would be expected. In this case, the CVs of each cluster should be evaluated as well.
13. GOLM metabolome retention indexed spectral library [17] and NIST Main EI MS Library (NIST SRD 1A v14 or v17) [18] are used for GC-MS. An in-house library [19] (in the "ac6b02479_si_001.zip" at https://osf.io/x8bw5/. Accessed 1 Feb 2018) and NIST MS/MS libraries (NIST SRD 1A v14 or v17) [18] are used for LC-MS.
14. The .mspLib file is generated by RAMClustR in steps in Subheading 3.4. It is written with a .mspLib file extension but is actually .msp format, and the extension can be modified to .msp to enable other programs (such as NIST MSSearch) to recognize it.
15. The InSource search options will perform a spectral search when the precursor ion is not defined. Precursor MS/MS will require that a precursor m/z is defined for each MS/MS spectra in the .mspLib file. RAMClustR does not utilize precursor selected MS/MS data at this point.
16. The combined similarity score is calculated as the product of mean of the forward dot product spectral match score (MF/Dot) and the reverse dot product spectral match score (Rev MF/Dot; scaled 0–999) and the retention time similarity score (range 0–1). Its search algorithm is based on NIST Mass Spectral Search Program [20]. For GC-MS data, one can use the **degolm** function in R to plot retention times against the GOLM retention index to increase confidence in the spectral match. A strong correlation will be reflected visually by sample names plotting roughly in an $x = y$ diagonal line and indicates satisfactory annotation effort. If not, the annotation would need to be re-evaluated. LC-MS data are annotated in a similar approach and can use retention time when available in the library.
17. Statistical analysis in this step is not specific to the current workflow. As with any statistical analysis, always check applicable assumptions regarding normality of distribution and equality of variance or use transformations, as appropriate. Any statistical R functions appropriate to the dataset or other statistical tools such as MetaboAnalyst [21] can be used to conduct the analysis on the dataset processed by this workflow, which are not limited to the examples given in Subheading 3.8. Publications emphasized on the statistical analysis of metabolomics data are also quite available [22–24].

References

1. Hall R, Beale M, Fiehn O, Hardy N, Sumner L, Bino R (2002) The missing link in functional genomics strategies. Plant Cell 14:1437–1440
2. Rhee EP, Gerszten RE (2012) Metabolomics and cardiovascular biomarker discovery. Clin Chem 58:139–147
3. Newgard CB (2017) Metabolomics and metabolic diseases: where do we stand? Cell Metab 25:43–56
4. Gibney MJ, Walsh M, Brennan L, Roche HM, German B, van Ommen B (2005) Metabolomics in human nutrition: opportunities and challenges. Am J Clin Nutr 82:497–503
5. Plumb RS, Johnson KA, Rainville P, Smith BW, Wilson ID, Castro-Perez JM, Nicholson JK (2006) UPLC/MS^E; a new approach for generating molecular fragment information for biomarker structure elucidation. Rapid Commun Mass Spectrom 20:1989–1994
6. Smith CA, Want EJ, O'Maille G, Abagyan R, Siuzdak G (2006) XCMS: processing mass spectrometry data for metabolite profiling using nonlinear peak alignment, matching, and identification. Anal Chem 78:779–787
7. Broeckling CD, Afsar FA, Neumann S, Ben-Hur A, Prenni JE (2014) RAMClust: a novel feature clustering method enables spectral-matching-based annotation for metabolomics data. Anal Chem 86:6812–6817
8. Broeckling CD, Ganna A, Layer M, Brown K, Sutton B, Ingelsson E, Peers G, Prenni JE (2016) Enabling efficient and confident annotation of LC-MS metabolomics data through MS1 spectrum and time prediction. Anal Chem 88:9226–9234
9. R Core Team (2017) R: a language and environment for statistical computing. R Foundation for Statistical Computing, Vienna. https://www.R-project.org/. Accessed 1 Feb 2018
10. Stacklies W, Redestig H, Scholz M, Walther D, Selbig J (2007) pcaMethods—a bioconductor package providing PCA methods for incomplete data. Bioinformatics 23:1164–1167
11. Ejigu BA, Valkenborg D, Baggerman G, Vanaerschot M, Witters E, Dujardin JC, Burzykowski T, Berg M (2013) Evaluation of normalization methods to pave the way towards large-scale LC-MS-based metabolomics profiling experiments. OMICS 17:473–485
12. Zelena E, Dunn WB, Broadhurst D, Francis-McIntyre S, Carroll KM, Begley P, O'Hagan S, Knowles JD, Halsall A, HUSERMET Consortium, Wilson ID, Kell DB (2009) Development of a robust and repeatable UPLC-MS method for the long-term metabolomic study of human serum. Anal Chem 81:1357–1364
13. Benjamini Y, Hochberg Y (1995) Controlling the false discovery rate: a practical and powerful approach to multiple testing. J R Stat Soc Series B 57:289–300
14. Chambers MC, MacLean B, Burke R, Amode D, Ruderman DL, Neumann S, Gatto L, Fischer B, Pratt B, Egertson J, Hoff K, Kessner D, Tasman N, Shulman N, Frewen B, Baker TA, Brusniak M-Y, Paulse C, Creasy D, Flashner L, Kani K, Moulding C, Seymour SL, Nuwaysir LM, Lefebvre B, Kuhlmann F, Roark J, Rainer P, Detlev S, Hemenway T, Huhmer A, Langridge J, Connolly B, Chadick T, Holly K, Eckels J, Deutsch EW, Moritz RL, Katz JE, Agus DB, MacCoss M, Tabb DL, Mallick P (2012) A cross-platform toolkit for mass spectrometry and proteomics. Nat Biotechnol 30:918–920
15. Li B, Tang J, Yang Q, Li S, Cui X, Li Y, Chen Y, Xue W, Li X, Zhu F (2017) NOREVA: normalization and evaluation of MS-based metabolomics data. Nucleic Acids Res. https://doi.org/10.1093/nar/gkx449
16. Bolstad BM, Irizarry RA, Astrand M, Speed TP (2003) A comparison of normalization methods for high density oligonucleotide array data based on variance and bias. Bioinformatics 19:185–193
17. Kopka J, Schauer N, Krueger S, Birkemeyer C, Usadel B, Bergmuller E, Dormann P, Weckwerth W, Gibon Y, Stitt M, Willmitzer L, Fernie AR, Steinhauser D (2005) GMD@CSB.DB: the Golm Metabolome Database. Bioinformatics 21:1635–1638
18. Linstrom PJ, Mallard WG (eds) (2005) NIST chemistry webbook, NIST standard reference database number 69. National Institute of Standards and Technology, Gaithersburg, MD. https://doi.org/10.18434/T4D303. Accessed 1 Feb 2018
19. Broeckling CD, Heuberger AL, Prince JA, Ingelsson E, Prenni JE (2013) Assigning precursor-product ion relationships in indiscriminant MS/MS data from non-targeted metabolite profiling studies. Metabolomics 9:33–43
20. Stein SE (1994) Optimization and testing of mass spectral library search algorithms for compound identification. J Am Soc Mass Spectrom 5:859–865
21. Xia J, Psychogios N, Young N, Wishart DS (2009) MetaboAnalyst: a web server for metabolomic data analysis and interpretation. Nucleic Acids Res 37:W652–W660
22. Smilde AK, Jansen JJ, Hoefsloot HC, Lamers RJA, Van Der Greef J, Timmerman ME

(2005) ANOVA-simultaneous component analysis (ASCA): a new tool for analyzing designed metabolomics data. Bioinformatics 21:3043–3048

23. Wiklund S, Johansson E, Sjöström L, Mellerowicz EJ, Edlund U, Shockcor JP, Gottfries J, Moritz T, Trygg J (2008) Visualization of GC/TOF-MS-based metabolomics data for identification of biochemically interesting compounds using OPLS class models. Anal Chem 80:115–122
24. Worley B, Powers R (2013) Multivariate analysis in metabolomics. Curr Metabolomics 1:92–107

Chapter 19

El-MAVEN: A Fast, Robust, and User-Friendly Mass Spectrometry Data Processing Engine for Metabolomics

Shubhra Agrawal, Sahil Kumar, Raghav Sehgal, Sabu George, Rishabh Gupta, Surbhi Poddar, Abhishek Jha, and Swetabh Pathak

Abstract

Analysis of large metabolomic datasets is becoming commonplace with the increased realization of the role that metabolites play in biology and pathophysiology. While there are many open-source analysis tools to extract peaks from liquid chromatography-mass spectrometry (LC-MS), gas chromatography-mass spectrometry (GC-MS), and tandem mass spectrometry (LC-MS/MS) data, these tools are not very interactive and are suboptimal when a large number of samples are to be analyzed. El-MAVEN is an open-source analysis platform that extends MAVEN and provides fast, powerful, and interactive analysis capabilities especially for datasets containing over 100 samples. The El-MAVEN workflow is easy to use with just four steps from loading data to exporting of the results. Advanced analysis and software techniques such as multiprocessing, machine learning, and reduction of memory leaks are implemented so as to provide a seamless and interactive user experience. Results from El-MAVEN can be exported in a range of formats allowing continued analysis on other platforms. Additionally, El-MAVEN is also fully integrated with Polly™, a cloud-based analysis platform that provides a range of tools for flux analysis and integrative-omics analysis. El-MAVEN is a powerful tool that enables fast and efficient analysis of large metabolomic datasets to accelerate the process of gaining insight from raw data.

Key words Mass spectrometry, Data processing, Metabolomics, Bioinformatics, Data analysis, Metabolic pathways, Liquid chromatography-mass spectrometry, MAVEN

1 Introduction

Metabolomics is an important data type that augments genomic, proteomic, and other -omic datasets to provide increased understanding of biology and disease at a systems level [1–8]. Liquid chromatography-mass spectrometry (LC-MS) and gas chromatography-mass spectrometry (GC-MS) are techniques commonly used in metabolomic studies to identify and quantify metabolites important in different biological, chemical, or disease conditions. Many such studies involve a large number of samples probed under various experimental conditions resulting in large datasets that cannot be reliably analyzed or quantified using tedious

Angelo D'Alessandro (ed.), *High-Throughput Metabolomics: Methods and Protocols*, Methods in Molecular Biology, vol. 1978, https://doi.org/10.1007/978-1-4939-9236-2_19,

manual processes. The volume and size of data generated using these techniques necessitates the use of robust, stable, and accurate analysis software that enables scientist to explore available data in an interactive, efficient, and reliable manner. Access to such analysis software accelerates the scientific process of measurement, analysis, and hypothesis generation.

Due to the need for automated analysis techniques for LC-MS and GC-MS data, multiple data analysis software platforms have been developed, including open-source software like MAVEN [9, 10], XCMS [11–13], MZmine [14, 15], or proprietary software such as SCIEX MultiQuant among others. While open-source software platforms facilitate mass spectrometry data analysis, there are well-documented issues with these tools, such as a large number of false positives and inaccuracies in peak detection [16–18]. Because of such inaccuracies, data analysts often have to review results and adjust analysis parameters, to ensure accurate results. However, most of the available tools have poor user-interface capabilities, with little or no interactive visualizations, thus requiring many rounds of analysis, exporting, viewing of results using separate software, and reanalysis to obtain accurate peak results [9]. This process becomes tedious, forcing some users to create ad hoc pipelines to achieve easier workflows [19]. Additionally, many available software platforms become slow and unreliable when used on high throughput datasets containing a large number of samples or metabolites. In certain cases, there is also a lack of an active support community and detailed documentation. These factors introduce delays and challenges when trying to gain biological insight from large -omics raw datasets. Proprietary software platforms can become cost-prohibitive to use, are restricted to vendor-specific file formats, and have the same data analysis and performance challenges when used with large datasets.

Here we describe El-MAVEN, an open-source, vendor-neutral software platform that allows interactive, fast, efficient, and reliable analysis of LC-MS, GC-MS, and LC-MS/MS datasets in just four steps from loading data to exporting results (*see* Fig. 1 for a summary of El-MAVEN features). With a highly interactive, widget-based user interface, El-MAVEN provides features that enable fast analysis for 100s of samples, each containing 100–1000s of metabolites, in just a few mouse clicks. Compared to other open-source analysis tools, the El-MAVEN analysis workflow is intuitive even for large datasets. For example, when using other analysis tools, an experiment tracking 10 metabolites over 100 samples would require 1000 (10×100) peaks to be curated one at a time. With El-MAVEN, peaks from all 100 samples can be grouped, so that only 10 peaks groups (one for each metabolite) need to be curated, thus drastically reducing analysis time. Table 1 provides a complete list of features that make El-MAVEN a powerful mass spectrometry data processing engine. This table also provides a comparison of El-MAVEN with other commonly used open-source tools.

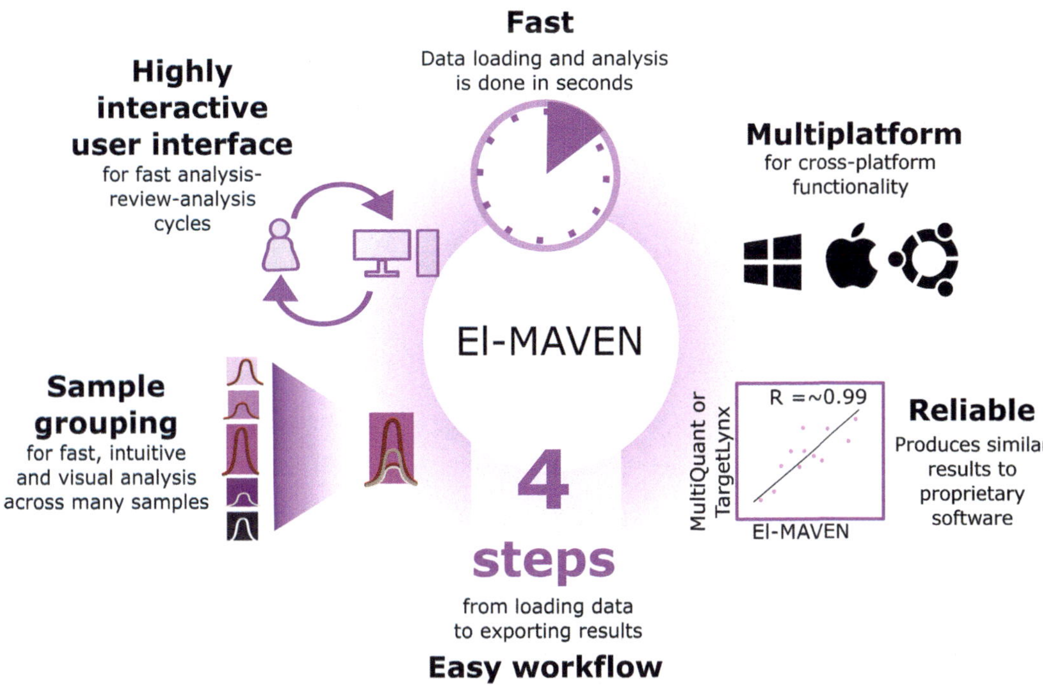

Fig. 1 Features that make El-MAVEN a powerful tool for analysis of large and small mass spectrometry data metabolomics datasets. El-MAVEN has many desirable features such as a highly interactive user interface, fast processing times, availability on multiple platforms, peaks comparable to commercially available software, sample grouping for fast comparison across samples, and an easy four-step analysis workflow

El-MAVEN was developed by extending the open-source software MAVEN [9, 10], to provide additional features that increase ease of use and provide advanced functionality for the analysis of large MS datasets beyond the standard MS data analysis features in MAVEN. Some of the additional features that El-MAVEN provides include (1) multiprocessing for fast data loading, (2) automated peak detection and scoring for an almost fully automated analysis workflow, (3) more options and algorithms to perform alignment and grouping to analyze a range of different datasets, (4) enhanced LC-MS/MS data analysis capability, and (5) additional visualizations and options to save and load settings. These additional features result in an overall improved interactive analysis workflow.

In addition to providing advanced analysis capabilities, a core requirement for the development of El-MAVEN was the need to address certain stability and usability issues encountered when using MAVEN for the analysis of datasets containing a large number of samples (~100 samples or more) each containing 100–1000s of metabolites. While MAVEN performs well for small datasets, data load times in MAVEN for large datasets (e.g., those containing large number of samples or metabolites) were on the order of minutes. Additionally, during analysis of such

Table 1
Comparison of features of El-MAVEN, MAVEN, and XCMS

Property	El-MAVEN	MAVEN	XCMS
Targeted and untargeted metabolite detection	✓	✓	✓
Automated peak detection	✓	✓	✓
Automated grouping of peaks across samples (accelerates analysis time) Isotope detection for analyzing labeled data	✓	✓	✓
Support for isotopic labels	✓ C13, N15, S34, D2 and dual-labeled	✓ C13, N15, S34, D2	✓ C13 as part of X13CMS
Peak alignment (to account for column drift)	✓	✓	✓
Data export	CSV, JSON, PNG, PDF, XML, .mzroll	CSV, PNG, PDF, XML, .mzroll	PNG, PDF, TSV, TXT, XLSX
Data import	.mzML, .mzXML, .cdf, .mzroll	.mzXML, .mzroll	.mzXML, .mzData, .cdf, .d, .wiff, . RAW
Widget-based interactive UI	✓	✓	
Multiprocessing for fast data load and analysis times	✓		
Automated peak scoring (quality indicators)	✓		
Save/load user options for reproducible analysis	✓		
Reset default options	✓		

large datasets, MAVEN performed poorly with high memory usage, long analysis durations, being unresponsive to user input, and crashing, thus making it impossible to analyze these large datasets. With the improvements made in El-MAVEN, large -omics datasets containing data from over 100 samples are loaded in a matter of seconds (~4× improvement compared with MAVEN), and analyses such as peak detection and isotope detection were completed within seconds. *See* Fig. 2 and Table 2 for detailed comparison of the performance on the same dataset. Overall, El-MAVEN provides a stable and interactive platform that enables rapid automated analysis of large metabolomic datasets.

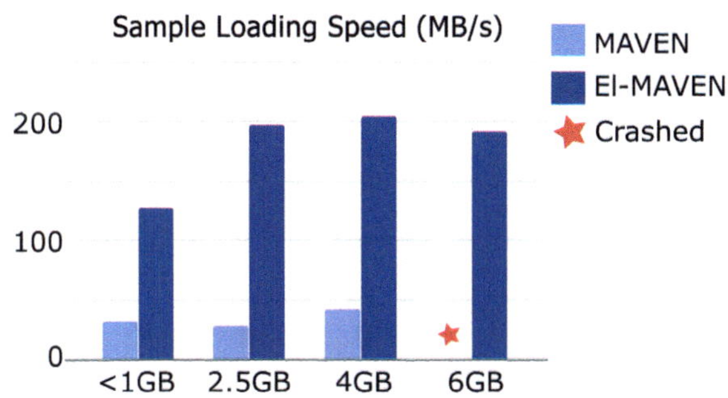

Fig. 2 Comparison of El-MAVEN and MAVEN sample load times for different sizes of datasets. El-MAVEN has faster sample load times for datasets of all sizes, seen by the higher MB/s bars in the figure. Note that for the 6 GB data set, MAVEN crashed (red star)

Table 2
A summary of metrics comparing the performance of MAVEN and El-MAVEN in data loading and analysis for two different datasets

Parameter	Features	MAVEN	El-MAVEN	Improvement
2.5 GB LC-MS data (high-resolution data)				
Time	Upload	93 s	13 s	7×
	Db peak detection	Crashed	1007 groups in 14 s (72 groups/s)	
	Isotope detection	–	3347 groups in 3 m 18 s (17 groups/s)	
Memory used	Upload	1806	2226	1.2× increase
	Db peak detection	Crashed	30	
	Isotope detection	–	35	
Processors used	Upload	1	4	
	Db peak detection	1	4	
	Isotope detection	–	4	
191 MB LCMS data (example dataset)				
Time	Upload	6	1.5	4×
	Db peak detection	213 groups in 2.9 s (73 groups/s)	219 groups in 1.6 s (137 groups/s)	1.9×
	Isotope detection	1311 groups in 6.9 s (190 groups/s)	1076 groups in 5 s (215 groups/s)	1.13×
Memory used	Upload	158	162	1.02× increase
	Db peak detection	11	115	10× increase
	Isotope detection	610	13	46×
Processors used	Upload	1	4	
	Db peak detection	1	4	
	Isotope detection	1	4	

2 Materials

El-MAVEN is a platform agnostic software that can be installed on Windows [7, 8, 10], Mac, and Ubuntu operating systems. As such, a computer capable of running these operating systems is needed. Additionally, a complete dataset from a mass spectrometry run is required for the analysis. Installers and a sample dataset are available at https://elucidatainc.github.io/ElMaven/Downloads/ to follow through the workflow. The steps described below are up-to-date for the current latest version of El-MAVEN (v.0.4.1) at the time of writing this paper.

3 Methods

Before beginning the analysis, El-MAVEN should be downloaded and installed. El-MAVEN installers are available for Windows [7, 8, 10] and Mac. Download the latest version of El-MAVEN from https://elucidatainc.github.io/ElMaven/Downloads/ for your preferred environment. Run the installer and launch the program once installation is complete.

Figure 3 shows a typical workflow used when analyzing data using El-MAVEN. There are four major steps in the workflow: (1) setting up the workspace (adjusting global settings, loading data, and compound databases), (2) performing peak detection (includes peak grouping and scoring), (3) peak curation (optional step in case the automated peaks need manual adjustment), and (4) data export. Because the workflow is interactive, the steps can be repeated with different options and settings, so as to perform the most appropriate analysis for the dataset at hand. If labeled data is being analyzed, isotope detection is also available as a component of the peak detection step.

All analysis steps described here are performed using the user interface. Several windows, menu bars, options, and visualizations are available for the user to perform interactive analysis on the dataset. Figure 4 highlights the main components of the user interface. The main visualization is the extracted ion chromatogram (EIC) plot of the selected metabolite. Samples can be seen in the sample window and a list of metabolites in the compound window. Samples are color-coded, and all peaks from a sample will have the same color in the EIC window. The main EIC window also shows the peak area for each sample for the given RT window. Using the widget bar to the right of the window, users can select a range of visualizations to be seen at the bottom of the user interface. Useful visualizations include alignment visualizations, isotope plots, and peak table views.

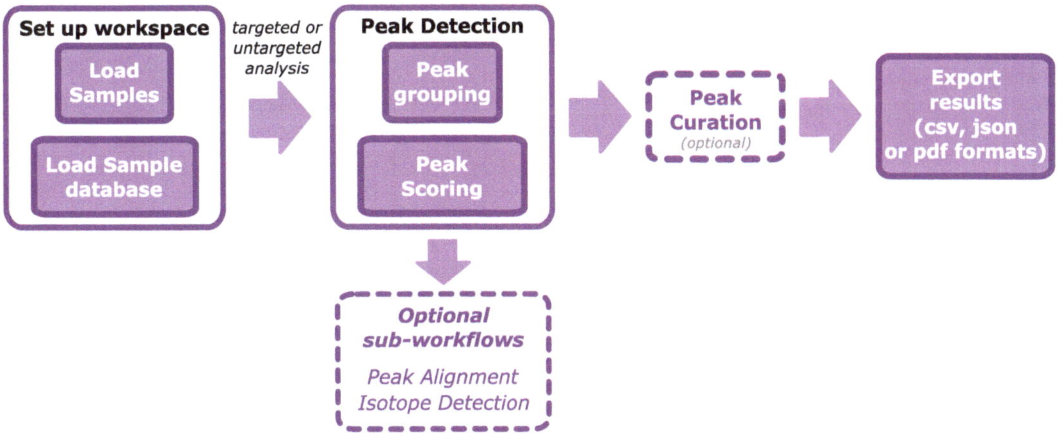

Fig. 3 A typical workflow in El-MAVEN from raw data to exported results. El-MAVEN has an extremely simplified workflow that allows users to analyze hundreds of samples in just a few steps. The first step involves setting up the workspace including loading data files and a reference compound. Global settings should also be set according to the experimental conditions. The next step involved peak detection, grouping, and scoring, all of which are performed by clicking on the "peaks" icon on the main menu. The machine learning algorithm automatically scores groups based on their quality, providing the user with tools to assess the quality of the resulting peaks. During peak detection, various options and settings can be changed to create sub-workflows such as alignment and isotope detection depending on the data and experimental setup. The next step is the optional step of manual peak curation, shown in the box with a dotted line. This step is only needed if the automated peak detection is unable to find certain metabolites due to strict filtering parameters or multiple groups are found for a metabolite. All peaks can then be exported in .csv, .xml, .json, or .pdf formats for further analysis

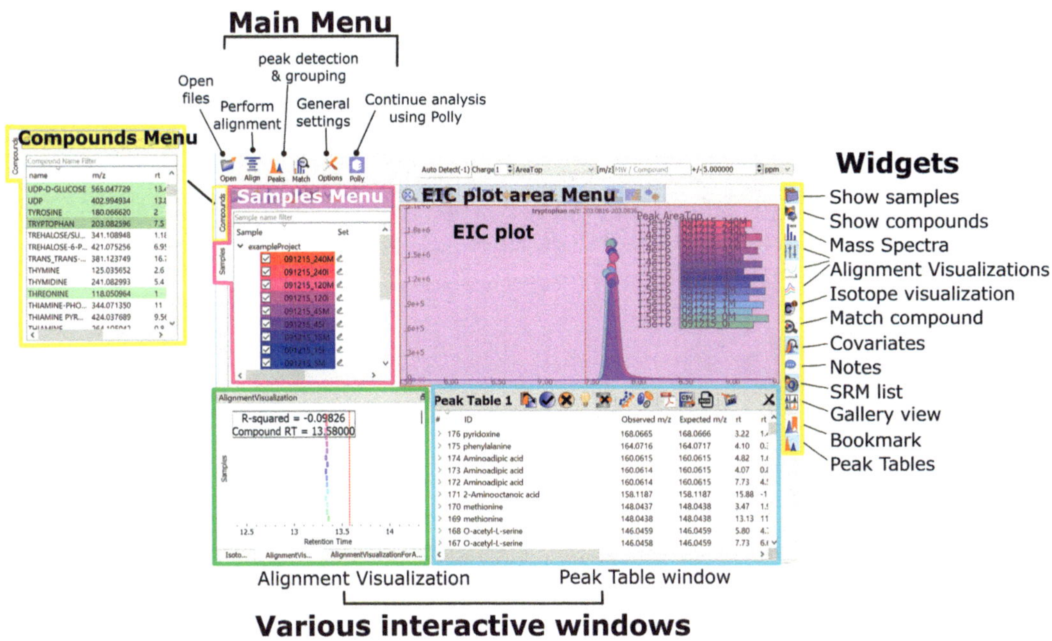

Fig. 4 El-MAVEN user-interface panels, widgets, and visualizations. This figure provides a map of the El-MAVEN user interface. At the top is the main menu, which contains buttons for sample loading, alignment, peaks detection, and global settings. The main visualization panel shows the EIC plots for a selected compound. The EIC menu provides options to show or hide various metrics or visualizations on the EIC panel. To the left is the Sample and Compound panel, which can be toggled using the tabs. Each of these panels has its own menu bar for Sample- and Compound-related options. On the right is the vertical tray containing various widget buttons. These buttons are used to show or hide different types of visualizations on the bottom panel of the user interface. For example, the "alignment button" shows or hides alignment results for the highlighted compound, and the "peaks table" button shows or hides peak detection results

Note that in addition to the user interface, a command-line version of El-MAVEN is also available, which allows the user to adjust a handful of parameters. The command-line version can be useful to automate the entire analysis process if an optimal analysis flow has been identified.

The four steps in the data analysis workflow are highlighted below. These steps are up-to-date for the latest current version of El-MAVEN (v0.4.1 at the time of writing this chapter). Changes in features, options, or settings might be introduced in future versions of El-MAVEN. Certain practical tips, details, and special points to note are provided in the Subheading 4. El-MAVEN also has an active community on GitHub that can be found at https://github.com/ElucidataInc/ElMaven, where source code, additional documentation, and support are available.

3.1 Setting Up the Workspace

3.1.1 Launch El-MAVEN and Adjust Global Settings

Launch El-MAVEN by clicking on the desktop shortcut. Once the program is fully loaded, click on the options button (the screw driver and hammer button as shown in Fig. 4) and adjust the global settings to capture experimental details for the data being analyzed. This step is extremely important to ensure that the appropriate analysis is performed on the dataset. For example, the settings should capture the correct ionization mode and type used in the experiment, the subset of data to be included in the analysis based on certain scan criteria, settings for peak detection and grouping, e.g., what type of smoothing algorithm to use, and so on. A complete list of the available global settings options is provided in Table 3. All subsequent analyses depend on these global settings, so it is imperative to adjust the settings correctly, although the default settings should work well for most good quality LC-MS and LC-MS/MS datasets.

Table 3
List of global settings and a description of the available options

Options	Description
Instrumentation tab	
Polarity/ionization mode	Polarity information is required for *m/z* calculation. User can set the polarity of the metabolites in their experiment from the drop-down list or set it to auto-detect
Ionization type	Ionization methods can affect *m/z* calculation. Drop-down list shows most popular ionization types
Q1 accuracy	For LC-MS/MS data only: Mass resolution in amu of the first quadrupole

(continued)

Table 3 (continued)

Options	Description
Q3 accuracy	For LC-MS/MS data only: Mass resolution in amu of the third quadrupole
Filterline	The drop-down list shows different scan types used in the experiment
File import	
Centroid scans	Centroid acquisition is an acquisition method where only centroid m/z and intensity are stored. Centroid m/z is calculated based on the average m/z value weighted by the intensity, and m/z values are assigned based on a calibration file. Leave the box unchecked if data is centroid data. Check the box if centroiding has to be done in El-MAVEN
Scan filter polarity	Provides option to import scans based on the polarity of ions in the scan. Especially helpful in polarity-switching experiments
Scan filter MS level	Provides option to import only MS1 or MS2 scans. This feature can be used with MS/MS data
Scan filter minimum intensity	Provides option to filter out scans with low intensity
Scan filter intensity minimum quantile cutoff	Scans with x% of their intensity values below the threshold will be filtered out during import
Enable multiprocessing	In order to reduce sample load time, El-MAVEN uses multiprocessing. Multiprocessing can be disabled if required. NOTE: Disabling multiprocessing will increase data load times
Peak detection	
Peak detection and grouping settings	
EIC smoothing algorithm	Smoothing of data points helps in increasing the signal/noise ratio. There are three algorithms provided for EIC smoothing: (a) Savitzky-Golay—It preserves the original shape and features of the signal [22]. (b) Gaussian—It reduces noise by averaging over the neighborhood and successfully preserves sharp edges. (c) Moving average—It takes the simple average of all points over time. Signal behavior is not natural. Least preferred method for smoothing
EIC smoothing window	Number of scans used for fitting in the smoothing algorithm can be adjusted here
Max retention time difference between peaks	Set a limit to RT difference between peaks in a group. Increase the value if alignment fails to center peaks satisfactorily
Baseline calculation	
Drop top x% intensities from chromatogram	Set the baseline for every peak. Baseline is obtained once x% of the highest intensities in a peak are removed from consideration. Baseline should be set high when there is more noise in the data
Baseline smoothing	Number of scans used for fitting in the smoothing algorithm can be adjusted here

(continued)

Table 3
(continued)

Options	Description
Peak filtering	
Parent peak filtering	
Min. signal baseline difference	Set the minimum difference between intensity and baseline to detect any signal as a valid peak for the parent
Min. peak quality	Set the minimum peak quality to be accepted. If the peak quality is lower than the selected value, the peak will be dropped for the parent
Isotope peak filtering	
Min. signal baseline difference	Set the minimum difference between intensity and baseline to detect any signal as a valid peak for the isotopes
Min. peak quality	Set the minimum peak quality to be accepted. If the peak quality is lower than the selected value, the peak will be dropped for the isotopes
Isotope detection	
Are samples labeled?	
Bookmarks, peak detection, file export	Select the labeled atoms that should be used in bookmarking, peak detection, and export. D2, deuterium; C13, labeled carbon; N15, labeled nitrogen; S34, labeled sulfur
Isotopic widget	Select the labeled atoms that should be displayed in the isotopic widget. D2, deuterium; C13, labeled carbon; N15, labeled nitrogen; S34, labeled sulfur
Abundance threshold	Set the minimum threshold for isotopic abundance. Isotopic abundance is the ratio of intensity of isotopic peak over the parent peak
Filter isotopic peaks based on these criteria	
Minimum isotope-parent correlation	Set the minimum threshold for isotope-parent peak correlation. This correlation is a measure of how often they appear together
Isotope is within [X] scans of parent	Set the maximum scan difference between isotopic and parent peaks. This is a measure of how closely they appear together on the RT scale
Maximum % error to natural abundance	Set the maximum natural abundance error expected. Natural abundance of an isotope is the expected ratio of amount of isotope over the amount of parent molecule in nature. Error is the difference between observed and natural abundance as a fraction of natural abundance
Correct for natural C13 isotope abundance	Check the box to correct for natural C13 abundance
EIC (XIC) [BETA]	
EIC type	Select a method to merge EICs over *m/z*. There are two options: (a) MAX—Merged EIC is created by taking the maximum intensity across the *m/z* window at a particular scan. (b) SUM—Merged EIC is created by taking the sum average of intensities across the *m/z* window at a particular scan

(continued)

Table 3 (continued)

Options	Description
Peak grouping	
Peak grouping score	Peaks are assigned a grouping score to determine whether they should be grouped together. There are two formulas for grouping score calculation: (a) score = 1.0/((distX*A) + 0.01)/((distY*B) + 0.01) (C*overlap), (b) score = 1.0/((distX*A) + 0.01)/((distY*B) + 0.01) The score depends on the following three parameters and their weights: (a) RT difference or DistX—Difference in RT between the peaks under comparison. Closer peaks are assigned a higher score (b) Intensity difference or DistY—Difference in intensity between peaks under comparison. Smaller difference accounts for a higher score (c) Overlap—Fraction of RT overlap between the peaks under comparison. Greater overlap accounts for a higher score.
Consider overlap-	Uncheck this box to calculate grouping score without overlap
Sliders	Are provided to adjust the weights (A, B, C) attached to each of the three parameters
Group rank	
Group rank formula	Group rank is one of the parameters for group filtering. There are two formulas below for group rank calculation (a) Group rank = $((1.1 - Q)^A) \times (1/(\log(I + 1))^B)$(b) Group rank = $((1.1 - Q)^A) \times (1/(\log(I + 1))^B) \times (dRT)^{(2\times C)}$ The score depends on the following three parameters and their respective weights A, B, and C (1) Q or group quality—Maximum peak quality of a group. Peaks are assigned a quality score by a machine learning algorithm in El-MAVEN. Better quality leads to a higher rank(2) I or group intensity—Maximum intensity of a group. Better intensity gives higher rank(3) dRT or RT difference—Difference between expected RT and group mean RT
Consider retention time	Check the box to use formula (b) for group rank calculation. Formula (a) is used by default
Sliders	Sliders are provided to adjust the weights (A, B, C) for group quality, intensity, and RT difference for group rank calculation

3.1.2 Load Data

After adjusting the global settings, data can be loaded by clicking on *File> Load Samples|Projects|Peaks* from the main Menu bar. Navigate to the folder containing sample files in .mzXML, .mzML, or .cdf format, select all sample files to be loaded for analysis, and click "Open." Note that if the dataset is not in the accepted format, the data files need to be converted before loading them in El-MAVEN (*see* **Note 1**). A loading bar displays the progress at the bottom right of the El-MAVEN window.

Once all samples are loaded, the samples should appear on the sample panel on the left of the user interface. If the sample panel is not visible, click the "Show Samples" button from the widget bar on the right. Samples can be sorted by clicking on the column header and recolored using the "recolor" button to aid in visualizations. Samples can also be selected/deselected by clicking on the checkbox to the left of the sample, to include or exclude them from the analysis.

Additional metadata about the samples (e.g., samples that are part of a cohort) can be uploaded by clicking on the "sample metadata" button or directly edited in the sample window. Metadata can be used for further comparisons and statistical analyses (*see* **Note 2**).

3.1.3 Select Reference Compound Database

A reference database provides information on metabolite names, chemical formula, Human Metabolome Database (HMDB) ID, and the class/category of compounds they belong to (if known). Two standard set of references are preloaded with the standard installation of El-MAVEN (1) KNOWNS.csv (for use with LC-MS data) and (2) SRM2.csv (for use with LC-MS/MS data). These references are similar to the ones available in MAVEN and are derived from published data [20, 21]. However, custom reference compound databases can also be uploaded and used for peak detection and other analysis, provided they comply with the format defined in the sample databases. To generate a custom reference database, it is recommended that you download the KNOWNS.csv or SRM2.csv files from https://elucidatainc.github.io/ElMaven/Downloads/ and use the headers and format of these files to create a custom reference database using a .csv editor. To select a reference compound database, click on the "*Compounds*" tab on the left. A list of available reference databases will appear in the drop-down field. If the desired reference database is not visible in the drop-down list, a new file can be uploaded by clicking on the "load database" button. In the file explorer that opens up, navigate to the folder containing the reference database file, select the appropriate .csv file, and click "*Open.*"

3.2 Peak Detection

Once the data and reference compounds have been loaded, the next general step is to perform peak detection. Note that the data should be visualized to see if peak alignment should be performed (*see* **Note 3**). Automated peak detection and grouping is an important part of the El- MAVEN workflow. The peak detection algorithm pulls the EICs, detects peaks, and performs grouping and filtering based on parameters controlled by the user. Identical peaks across samples are grouped together, and a quality score is calculated for every peak using a machine learning algorithm. The formula for the group score incorporates the difference in retention time (RT) and intensities between peaks (smaller difference leads to a better score) and any existing overlap between them (higher extent of overlap leads to better score).

To perform peak detection and grouping, click on the "Peaks" button on the main Menu. This window provides options to adjust settings for peak detection and group filtering in separate tabs (*see* Fig. 5).

The first tab allows the user to select between Automated Feature Detection (untargeted analysis) and Compound Database Search (targeted analysis). If the "Automated Feature Detection" option is selected, thousands of mass slices are created across the whole *m/z* and retention time space to find all peaks present in the sample. This strategy is used when looking for new/unknown metabolites in the samples or during untargeted analysis. Instead, if the reference compound database is used to search for compounds, the "Compound Database Search" should be selected. With this option, the compound database is used to search for compounds listed in the reference file using their *m/z* information. For better accuracy, retention time information can also be used for the search.

If labeled details are being analyzed, check the "Report Isotopic Peaks" option and adjust the "Isotopic Detection Options" to match the conditions of the experiment. For example, if the samples have labeled carbon and nitrogen, then boxes corresponding to C13 and N15 should be checked. This will enable isotopic detection that returns a list of isotopologues along with the detected group for every metabolite in a peak table. It is important to note that isotope detection will only work if the chemical formulae of the metabolites have been provided in the compound database.

The second "group filtering" tab provides additional options for filtering out groups. After grouping is done, groups that do not pass the selected criteria are filtered out. There are various options that users can set to filter out undesirable peaks. It is recommended to start analysis using default values and adjust these settings after viewing the results. The options allow users to set minimum threshold values for peak intensity, peak quality, signal/blank ratio, signal/baseline ratio, and peak width.

Once all the desired options are set, click on "Find Peaks" to perform peak detection and scoring. The result will be a peak table at the bottom of the user interface that provides a summary of the results of the peaks and groups identified. The peaks can be explored and visualized and marked as good or bad using buttons on the "peak table menu" in preparation for export (*see* **Note 4**). If labeled data is analyzed, the "isotope plots" button on the EIC menu can be used to toggle the isotope plots visualization and explore results.

Note that the results from peak detection can be visually reviewed, and if the peaks are not optimal, peak detection can be rerun by following the steps above and adjusting parameter values for improved analysis. Each run of the peak detection algorithm creates a new peak table that can be reviewed interactively. The process can be repeated multiple times until satisfactory peaks are obtained. Settings can also be saved and reloaded at a later time from the *File>Save Settings* option.

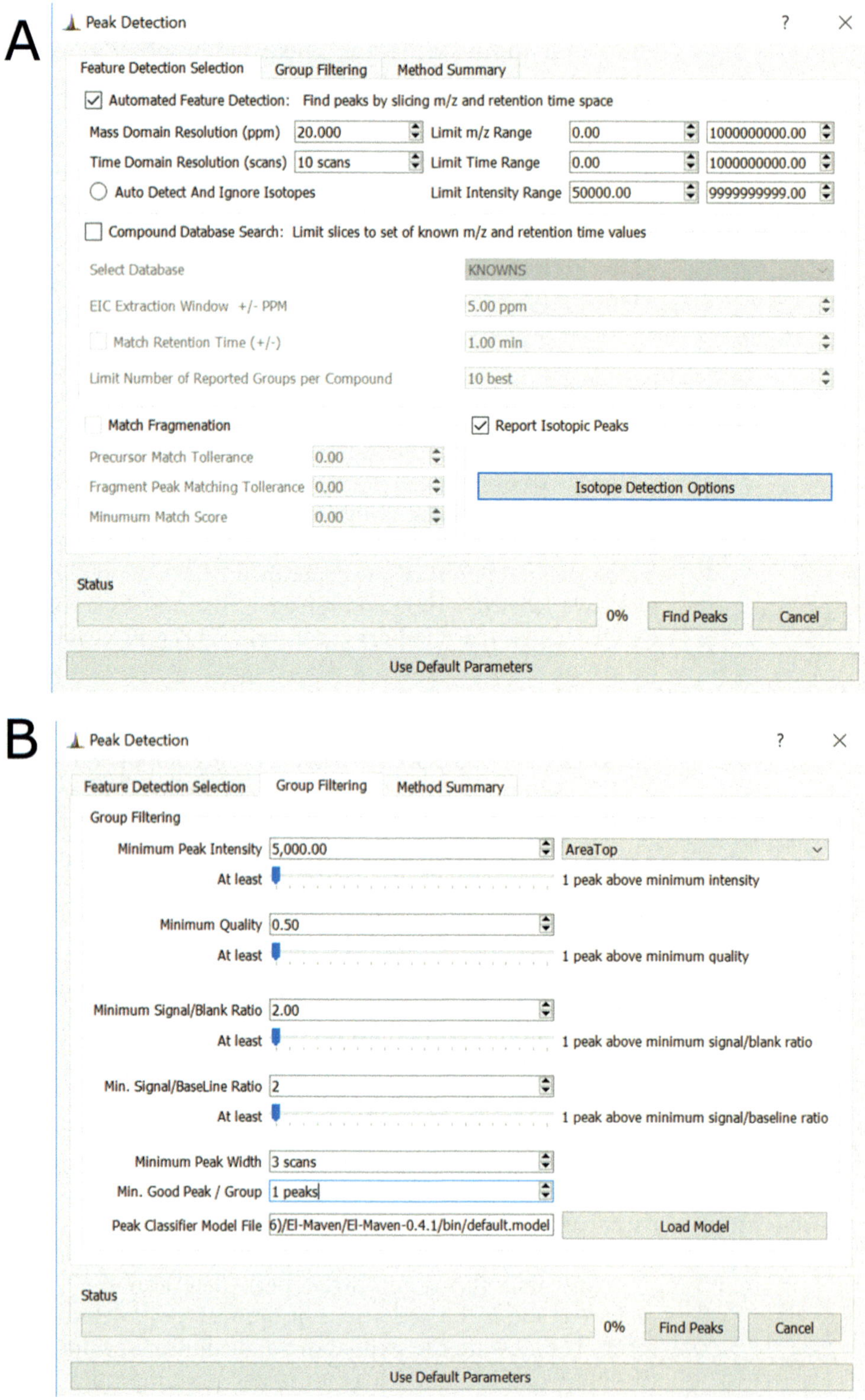

Fig. 5 Peak detection settings window. This figure shows the peak detection settings window. The first tab (**a**) is used to adjust settings for peak detections such as selecting between automated peak detection and compound database search. The second tab (**b**) allows the user to select various options to filter groups

3.3 Peak Curation (Optional)

In general, the automated peak detection algorithm should provide peaks that can directly be exported for further analysis. However, sometimes, a user might want to export only a subset of peaks for further analysis. In this case, peaks from the peak table can be curated by marking them as good or bad and exported as needed.

In certain other scenarios, if automated peak detection does not yield appropriate results, manual peak-picking and curation might be required (*see* **Note 5**). If manual picking and curation is to be performed, the compound DB widget is used to iterate overall the compounds in the compound DB on the extreme left of the window. For a selected compound, the EIC for the highest ranked group for that m/z is shown in the EIC panel. Double-clicking on peak tops (the circles at the top of the peaks) in the EIC panel will add the metabolite to the bookmarks table in the bottom panel. Once a metabolite is in the bookmarks table, it can be moved to another peak table (using the merge option on the "peak table" menu) or exported as needed.

3.4 Exporting Results

Once the desired peaks are curated or marked as "good," there are multiple export options available for storing marked peak data. Data can be exported in CSV format either as a group summary (default option) or as a more detailed peak-level report (using the "Peak Detailed Format"). Additionally, the EICs and group metrics can be exported to a JSON file, which can be used to recreate visualizations outside El-MAVEN. A PDF report can also be exported, with the EIC for every metabolite. The *Export Groups to CSV* option lets the user save all, good, bad, or selected peaks. All peak data and analysis can be stored as an El-MAVEN project (.mzroll file) that can be directly loaded into El-MAVEN at another date.

3.5 Additional Optional Analyses

The steps outlined above are the general basic workflow for analyzing data with El-MAVEN. However, there are additional analyses that can be performed in certain cases to gain more insight from the dataset.

3.5.1 Statistical Analysis

In certain cases, especially when comparing two experimental conditions, for example, statistical analysis is required to explore the differences between samples from the two conditions. El-MAVEN comes equipped with a statistics module for comparing data across different cohorts, such as comparing samples with blanks or standards, or comparing diseased vs. healthy samples. To do this, the "compare samples" button should be clicked from the "Peak table" menu. The samples must be labeled as belonging to two different cohorts either by editing the "Set" column in the Sample space or uploading a Metafile with sample and cohort names.

Multiple types of plots can be generated to identify metabolites of interest, including a scatter plot and a volcano plot. Metabolites with significant differences in peak area between the two sets being compared are colored in the plots. The plot is interactive, and so clicking on a metabolite will show the peaks in the EIC panel. Additionally, a set of data points on the scatter plot can be selected by drawing a box around them. The groups associated with the selected points will become available in a new peak table that can be exported like any other peak table.

3.5.2 Integration with Polly™ or Other Downstream Tools

Once peak detection and grouping has been performed, the .csv or .json result files can be used as an input to other software platforms for further downstream analysis like flux analysis and pathway analysis. One such downstream analysis tool is Polly™, a cloud-based proprietary software developed by Elucidata, the developers of El-MAVEN, that allows for multiple different post-processing workflows of mass spectrometry data. Please note that the results exported from El-MAVEN can be used with any other tool (open-source or proprietary) that provides the required analysis features—Polly™ is just one such downstream analysis tool. Polly™ provides a suite of tools for advanced analysis such as absolute quantitation (Polly™ QuantFit), natural abundance isotope correction (Polly™ IsoCorrect), flux analysis (PollyPhi™ Relative LC-MS), and integrative-omics analysis (Polly™ Intomix). Polly™ is fully integrated with El-MAVEN, and users with a Polly™ account can seamlessly click on the "Polly" button on the main menu in El-MAVEN, log-in to Polly™, and continue the data analysis process from a dashboard within Polly™. The integration of El-MAVEN with Polly™ provides a means for data analysts to go further in the process to gain insight into their datasets.

4 Notes

1. *Data preparation*: El-MAVEN accepts data in .mzXML, .mzML, or .cdf format. If data is in any other format, it must be converted to one of these formats accepted by El-MAVEN. msConvert (developed and maintained by proteoWizard) is a command line/ GUI tool that can be used to convert between various common mass spectrometry data formats such as .mzXML, .mzML, .RAW (ThermoFisher, Waters), .d (Agilent), and .wiff (ABSciex).

   When using msConvert, make sure to uncheck the Zlib compression. El-MAVEN in its current form does not support Zlib compression. Additionally, the following options should be checked: 64-bit encoding, write index, and TPP compatibility. For an example of a screenshot showing the desired settings for msConvert, please see https://github.com/ElucidataInc/ElMaven/wiki/Unlabeled-LCMS-Workflow#preprocessing. Once the files have been converted to the desired format, they can be loaded in El-MAVEN for analysis.

2. *Sample loading*: When loading samples, information on sample sets can be uploaded using the "Sample metadata" button. An example of sample metadata can be seen at https://github.com/ElucidataInc/ElMaven/wiki/Introduction-to-ElMaven-UI#3-load-sample-files. The cohort or group information about a sample can also be directly edited in the samples window by clicking on the pencil icon. When samples are loaded in El-MAVEN, samples with the name "Blank" are automatically marked as blanks. If blank samples are not named as such, they can be marked manually by selecting the blank samples using the checkbox on the left of the sample and clicking on the "Mark as Blank" button. Multiple blanks can be marked together, and the blanks will be colored black.
3. *Peak alignment*: It is common for a drift to be observed in RT across samples with prolonged use of the LC column. To correct for this drift, El-MAVEN provides an "Alignment" algorithm, which shifts the peak RTs in every sample and brings the peaks closer to median RT of the group. Alignment is not necessary for all analyses. A visual check can be performed to determine whether alignment should be performed. The "Alignment visualization" widget on the right widget panel opens a visualization window at the bottom of the user interface. This visualization depicts each peak from the selected group as a box at its current RT. Samples are perfectly aligned when all peak boxes in a group lie on the same vertical axis. If there is a spread in the peak boxes, the "Align" button and the main Menu should be used to adjust alignment settings, to select the alignment algorithm, and to perform sample alignment. After alignment, the peak boxes in the alignment visualization panel should be considerably well aligned. Figure 6

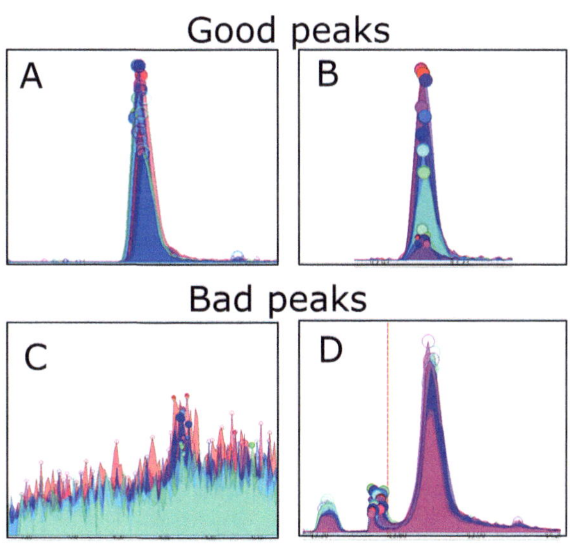

Fig. 7 Examples of good and bad peaks. This figure shows a few examples of good peaks (**a**, **b**) and bad peaks (**c**, **d**), based on the criteria specified in the text

shows an example of a sample pre- and post-alignment using default settings and a Loess fit. The peak boxes are clearly scattered pre-alignment, and aligned well after the alignment algorithm is applied. The alignment can be performed multiple times for a peak, using different settings, until the peaks alignment looks acceptable.

4. *Guidelines for peak-picking*: Several factors should be considered when determining whether a peak is good or bad. Some of the factors to consider include a peak's shape, width, and intensity compared with blanks among others. In general, narrow, Gaussian-shaped peaks that have a high intensity compared with blanks and that are consistent with respect to metabolites and standards are generally considered good peak. Deviations from these features constitute a bad peak. Sometimes, El-MAVEN assigns multiple groups to one metabolite either when no appropriate RT information is available in the compound DB or when there are multiple groups found within a user-defined RT window for the metabolite. In such cases, if there is no contextual information to help select the appropriate group for that particular metabolite, it is recommended to export all the groups and decide which group to keep after downstream analysis when more information about the sample and data is available. Examples of good and bad peaks are shown in Fig. 7.

5. *Scenarios when manual peak curation should be performed*: In certain scenarios, automated peak detection does not yield appropriate results, and manual peak curation should be performed. Below are a few examples of scenarios where manual curation should be performed:

   (a) If some metabolites have not been detected. If the metabolite is not marked in green, in the compound DB after automated peak detection, it is not present in the peak table. In this case if additional metabolites are to be curated, the user can go through the metabolites one at a time from the compound DB list and bookmark the groups to be selected. The selected groups will be added to the bookmark table, and using the merge table option (from the bookmark menu), the selected groups can be merged into the desired peak table.

   (b) If multiple groups have been detected for a metabolite. To check this, the peak table can be sorted by the compound name (by clicking on the compound header). When multiple peaks have been detected for a metabolite, each peak can be visually judged, and the desired peak can be manually selected and marked as "good."

   (c) If groups are noisy. If groups look noisy and ill-defined, they should be marked as bad, or the settings for group

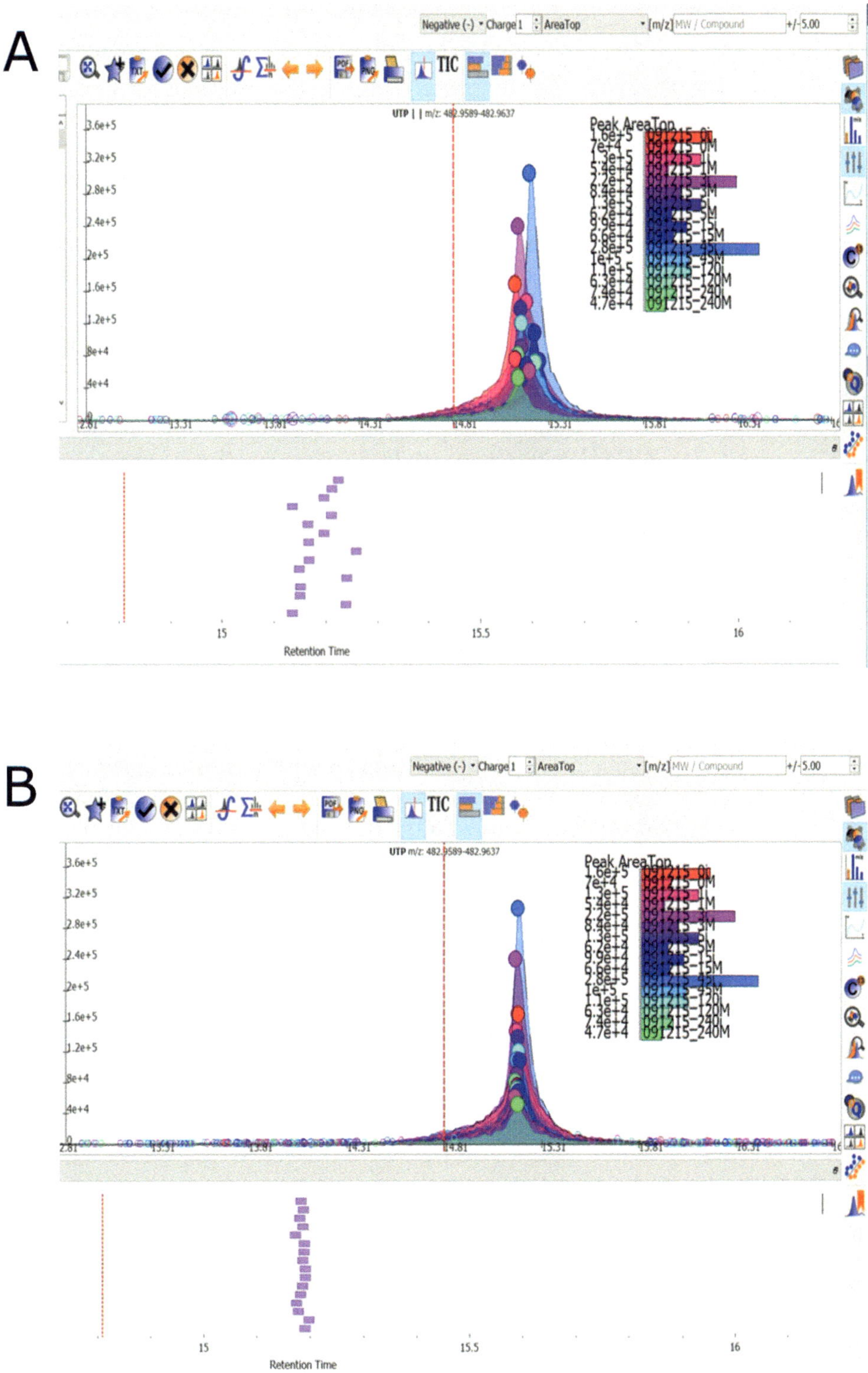

Fig. 6 Peak groups visualized pre- and post-alignment. This figure shows an example of a peak group pre- and post-alignment using default settings and a Loess fit. Each box represents a peak from the selected group at its current RT. Samples pre-alignment (**a**) show a spread in the vertical axis, whereas post-alignment (**b**), the peak boxes lie on the same vertical axis, indicating a good peak alignment

filtering (in the "Peak detection" window) should be changed to be more stringent to filter such noisy peaks out.

(d) If undesirable grouping is performed for certain metabolites during automated peak detection. In this case, the peaks that should be grouped together can be selected by pressing the "Shift" button and dragging the mouse across all the peaks to be grouped. When this is done, all the selected peaks are integrated, essentially creating a new group. The grouping algorithm can also be adjusted from the general options window.

5 Conclusion and Future Directions

El-MAVEN is a fast, stable, and interactive software platform that accelerates the analysis of LC-MS, GC-MS, and LC-MS/MS datasets. It is especially powerful when used on datasets with a large number of samples or metabolites, with fast data load times and analysis times, owing to the improved backend implementation that leverages multiprocessing and has minimal memory leaks. With El-MAVEN, 100s of samples, each containing hundreds of metabolites, can be analyzed with just a few clicks to go from raw data to peak detection and export. The interactive visualizations provided in the user interface allow analysts to review analysis results at each stage. If the results are not desirable, El-MAVEN provides the user with a range of options that can be fine-tuned including the use of different algorithms for alignment and settings for peak detection and grouping to identify the analysis method that works best with a particular dataset. Final analysis results from El-MAVEN can be easily integrated with other platforms that accept data in CSV, XML, or JSON formats. Alternatively, El-MAVEN analysis results can directly be transferred to Polly™, a cloud-based platform that allows for a range of post-processing techniques to be applied on results from El-MAVEN. Overall, El-MAVEN provides a range of features in an easy-to-use interface to enable fast, efficient, and rich analysis of large metabolomic datasets to accelerate the process from raw data to scientific insight.

Acknowledgments

El-MAVEN is an extension of MAVEN, and as such we would like to acknowledge the creators of MAVEN. We would also like to acknowledge El-MAVEN contributors and community on GitHub, especially Victor Chubukov, Lance Parsons, and Eugene Melamud for their help in identifying bugs and fixes. The content and figures in this paper were structured, written, edited, and formatted by Chandni Valiathan at Illumetis, LLC.

References

1. Xia J, Broadhurst DI, Wilson M, Wishart DS (2013) Translational biomarker discovery in clinical metabolomics: an introductory tutorial. Metabolomics 9:280–299. https://doi.org/10.1007/s11306-012-0482-9
2. Mathew AK, Padmanaban VC (2013) Metabolomics: the apogee of the omics trilogy. Int. J. Pharm. Pharm. Sci. 5:45–48
3. Puchades-Carrasco L, Pineda-Lucena A (2015) Metabolomics in pharmaceutical research and development. Curr. Opin. Biotechnol. 35:73–77
4. Holmes E, Wilson ID, Nicholson JK (2008) Metabolic phenotyping in health and disease. Cell 134:714–717
5. Wei R (2011) Metabolomics and its practical value in pharmaceutical industry. Curr Drug Metab 12:345–358. https://doi.org/10.2174/138920011795202947
6. Nicholson JK, Lindon JC (2008) Systems biology: metabonomics. Nature 455:1054–1056
7. Vinayavekhin N, Homan EA, Saghatelian A (2010) Exploring disease through metabolomics. ACS Chem. Biol. 5:91–103
8. Fiehn O (2001) Combining genomics, metabolome analysis, and biochemical modelling to understand metabolic networks. Comp. Funct. Genomics 2:155–168
9. Clasquin MF, Melamud E, Rabinowitz JD (2012) LC-MS data processing with MAVEN: a metabolomic analysis and visualization engine. Curr Protoc Bioinforma. https://doi.org/10.1002/0471250953.bi1411s37
10. Melamud E, Vastag L, Rabinowitz JD (2010) Metabolomic analysis and visualization engine for LC-MS data. Anal Chem 82:9818–9826. https://doi.org/10.1021/ac1021166
11. Smith CA, Want EJ, O'Maille G et al (2006) XCMS: processing mass spectrometry data for metabolite profiling using nonlinear peak alignment, matching, and identification. Anal Chem 78:779–787. https://doi.org/10.1021/ac051437y
12. Smith CA, Want EJ, O'Maille G, Abagyan R, Siuzdak G (2006) LC/MS preprocessing and analysis with xcms. Anal Chem 78:779–787
13. Tautenhahn R, Patti GJ, Rinehart D, Siuzdak G (2012) XCMS online: a web-based platform to process untargeted metabolomic data. Anal Chem 84:5035–5039. https://doi.org/10.1021/ac300698c
14. Pluskal T, Castillo S, Villar-Briones A, Orešič M (2010) MZmine 2: modular framework for processing, visualizing, and analyzing mass spectrometry-based molecular profile data. BMC Bioinformatics 11. https://doi.org/10.1186/1471-2105-11-395
15. Katajamaa M, Miettinen J, Orešič M (2006) MZmine: toolbox for processing and visualization of mass spectrometry based molecular profile data. Bioinformatics 22:634–636. https://doi.org/10.1093/bioinformatics/btk039
16. Coble JB, Fraga CG (2014) Comparative evaluation of preprocessing freeware on chromatography/mass spectrometry data for signature discovery. J Chromatogr A 1358:155–164. https://doi.org/10.1016/j.chroma.2014.06.100
17. Myers OD, Sumner SJ, Li S et al (2017) Detailed investigation and comparison of the XCMS and MZmine 2 chromatogram construction and chromatographic peak detection methods for preprocessing mass spectrometry metabolomics data. Anal Chem 89:8689–8695. https://doi.org/10.1021/acs.analchem.7b01069
18. Myers OD, Sumner SJ, Li S et al (2017) One step forward for reducing false positive and false negative compound identifications from mass spectrometry metabolomics data: new algorithms for constructing extracted ion chromatograms and detecting chromatographic peaks. Anal Chem 89:8696–8703. https://doi.org/10.1021/acs.analchem.7b00947
19. Libiseller G, Dvorzak M, Kleb U et al (2015) IPO: a tool for automated optimization of XCMS parameters. BMC Bioinformatics 16. https://doi.org/10.1186/s12859-015-0562-8
20. Lu W, Bennett BD, Rabinowitz JD (2008) Analytical strategies for LC-MS-based targeted metabolomics. J Chromatogr B Anal Technol Biomed Life Sci 871:236–242. https://doi.org/10.1016/j.jchromb.2008.04.031
21. Lu W, Clasquin MF, Melamud E et al (2010) Metabolomic analysis via reversed-phase ion-pairing liquid chromatography coupled to a stand alone orbitrap mass spectrometer. Anal Chem 82:3212–3221. https://doi.org/10.1021/ac902837x
22. Savitzky A, Golay MJE (1964) Smoothing and differentiation of data by simplified least squares procedures. Anal Chem 36:1627–1639. https://doi.org/10.1021/ac60214a047

Chapter 20

Pre-analytic Considerations for Mass Spectrometry-Based Untargeted Metabolomics Data

Dominik Reinhold, Harrison Pielke-Lombardo, Sean Jacobson, Debashis Ghosh, and Katerina Kechris

Abstract

Metabolomics is the science of characterizing and quantifying small molecule metabolites in biological systems. These metabolites give organisms their biochemical characteristics, providing a link between genotype, environment, and phenotype. With these opportunities also come data challenges, such as compound annotation, missing values, and batch effects. We present the steps of a general pipeline to process untargeted mass spectrometry data to alleviate the latter two challenges. We assume to have a matrix with metabolite abundances, with metabolites in rows and samples in columns. The steps in the pipeline include summarizing technical replicates (if available), filtering, imputing, transforming, and normalizing the data. In each of these steps, a method and parameters should be chosen based on assumptions one is willing to make, the question of interest, and diagnostic tools. Besides giving a general pipeline that can be adapted by the reader, our goal is to review diagnostic tools and criteria that are helpful when making decisions in each step of the pipeline and assessing the effectiveness of normalization and batch correction. We conclude by giving a list of useful packages and discuss some alternative approaches that might be more appropriate for the reader's data.

Key words Mass spectrometry, Untargeted, Metabolomics, Processing, Pre-analytic, Technical replicates, Filtering, Imputation, Normalization

1 Introduction

Metabolomics is concerned with the comprehensive characterization and quantification of small molecule metabolites in biological systems; the metabolites present in a sample or system are referred to as the metabolome (*see* [1, 2] and http://www.metabolomics-society.org/). Metabolomics allows for an assessment of a cellular state within the context of the environment (*see* [2–5] and references therein) and gives complementary insight into biological processes as compared to genomics and proteomics (*see* [2, 3, 6]). Metabolites can also serve as markers of disease severity and drug sensitivity [2, 7–10]. Thus, metabolites give organisms their

Angelo D'Alessandro (ed.), *High-Throughput Metabolomics: Methods and Protocols*, Methods in Molecular Biology, vol. 1978, https://doi.org/10.1007/978-1-4939-9236-2_20, © Springer Science+Business Media, LLC, part of Springer Nature 2019

biochemical characteristics, providing a link between genotype, environment, and phenotype.

Metabolomics can be classified as targeted or untargeted. Targeted metabolomics measures only smaller, defined groups of chemically characterized and biochemically annotated metabolites [11], while untargeted metabolomics is designed to comprehensively measure analytes in a sample, including chemical unknowns. In addition, internal standards can be included in both approaches for absolute quantification and/or to evaluate technical variation. In the following, we focus on untargeted metabolomics and will refer to it simply as metabolomics. We will also use the terms metabolites and compounds interchangeably.

We focus on liquid chromatography mass spectrometry (LC-MS), but much of the discussion applies to other MS applications, such as gas chromatography and LC-MS-based proteomics data. The main alternative to MS is nuclear magnetic resonance (NMR) spectroscopy [12, 13]. Both MS and NMR technologies have their strengths and weaknesses, supplementing and complementing each other [12–14]: For example, MS is more sensitive than NMR and can detect lower-abundance metabolites. NMR, on the other hand, is better at determining the structure of unknown metabolites, is nondestructive, and can be used in vivo.

While MS-based metabolomics brings many opportunities, the complexity of the metabolome and technical limitations also bring challenges. The challenges we focus on here are missing values and unwanted variation, such as batch effects; annotation of compounds is another challenge that we only discuss in the context of missing values. These challenges occur also in other omics fields such as transcriptomics, and tools that have been developed there are often adapted or directly used for metabolomics data. For example, ComBat was developed to "combat" batch effects in gene expression microarray data [15–17], but it has been used in a variety of fields, including metabolomics (*see* [18] and references therein). On the other hand, there are also many MS-specific tools for metabolomics data. An example for a normalization method using internal standards is Cross-contribution Compensating Multiple standard Normalization (CCMN) [19]. Another MS-specific normalization method, using quality control samples, is the mixture model normalization implemented in Metabomxtr [20, 21].

Our goal is to make the reader aware of some of the challenges in MS-based metabolomics data and ways to evaluate and/or alleviate their impact on downstream analyses and results. In particular, we aim to provide the reader with a pipeline for "cleaning" the data for statistical analysis.

The challenges in MS-based metabolomics data we discuss here are missing values, right-skewness of data, batch and run-order effects, and other unwanted variations. Batches often refer to the set of samples run on a particular date, and signal intensities can vary between these batches. Within a batch, run-order effects refer to changes in signal intensities over the different MS runs within a batch. We will return to this when discussing normalization methods.

To address these challenges, we use the pre-analytic pipeline outlined in Fig. 1. We think of pre-analytic processing as the steps starting with a data matrix containing metabolites or compounds in rows and samples, including technical replicates if available, in columns. Note that compounds are characterized by two measurements, retention time and mass. Retention time is the time from injection of the sample to the top of the peak for each compound of interest [22]. The mass of a compound can be used to determine the kind and number of atoms in the molecule. The same compound can have different retention times in different samples, resulting in uncertainty of whether signals from two or more samples correspond to the same metabolite. For a discussion of this peak-shift problem, causes, and alignment methods, *see*, e.g., [23, 24].

Pre-analytic processing pipeline: The steps we consider to be important parts of a pre-analytic processing pipeline are as follows:

1. Summarization of technical replicates (if available).
2. Filtering of metabolites.
3. Imputation of missing values.
4. Transformation of data.
5. Normalization and batch correction.

Note that these steps are often referred to as preprocessing. To avoid confusion with the preprocessing upstream (peak finding, annotation, etc.), we use the term pre-analytic. The individual challenges and approaches to alleviate them are discussed in Subheading 2.

Example data set: For illustration purposes, we use data generated from subjects from the COPDGene genetic epidemiology study [25]. Chronic obstructive pulmonary disease (COPD) is characterized by several clinical phenotypes, including airflow obstruction, emphysema, chronic bronchitis, and frequent exacerbations (*see* [9, 10] and references therein). Smoking is the major risk factor, but most smokers do not develop COPD [10]. Links between alterations in sphingolipid metabolism and disease susceptibility were suggested [9, 26, 27]. For 131 plasma samples from current and former smokers in COPDGene, the lipid fraction was analyzed, using an ultrahigh performance liquid chromatograph (Agilent 1290 Series) and a time-of-flight mass spectrometer (Agilent 6210), which resulted in data on 6166 features. For more detailed information about the cohort and data generation, *see* [10]. Data are available at Metabolomics Workbench (Project ID PR000438).

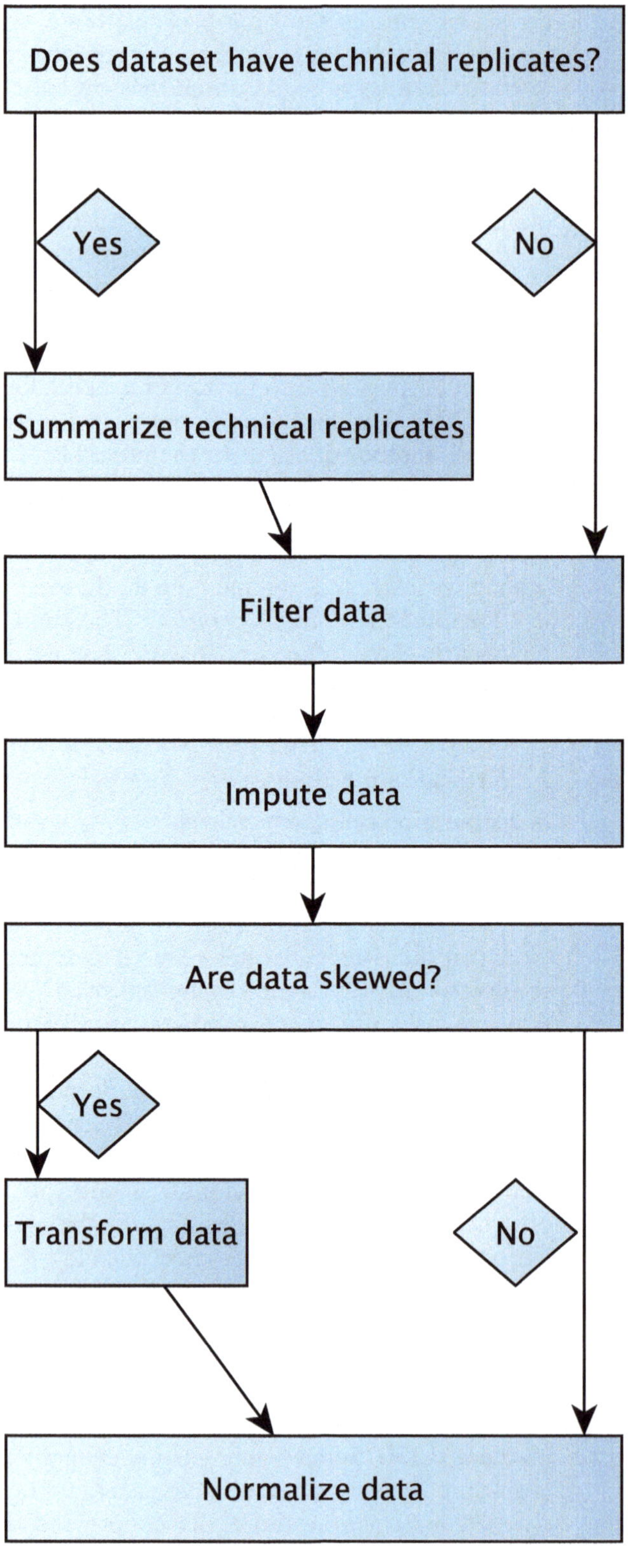

Fig. 1 Pre-analytic pipeline. The five individual pipeline steps are discussed in more detail in Subheadings 2.1–2.5, with each section corresponding to one step

2 Methods

In this section, we go through the five pipeline steps (Fig. 1) in detail. We use the MSPrep package [28], which implements or uses existing packages of the discussed methods; however, note that all steps can be performed without using MSPrep. Within certain steps, we illustrate a specific method using diagnostic plots and our motivating data set but note that different options may be appropriate for other data sets. We also give short explanations for major sources of challenges as well as our rationale for the chosen pipeline steps. We will focus more on the importance of these pre-analytic steps rather than on the specific method used at individual steps (*see* **Notes 1** and **2**). For reviews, evaluations, and discussions of the methods, *see*, e.g., [20, 29–34] and references therein. Finally, in Subheading 3, we will discuss alternatives and challenges.

2.1 Summarization of Technical Replicates

Many studies use technical replicates, which are repeated runs of the same sample through the mass spectrometer. The benefits of technical replicates include the ability to obtain a more precise and robust measurement of a biological sample if we run it through the mass spectrometer multiple times and use the median or mean intensity as the final measurement. In particular, this could help with missing values if a metabolite was measured in a certain number of technical replicates; details and references are provided in our example summarization below. The variability of the technical replicates can also be used to determine if the metabolite could be measured with high or low accuracy, as measured by the coefficient of variation.

However, technical replicates are not always performed due to cost or lack of available material. The downside of technical replicates is the increased complexity in the peak alignment process, as discussed above, which could result in more "duplicated" metabolites (i.e., rows in our data with metabolites having same mass and estimated chemical formula, but retention times that are sufficiently different for the software to consider the metabolites as being separated). Other disadvantages include fewer biological samples that can fit in a batch and the need for additional analyses steps.

Example of summarization of technical replicates: In our data, three technical replicates were randomized within each batch. We examined the missingness and variability of metabolite measurements using the coefficient of variation (CV) in the technical replicates. For a sample, the CV is defined as the standard deviation divided by the mean. When using the CV as criterion for acceptable variation in the measurements, we account for the fact that the standard deviation naturally increases with the mean. Based on missingness and CV, technical replicates were summarized (*see* Fig. 2). For the COPD data, we counted how often there were 0,

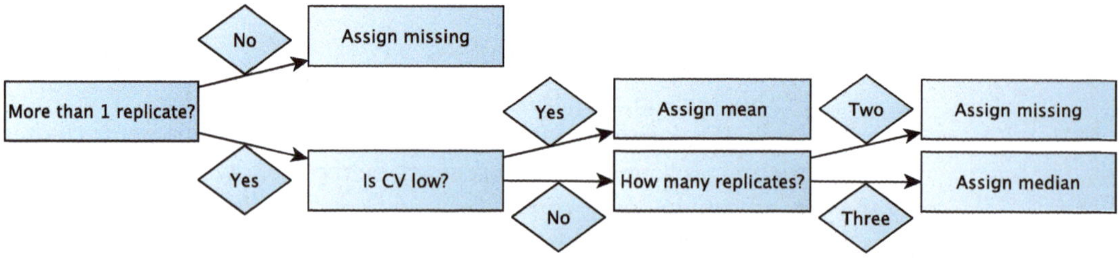

Fig. 2 Example of summarization of technical replicates. For each metabolite and sample, we calculate the coefficient of variation (CV). If two or more replicates have a missing value, we assign a missing value to the metabolite for this sample. If at most one of the three values is missing, and the CV is low (≤0.5), we summarize the replicates with the mean. If the CV is high (>0.5) and we have a missing value, we conclude that the metabolite cannot be measured accurately, and we assign a missing value. With a high CV and no missing values, we choose the median as a robust summary of the replicates

Table 1
Missingness in technical replicates across all samples and metabolites. The number of missing values across the three technical replicates is summarized by the count of metabolite-sample combinations and by the overall percent

# Missing values	Count	Percent
0	373,965	46.19
1	98,457	12.16
2	95,945	11.85
3	241,213	29.79

1, 2, or 3 missing values in the respective technical replicates across all samples and metabolites (Table 1). Across all metabolite and sample combinations, approximately 60% had less than two missing values. If two out of three replicates have a missing value, then we assign a missing value to the metabolite in that sample; for a discussion, *see*, e.g., [35, 36]. When there were at least two non-missing values, we then examine the CV as summarization criterion because our rationale is that low variability indicates that the metabolite could be measured accurately and that in this case the missing value (if any) might well be due to technical issues. In this setting, summarizing the replicates as mean or median depending on the CV will reduce the run-order effect. In **Notes 3** and **4**, we discuss alternatives to this approach.

2.2 Filtering

In this work, we refer to filtering as the removal of metabolites if there are too many missing values across samples. Using our example data set, after summarization of the technical replicates, we examined the distribution of missingness in our COPD data and found that although there was a large number of metabolites with

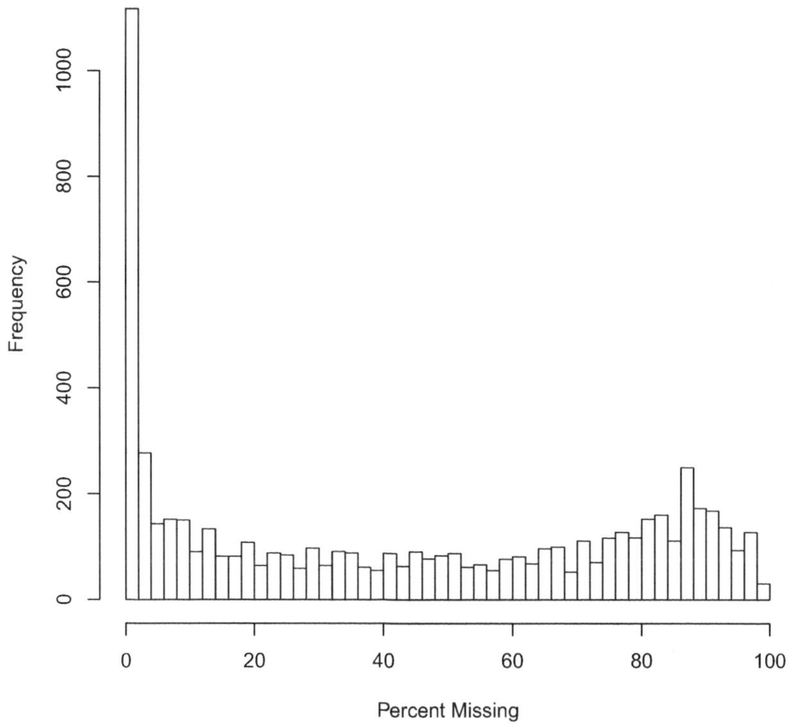

Fig. 3 Example of Missingness. After summarization of replicates, histogram of missingness across samples for each of the 6166 metabolite features

very few missing values, 2682 (43.5%) of the metabolites had more than 50% missing (Fig. 3).

To understand the need for filtering and the impact of imputation for downstream steps in the pipeline, we briefly discuss the reasons for missing values. Missingness is often classified as *missing completely at random* (MCAR), *missing at random* (MAR), and *missing not at random* (MNAR) (*see* [37, 38]): MCAR means that each measurement is equally likely to have a missing value, that is, there is no systematic difference between missing and non-missing data. MAR means that the probability that a value is missing only depends on observed data of other variables measured, that is, any systematic difference between missing and non-missing data can be explained by differences in observed data. MNAR means that after the observed data (of all measured variables) are taken into account, systematic differences remain between the missing and non-missing values.

Missingness in untargeted MS data can have the following reasons:

1. Metabolite is not present.
2. Metabolite is below detection threshold.
3. Other technical reasons, such as misalignment of the detected peaks.

These reasons can partially be classified using the "missingness mechanisms" described above: MCAR, MAR, and MNAR. For metabolites that are not present or below the detection threshold, the data are MNAR. If the data is missing for other technical reasons, the probability of missingness can, for example, depend on m/z [33], which is a measureable quantity. If the probability of missingness were completely determined by m/z, then this would be attributed to MAR. However, missingness is also intensity (abundance) dependent [33], which means there is an MNAR component. This is in contrast to MCAR, where the probability of missingness does not depend on measureable or unmeasureable quantities. It is often difficult or impossible to determine what mechanism caused the missing value. However, understanding the sources of missingness helps to understand the potential biases introduced by imputation, which is discussed below.

Filtering: If a metabolite has too many missing values, it might not be present in many samples, is very low abundant, or is generally hard to quantify. In that case, it might be best to remove that metabolite. There is no consensus on the right filter, but in our experience filtering out the metabolites that have less than 80% non-missing values strikes a good balance of obtaining a relatively "clean" data set for which imputation and normalization will work reasonably well; this was termed "80%" rule in [39] where 80% of the metabolites were required to be present within each group. When there are too many missing values, imputation might not work very well. Restricting ourselves to allow fewer missing values could result in losing too many potentially important metabolites (*see* [36] and references therein). In our COPD data, we have 6166 metabolite features before filtering. Allowing for weak filtering (up to 40% missing values), this reduces the number of metabolites to 3084 (~50% of the original metabolites). More stringent filtering (up to 20% missing values) results in 2333 metabolites (~38% of original metabolites). *See* **Notes 5–7** for alternative approaches.

2.3 Imputation

Many downstream analyses, such as classification, require complete data. Imputation refers to the estimation of missing values. There is a wide selection of imputation strategies available, each of which has their own assumptions about the nature of missing values. The impact on downstream analyses, such as differential abundance, has been noted in the literature (*see*, e.g., [33]). For example, a simple imputation method is to replace missing values with a small value, such as zero, the minimum observed value in other samples, or half the minimum. These imputation methods might be appropriate if missingness is due to low abundance or absence of the metabolite. If missingness is due to other technical issues and the metabolite is present at high abundance but not detected, these imputation methods cause bias and can cause high variability if the metabolite is abundantly present and detected in the other samples.

As an example application, we use k-nearest neighbor (kNN) imputation. Roughly, for the metablites that is missing a particular metabolite M at a sample, the k metabolites that have the most similar metabolite values (profile) will be considered, and the median of their abundances across other samples for metabolite M is used to impute the missing value [40]. Other popular imputation methods are Bayesian PCA (BPCA) [41] and random forest [42]. Recent articles reviewing and evaluating these and other methods are [32, 34]. *See* also **Notes 8** and **9** for additional discussion.

2.4 Transforming the Data

Metabolomics data is typically right-skewed and heteroscedastic. The latter refers to unequal variance of observations across samples; in right-skewed data, variance often increases with the (average) value of the measurement. These characteristics of the data may violate assumptions for standard statistical analyses. A simple and widely used transformation to make data more symmetric and homoscedastic is the log transformation (*see*, e.g., [43]). We recommend this as the default choice and discuss alternatives and evaluation tools in **Notes 10** and **11** (*see* also [43]). An illustrative example of the effect of a $\log_2$ transformation for an example metabolite is presented in Fig. 4. We used the $\log_2$ transformation on our COPDGene data.

2.5 Normalization and Batch Correction

Batch and run-order effects: Similar to other omics technologies, MS suffers from batch effects. As mentioned above, batches often refer to the set of samples run on a particular date. In principle, one could run the machine longer and increase the "batch" size. Besides batch effects, run-order effects within a batch can be observed, and both batch and run-order effects can be metabolite specific (*see*, e.g., [20]).

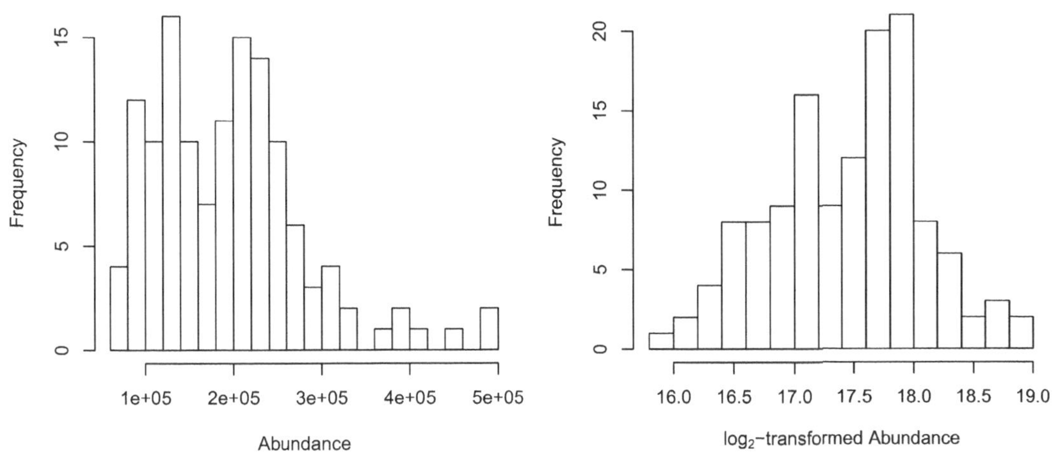

Fig. 4 Distribution of example metabolite (sphingosine) abundance before (left panel) and after $\log_2$ transformation (right panel)

Normalization: To alleviate batch effects and other unwanted variation, the data should be normalized so that different samples can be compared. Similar to imputation strategies, a variety of normalization methods are available, again each with their own assumptions. For recent discussions of methods, *see*, e.g., [20, 29–32] and references therein. A simple yet effective first step is typically a median normalization, which aligns the median signal of all metabolites across samples. The assumption is that most of the metabolites do not change considerably across samples (*see* also **Notes 12–16**). While median normalization typically removes some unwanted variation, other unwanted variations usually remain, some of which are often due to batch effects. The latter can be reduced using, for example, ComBat [15]. Some other popular methods to reduce unwanted variation are Surrogate Variable Analysis (SVA) [16, 44], Remove Unwanted Variation (RUV) [45], and Cross-contribution Compensating Multiple standard Normalization (CCMN) [19]. Normalization methods can also be based on quality control (QC) samples, as, for example, in [20, 21]. *See* **Notes 12–16** for additional discussions. Diagnostic tools, discussed next, can guide the choice of normalization methods.

Diagnostic tools: Diagnostic tools allow us to evaluate the effect of different normalization methods and can help guide our choice of normalization methods. We will discuss relative log expression (RLE) plots and PCA plots. Limitations of these tools, as well as heatmaps combined with hierarchical clustering, will be discussed in **Note 16**.

RLE plots: Relative log expression plots were originally developed to analyze microarray gene expression data [46–48]. These are boxplots of transformed and centered abundance data. More precisely, after log transforming the abundances, for a given sample and metabolite, we subtract the median of the log abundances across all samples. These plots are often better at revealing heterogeneity between samples and batches than boxplots based on $\log(y_{ij})$ without subtracting the median. A discussion using real and simulated data can be found in [46]. We adapt their discussion to our situation. Assume the following:

1. The abundances of a majority of metabolites are unaffected by the biological factors of interest.
2. For each metabolite i that satisfies assumption [1], the distribution of $\log(y_{ij})$ is approximately symmetric.

For a metabolite i satisfying these assumptions, the distribution of the transformed abundances is approximately centered at zero and has constant variance for $j = 1, \ldots, n$, where n is the sample size. In particular, the RLE plots should be approximately centered at 0 and have comparable interquartile ranges if the above assumptions are satisfied.

PCA plots: Principal component analysis determines, roughly speaking, the orthogonal directions of maximal variation in the data. Often, the data will then be projected onto the two- or three-dimensional space spanned by these directions of maximal variation or principal components (PCs). For an overview article of PCA, *see* [49]. If the samples cluster by batches in these plots, that means a high amount of variation in the data stems from batch effects. These clustered batches can often be seen in the data before normalization. After normalization, these clusters should be closer together and their difference minimized.

Illustrative example: In our COPDGene data, we observe batch effects before normalization. After a median and ComBat normalization, these effects are alleviated (Figs. 5 and 6). In particular, we observe in the RLE plots that before normalization, the boxplots of some batches tend to be centered above or below zero, while after normalization, all boxplots are roughly centered at zero (Fig. 5). Similarly, in the PCA plots, we observe strong batch effects before normalization. The batch effect is alleviated after normalization (Fig. 6).

2.6 Concluding Remarks

The pre-analytic pipeline is used to prepare the data for downstream analyses, such as differential abundance analysis of metabolites between two groups, regression analysis to determine associations between phenotypes and metabolites, classification using metabolite profiles, and pathway analysis, to name a few. While the downstream analysis is beyond the scope of this chapter, we want to emphasize that the chosen pre-analytic steps and methods will have an impact on downstream analyses. Sensitivity analyses can be per-

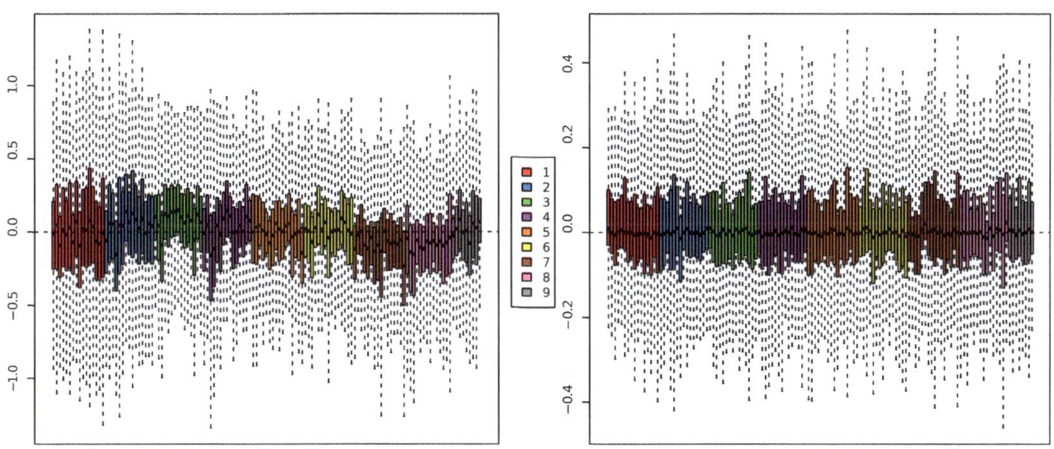

Fig. 5 RLE plots. The kNN-imputed data before normalization (left panel). The kNN-imputed data after median and ComBat normalization (right panel). The nine batches are color-coded. The kNN imputation was done using the VIM R package, with $k = 5$ [40]. The RLE plot was produced using plotRLE in the EDASeq package [57]. Note the different range on the y-axis

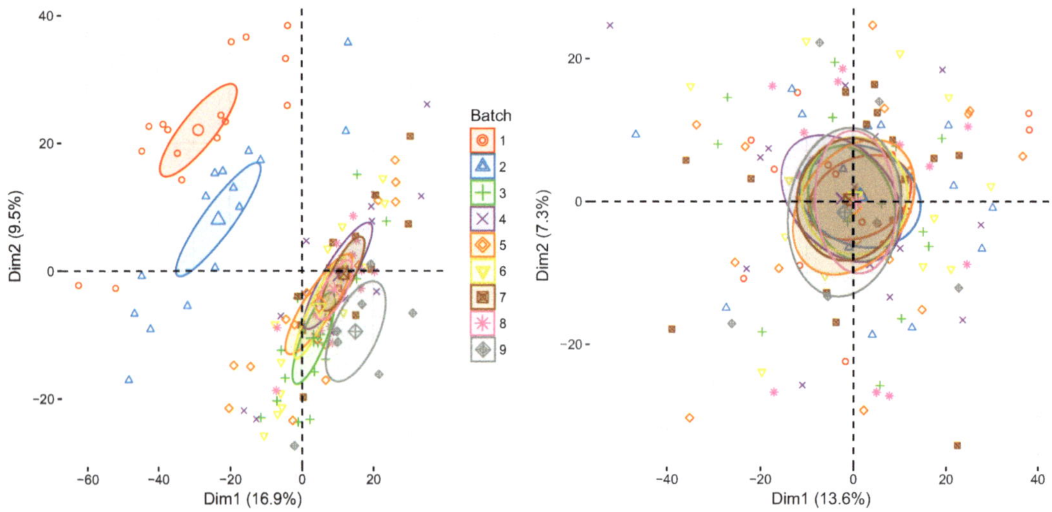

Fig. 6 PCA plots. The kNN-imputed data before normalization (left panel) and after median and ComBat normalization (right panel). The PCA plot was produced using the factoextra package [58]

formed to evaluate the impact of various pre-analytic steps on downstream results. For reviews, evaluations, and discussions of pre-analytic steps, as well as their impact on downstream results, *see*, e.g., [20, 29–34] and references therein.

We conclude with a table of selected free and/or open-source tools that implement normalization methods (Table 2).

3 Notes

Following some general notes, we provide specific comments corresponding to the individual steps in our pipeline.

1. MS technologies and data processing methods constantly evolve, and we encourage the user to read the most current methods and review articles.
2. It is not clear if imputation or normalization should be performed first. Imputation and normalization affect each other, and while imputation works better if there is less unwanted variation such as batch effects, some normalization methods need complete data. We present imputation followed by normalization for the methods applied to our example data set. However, before proceeding with an analysis, we recommend that a user check whether a normalization method can handle missing values and also be aware of the assumptions regarding the form of the data for an imputation method.

Table 2
Selected free and/or open-source tools and packages for imputation and normalization. CRAN packages are available at https://cran.r-project.org, and BioConductor packages are available at https://www.bioconductor.org

Method/software	Link and/or package	References
Imputation methods:		
kNN	CRAN (VIM)	[40]
Bayesian PCA	Bioconductor (pcaMethods)	[41, 59, 60]
Probabilistic PCA	Bioconductor (pcaMethods)	[59]
Random forest imputation	CRAN (randomForest)	[42]
MetaboAnalyst	http://www.metaboanalyst.ca/	[53]
Normalization methods:		
ComBat	Bioconductor (sva)	[15–17, 44]
Cross-contribution robust multiple standard normalization	CRAN (crmn)	[19]
EigenMS	https://sourceforge.net/projects/eigenms/ (EigenMS R package)	[61, 62]
Metabomxtr	Bioconductor (Metabomxtr)	[20, 21]
MetaboQC	CRAN (MetaboQC)	[63]
MetNorm	CRAN (MetNorm)	[64, 65]
MSPrep	https://sourceforge.net/projects/msprep/ (MSPrep R package)	[28]
Remove unwanted variation	CRAN (RUV)	[45]
Support vector regression normalization	http://www.metabolomics-shanghai.org/ (MetNormalizer R package)	[66]
Surrogate variable analysis	CRAN (SVA)	[16, 44]
MetaboAnalyst	http://www.metaboanalyst.ca/	[53]

3. For the pipeline in Fig. 2, we required the metabolite to be present in at least two of the three technical replicates. An alternative is to require it to be present in all three replicates (*see* [36, 50, 51]).
4. When summarizing the technical replicates that are randomized within batches or across batches, we also lose the ability to model the run-order or batch effect using these replicates. In case the technical replicates were run back to back, the run-order is preserved.
5. We discussed and used a "global" filtering approach, in which we removed any metabolite that has at least 80% non-missing

values. *See* [36] for a review of other proposed filters, of requiring 50% [52] and 60–75% [35] presence.

6. If there are specified subject groups of interest, an alternative approach is to filter within groups [39] or consider a present/absent analysis. For example, given two groups (e.g., cases and controls or treatment and placebo), it could be of interest whether metabolites are present in one group but not in the other group. In a present/absent analysis, the data are dichotomized, for example, using a logistic regression framework. This would only work for larger sample sizes.
7. In metabolomics and other omics data sets, investigators may also filter by low variance, low overall abundance, or other characteristics; *see*, e.g., [53] and references therein.
8. As discussed, imputing missing values with a small value can cause bias if the metabolite is highly abundant but not detected. Other methods, such as kNN, can cause bias as well, for example, if the metabolite is indeed not present but a larger value is imputed.
9. Nearest neighbor imputation can be based on different distances to determine nearest neighbors, and the imputed value could take different forms, such as median or mean. A modified version of kNN that accounts for truncation at the minimum value was introduced in [54]. In cases where there are not as many metabolites, the neighbors may also be defined to be at the sample level.
10. While a log transformation typically results in more symmetrically distributed data, there are alternative methods to transform the data to make the data more symmetric and homoscedastic. A powerful tool is the Box-Cox transformation [55]. From a one- or two-parameter family of power transformations, with the log transformation as limit, a parameter is estimated. The data is transformed accordingly.
11. If the sample size is small, determining the most appropriate transformation might not be possible using statistical tests, and even visual assessment of the distributions of metabolites before and after a transformation might not result in a clear decision on the transformation. In this case, we recommend using the log transformation as discussed above.
12. An alternative to median normalization is third quartile normalization, which might be more robust if there are many missing values.
13. The assumption for median normalization and other methods is that most metabolites do not change considerably across samples. Although this may be appropriate for our example study, this assumption may not hold for other studies that have substantial global changes across the metabolome (e.g., perturbation studies).

14. Quality control samples, which are pooled samples and thus should only show technical variability across MS runs, can be used for normalization. In particular, if QC samples are run in different batches and at different run-order positions, they can be used to estimate run-order and batch effects [20, 21].
15. Assumptions for normalization methods are often not verifiable, and one has to choose the one that seems plausible and "works well" for the data. While diagnostic tools can help us determine if we were successful in removing unwanted variation, they cannot tell us if we also removed biological variation of interest or introduced other artifacts. *See*, e.g., [46], for a more detailed discussion.
16. Heatmaps and hierarchical clustering of the samples can also be used to look for similarities and differences in metabolite profiles. This can be used as a diagnostic tool to assess normalization methods. If samples cluster by batches (similar to the clusters in PCA), then further normalization might be needed. For a review of heatmaps, *see*, e.g., [56].

Acknowledgments

D.R., S.J., and K.K. were supported by NIH/NHLBI Grant Number P20 HL113445. D.R., D.G., and K.K. were supported by NIH/NCATS Colorado CTSA Grant Number UL1 TR002535. H.P.-L. was supported by training grant T15 LM009451. Contents are the authors' sole responsibility and do not necessarily represent official NIH views.

References

1. Jordan KW, Nordenstam J, Lauwers GY, Rothenberger DA, Alavi K, Garwood M, Cheng LL (2009) Metabolomic characterization of human rectal adenocarcinoma with intact tissue magnetic resonance spectroscopy. Dis Colon Rectum 52(3):520–525. https://doi.org/10.1007/DCR.0b013e31819c9a2c. PubMed PMID: 00003453-200903000-00024
2. Spratlin JL, Serkova NJ, Eckhardt SG (2009) Clinical applications of metabolomics in oncology: a review. Clin Cancer Res 15(2):431
3. Griffin JL, Shockcor JP (2004) Metabolic profiles of cancer cells. Nat Rev Cancer 4:551. https://doi.org/10.1038/nrc1390
4. Mendes P, Kell DB, Westerhoff HV (1996) Why and when channelling can decrease pool size at constant net flux in a simple dynamic channel. Biochim Biophys Acta 1289(2):175–186. https://doi.org/10.1016/0304-4165(95)00152-2
5. Mendes P, Kell DB, Westerhoff HV (2005) Channelling can decrease pool size. Eur J Biochem 204(1):257–266. https://doi.org/10.1111/j.1432-1033.1992.tb16632.x
6. Boros LG, Lerner MR, Morgan DL, Taylor SL, Smith BJ, Postier RG, Brackett DJ (2005) [1,2-13C2]-D-glucose profiles of the serum, liver, pancreas, and DMBA-induced pancreatic tumors of rats. Pancreas 31:4
7. El-Deredy W, Ashmore SM, Branston NM, Darling JL, Williams SR, Thomas DGT (1997) Pretreatment prediction of the chemotherapeutic response of human glioma cell cultures using nuclear magnetic resonance spectroscopy and artificial neural networks. Cancer Res 57(19):4196

8. Griffin JL, Pole JCM, Nicholson JK, Carmichael PL (2003) Cellular environment of metabolites and a metabonomic study of tamoxifen in endometrial cells using gradient high resolution magic angle spinning 1H NMR spectroscopy. Biochim Biophys Acta 1619(2):151–158. https://doi.org/10.1016/S0304-4165(02)00475-0
9. Bahr TM, Hughes GJ, Armstrong M, Reisdorph R, Coldren CD, Edwards MG, Schnell C, Kedl R, LaFlamme DJ, Reisdorph N, Kechris KJ, Bowler RP (2013) Peripheral blood mononuclear cell gene expression in chronic obstructive pulmonary disease. Am J Respir Cell Mol Biol 49(2):316–323. https://doi.org/10.1165/rcmb.2012-0230OC. PubMed PMID: 23590301; PMCID: PMC3824029
10. Bowler RP, Jacobson S, Cruickshank C, Hughes GJ, Siska C, Ory DS, Petrache I, Schaffer JE, Reisdorph N, Kechris K (2015) Plasma sphingolipids associated with chronic obstructive pulmonary disease phenotypes. Am J Respir Crit Care Med 191(3):275–284. https://doi.org/10.1164/rccm.201410-1771OC. PubMed PMID: 25494452; PMCID: PMC4351578
11. Roberts LD, Souza AL, Gerszten RE, Clish CB (2012) Targeted metabolomics. Curr Protoc Mol Biol Chapter 30:Unit30.2. https://doi.org/10.1002/0471142727.mb3002s98
12. Gowda GAN, Raftery D (2017) Recent advances in nmr-based metabolomics. Anal Chem 89(1):490–510. https://doi.org/10.1021/acs.analchem.6b04420
13. Markley JL, Brüschweiler R, Edison AS, Eghbalnia HR, Powers R, Raftery D, Wishart DS (2017) The future of NMR-based metabolomics. Curr Opin Biotechnol 43:34–40. https://doi.org/10.1016/j.copbio.2016.08.001
14. Gowda GAN, Djukovic D (2014) Overview of mass spectrometry-based metabolomics: opportunities and challenges. Methods Mol Biol 1198:3–12. https://doi.org/10.1007/978-1-4939-1258-2_1
15. Johnson WE, Li C, Rabinovic A (2007) Adjusting batch effects in microarray expression data using empirical Bayes methods. Biostatistics 8(1):118–127. https://doi.org/10.1093/biostatistics/kxj037
16. Leek JT, Storey JD (2007) Capturing heterogeneity in gene expression studies by surrogate variable analysis. PLoS Genet 3(9):e161. https://doi.org/10.1371/journal.pgen.0030161
17. Leek JT, Storey JD (2008) A general framework for multiple testing dependence. Proc Natl Acad Sci 105(48):18718
18. Fernández-Albert F, Llorach R, Garcia-Aloy M, Ziyatdinov A, Andres-Lacueva C, Perera A (2014) Intensity drift removal in LC/MS metabolomics by common variance compensation. Bioinformatics 30(20):2899–2905. https://doi.org/10.1093/bioinformatics/btu423
19. Redestig H, Fukushima A, Stenlund H, Moritz T, Arita M, Saito K, Kusano M (2009) Compensation for systematic cross-contribution improves normalization of mass spectrometry based metabolomics data. Anal Chem 81(19):7974–7980. https://doi.org/10.1021/ac901143w
20. Reisetter AC, Muehlbauer MJ, Bain JR, Nodzenski M, Stevens RD, Ilkayeva O, Metzger BE, Newgard CB, Lowe WL Jr, Scholtens DM (2017) Mixture model normalization for non-targeted gas chromatography/mass spectrometry metabolomics data. BMC Bioinformatics 18(1):84. https://doi.org/10.1186/s12859-017-1501-7. PubMed PMID: 28153035; PMCID: PMC5290663
21. Nodzenski M, Muehlbauer MJ, Bain JR, Reisetter AC, Lowe WL, Scholtens DM (2014) Metabomxtr: an R package for mixture-model analysis of non-targeted metabolomics data. Bioinformatics 30(22):3287–3288. https://doi.org/10.1093/bioinformatics/btu509
22. Snyder LR, Kirkland JJ, Dolan JW (2010) Introduction to modern liquid chromatography, 3rd edn. Wiley, Hoboken, NJ
23. Åberg KM, Alm E, Torgrip RJO (2009) The correspondence problem for metabonomics datasets. Anal Bioanal Chem 394(1):151–162. https://doi.org/10.1007/s00216-009-2628-9
24. Zhou B, Xiao JF, Tuli L, Ressom HW (2012) LC-MS-based metabolomics. Mol BioSyst 8(2):470–481. https://doi.org/10.1039/c1mb05350g
25. Regan EA, Hokanson JE, Murphy JR, Make B, Lynch DA, Beaty TH, Curran-Everett D, Silverman EK, Crapo JD (2010) Genetic epidemiology of COPD (COPDGene) study design. COPD 7(1):32–43. https://doi.org/10.3109/15412550903499522. PubMed PMID: 20214461; PMCID: PMC2924193
26. Petrache I, Petrusca DN, Bowler RP, Kamocki K (2011) Involvement of ceramide in cell death responses in the pulmonary circulation. Proc Am Thorac Soc 8(6):492–496. https://doi.org/10.1513/pats.201104-034MW. PubMed PMID: 22052925; PMCID: PMC3359077
27. Ahmed FS, Jiang XC, Schwartz JE, Hoffman EA, Yeboah J, Shea S, Burkart KM, Barr RG (2014) Plasma sphingomyelin and longitudinal change in percent emphysema on CT. The MESA lung study. Biomarkers 19(3):207–213. https://doi.org/10.3109/

1354750X.2014.896414. PubMed PMID: 24649875; PMCID: PMC4088962

28. Hughes G, Cruickshank-Quinn C, Reisdorph R, Lutz S, Petrache I, Reisdorph N, Bowler R, Kechris K (2014) MSPrep—summarization, normalization and diagnostics for processing of mass spectrometry-based metabolomic data. Bioinformatics 30(1):133–134. https://doi.org/10.1093/bioinformatics/btt589
29. Ejigu BA, Valkenborg D, Baggerman G, Vanaerschot M, Witters E, Dujardin JC, Burzykowski T, Berg M (2013) Evaluation of normalization methods to pave the way towards large-scale LC-MS-based metabolomics profiling experiments. OMICS 17(9):473–485. https://doi.org/10.1089/omi.2013.0010. PubMed PMID: 23808607; PMCID: PMC3760460
30. Han TL, Yang Y, Zhang H, Law KP (2017) Analytical challenges of untargeted GC-MS-based metabolomics and the critical issues in selecting the data processing strategy. F1000Res 6:967. https://doi.org/10.12688/f1000research.11823.1. PubMed PMID: 28868138; PMCID: PMC5553085
31. Chen J, Zhang P, Lv M, Guo H, Huang Y, Zhang Z, Xu F (2017) Influences of normalization method on biomarker discovery in gas chromatography-mass spectrometry-based untargeted metabolomics: what should be considered? Anal Chem 89(10):5342–5348. https://doi.org/10.1021/acs.analchem.6b05152
32. Di Guida R, Engel J, Allwood JW, Weber RJ, Jones MR, Sommer U, Viant MR, Dunn WB (2016) Non-targeted UHPLC-MS metabolomic data processing methods: a comparative investigation of normalisation, missing value imputation, transformation and scaling. Metabolomics 12:93. https://doi.org/10.1007/s11306-016-1030-9. PubMed PMID: 27123000; PMCID: PMC4831991
33. Hrydziuszko O, Viant MR (2012) Missing values in mass spectrometry based metabolomics: an undervalued step in the data processing pipeline. Metabolomics 8(1):161–174. https://doi.org/10.1007/s11306-011-0366-4
34. Wei R, Wang J, Su M, Jia E, Chen S, Chen T, Ni Y (2018) Missing value imputation approach for mass spectrometry-based metabolomics data. Sci Rep 8(1):663. https://doi.org/10.1038/s41598-017-19120-0
35. Han J, Danell RM, Patel JR, Gumerov DR, Scarlett CO, Speir JP, Parker CE, Rusyn I, Zeisel S, Borchers CH (2008) Towards high-throughput metabolomics using ultrahigh-field Fourier transform ion cyclotron resonance mass spectrometry. Metabolomics 4(2):128–140. https://doi.org/10.1007/s11306-008-0104-8
36. Payne TG, Southam AD, Arvanitis TN, Viant MR (2009) A signal filtering method for improved quantification and noise discrimination in Fourier transform ion cyclotron resonance mass spectrometry-based metabolomics data. J Am Soc Mass Spectrom 20(6):1087–1095. https://doi.org/10.1016/j.jasms.2009.02.001
37. Sterne JAC, White IR, Carlin JB, Spratt M, Royston P, Kenward MG, Wood AM, Carpenter JR (2009) Multiple imputation for missing data in epidemiological and clinical research: potential and pitfalls. BMJ 338:b2393. https://doi.org/10.1136/bmj.b2393
38. Little R, Rubin D (2002) Statistical analysis with missing data, 2nd edn. Wiley, New York
39. Bijlsma S, Bobeldijk I, Verheij ER, Ramaker R, Kochhar S, Macdonald IA, van Ommen B, Smilde AK (2006) Large-scale human metabolomics studies: a strategy for data (pre-) processing and validation. Anal Chem 78(2):567–574. https://doi.org/10.1021/ac051495j
40. Kowarik A, Templ M (2016) Imputation with the R package VIM. J Stat Software 74(7):16. https://doi.org/10.18637/jss.v074.i07
41. Oba S, Sato M-a, Takemasa I, Monden M, Matsubara K-i, Ishii S (2003) A Bayesian missing value estimation method for gene expression profile data. Bioinformatics 19(16):2088–2096. https://doi.org/10.1093/bioinformatics/btg287
42. Breiman L (2001) Random forests. Mach Learn 45(1):5–32. https://doi.org/10.1023/A:1010933404324
43. van den Berg RA, Hoefsloot HCJ, Westerhuis JA, Smilde AK, van der Werf MJ (2006) Centering, scaling, and transformations: improving the biological information content of metabolomics data. BMC Genomics 7:142. https://doi.org/10.1186/1471-2164-7-142
44. Leek JT, Johnson WE, Parker HS, Jaffe AE, Storey JD (2012) The sva package for removing batch effects and other unwanted variation in high-throughput experiments. Bioinformatics 28(6):882–883. https://doi.org/10.1093/bioinformatics/bts034
45. Gagnon-Bartsch JA, Speed TP (2012) Using control genes to correct for unwanted variation in microarray data. Biostatistics 13(3):539–552. https://doi.org/10.1093/biostatistics/kxr034. PubMed PMID: 22101192; PMCID: PMC3577104
46. Gandolfo LC, Speed TP (2018) RLE plots: visualising unwanted variation in high dimensional data. PLoS One 13(2):e0191629

47. Bolstad BM, Collin F, Brettschneider J, Simpson K, Cope L, Irizarry RA, Speed TP (2005) Quality assessment of affymetrix GeneChip data. In: Gentleman R, Carey VJ, Huber W, Irizarry RA, Dudoit S (eds) Bioinformatics and computational biology solutions using R and bioconductor. Springer, New York, NY, pp 33–47
48. Brettschneider J, Collin F, Bolstad BM, Speed TP (2008) Quality assessment for short oligonucleotide microarray data. Technometrics 50(3):241–264. https://doi.org/10.1198/004017008000000334
49. Abdi H, Williams LJ (2010) Principal component analysis. Wiley Interdiscip Rev Comput Stat 2(4):433–459. https://doi.org/10.1002/wics.101
50. Dunn WB, Overy S, Quick WP (2005) Evaluation of automated electrospray-TOF mass spectrometryfor metabolic fingerprinting of the plant metabolome. Metabolomics 1(2):137–148. https://doi.org/10.1007/s11306-005-4433-6
51. Overy SA, Walker HJ, Malone S, Howard TP, Baxter CJ, Sweetlove LJ, Hill SA, Quick WP (2005) Application of metabolite profiling to the identification of traits in a population of tomato introgression lines. J Exp Bot 56(410):287–296. https://doi.org/10.1093/jxb/eri070
52. Smith CA, Want EJ, O'Maille G, Abagyan R, Siuzdak G (2006) XCMS: processing mass spectrometry data for metabolite profiling using nonlinear peak alignment, matching, and identification. Anal Chem 78(3):779–787. https://doi.org/10.1021/ac051437y
53. Xia J, Wishart David S (2016) Using MetaboAnalyst 3.0 for comprehensive metabolomics data analysis. Curr Protoc Bioinformatics 55(1):14.0.1–14.0.91. https://doi.org/10.1002/cpbi.11
54. Shah JS, Rai SN, DeFilippis AP, Hill BG, Bhatnagar A, Brock GN (2017) Distribution based nearest neighbor imputation for truncated high dimensional data with applications to pre-clinical and clinical metabolomics studies. BMC Bioinformatics. 18:114. https://doi.org/10.1186/s12859-017-1547-6
55. Box GEP, Cox DR (1964) An Analysis of Transformations. J R Stat Soc Series B 26(2):211–252
56. Bojko A (ed) (2009) Informative or misleading? Heatmaps deconstructed. Human-computer interaction new trends. Springer, Berlin
57. Risso D, Schwartz K, Sherlock G, Dudoit S (2011) GC-content normalization for RNA-Seq data. BMC Bioinformatics 12(1):480. https://doi.org/10.1186/1471-2105-12-480
58. Kassambara A, Mundt F (2017) Factoextra: extract and visualize the results of multivariate data analyses. https://cran.r-project.org/web/packages/factoextra/index.html
59. Stacklies W, Redestig H, Scholz M, Walther D, Selbig J (2007) pcaMethods—a bioconductor package providing PCA methods for incomplete data. Bioinformatics 23(9):1164–1167. https://doi.org/10.1093/bioinformatics/btm069
60. Bishop CM (ed) (1999) Variational principal components. 1999 ninth international conference on artificial neural networks ICANN 99 (Conf Publ No 470)
61. Karpievitch YV, Nikolic SB, Wilson R, Sharman JE, Edwards LM (2015) Metabolomics data normalization with EigenMS. PLoS One 9(12):e116221. https://doi.org/10.1371/journal.pone.0116221
62. Karpievitch YV, Taverner T, Adkins JN, Callister SJ, Anderson GA, Smith RD, Dabney AR (2009) Normalization of peak intensities in bottom-up MS-based proteomics using singular value decomposition. Bioinformatics 25(19):2573–2580. https://doi.org/10.1093/bioinformatics/btp426
63. Calderón-Santiago M, López-Bascón MA, Peralbo-Molina Á, Priego-Capote F (2017) MetaboQC: a tool for correcting untargeted metabolomics data with mass spectrometry detection using quality controls. Talanta 174:29–37. https://doi.org/10.1016/j.talanta.2017.05.076
64. De Livera AM, Dias DA, De Souza D, Rupasinghe T, Pyke J, Tull D, Roessner U, McConville M, Speed TP (2012) Normalizing and integrating metabolomics data. Anal Chem 84(24):10768–10776. https://doi.org/10.1021/ac302748b
65. De Livera AM, Sysi-Aho M, Jacob L, Gagnon-Bartsch JA, Castillo S, Simpson JA, Speed TP (2015) Statistical methods for handling unwanted variation in metabolomics data. Anal Chem 87(7):3606–3615. https://doi.org/10.1021/ac502439y
66. Shen X, Gong X, Cai Y, Guo Y, Tu J, Li H, Zhang T, Wang J, Xue F, Zhu Z-J (2016) Normalization and integration of large-scale metabolomics data using support vector regression. Metabolomics 12(5):89. https://doi.org/10.1007/s11306-016-1026-5

Part VI

Metabolic Measurements with Techniques Orthogonal to Mass Spectrometry

Chapter 21

Temporal Metabolite, Ion, and Enzyme Activity Profiling Using Fluorescence Microscopy and Genetically Encoded Biosensors

Douglas A. Chapnick, Eric Bunker, Xuedong Liu, and William M. Old

Abstract

Living cells employ complex and highly dynamic signaling networks and transcriptional circuits to maintain homeostasis and respond appropriately to constantly changing environments. These networks enable cells to maintain tight control on intracellular concentrations of ions, metabolites, proteins, and other biomolecules and ensure a careful balance between a cell's energetic needs and catabolic processes required for growth. Establishing molecular mechanisms of genetic and pharmacological perturbations remains challenging, due to the interconnected nature of these networks and the extreme sensitivity of cellular systems to their external environment. Live cell imaging with genetically encoded fluorescent biosensors provides a powerful new modality for nondestructive spatiotemporal tracking of ions, small molecules, enzymatic activities, and molecular interactions in living systems, from cells, tissues, and even living organisms. By deploying large panels of cell lines, each with distinct biosensors, many critical biochemical pathways can be monitored in a highly parallel and high-throughput fashion to identify pharmacological vulnerabilities and combination therapies unique to a given cell type or genetic background. Here we describe the experimental and analytical methods required to conduct multiplexed parallel fluorescence microscopy experiments on live cells expressing stable transgenic synthetic protein biosensors.

Key words Fluorescent biosensor, FRET, Fluorescence microscopy, Mechanism of action, Profiling

1 Introduction

Fluorescence microscopy is a powerful analytical tool for rapidly profiling biochemical changes within cells in response to chemical stimuli. Using a relatively standard wide-field microscope, researchers can simultaneously excite and measure the emission of fluorescent small molecules or genetically encoded protein biosensors, delivered intracellularly, and thus monitor a growing list of possible intracellular activities or molecules important for cellular function and drug response. A large advantage of this indirect approach of using biochemical reporters over other destructive methods is the

Angelo D'Alessandro (ed.), *High-Throughput Metabolomics: Methods and Protocols*, Methods in Molecular Biology, vol. 1978, https://doi.org/10.1007/978-1-4939-9236-2_21,

ability to continually monitor the same population of cells over time in an automated fashion. Fluorescent reporters are commercially available in two major forms: synthetic small molecules containing a conjugated fluorophore [1] and genetically encoded fluorescent biosensors. Genetically encoded fluorescent biosensors most frequently incorporate chimeric fluorescent proteins (FPs), for example, from jellyfish (*Aequorea victoria*) or various coral species [2] or luciferase enzymes derived from various species of firefly or sea pansy (*Renilla reniformis*) [3]. Fluorescent biosensors enable the tracking of biochemical events such as enzymatic activities and binding events not accessible to orthogonal methods such as mass spectrometry or RNA-seq.

One strategy for genetically encoded biosensor design exploits the phenomena of Forster resonance energy transfer (FRET) [4], where a donor fluorophore is coupled to an accepter fluorophore through a functional chimeric protein sequence that interacts with one or many chemicals in cells [5]. The role of the functional chimeric protein sequence is to couple the distance and/or orientation between FRET donor and acceptor FPs to a biochemical event in the cell. For example, the EKAR sensor (Fig. 1) contains a chimeric protein sequence comprised of an ERK substrate peptide derived from the protein Cdc25C, a phosphopeptide binding domain derived from the protein Pin1, and a consensus ERK docking domain with the amino acid sequence FQFP [6]. When ERK kinase activity is high in biosensor-expressing cells, the Cdc25C substrate motif is phosphorylated by ERK resulting in intramolecular binding by the WW domain. As a result, the EKAR sensor adopts a constrained conformation, decreasing the average distance between FRET donor and acceptor fluorophores and increasing the observed FRET signal. In this manner, a designed synthetic protein allows for fluorescent measurements to report biochemical ERK activity throughout a population of cells. Such FRET-based biosensors have been constructed for hundreds of metabolites, enzymes, and intermolecular events within cells [5].

We demonstrate a method of delivering multiple targeted biosensors to mammalian cells and a multiplexed parallel measurement experiment to simultaneously measure small molecule metabolites, ion compartmentalization, and enzyme activity modulation throughout the same cell-based model system and in response to chemical perturbation with small molecule compounds. This method can be used to increase the efficiency of cell biology research by providing a rapid means of determining the biochemical pathways that are perturbed within cells in response to stimuli. This approach enables a researcher to identify and focus on molecular and cellular components relevant for further study of the mechanism of action to perturbations in a largely literature-independent manner. A common problem encountered is that molecular mechanisms of action reported in the literature exhibit

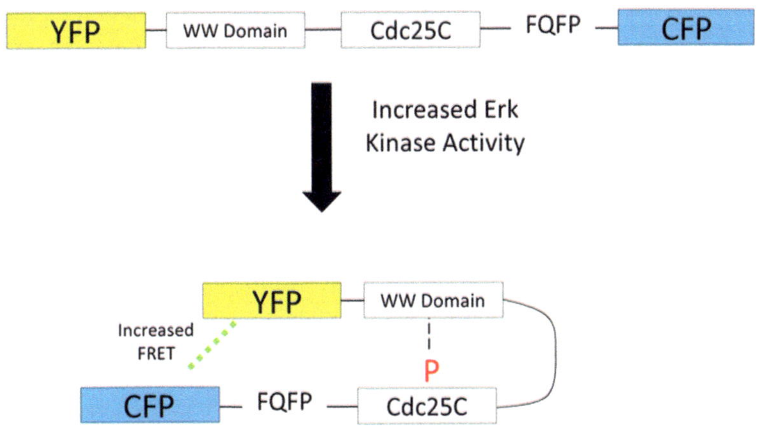

Fig. 1 A schematic diagram of the EKAR biosensor for ERK kinase activity developed by Svoboda and colleagues. In the presence of high ERK kinase activity within a cell, the synthetic protein adopts a constrained conformation that increases FRET between the CFP and YFP fluorophores

cell-type specificity and thus are not generalizable to a new cell type or genetic background. The overall technique displayed here circumvents this issue, allowing for highly parallel measurement of an array of biochemical activities, metabolites and biomolecules in any cell type of interest, which facilitates the characterization of unknown mechanisms of action. The ability to screen large numbers of drug perturbations over multiple doses and timepoints enables empirical search over large combinatorial spaces for drug interactions in a given cellular model.

2 Materials

Prepare all reagents that come in contact with cells in a biosafety cabinet and with solvents known to be sterile filtered.

2.1 Mammalian Cell Culture

1. Phenol red-free and riboflavin-free DMEM is commercially available as FluoroBrite™ DMEM from Thermo Fisher (A1896701).
2. Fetal bovine serum (FBS) is used at a final concentration in DMEM at 10% by volume.
3. Penicillin-streptomycin used in DMEM at a final concentration of 50 U/ml.
4. GlutaMAX supplement (Thermo Fisher, #35050061) is supplied as a 100× stock and is used at a dilution of 1:100 in DMEM.
5. Phosphate buffered saline (PBS).
6. Trypsin-EDTA (0.05%).

7. Ouabain was obtained from Selleck Chemicals (Cat. # S4016) and reconstituted in a 10 mM DMSO (Thermo Fisher, # D12345) stock solution.
8. Phorbol 12-myristate 13-acetate (PMA) was obtained from Sigma-Aldrich (Cat. # P8139-1MG) and reconstituted in a 1 mg/ml DMSO (Thermo Fisher, # D12345) stock solution.
9. 35 mm × 10 mm cell culture petri dishes.
10. 12 well cell culture dishes.
11. SK-N-BE(2) neuroblastoma cells (ATCC CRL-2271).
12. 10 ml serological pipettes.
13. Hemacytometer (Fisher Scientific, 02-671-6).
14. A centrifuge capable of 200 × *g* force with a swinging bucket rotor capable of spinning 384-well plates.

2.2 Biosensor Gene Delivery

1. Opti-MEM™ I Reduced Serum Medium (Thermo Fisher, #31985062).
2. Polyethylenimine (PEI), Linear, MW 25000, Transfection Grade (Polysciences, Inc. # 23966-1).
3. Cellulose Acetate Syringe Filters (Thermo Fisher, #F2500-16).
4. PiggyBac Transposase Expression Vector (referred to below as mPB) (System Biosciences, PB210PA-1).
5. PiggyBac compatible vector containing a biosensor gene. (Note: The biosensor of interest can be cloned into the expression vector system PB-CMV-MCS-EF1a-Puro (System Biosciences, PB510B-1) according the manufacturer's instructions.)

2.3 Image Acquisition

1. Corning 384 well optical imaging plates (#3985).
2. ImageXpress Micro XL high-content imaging microscope (Molecular Devices). This microscope is equipped with the following filter sets for measurement of CFP/YFP FRET or GFP/RFP FRET biosensors:

CFP/YFP FRET: FRET Ex. 438/25, Dichroic 520LP, FRET Em. 542/25 (Semrock Filter Set MOLE-0189); CFP Ex. 438/25, Dichroic 458LP, CFP Em. 483/25 (Semrock Filter Set CFP-2432B-NTE-Zero).

GFP/RFP FRET: FRET Ex. 472/25, Dichroic 495LP, FRET Em. 641/25 (Semrock Filter Set MOLE-0190); GFP Ex. 472/25, Dichroic 495LP, GFP Em. 520/25 (Semrock Filter Set GFP-3035D-NTE-Zero).

2.4 Image Analysis/ Data Display

1. A MATLAB product license (MathWorks).

3 Methods

3.1 Cell Culture and Manufacturing Stable Transgenic Mammalian Cell Lines with Biosensors

1. For 10^6 cells in one 35 mm cell culture petri dish, wash a passage plate of SK-N-BE(2) adherent cells twice with 5 ml room temperature PBS.
2. Remove all PBS and add 1 ml of prewarmed (37 °C) Trypsin-EDTA solution, and incubate at 37 °C for 5 min.
3. After 5 min, gently homogenize the cell/solution slurry with a 5 ml serological pipette to release all cells from the bottom surface of the dish and to reduce any clumps of cells to single cells. Quench the trypsin reaction by addition of 2 ml DMEM supplemented with 10% FBS, 50 U/ml penicillin-streptomycin, and 1× GlutaMAX (hereafter referred to as DMEM Media).
4. Spin cells in a 15 ml conical tube at 200 × *g* for 5 min.
5. Remove and discard the supernatant, and resuspend cellular pellet in 1 ml DMEM Media. Homogenize with a 5 ml serological pipette to reduce any clumps of cells to single cells.
6. Add all cells and slurry to a single well of a 12-well plate.
7. Prepare a 10 ml of 1 mg/ml stock of PEI by combining 10 mg 25 kD PEI with 5 ml ddH_2O adjusted to pH 2 with approximately 20 μl concentrated HCl. Mix and incubate at 50 °C until all solids dissolve, which takes approximately 60 min.
8. Adjust the pH of the dissolved PEI solution to 7.4 with NaOH (approximately 75 μl of 1 M NaOH).
9. Bring the PEI solution to a total volume of 10 ml with ddH_2O, and aliquot 50 μl aliquots for future use.
10. In a 1 ml Eppendorf tube, combine 1 ml prewarmed (37 °C) Opti-MEM I media with 10 μl of 1 mg/ml PEI solution, and mix by inversion, twice.
11. Add 1 μg of total plasmid DNA for both the mPB plasmid and the pBSR2 plasmid containing the synthetic biosensor gene (*see* **Note 1**). Mix by inversion twice and incubate at 37 °C for 20 min.
12. Add PEI/plasmid solution to the 12-well plate of 1 ml DMEM Media and cells. Incubate for 6 h.
13. After 6 h remove all media from cells and add 1 ml fresh prewarmed DMEM Media.
14. After 12–24 h, trypsinize and seed all cells into a 35 mm petri dish with 5 ml DMEM Media using the same procedure as **steps 1–5** above (*see* **Note 2**).

3.2 Preparation of 384 Well Imaging Dishes with Biosensor Expressing Cells

1. Repeat the trypsinization procedure described in **steps 1–5** of Subheading 3.1 to recover each biosensor-expressing cell population in separate 15 ml conical tubes.
2. Resuspend cell pellets in DMEM Media to a final concentration of 300,000 cells per ml (*see* **Note 3**).
3. For each sensor-expressing cell population, add 50 μl of resuspended cell slurry to each desired well of a 384-well plate (*see* **Notes 4–6**).
4. Spin the 384-well plate at 200 × *g* in a microplate centrifuge for 2 min (*see* **Note 7**).
5. Incubate the 384-well plate at 37 °C, 5% CO_2, and 95% relative humidity overnight to allow cells to adhere.

3.3 Parallel Acquisition of Biosensor Experiments

We do not describe acquisition parameters specific to any microscope model, due to the diverse array of available wide-field microscope models and accompanying software. Instead, the following describes to steps and parameters that we utilize in a frequently conducted experiment:

1. Insert the 384-well plate into the microscope chamber.
2. Apply autofocusing parameters according to the microscope software and design specifications of the plate manufacturer.
3. Set the following parameters:

Objective: 10×.

Camera binning: 4 × 4.

Number of wavelengths: 2.

Wavelength 1: FRET Ex. 438/25, Dichroic 520LP, FRET Em. 542/25.

Wavelength 2: CFP Ex. 438/25, Dichroic 458LP, CFP Em. 483/25.

Acquisition duration: 24 h.

Acquisition frequency: 10 min.

4. Perform a pretreatment duty cycle (one timepoint for each well to be acquired).
5. Add the appropriate stock solutions to the appropriate wells to achieve the desired final concentrations of both vehicle and chemical solutes.
6. Restart the acquisition and allow to continue for 24 h.

3.4 Image Processing Using MATLAB

1. Using the acquisition settings described above, each well yields an image, represented in MATLAB as a two-dimensional array (frequently referred to as pictures) for each timepoint. Processing of these images is summarized in Fig. 2. Wavelength

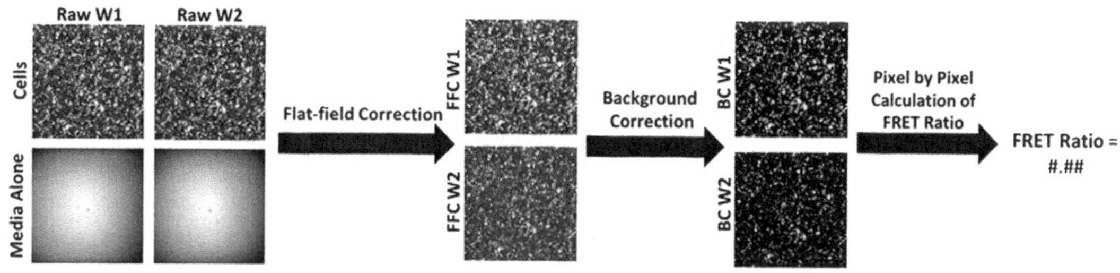

Fig. 2 A diagram describing the steps required to convert raw images into a single value of activity (FRET ratio) for each well, for each timepoint. These steps included flat-field correction, background correction, and pixel by pixel calculations. W1 = wavelength 1, W2 = wavelength 2, FFC = flat-field corrected, BC = background corrected

1 (W1) is the FRET channel, while wavelength 2 (W2) is the CFP channel. Flat-field correction (FFC) of the raw W1 and W2 images can be achieved by division of the W1 array by the W1 array of the media only well (W1B) (*see* **Note 6**) using the following MATLAB (Version R2015b) code:

```
FFCW1= W1./W1B;
FFCW2= W2./W2B;
```

2. For each FFCW# image, where # can be either 1 or 2 for W1 or W2, respectively, background correction can be achieved with the following MATLAB code. The background variable "Background" represents the bottom quantile fraction used to threshold pixel intensities for computing the mean background intensity, which is then subtracted from the FRET and CFP image intensities.

```
W1hist=sort(reshape(FFCW1,1,numel(FFCW1)));
W2hist=sort(reshape(FFCW2,1,numel(FFCW2)));
Background=0.05; %Value can be adjusted
BackgroundvalueW1=mean(W1hist(1:round(numel(W1
hist)*Background)));
BackgroundvalueW2=mean(W2hist(1:round(numel(W2
hist)*Background)));
BCW1=FFCW1-BackgroundvalueW1;
BCW2=FFCW2-BackgroundvalueW2;
```

3. For each BCW# image, where # can be either 1 or 2 for W1 or W2, respectively, the mean FRET ratio (FRETRatio) for the population of pixels can be calculated with the following MATLAB code:

```
Thresholdvalue=500; % a pixel must be 500
units above background to be considered signal
BCW1(BCW1<Thresholdvalue)=0;
BCW1(BCW2<Thresholdvalue)=0;
BCW2(BCW1<Thresholdvalue)=0;
BCW2(BCW2<Thresholdvalue)=0;
```

```
RatioImage=double(BCW1)./double(BCW2);
FRETRatio=sum(sum(RatioImage))/nnz(RatioImage);
```

4. Repeat this calculation for each well and timepoint to construct a three-dimensional array (PlateRatioData(t,x,y)), where t is the time dimension, x is the column number in a 384-well plate, and y is the row number in a 384-well plate. The resulting PlateRatioData(t,x,y) array can be used to construct a heatmap-style graphical representation of the entire 384 well experiment using the following MATLAB code. This code is custom tailored to a plate design where each column represents a different sensor):

```
times=1:size(PlateRatioData,1);
columns=1:size(PlateRatioData,2);
rows=1:size(PlateRatioData,3);
PlateRatioData2= PlateRatioData(times,columns,
rows);
For x=1:size(PlateRatioData2,2)
temp=PlateRatioData2(:,x,:);
FracSat(:,x,:)=(temp-nanmin(nanmin(temp)))/
(nanmax(nanmax(temp))-nanmin(nanmin(temp)));
FracSat(isnan(FracSat))=0;
end
breaks=[0:size(PlateRatioData2,1)+1:(size(Plat
eRatioData2,1)*(size(PlateRatioData2,3)+1))+1];
for y=1:(numel(breaks)-1)
data_rs(breaks(y)+1:breaks(y+1)-
1,:)=datan(:,:,y);
end
scale1=[0.2 1.0];
namesx2=namesx(columns);
namesy2=namesy(rows);
imagesc(data_rs,scale1)
ytickarray=1:size(namesy2,2);
set(gca,'xtick',1:size(data2,2),'xticklabel',
namesx2,'xticklabelrotation',...
90,'ytick',ytickarray*(breaks(2))-(breaks(
2))/2,'yticklabel',namesy2,'ticklength',[0
0],'xaxislocation','top')
```

4 Notes

1. We describe here the construction of adam17 biosensor-expressing cells (activity shown in the sixth column of heatmap in Fig. 3), which requires cotransfection of a plasmid containing the PB transposase and the adam17 biosensor-expressing plasmid TSEN Bsr2 [7]. The mode of gene delivery for each biosensor will depend on the vector in which it is supplied.

Whereas some biosensor genes are supplied in retroviral vectors, this transfection protocol can be adapted to aid in the viral packaging steps of such retroviral gene delivery protocols.

2. The TSEN Bsr2 biosensor delivery that is used as an example biosensor here contains a selectable gene that is simultaneously delivered. In this case, the cytotoxic compound Blasticidin S can be used to selectively kill cells in the population that did not exhibit gene transfer and expression. Such selectable markers are highly specific to the plasmid containing the biosensor gene, and researchers should consult the original source of biosensor-expressing plasmids for the proper selection protocol.
3. A hemacytometer should be used for cell counting according to the manufacturer's instructions.
4. When adding media to wells in a 384-well plate, be sure to pipette into a bottom corner of the well. Additionally, do not

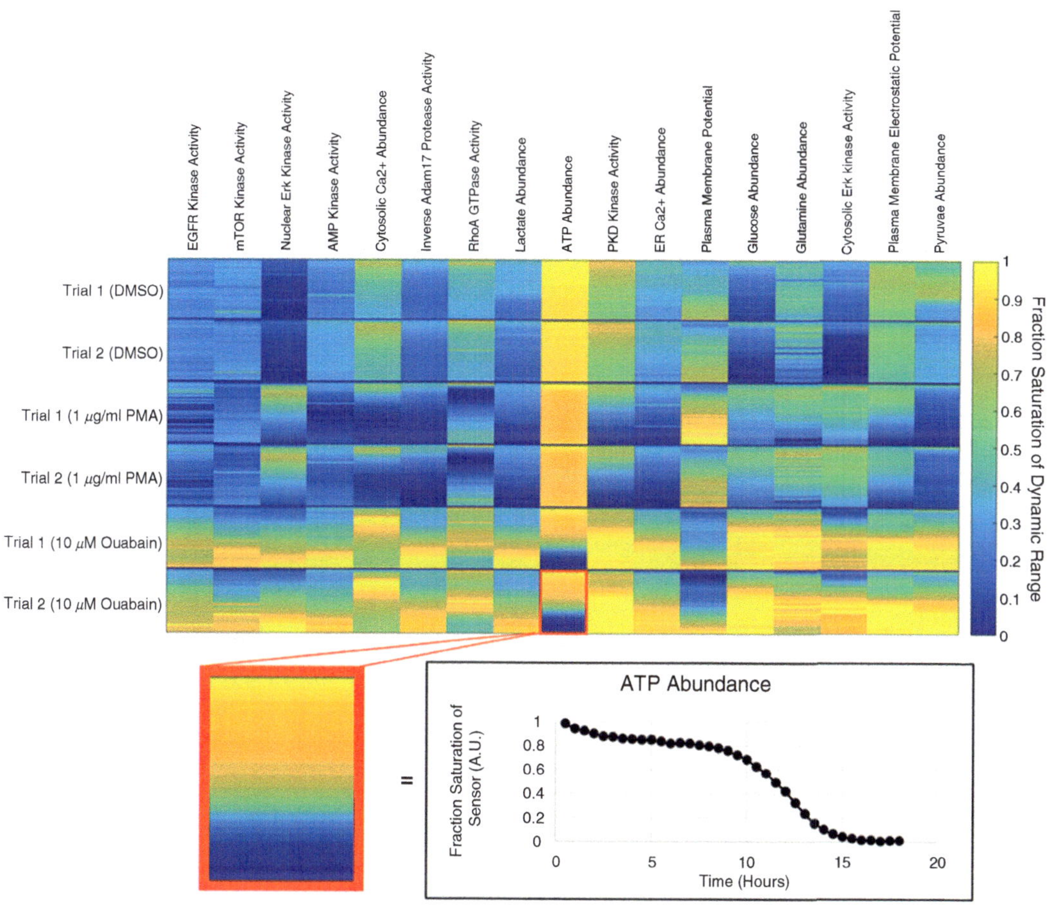

Fig. 3 A graphical output of an experiment utilizing 17 biosensors. SK-N-BE(2) biosensor-expressing cells were exposed to either vehicle, ouabain, or phorbol myristate acetate (PMA). Each tile represents a heatmap of a time-dependent plot of the relative changes in FRET ratio of each sensor within the dynamic range observed for that sensor throughout the experiment (fraction saturation) (*see* **Note 8**)

clear all liquid from the pipette tip, which can frequently result in introduction of air bubbles.

5. Positioning of biosensors and drug treatments in such a multiplexed experiment as is described here is largely flexible. To remain compatible with the analytical pipeline described, below, one should seed biosensors in whole plate columns and conduct variable chemical stimuli perturbations in plate rows.
6. Be sure to include at least one well with media alone, which is used in image acquisition to facilitate flat-field correction for each wavelength.
7. Spinning whole plates requires a swinging bucket rotor and centrifuge. This spinning is not absolutely required, but removes any air bubbles that arise from pipetting. Degassing of media can also result in air bubbles and can be removed using this recommended centrifugation.
8. The following biosensors were used for this demonstration:

Targeted analyte	Biosensor
EGFR kinase activity	Picchu [8]
mTOR kinase activity	TORCAR [9]
Nuclear ERK kinase activity	EKAR-NLS [8]
AMP kinase activity	AMPKAR [10]
Cytosolic Ca^{2+} abundance	D3-cpv [11]
Adam17 protease activity	TSEN [7]
RhoA GTPase activity	Raichu RhoA [12]
Lactate abundance	Laconic [13]
ATP abundance	ATEAM [14]
PKD kinase activity	DKAR [15]
ER Ca^{2+} abundance	D1ER [16]
Plasma membrane potential	VSFP-CR [17]
Glucose abundance	FLIPglu-30uDelta13V [18]
Glutamine abundance	FLIPQTV3.0 8 m [19]
Cytosolic ERK kinase activity	EKAR-NES [8]
Plasma membrane electrostatic potential	MCS+ [20]
Pyruvate abundance	Pyronic [21]

Acknowledgments

This work was supported by NIH grant R01GM113141 and the DARPA cooperative agreement W911NF-14-2-0019.

References

1. Terai T, Nagano T (2013) Small-molecule fluorophores and fluorescent probes for bioimaging. Pflugers Arch 465:347–359
2. Mohsin M, Ahmad A, Iqbal M (2015) FRET-based genetically-encoded sensors for quantitative monitoring of metabolites. Biotechnol Lett 37:1919–1928
3. Greer LF 3rd, Szalay AA (2002) Imaging of light emission from the expression of luciferases in living cells and organisms: a review. Luminescence 17:43–74
4. Padilla-Parra S, Tramier M (2012) FRET microscopy in the living cell: different approaches, strengths and weaknesses. BioEssays 34:369–376
5. Sanford L, Palmer A (2017) Recent advances in development of genetically encoded fluorescent sensors. Methods Enzymol 589:1–49
6. Harvey CD, Ehrhardt AG, Cellurale C et al (2008) A genetically encoded fluorescent sensor of ERK activity. Proc Natl Acad Sci U S A 105:19264–19269
7. Chapnick DA, Bunker E, Liu X (2015) A biosensor for the activity of the "sheddase" TACE (ADAM17) reveals novel and cell type-specific mechanisms of TACE activation. Sci Signal 8:rs1
8. Komatsu N, Aoki K, Yamada M et al (2011) Development of an optimized backbone of FRET biosensors for kinases and GTPases. Mol Biol Cell 22:4647–4656
9. Zhou X, Clister TL, Lowry PR et al (2015) Dynamic visualization of mTORC1 activity in living cells. Cell Rep 10:1767–1777
10. Tsou P, Zheng B, Hsu CH et al (2011) A fluorescent reporter of AMPK activity and cellular energy stress. Cell Metab 13:476–486
11. Ravier MA, Cheng-Xue R, Palmer AE et al (2010) Subplasmalemmal Ca(2+) measurements in mouse pancreatic beta cells support the existence of an amplifying effect of glucose on insulin secretion. Diabetologia 53: 1947–1957
12. Yoshizaki H, Ohba Y, Kurokawa K et al (2003) Activity of Rho-family GTPases during cell division as visualized with FRET-based probes. J Cell Biol 162:223–232
13. San Martin A, Ceballo S, Ruminot I et al (2013) A genetically encoded FRET lactate sensor and its use to detect the Warburg effect in single cancer cells. PLoS One 8:e57712
14. Imamura H, Nhat KP, Togawa H et al (2009) Visualization of ATP levels inside single living cells with fluorescence resonance energy transfer-based genetically encoded indicators. Proc Natl Acad Sci U S A 106:15651–15656
15. Kunkel MT, Toker A, Tsien RY et al (2007) Calcium-dependent regulation of protein kinase D revealed by a genetically encoded kinase activity reporter. J Biol Chem 282:6733–6742
16. Palmer AE, Jin C, Reed JC et al (2004) Bcl-2-mediated alterations in endoplasmic reticulum Ca2+ analyzed with an improved genetically encoded fluorescent sensor. Proc Natl Acad Sci U S A 101:17404–17409
17. Lam AJ, St-Pierre F, Gong Y et al (2012) Improving FRET dynamic range with bright green and red fluorescent proteins. Nat Methods 9:1005–1012
18. Takanaga H, Frommer WB (2010) Facilitative plasma membrane transporters function during ER transit. FASEB J 24:2849–2858
19. Gruenwald K, Holland JT, Stromberg V et al (2012) Visualization of glutamine transporter activities in living cells using genetically encoded glutamine sensors. PLoS One 7:e38591
20. Ma Y, Yamamoto Y, Nicovich PR et al (2017) A FRET sensor enables quantitative measurements of membrane charges in live cells. Nat Biotechnol 35:363–370
21. San Martin A, Ceballo S, Baeza-Lehnert F et al (2014) Imaging mitochondrial flux in single cells with a FRET sensor for pyruvate. PLoS One 9:e85780

Chapter 22

Microplate Assays for Spectrophotometric Measurement of Mitochondrial Enzyme Activity

Rachel C. Janssen and Kristen E. Boyle

Abstract

Spectrophotometric analysis of metabolic enzyme activity from homogenized tissues is a valuable method for investigating mitochondrial content and capacity. Enzyme activity is normally measured in single cuvette spectrophotometers, requiring a large sample volume and low throughput. Here, we describe microplate assays for high-throughput analysis of mitochondrial enzymes citrate synthase, β-hydroxyacyl CoA dehydrogenase, aconitase, and mitochondrial electron transport system (ETS) complexes I, II, III, and IV.

Key words Electron transport chain, High throughput, Citrate synthase, Mitochondrial complexes, Mitochondrial function

1 Introduction

Spectrophotometric assessment of mitochondrial enzyme activity is a simple, useful means to assess overall mitochondrial content and/or health and is used in both clinical diagnostics and research applications. Traditionally, these assays were performed in single cuvettes, limiting high-throughput capabilities. We have adapted these assays for higher-throughput measures in 96-well microplates. While traditional oxygen consumption measures in isolated mitochondria are considered a gold standard for assessment of mitochondrial function, advantages of microplate enzyme activity assays are their high-throughput nature and that enzyme activity can be measured in small quantities of previously frozen tissue or cell lysates. Likewise, though mass spectrometry metabolomic measures provide excellent precision and breadth of metabolite content, interpretation of these data will often benefit from assessment of directional changes in metabolism, which can be provided by assessment of mitochondrial enzyme activity. However, because these assays are performed in vitro with an abundance of substrate present in a lysed sample, they are considered an index of maximal

Angelo D'Alessandro (ed.), *High-Throughput Metabolomics: Methods and Protocols*, Methods in Molecular Biology, vol. 1978, https://doi.org/10.1007/978-1-4939-9236-2_22,

enzymatic capacity and should not be interpreted as physiological activation.

Over the past 75 years, many respected laboratories have developed and modified assays for each of the mitochondrial enzymes described here, using varied technologies such as radiochemical or spectrophotometric assessment. Here, we have adapted single cuvette measures of mitochondrial enzyme activity for use in spectrophotometer microplate readers. The citrate synthase assay is adapted from Frazier and Thorburn [1] and is often used as an index of mitochondrial content. Typically, activity of the other mitochondrial enzymes is referenced to citrate synthase activity to account for differences in mitochondrial content between samples. β-hydroxyacyl CoA dehydrogenase (β-HAD) assay is adapted from Lynen and Wieland [2] for estimating activity of β-oxidation. The iron-sulfur cluster of aconitase renders it sensitive to superoxide damage. Thus, diminished activity can be used as an index of oxidative stress. This microplate assay is adapted from Gardner et al. [3]. Singly, or in combination, mitochondrial ETS complex activity assays are useful for estimating mitochondrial function. Complex I and II assays are adapted from Bindoff et al. [4], while complex III and IV assays are adapted from Spinazzi et al. [5]. Together, these assays can give a general assessment of mitochondrial content, health, and activity. Each laboratory using these assays should adapt a preferred method by optimizing sample preparation and assay conditions to achieve a linear reaction with a high signal-to-background ratio.

2 Materials

2.1 Materials and Equipment for Sample Preparation

1. Reagents.
   (a) Mannitol.
   (b) Sucrose.
   (c) HEPES.
   (d) EGTA.
   (e) Potassium hydroxide (KOH).
   (f) 70% Ethanol.
   (g) Methanol.
   (h) Dry ice.
   (i) Bicinchoninic acid (BCA) protein quantitation kit (Pierce 23227).
2. Equipment.
   (a) All-glass tissue grinder, size 20, for use with muscle samples (e.g., Kimble DUALL).

(b) Glass tube/PTFE pestle tissue grinder, size 20, for use with liver samples (e.g., Kimble DUALL).
(c) Overhead stirrer motor (e.g., Caframo).
(d) Refrigerated microcentrifuge.
(e) Scalpel blades or surgical scissors.
(f) Water-bath sonicator (e.g., Fisher Scientific M1800) for muscle samples.

2.2 Materials and Equipment for Activity Assays

1. Reagents.
   (a) *See* Table 1.
   (b) Ethanol, absolute.
   (c) Hydrochloric acid (HCl); concentrated, 1 M and 0.1 M.
   (d) Potassium borohydride.
   (e) Sodium hydrosulfite.
2. Equipment.
   (a) 12-channel pipette (2–20 μL).
   (b) UV-Vis spectrophotometer microplate reader with temperature control (e.g., Molecular Devices SpectraMax 190).
   (c) Flat-bottom, clear, untreated 96-well plates.
   (d) UV-transparent 96-well plates (e.g., Corning 3635).
   (e) Polypropylene tubes (15 and 50 mL).

3 Methods

3.1 Reagents for Sample Preparation and Activity Assays

1. Prepare 50 mL of Sample Lysis Buffer: 210 mM mannitol, 70 mM sucrose, 5 mM HEPES, 1 mM EGTA, pH 7.2 with KOH. Filter sterilize. Store at 4 °C, stable for 6 months. Keep Sample Lysis Buffer on ice during use.
2. Prepare potassium phosphate buffers (Kpi):
   (a) 0.5 M Kpi, pH 7.4: Mix 20% 0.5 M KH_2PO_4 with 80% 0.5 M K_2HPO_4, pH to 7.4 (use KH_2PO_4 if pH > 7.4, K_2HPO_4 if < 7.4). Filter sterilize. Store at 4 °C, stable for 2 months.
   (b) 0.5 M Kpi, pH 7.0: Mix 40% 0.5 M KH_2PO_4 with 60% 0.5 M K_2HPO_4, pH to 7.0 (use KH_2PO_4 if pH > 7.0, K_2HPO_4 if < 7.0). Filter sterilize. Store at 4 °C, stable for 2 months.
3. *See* Table 1 for working stock solutions of reagents for activity assays.

Table 1
Preparation of stock reagents for mitochondrial enzyme activity assays

Reagent	Sigma cat. #	Stock solution	Storage	Stability[a]	Use
Potassium phosphate, dibasic (K_2HPO_4)	P2222	0.5 M in dH_2O	RT	1 year	CS, CI, CII, CIII, CIV
Potassium phosphate, monobasic (KH_2PO_4)	P5655	0.5 M in dH_2O	RT	1 year	CS, CI, CII, CIII, CIV
Acetyl coenzyme A (ACoA)	A2181	10 mM in dH_2O	−80 °C in aliquots[b]	6 months	CS
5,5′-Dithiobis-(2-nitrobenzoic acid) (DTNB)	D8130	1 mM in 100 mM Tris, pH 8.0	Freshly prepared		CS
Oxaloacetate (OAA)	O4126	5 mM in dH_2O	Freshly prepared		CS
BSA, essentially fatty acid-free	A6003	5% in dH_2O	4 °C	1 month	CI
Coenzyme Q_1 (CoQ_1)	C7956	5 mM in ethanol	−20 °C	6 months	CI, CII
β-Nicotinamide adenine dinucleotide (NADH)	N4505	10 mM in dH_2O	Freshly prepared		CI, β-HAD
Potassium cyanide (KCN)[c]	60178	200 mM in dH_2O	Freshly prepared		CI, CII, CIII, CIV
Rotenone[c]	R8875	0.125 mM in ethanol	−20 °C in aliquots[d]	1 month	CI, CII, CIII
Antimycin A (AntiA)[c]	A8674	10 mg/mL in ethanol	Stock: 1 month. Freshly prepare 1 mg/mL in ethanol		CI, CII, CIII
Succinic acid	S7501	400 mM in dH_2O, pH to 7.4 with 5 M KOH	−20 °C in aliquots	6 months	CII
Cytochrome c (oxidized CytC)	C2506	1 mM in dH_2O	−20 °C in aliquots	6 months	CIII
Decylubiquinone (DQ)	D7911	10 mM in ethanol	−20 °C in aliquots[d]	3 months	CIII
Tween 20	P7949	2.5% in dH_2O	Freshly prepared		CIII
EDTA	E1644	5 and 200 mM in dH_2O	RT	1 year	CIII, β-HAD
Tris–HCl	T3253	1 M in dH_2O, pH to 7.0 with HCl	RT	1 year	β-HAD
Triton X-100	T8787	10% in dH_2O	RT	1 year	β-HAD

(continued)

Table 1 (continued)

Reagent	Sigma cat. #	Stock solution	Storage	Stability[a]	Use
Acetoacetyl coenzyme A (AACoA)	A1625	5 mM in dH_2O	−80 °C in aliquots[b]	3 months	β-HAD
Manganese chloride	M3634	50 mM in dH_2O	RT	1 year	Aconitase
Tris–HCl	T3253	1 M in dH_2O, pH to 7.4 with HCl	RT	1 year	Aconitase
Sodium citrate	S4641	500 mM in dH_2O	RT	1 year	Aconitase
β-Nicotinamide adenine dinucleotide phosphate ($NADP^+$)	N5755	100 mM in dH_2O	−80 °C in aliquots[b]	1 year	Aconitase
Isocitrate dehydrogenase (IDH)	94596	16 mU/mL in dH_2O	Freshly prepared		Aconitase

BSA bovine serum albumin, *RT* room temperature, *CS* citrate synthase, *C* ETS complex, *β-HAD* β-hydroxyacyl CoA dehydrogenase

[a]Stability periods are estimates

[b]Avoid repeated freeze/thaw

[c]Highly toxic—wear appropriate PPE

[d]Protect from light

4. Freshly prepare reduced decylubiquinone (decylubiquinol) for complex III assay: to 250 μL of 10 mM decylubiquinone (DQ) in ethanol in a 1.5 mL microfuge tube, add a few grains of potassium borohydride (highly toxic, wear appropriate personal protective equipment [PPE]). Add 0.1 M HCl in 5 μL aliquots until the solution becomes colorless. Spin at 10,000 × *g* for 1 min, and transfer the solution to a new 1.5 mL tube, leaving behind the grains of potassium borohydride. Adjust the solution to a pH between 2 and 3 using 1 M HCl in 5 μL aliquots. Keep the reduced DQ on ice and protected from light. Scale up the reduced DQ preparation if assaying more than 25 samples.
5. Freshly prepare reduced cytochrome c (CytC) for complex IV assay: make a 1 mM solution of oxidized CytC in 20 mM Kpi, pH 7.0 in a 1.5 mL microfuge tube. Add a few grains of sodium hydrosulfite to reduce the CytC solution. Vortex well. The solution will change color from brown to orange/pink. If no change in color is observed, a few more grains of sodium hydrosulfite might be required, although the least amount of sodium hydrosulfite necessary for a color change should be used. Measure the ratio of absorbance at 550–565 nm on an

aliquot of the reduced CytC solution diluted 50-fold. The 550/565 nm ratio of reduced CytC should be > 6.

3.2 Sample Preparation

Tissue samples should be stored at −80 °C prior to preparation of homogenates (*see* **Note 1**).

This preparation of sample from frozen tissue results in a mitochondria-“enriched” sample, though it is recommended to reference other mitochondrial enzyme activity measures to citrate synthase activity as an index of mitochondrial content. Activity assays can also be performed in isolated mitochondria, though these would need to be prepared from freshly collected samples (*see* **Note 2**), and in cultured cells (*see* **Note 3**).

1. Weigh 20–30 mg of frozen liver or 30–40 mg of frozen skeletal muscle in a 1.5 mL microfuge tube, and keep on dry ice or in liquid nitrogen until homogenization. For skeletal muscle, remove any visible connective tissue or fat with a scalpel blade. For liver and skeletal muscle, fine-mince on dry ice near thawing with scissors. Return to appropriate microfuge tube, and keep on dry ice until all samples have been minced. One sample at a time, transfer minced sample to a pre-chilled glass grinder. Add ice-cold Sample Lysis Buffer to the glass tube at 5% wet tissue weight to volume (e.g., to 30 mg tissue, add 600 μL Sample Lysis Buffer).
2. Disrupt tissue using a PTFE pestle for liver and a glass pestle for skeletal muscle. Keep glass tube on ice, and homogenize the tissue with 15 slow up and down strokes at 300 rpm using an overhead stirrer motor. Transfer the homogenized tissue to a chilled microfuge tube, and keep on ice. Clean tissue grinder and repeat with remaining samples (*see* **Note 4**).
3. Once all samples have been homogenized, centrifuge the solubilized tissue at 600 × *g* for 10 min at 4 °C. Transfer the supernatant to a new, pre-chilled microfuge tube, and keep on ice. Freeze/thaw three times by snap-freezing the samples in a dry ice/methanol slurry and thawing in a room temperature water bath. Keep samples on ice after freeze/thaw is completed. Skeletal muscle samples used for the citrate synthase and complex I assays must also be sonicated on ice just prior to use (*see* **Note 5**).
4. Perform BCA protein quantitation assay according to manufacturer’s instructions (*see* **Note 6**). Dilute samples 1:10 in Sample Lysis Buffer, and use albumin standards from BCA kit prepared in Sample Lysis Buffer.

3.3 Protocol

Protocol is for each assay to be run with duplicate measures on each unknown sample at an assay volume of 200 μL in a 96-well microplate. Most of the assays are time-sensitive and require measures of 1 row at a time.

3.3.1 Citrate Synthase

1. Calculate the total number of wells required for the assay: (# samples +1 blank) × 2. Make the Assay Solution of 50 mM Kpi buffer (pH 7.4), 100 μM DTNB, 115 μM ACoA in dH_2O. Total volume should allow for 186 μL/well, plus 10% for pipetting error. Warm Assay Solution to 30 °C. Set spectrophotometer temperature to 30 °C.
2. Pipette 10 μL Sample Lysis Buffer for blank or 10 μL sample into 2 wells each of a 96-well plate, filling only 1 row at a time.
3. Immediately pipette 186 μL of Assay Solution into each of the sample wells, 1 row at a time, and read every 15 s for 2 min at 412 nm for baseline rate.
4. Add 4 μL of 5 mM OAA (100 μM final concentration) to each blank/sample well using a multichannel pipette, 1 row at a time.
5. Immediately return to spectrophotometer, and read again every 15 s for 3 min at 412 nm to measure the rate of thionitrobenzoate anion (TNB) appearance. Repeat **steps 2–5**, 1 row at a time, until all samples are complete.

3.3.2 β-Hydroxyacyl CoA Dehydrogenase

1. Calculate the total number of wells required for the assay: (# samples +1 blank) × 2. Make the Assay Solution of 50 mM Tris–HCl buffer (pH 7.0), 250 μM NADH, 2 mM EDTA, 0.2% Triton X-100 in dH_2O. The total volume should allow for 186 μL/well, plus 10% for pipetting error. Warm solution to 30 °C. Set spectrophotometer temperature to 30 °C.
2. Pipette 10 μL Sample Lysis Buffer for blank or 10 μL sample into 2 wells each of a 96-well plate, filling only 1 row at a time.
3. Pipette 186 μL of Assay Solution into each of the blank/sample wells, 1 row at a time. Then add 4 μL of 5 mM AACoA (100 μM final concentration) to each well using a multichannel pipette, 1 row at a time.
4. Immediately place in spectrophotometer, and read every 15 s for 4 min at 340 nm to measure the rate of NADH disappearance. Repeat **steps 2–4**, 1 row at a time, until all samples are complete.

3.3.3 Aconitase

1. Calculate the total number of wells required for the assay: (# samples +1 blank) × 2. Make the Assay Solution of 50 mM Tris–HCl buffer (pH 7.4), 5 mM sodium citrate, 0.2 mM $NADP^+$, 0.6 mM manganese chloride in dH_2O. The total volume should allow for 185 μL/well, plus 10% for pipetting error. Warm solution to 25 °C. Set spectrophotometer temperature to 25 °C.

2. Pipette 10 μL Sample Lysis Buffer for blank or 10 μL sample into 2 wells each of a 96-well plate. This assay is not time-sensitive, so more than 1 row can be completed at a time.
3. Pipette 185 μL of Assay Solution into each of the blank/sample wells. Then add 5 μL of 16 mU/mL IDH to each well using a multichannel pipette.
4. Immediately place in spectrophotometer, and read every 1 min for 30 min at 340 nm to measure the rate of NADPH appearance.

3.3.4 Complex I (NADH/CoQ_1 Oxidoreductase)

1. Calculate the total number of wells required for the assay: (# samples +1 blank) × 2. Make the Assay Solution of 50 mM Kpi buffer (pH 7.4), 50 μM NADH, 1 mM KCN, 5 μg/mL AntiA, 0.1% BSA, 50 μM CoQ_1 in dH_2O. The total volume should allow for 186 μL/well, plus 10% for pipetting error. Warm solution to 30 °C. Set spectrophotometer temperature to 30 °C.
2. Pipette 10 μL Sample Lysis Buffer for blank or 10 μL sample into 2 wells each of a 96-well plate, filling only 1 row at a time.
3. Pipette 186 μL of Assay Solution into each of the blank/sample wells, 1 row at a time.
4. Immediately place in spectrophotometer, and read every 15 s for 3 min at 340 nm to measure the rate of NADH disappearance.
5. Remove plate, and add 4 μL of 0.125 mM rotenone (2.5 μM final concentration) to each blank/sample well using a multichannel pipette.
6. Immediately return to spectrophotometer, and read again every 15 s for 3 min at 340 nm. Repeat **steps 2–6**, 1 row at a time, until all samples are complete.

3.3.5 Complex II (Succinate Dehydrogenase)

Important: this assay requires ultraviolet-transparent microplates.

1. Calculate the total number of wells required for the assay: (# samples +1 blank) × 2. Make the Assay Solution of 50 mM Kpi buffer (pH 7.4), 6 μg/mL AntiA, 10 mM succinate, 2.5 μM rotenone in dH_2O. The total volume should allow for 186 μL/well, plus 10% for pipetting error. Set spectrophotometer temperature to 30 °C.
2. Pipette 10 μL Sample Lysis Buffer for blank or 10 μL sample into 2 wells each of a 96-well plate, filling only 1 row at a time.
3. Pipette 186 μL of Assay Solution into each of the blank/sample wells, 1 row at a time.

4. Incubate the plate at 30 °C for 10 min inside the spectrophotometer.
5. Add 4 μL of 5 mM CoQ_1 (100 μM final concentration) to each blank/sample well using a multichannel pipette.
6. Immediately place in spectrophotometer, and read every 15 s for 4 min at 280 nm to measure the rate of CoQ_1 disappearance. Repeat **steps 2–6**, 1 row at a time, until all samples are complete.

3.3.6 Complex III (Decylubiquinol-Cytochrome C Reductase)

1. Calculate the total number of wells required for the assay: (# samples +1 blank) × 4. Make the Assay Solution of 25 mM Kpi buffer (pH 7.5), 0.5 mM KCN, 0.1 mM EDTA, 0.025% Tween 20, 75 μM oxidized CytC in dH_2O. The total volume should allow for 188 μL/well, plus 10% for pipetting error. Remove half of the Assay Solution to a separate tube, and add AntiA to a final concentration of 10 μg/mL. Warm solutions to 37 °C. Set spectrophotometer temperature to 37 °C.
2. Pipette 10 μL Sample Lysis Buffer for blank or 10 μL sample into 4 wells each of a 96-well plate, filling only 1 row at a time.
3. Pipette 188 μL of Assay Solution without AntiA into 2 wells per blank/sample and 188 μL of Assay Solution with AntiA into the other 2 blank/sample wells, 1 row at a time.
4. Immediately place in spectrophotometer and read every 15 s for 2 min at 550 nm.
5. Remove plate, and add 2 μL of 10 mM reduced DQ (100 μM final concentration) to each blank/sample well using a multichannel pipette.
6. Immediately return to spectrophotometer, and read again every 15 s for 2 min at 550 nm to measure the rate of reduced CytC appearance. Repeat **steps 2–6**, 1 row at a time, until all samples are complete.

3.3.7 Complex IV (Cytochrome C Oxidase)

1. Calculate the total number of wells required for the assay: (# samples +1 blank) × 2. Make the Assay Solution of 50 mM Kpi buffer (pH 7.0), 60 μM reduced CytC in dH_2O. Total volume should allow for 190 μL/well, plus 10% for pipetting error. Warm solutions to 37 °C. Set spectrophotometer temperature to 37 °C.
2. Pipette 190 μL of Assay Solution into 2 wells per blank/sample of a 96-well plate, filling only 1 row at a time.
3. Place in spectrophotometer and read every 15 s for 2 min at 550 nm.
4. Pipette 10 μL Sample Lysis Buffer for blank or 10 μL sample into 2 wells each using a multichannel pipette.

5. Immediately return to spectrophotometer, and read again every 15 s for 3 min at 550 nm to measure the rate of reduced CytC disappearance. Repeat **steps 2–5**, 1 row at a time, until all samples are complete.
6. Test the specificity of complex IV activity by adding 0.3 mM KCN to the Assay Solution in a separate set of reactions as in **steps 1–5**. However, specificity (percentage inhibition by KCN) is typically close to maximum.

3.4 Data Analysis

1. Obtain the absolute value for each slope measured in the linear range: change in absorbance (OD reading) per minute; ΔA/min (Fig. 1). Calculate the slope mean for each duplicate (*see* **Note** 7). Calculate the corrected slope for each sample:

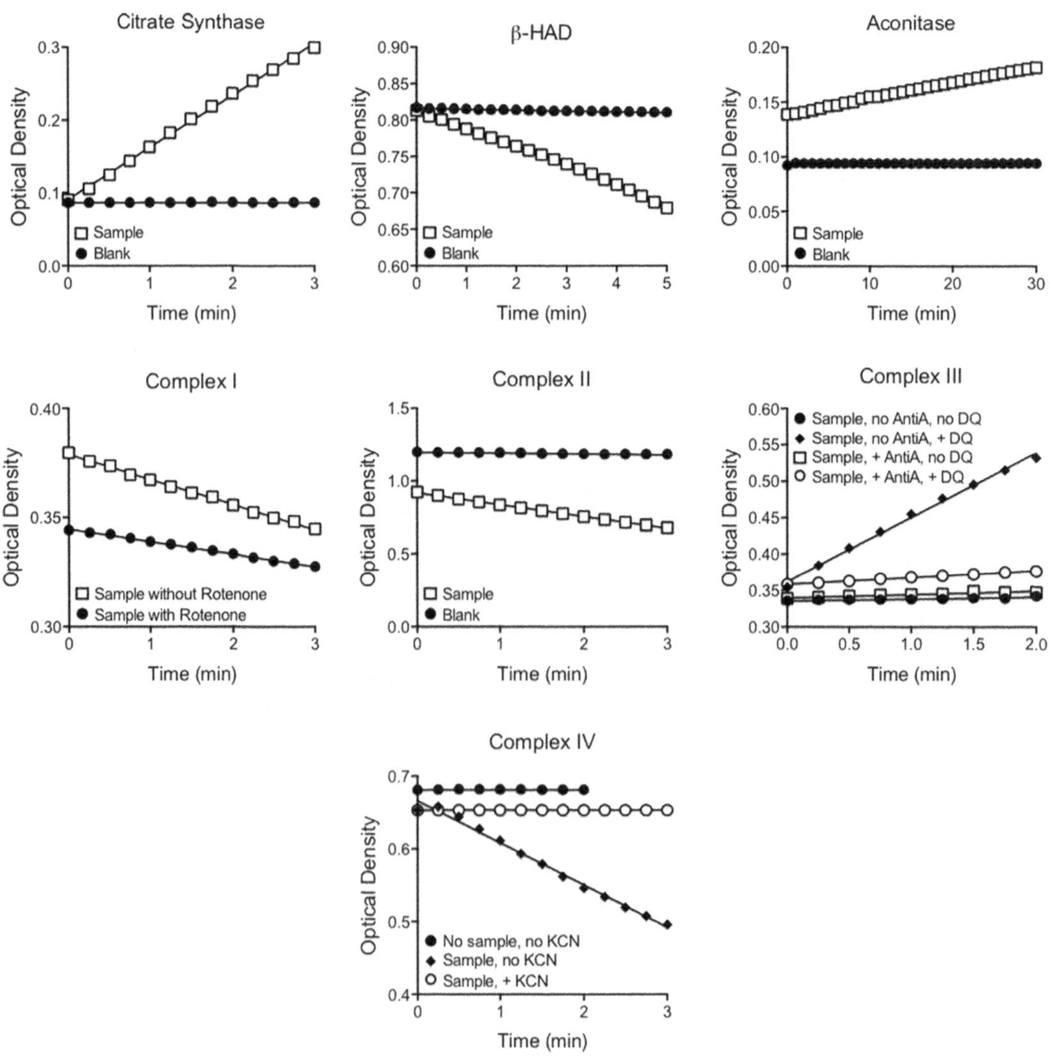

Fig. 1 Sample data for each assay

(a) *Citrate synthase*: Subtract the slope mean without sample from the corresponding slope mean with sample added.

(b) *β-hydroxyacyl CoA dehydrogenase*: Subtract the slope mean without sample from the corresponding slope mean with sample added.

(c) *Aconitase*: Subtract the slope mean without sample from the corresponding slope mean with sample added. Typically, a good linear range for measuring the slope is between 10 and 25 min.

(d) *Complex I*: Subtract the slope mean without rotenone from the corresponding slope mean with rotenone added to calculate the specific complex I activity (rotenone-sensitive rate; *see* **Note 8**).

(e) *Complex II*: Subtract the slope mean without sample from the corresponding slope mean with sample added.

(f) *Complex III*: In the samples without AntiA added, subtract the slope mean without sample added from the slope mean with sample added. Do the same for the samples with AntiA added. Finally, subtract the corrected slope of the samples with AntiA added from the corrected slope of the samples without AntiA added to calculate the AntiA-sensitive rate.

(g) *Complex IV*: Subtract the slope mean without sample from the corresponding slope mean with sample added.

2. According to extinction coefficients listed in Table 2, calculate the activity for each sample using the following equation (*see* **Note 9**). Sample dilution is sample volume/total assay volume (0.05 μL/μL):

$$\text{Enzyme activity}\left(\frac{\text{nmol}}{\text{min}}\right)=\frac{\text{Corrected slope}\left(\frac{\Delta A}{\text{min}}\right)\times\left(\frac{1{,}000{,}000\,\mu\text{mol}}{\text{mol}}\right)}{\text{Sample dilution}\times\varepsilon\times\text{Pathlength (cm)}}$$

3. Finally, divide the calculated activity by the protein concentration for each sample (mg/mL). Reporting units for all assays are nmol/min/mg.

4 Notes

1. The Sample Lysis Buffer, samples, tissue grinders, and microfuge tubes must be kept on ice during sample preparation to ensure optimum enzymatic activities. Pre-cool the tissue grinder tube and pestle and three 1.5 mL microfuge tubes per sample on ice at least 5 min prior to use.

Table 2
Extinction coefficients for assay calculations

Assay	Extinction coefficient (ε)
CS	TNB: 13,600 $L \cdot mol^{-1} \cdot cm^{-1}$
β-HAD	NADH: 6220 $L \cdot mol^{-1} \cdot cm^{-1}$
Aconitase	NADPH: 6220 $L \cdot mol^{-1} \cdot cm^{-1}$
CI	NADH: 6220 $L \cdot mol^{-1} \cdot cm^{-1}$
CII	CoQ_1: 12,000 $L \cdot mol^{-1} \cdot cm^{-1}$
CIII	Reduced CytC: 18,500 $L \cdot mol^{-1} \cdot cm^{-1}$
CIV	Reduced CytC: 18,500 $L \cdot mol^{-1} \cdot cm^{-1}$

2. Mitochondria can be isolated from fresh tissue samples or cultured cells to be used in the enzyme assays. Based on the protocol of Bieber et al. [6], prepare a mitochondrial isolation buffer of 0.25 M sucrose and 0.2 mM EDTA in dH_2O; adjust pH to 7.5 using KOH. All reagents, tubes, and equipment must be pre-chilled on ice. Mince and homogenize tissues in this solution at 10% wet tissue weight to volume as for frozen tissues (Subheading 3.2, **step 2**). For cells, harvest in 500 μL mitochondrial isolation buffer using a cell scraper. Cells can be homogenized by gently pulling 5 times through a small-bore needle (e.g., insulin syringe). Next, centrifuge homogenized tissues or cells at 4 °C, 700 × *g* for 10 min. Reserve supernatant. Gently resuspend pellet in 500 μL mitochondrial resuspension buffer (70 mM sucrose, 1 mM EDTA, 220 mM mannitol, 2 mM HEPES in dH_2O; adjust pH to 8.0 with KOH), and repeat centrifugation, combining supernatants. Centrifuge combined supernatants at 4 °C, 6800 × *g* for 10 min. Discard supernatant, and gently resuspend pellet in mitochondrial resuspension buffer. Mitochondrial isolates can then be immediately used for enzyme activity assay or flash frozen in liquid nitrogen and stored at −80 °C until assay.
3. For cultured cells, we recommend harvesting from 1 to 2 10 cm dishes, depending on cell type. For cell harvest, rinse twice in ice-cold phosphate buffered saline (PBS), and collect cells with cell scraper in ice-cold PBS. Pellet the sample and remove the supernatant. Flash freeze the pellet in liquid nitrogen and store at −80 °C. Continue with sample preparation as for frozen tissue (Subheading 3.2, **step 2**). Mitochondria can also be isolated from freshly harvested cells, as described in **Note 2**.
4. As detailed by Spinazzi et al. [5], a tight compliance between the glass grinder tube and pestle is vital. Tissue grinder must be

cleaned well between samples using detergent and a small bottle brush. Rinse thoroughly, using distilled water for the final rinses. Remove residual water from tube, and place on ice before next sample.

5. Remove 1–100 μL aliquot of sample to new, pre-chilled microfuge tube. Add ice to water-bath sonicator to create a slurry. Place samples in sonicator, turn on (one setting, 40 kHz sonicator) for 2 s burst and turn off for 5 s, and repeat 7 times. Sonicated samples must be used within 1 day.

6. Total concentration of sample used in each assay can have a wide range, depending on tissue or sample type used. For example, in the citrate synthase assay, ~5 μg/well of skeletal muscle is sufficient, but ~20–25 μg/well of liver or cell lysate might be required. The table below gives approximate sample amounts required for each assay. Depending on initial sample concentrations, samples may be diluted in Sample Lysis Buffer to achieve appropriate assay sample volumes. Likewise, sample volumes greater than 10 μL may be necessary. In this case, increase sample/blank volume, and reduce Assay Solution volume accordingly to maintain a 200 μL total assay volume. Likewise, adjust the sample dilution calculation for the final enzyme activity calculation (Subheading 3.4, **step 2**).

Assay	Approximate sample/well (μg)
CS	5–25
β-HAD	10–30
Aconitase	60–80
CI	10–40
CII	10–50
CIII	10–25
CIV	5–25

7. For each slope, the first few measures might be variable. If necessary, eliminate these values consistently across all samples and controls to assure a stable slope. It is useful to graph the slope of each sample to flag these variable time points. Subtraction of the slope mean of the blank (Sample Lysis Buffer only) might be necessary unless the rate is negligible.

8. Rotenone is a complex I inhibitor. The rate without rotenone is the total activity, which includes a substantial slope for nonspecific rotenone-insensitive activity. Therefore, rates both with and without rotenone must be measured for each individual well. The difference between the rotenone-sensitive rate and the total rate (without rotenone) represents the complex I-specific activity.

9. This equation uses Beer's law, $A_\lambda = \varepsilon \cdot L \cdot c$, where A_λ is the absorbance (OD) at the specified wavelength (λ), ε is the extinction coefficient, L is the pathlength, and c is the molar concentration of substrate or product. The pathlength refers to the distance the light travels through the sample. The enzyme activity is equal to the change in concentration (ΔC). Therefore:

$$\Delta C = \frac{\Delta A_\lambda / \text{min}}{\varepsilon \frac{L \cdot \text{mol}^{-1}}{\text{cm}} \times \text{Pathlength (cm)}}$$

Pathlength may be determined for each individual well using a pathlength feature available on some spectrophotometer microplate readers. For example, the SpectraMax 190 from Molecular Devices uses a PathCheck feature that automatically determines the pathlength for each well after the kinetic assay is complete. Pathlength is variable based on spectrophotometer and microplate used, buffer composition, sample concentration, and assay volume. Pathlengths generally range from 0.5 to 0.75 cm for a typical microplate assay with 200 μL assay volume. In the absence of empirical pathlength data, an estimate of 0.625 cm can be used provided the same microplates, reader, and total assay volume are used for all samples and the approximate sample concentration is consistent between samples.

References

1. Frazier AE, Thorburn DR (2012) Biochemical analyses of the electron transport chain complexes by spectrophotometry. Methods Mol Biol 837:49–62
2. Lynen F, Wieland O (1955) β-Ketoreductase. Meth Enzymol 1:566–573
3. Gardner PR, Nguyen DD, White CW (1994) Aconitase is a sensitive and critical target of oxygen poisoning in cultured mammalian cells and in rat lungs. Proc Natl Acad Sci U S A 91:12248–12252
4. Bindoff LA, Desnuelle C, Birch-Machin MA, Pellissier JF, Serratrice G, Dravet C et al (1991) Multiple defects of the mitochondrial respiratory chain in a mitochondrial encephalopathy (MERRF): a clinical, biochemical and molecular study. J Neurol Sci 102:17–24
5. Spinazzi M, Casarin A, Pertegato V, Salviati L, Angelini C (2012) Assessment of mitochondrial respiratory chain enzymatic activities on tissues and cultured cells. Nat Protoc 7:1235–1246
6. Bieber LL, Abraham T, Helmrath T (1972) A rapid spectrophotometric assay for carnitine palmitoyltransferase. Anal Biochem 50:509–518

Chapter 23

Quantitative NMR-Based Metabolomics on Tissue Biomarkers and Its Translation into In Vivo Magnetic Resonance Spectroscopy

Natalie J. Serkova, Denise M. Davis, Jenna Steiner, and Rajesh Agarwal

Abstract

Nuclear magnetic resonance (NMR) spectroscopy is an established analytical platform for analyzing metabolic profiles of cells, tissues, and body fluids. There are several advantages in introducing an NMR-based study design into metabolomics studies, including a fast and comprehensive detection, characterization, and quantification of dozens of endogenous metabolites in a single NMR spectrum. Quantitative proton ^{1}H-NMR is the most useful NMR-based platform for metabolomics. The frozen tissues can be analyzed noninvasively using a high-resolution magic angle spinning (HR-MAS) ^{1}H-NMR spectroscopy; or several extraction techniques can be applied to detect additional metabolites using a conventional liquid-based NMR technique. In this chapter, we report on tissue collection, handling, extraction methods, and ^{1}H-NMR acquisition protocols developed in the past decades for a precise and quantitative NMR-metabolomics approach. The NMR acquisition protocols (both HR-MAS and conventional ^{1}H-NMR spectroscopy) and spectral analysis steps are also presented. Since NMR can be applied "in vivo" using horizontal bore MRI scanners, several in vivo sequences for localized ^{1}H-MRS (magnetic resonance spectroscopy) are presented which can be directly applied for noninvasive detection of brain metabolites.

Key words Nuclear magnetic resonance (NMR), Magnetic resonance spectroscopy (MRS), High-resolution magic angle spinning (HR-MAS), Endogenous metabolites, Chemical shift, Metabolite quantification, Tissue extraction

1 Introduction

Nuclear magnetic resonance (NMR) spectroscopy is one of the two principal analytical techniques in metabolomics (along with mass spectrometry (MS)-based platforms). The advantages of NMR-based metabolomics include its quantitative and nondestructive nature and high reproducibility, as well as the ability to detect all metabolites simultaneously in one snapshot. The disadvantages of NMR are related to low sensitivity (NMR detects only high-abundant metabolites in the micromole to millimole range) and signal overlap between individual metabolites, particularly on

Angelo D'Alessandro (ed.), *High-Throughput Metabolomics: Methods and Protocols*, Methods in Molecular Biology, vol. 1978, https://doi.org/10.1007/978-1-4939-9236-2_23,

one-dimensional proton NMR spectra. Proton NMR (^{1}H-NMR) is the major NMR application to quantify endogenous metabolites and to elucidate chemical structures of unknown metabolites. ^{1}H-NMR can identify and quantify up to 150 endogenous metabolites from a biological sample [1–4] (Table 1). Metabolomics applications with other nuclei (^{31}P-, ^{13}C-, and ^{19}F-NMR, mostly) remain limited due to low sensitivity or low natural abundance. Fortunately, the recent progress in the development of high-field magnets (750 MHz and above), cold probe (CryoProbe) technologies, and the use of dynamic nuclear polarization for hyperpolarized ^{13}C-NMR has dramatically improved the low limit of detection (LLOD) for multinuclear NMR spectroscopy.

The basic principle of NMR lies on the fact that each spinning proton will behave as a "mini-magnet" when placed into a strong magnetic field. The strong external magnetic field aligns the protons' magnetic moments parallel and antiparallel to the magnetic field (Fig. 1a). The population of spins in the parallel orientation (which are of a slightly lower energy level) is larger than in the antiparallel orientation as described by the Boltzmann distribution. This results in a net magnetic moment in the sample aligned with the direction of the external magnetic field (Fig. 1b). When a radio-frequency pulse is applied to the sample, the magnetization orientation is changed, and after the pulse is gone, the system relaxes to its original status (Fig. 1b). Hydrogen nuclei in different tissues have different relaxation properties, which can be detected by radio-frequency MR receivers. MR tissue relaxation characteristics following a radio-frequency pulse reveal information about the concentration, mobility, and chemical bonding of hydrogen and, less frequently, other tissue elements. The strongest NMR signals in a living system arise from water and fat protons due to their metabolic abundance. Other endogenous and exogenous metabolites also give signals on ^{1}H-NMR but with much weaker signal-to-noise ratios due to lower concentrations. A significant number of endogenous metabolite signals are usually obscured by the water signal (both ex vivo and in vivo), and their detection requires adequate techniques for water suppression.

Nuclei in molecules are surrounded by electrons which generate small magnetic fields in the vicinity of the nuclei due to the electron motion (Fig. 1c). This results in slightly different magnetic environments between nuclei. A nucleus with high electron density resonates at a lower frequency than a free proton. For this reason, this phenomenon is referred to as "the chemical shift" related to differences in resonance frequencies. The shift is expressed in terms of the dimensionless units of part per million (ppm) relative to the frequency of free protons in space (Fig. 2a). It is a unique "spectral signature" for a molecule/a metabolite (different protons in different molecules exhibit different absorption frequencies), and each compound can be defined by a unique pattern of chemical shifts (an NMR fingerprint) (Fig. 2b). For metabolic NMR applications,

Table 1
A typical ^{1}H-NMR quantitative data set in brain tumor extracts from three various patients with resected, pathology-confirmed glioblastomas (GBM, grade IV)

		GBM	GBM	GBM
Amino acid/protein metabolism	*Aromatic AAs*	19.79	11.72	14.67
Energy metabolism	*Adenosines*	5.47	2.80	3.41
Energy/carbohydrate storage	*Glycogens*	0.25	0.59	0.89
Glucose metabolism	*Glucose*	17.97	36.16	16.21
DNA/purine metabolism	*Nucleotides*	1.12	0.79	1.11
Unknown	*Metabolites X*	10.97	5.24	5.88
Osmolyte, phospholipid metabolism	*Inositol*	1.16	1.72	1.78
Osmolytes, carbohydrates, energy	*Polyols+sugars*	50.06	72.77	40.39
Amino acid/protein/puridine metabolism	*Glycine*	1.31	4.49	2.58
Osmolyte, transsulfuration/ methionine Met	*Taurine*	7.80	13.50	7.12
Phospholipid derivatives	*Total choline*	13.79	16.15	10.05
Energy metabolism	*Total creatine*	1.86	2.06	1.52
Amino acid/protein metabolism	*Tyrosine*	1.17	3.62	1.09
Amino acid/protein metabolism	*Aspartate*	0.23	0.07	0.11
Amino acid/protein metabolism	*Methionine*	1.02	0.34	0.67
ROS, oxidative stress	*Total glutathione*	2.85	1.38	1.29
Alternative fuel, mitochondrial metabolism	*Glutamine*	2.57	3.89	2.12
Krebs cycle, mitochondrial metabolism	*Succinate*	1.53	1.44	0.71
Amino acid, glutamate metabolsim	*Glutamate*	6.92	3.68	3.38
Ketine body metabolism, lipogenesis	*Ketone bodies*	0.28	0.45	0.12
General metabolic source	*Total CH2,CH3*	13.10	10.75	8.56
Lipogenesis, ketone body metabolism	*Acetate*	0.63	1.28	0.56
Amino acid/protein metabolism	*Lysine*	2.32	3.44	1.80
Amino acid metabolism, glycolysis	*Alanine*	1.06	0.73	0.52
Glycolysis, bioenergetics	*Lactate*	16.97	16.70	12.99
Ketione body metabolism, lipogenesis	*Hydroxybutyrate*	0.30	0.31	0.30
Amino acid/protein metabolism	*Essential AAs*	5.76	5.15	3.95
Signal transduction	*Sphingomyelin*	0.29	0.34	0.25
Lipid metabolism	*Total UFA*	6.15	15.70	14.18
Lipid metabolism	*TAG*	0.93	2.82	1.95
Lipid/phospholipid metabolism	*Glycerol-PL*	2.94	5.53	4.05
Phospholipid, cell membrane, apoptosis	*Ptd-serine*	0.25	0.17	0.08

(continued)

Table 1
(continued)

		GBM	GBM	GBM
Phospholipid, cell membrane	*Ptd-choline*	2.03	2.48	2.33
Phospholipid, cell membrane, signal trans	*Ptd-inositol*	0.32	0.86	0.20
Phospholipid, cell membrane	*Total cholines*	1.98	3.45	3.55
Phospholipid, cell membrane	*Ptd-ethanolamine*	0.29	0.64	0.58
Lipid metabolism	*PUFA*	2.50	6.23	5.75
Lipid metabolism	*Total FA#1*	3.18	6.47	6.29
Lipid metabolism	*Free FA*	0.45	0.76	1.10
Lipid metabolism	*Total lipids CH2n*	82.13	132.37	122.24
Lipid metabolism	*Total lipids*	17.57	22.92	18.89
Lipid metabolism, cholesterol	*Cholesterol*	0.84	2.08	1.93

All metabolite concentrations are given as micromole per gram tissue. Since NMR metabolite libraries are well established, from over 50 endogenous metabolites only 1 metabolite remained unidentified

AA amino acid, *FA* fatty acid, *GBM* glioblastoma, *PL* phospholipids, *Ptd* phosphatidyl (moiety), *UFA* unsaturated fatty acids

it is important to know the precise chemical shifts of each metabolite and to create “chemical shift libraries” (taking into account that pH changes directly influence chemical shift of metabolites) [5]. A web-based metabolite database has also already been established (http://www.hmdb.ca/) [6].

Even though ^{1}H-NMR has emerged into a high-throughput analytical platform with minimal sample preparation and no need for derivatization, the separation of hydrophilic from lipophilic followed by the water removal by freeze-drying will improve the signal overlap and subsequently increase the number of detectable metabolites. Several extraction procedures have been validated to enable (a) protein and their macromolecule precipitation and (b) separation of water-soluble lipid metabolites, followed by freeze-drying of the sample and its restoration in deuterium solvents in order to decrease the water signal in ^{1}H-NMR spectra (Fig. 2c).

The established ex vivo NMR metabolic biomarkers can then be translated into noninvasive magnetic resonance spectroscopy (MRS) [7, 8] or positron emission tomography (PET) using metabolic radiotracer [9] protocols. In this article, we provide step-by-step instructions and standard operating procedures (SOPs) for correct tissue collection and handling for NMR-based metabolic studies, specifically for HR-MAS ^{1}H-NMR. We provide two extraction protocols which are broadly used for quantitative

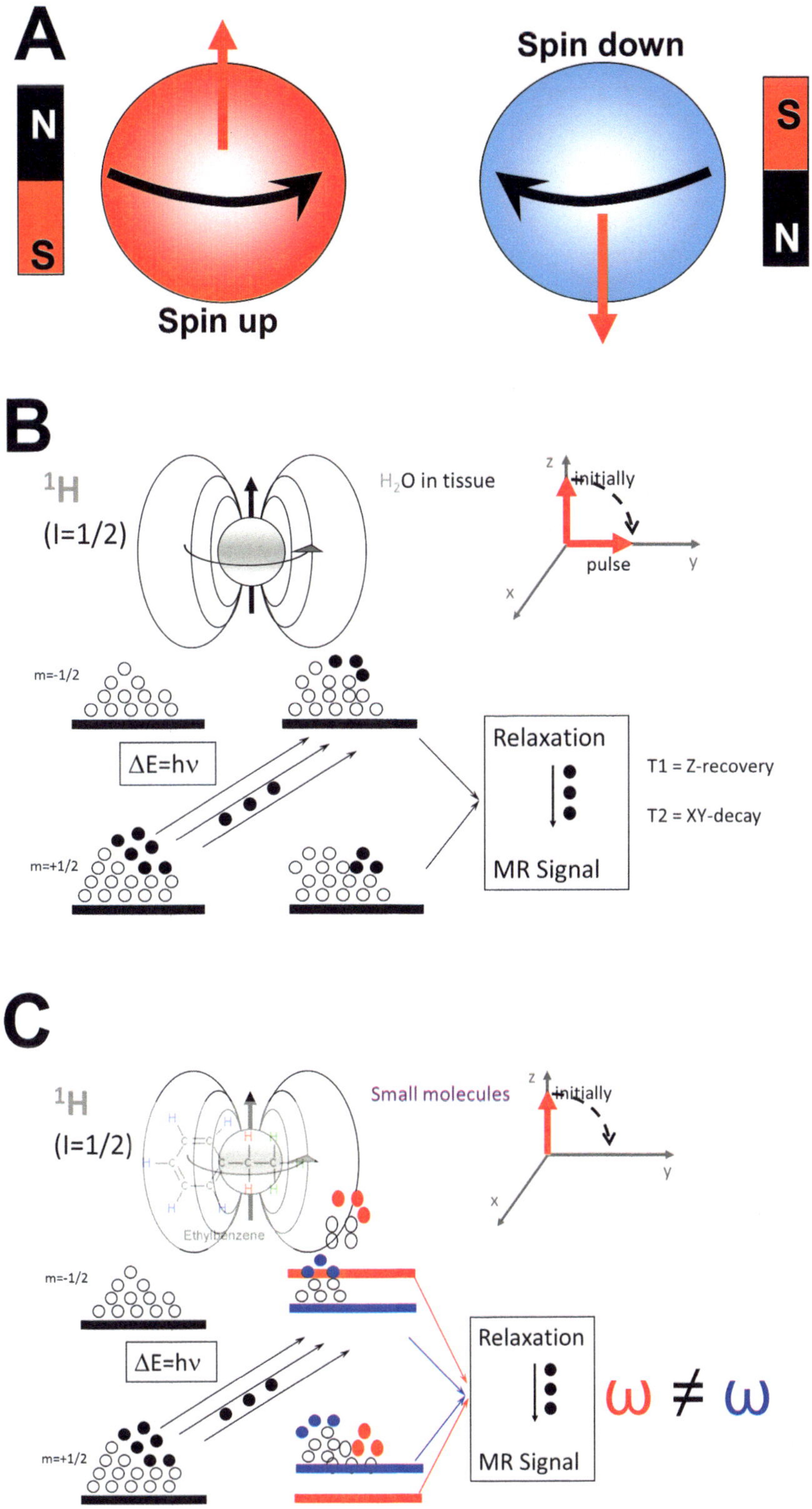

Fig. 1 Principles of NMR: (**a**) Each spinning proton behaves like a mini-magnet if placed into an NMR spectrometer; (**b**) an NMR experiment with magnetization and relaxation after a radio-frequency pulse; (**c**) the phenomenon of the chemical shift arises because of shielding (screening) of the proton from the external magnetic field by the neighboring electrons; as such each chemical group of each metabolite possesses a unique chemical shift

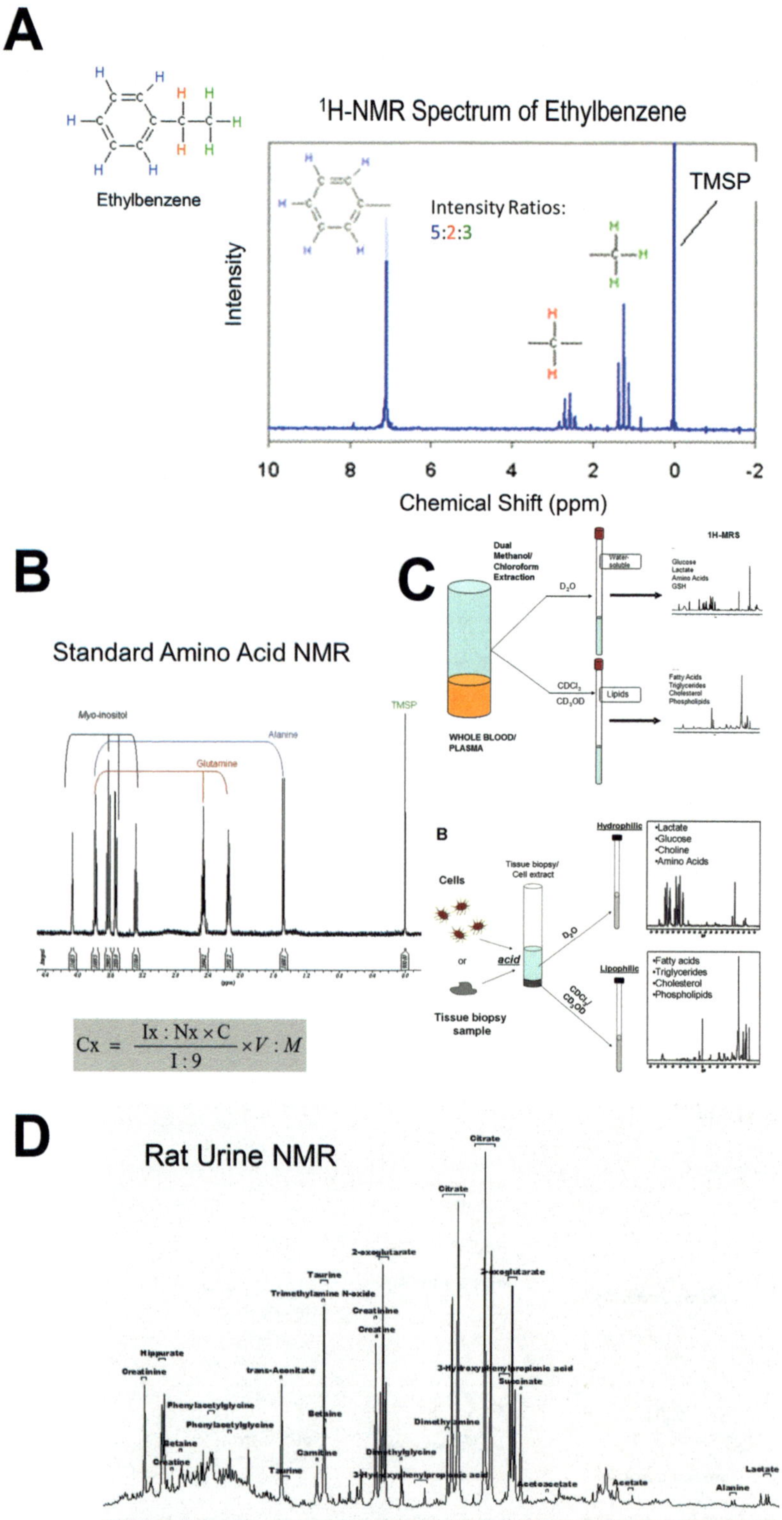

Fig. 2 ^{1}H-NMR spectra assignment and quantification: (**a**) assigning a simple ^{1}H-NMR spectrum of a pure compound (ethylbenzene); (**b**) chemical shifts and quantification of a metabolite standard (myoinositol, glutamine, alanine) solution using TMSP as an external standard; (**c**) lipophilic and hydrophilic NMR spectra yield more metabolites after dual methanol/chloroform (blood products) and acid (cells and tissues) extraction; (**d**) metabolite assignment on ^{1}H-NMR spectrum of rat urine (non-processed sample)

NMR-based metabolomics using conventional "liquid-state" NMR probe heads. Finally, some in vivo localized ^{1}H-MRS (magnetic resonance spectroscopy) sequences are presented which can be used to noninvasively obtain major brain metabolites in human or animal brain.

2 Materials

Since most metabolites are highly degradable at room temperatures (high-energy phosphate depletion, glucose utilization with lactic acid production, lipid peroxidation) [10], all collected specimens should be immediately placed in liquid nitrogen (*see* **Notes 1** and **2**), and ex vivo procedures should be performed on frozen tissue specimens using ice-cold solvents (ca. 4°) (*see* also **Notes 3–5**). This applies for both high-resolution conventional (liquid) NMR (after performing tissue extraction using Subheading 3.2 or 3.3) and HR-MAS (nondestructive tissue NMR, Subheading 3.4).

2.1 Tissue Sampling

- Pre-labeled aluminum foils (ca. 3′ × 3′).
- Liquid nitrogen N_2.
- Surgical bulldog clamps (precooled in liquid nitrogen).
- Safety goggles (*see* **Note 1**).

2.2 Perchloric Acid (PCA) Tissue Extraction

1. 100–400 mg of frozen study tissue (*see* **Note 2**).
2. Mortar, pestle, spatula.
3. Electric tissue homogenizer.
4. 50-mL centrifuge tubes.
5. Temperature-controlled centrifuge.
6. Vortex mixer.
7. pH meter with conventional and NMR electrodes.
8. 12% PCA (concentrated PCA is diluted with water to prepare 12%), ice cold for tissue extraction.
9. Bidistilled water, ice-cold for tissue extraction.
10. KOH—8 M, 1 M, and 0.1 M (44.8 g/100 mL for 8 M, 5.6 g/100 mL for 1 M, and 0.56/100 mL for 0.1 M), for pH adjustment, room temperature.
11. PCA—0.5 M, for pH adjustment, room temperature.
12. Ice.
13. Liquid nitrogen, N_2.
14. Freeze dryer (lyophilizer).
15. 50 mL fast-freeze flasks or glasses (Labconco).
16. Deuterium water D_2O, room temperature.

17. Deuterium DCl, room temperature.
18. Deuterium NaOD, room temperature.
19. Deuterium methanol/chloroform solution (1:2), room temperature.
20. NMR tubes (conventional 5 mm diameter).

2.3 Dual Methanol/Chloroform Tissue Extraction

1. 50–250 mg of frozen study tissue (*see* **Note 2**).
2. Mortar, pestle, spatula.
3. Electric tissue homogenizer.
4. 50 mL centrifuge tubes.
5. 12 mL glass test tubes with stoppers.
6. Serological glass pipettes.
7. Glass Pasteur pipettes.
8. Temperature-controlled centrifuge.
9. Vortex mixer.
10. 100% methanol, ice-cold for tissue extraction.
11. 100% Chloroform, Ice-Cold for Tissue Extraction.
12. Bidistilled water, ice-cold for tissue extraction.
13. Ice.
14. Liquid nitrogen, N_2.
15. Freeze dryer (lyophilizer).
16. 50 mL fast-freeze flasks or glasses (Labconco).
17. Deuterium water D_2O, room temperature.
18. Deuterium methanol/chloroform solution (1:2), room temperature.
19. NMR tubes (conventional 5 mm diameter).

2.4 HR-MAS Sample Preparation

1. 5–20 mg frozen tissue (*see* **Note 2**).
2. HR-MAS NMR probe head.
3. Zirconium HR-MAS rotor.
4. Deuterium water D_2O, ice cold.
5. Phosphate-buffered saline (PBS), ice cold.

2.5 NMR Spectrometers and MRI Scanners

1. 400–800 MHz high-resolution NMR spectrometers (vertical bores).
2. 1.5–7 T human MR scanners (vertical bores).
3. 4.7–11.4 T animal MR scanners (vertical bores).

3 Methods

The choice of tissue preparation for NMR acquisition depends on tissue properties and metabolites of interest. In general, all hydrophilic metabolites (water-soluble metabolites, such as amino acids, carbohydrates, osmolytes, high-energy phosphates) have the best recovery rate while undergoing acid extraction (Subheadings 2.2 and 3.2 using perchloric acid) [11]. For best hydrophobic/lipophilic metabolite recovery (fatty acids, triglycerides, phospholipids), dual-phase extraction (Subheadings 2.3 and 3.3 using methanol/chloroform) is preferable. Nevertheless, both methods will allow for qNMR analysis on both hydrophilic and lipophilic fractions from the same specimen. The choice of the method also depends on the lipid content of a particular tissue (such as the adipose tissue, fatty livers, lactating mammary glands; Table 1). If the specimens contain high-lipid levels, then the dual-phase methanol/chloroform is recommended Subheading 3.3 to prevent lipid contamination of the hydrophilic fraction [12, 13]. For the most tissues (the brain, heart, kidney, normal liver, lung, placenta, tumor biopsies), however, the acid extraction protocol in Subheading 3.2 [10, 13–23] or the nondestructive tissue preparation for HR-MAS NMR in Subheading 3.4 [8, 24–28] is recommended (Fig. 2c).

3.1 Tissue Specimen Collection and Storage

1. Prepare aluminum foils (ca. 3 × 3 in.) and label them.
2. Prepare liquid nitrogen; place the edges of bulldog clamps into liquid nitrogen (*see* **Note 1**).
3. Promptly remove tissue of interest.
4. Using a precooled bulldog clamp, immediately place the specimen into liquid nitrogen (*see* **Note 1**).
5. Cool the aluminum foil in liquid nitrogen for ca. 5 s.
6. Transfer frozen biopsy into pre-cooled aluminum foil.
7. Place wrapped sample in liquid nitrogen for another ca. 5 min (*see* **Note 1**).
8. Store frozen samples at −80 °C.

3.2 Perchloric Acid (PCA) Extraction on Human and Animal Tissues Specimens

1. Precool 12% PCA in an ice bath. Precool a mortar, pestle, and spatula by immersing them in a liquid nitrogen bath (*see* **Note 1**). Set and precool the centrifuge at 4 °C. Take 4 mL of 12% PCA in a 50 mL centrifuge tube, and precool it in an ice bath.
2. 0.1–0.5 g (check for tissue weight!) of tissue is placed in the precooled pestle and powdered by crushing with the mortar while periodically cooling the pestle and mortar in liquid nitrogen (*see* **Note 1**).

3. Using the precooled spatula, the powdered tissue is transferred to the tube containing 4 mL of 12% PCA (ice-cold). Add an additional 2 mL of 12% PCA (ice-cold). Mix well using a vortex mixer and then homogenize using an electric tissue homogenizer (always precool the knife in the ice-cold water bath between the samples).
4. Place the tube containing tissue/PCA extract in an ultrasound ice bath for approximately 5 min.
5. Centrifuge at 1300 × *g* for 20 min at 4 °C.
   (a) Transfer the supernatant into a new 50 mL tube and resuspend the pellet in 2 mL ice-cold 12% PCA in the old 50 mL tube. Vortex well to resuspend the pellet.
   (b) Centrifuge the resulting suspension under the same conditions as in **step a**.
   (c) Collect the supernatant into the same 50 mL tube as before, and save the pellet for **step 10**.
6. Neutralize the supernatant with KOH (first using 8 M, then 1 M, and 0.1 M) so that it is at pH 7.0.
7. Centrifuge at 1300 × *g* for 10 min at 4 °C.
8. Collect the supernatant into a 50-mL fast-freeze flask, and lyophilize it in a freeze dryer overnight.
9. The lyophilized sample will be redissolved in 0.5 mL D_2O, and its pH adjusted to 6.6 using DCl and NaOD, for NMR analysis. Centrifuge if required.
10. Redissolve the tissue pellet (which remains after centrifugation in **step 5**) in 4 mL ice-cold water, vortex well, and put it in an ultrasound ice bath for 10 min.
11. Neutralize the pellet suspension (containing tissue lipids) with KOH so that it is at pH 7.0, transfer into a fast-freeze flask, and lyophilize it in a freeze dryer overnight.
12. The lyophilized lipids will be redissolved in 0.6 mL deuterium methanol/chloroform (1:2 v/v) solution for NMR analysis. Centrifuge if required.

3.3 Dual Methanol/Chloroform Extraction for Human and Animal Tissues

1. For large volume samples (0.25–0.5 g), use 4 mL methanol, 4 mL chloroform, and 4 mL water.
2. For small volume samples (0.05–0.25 g), use 2 mL methanol, 2 mL chloroform, and 2 mL water.
3. Precool methanol, chloroform, and water on ice. Precool a mortar, pestle, and spatula by immersing them in a liquid nitrogen bath (*see* **Note 1**). Set and precool the centrifuge at 4 °C. Take 4 mL (2 mL) of methanol in a 50 mL centrifuge tube and precool it in an ice bath.

4. 0.05–0.5 g (check for tissue weight!) of frozen tissue is placed in the precooled pestle and powdered by crushing with the mortar while periodically cooling the pestle and mortar in liquid nitrogen (*see* **Note 1**).
5. Using the precooled spatula, the powdered tissue is transferred to the tube containing 4 mL (2 mL) methanol (ice-cold). Mix well using a vortex mixer and then homogenize using an electric tissue homogenizer.
6. Place the tube onto ice bath for 10 min. Precool glass tubes.
7. Transfer methanol extract into a glass test tube using a serological glass pipette.
8. Add 4 mL (2 mL) of ice-cold chloroform using another serological glass pipette. Shake it and then vortex. Leave on ice for 10 min.
9. Add 4 mL (2 mL) of ice-cold water. Shake and vortex it.
10. Leave samples overnight at 4 °C in the refrigerator for phase separation.
11. Next morning, centrifuge the samples for 30 min at (at least) 1400 × *g* at 4 °C.
12. Transfer upper methanol phase (water-soluble extract) into a lyophilizing fast-freeze flask using a glass Pasteur pipette and lyophilize overnight.
13. Transfer lower chloroform phase (lipid extract) into a new glass tube *without top white protein pellet* by using a glass Pasteur pipette. Evaporate chloroform under nitrogen stream.
14. Redissolve water-soluble extracts in 0.5 mL D_2O for NMR analysis. Centrifuge if required.
15. Redissolve lipid extracts in 0.6 mL deuterium methanol/chloroform (1:2 v/v) solution for NMR analysis. Centrifuge if required.

3.4 HR-MAS Sample Preparation

1. Place frozen tissue (5–20 mg) in a zirconium rotor with a 12–25 microL insert.
2. Fill the remaining space with D_2O-based phosphate-buffered saline (PBS, pH = 7.5).
3. Place the rotor in an NMR spectrometer equipped with a HR-MAS probe head.

3.5 NMR Spectra Acquisition

1. All water-soluble and lipid tissue extracts from Subheading 3.2 or 3.3 are analyzed by ^{1}H-NMR spectroscopy using a vertical bore high-resolution NMR spectrometer (the typical ^{1}H-resonance frequency for metabolomics studies varies between 400 and 750–800 MHz, with the corresponding field strength of 7.0–17.5 T) (*see* **Note 6**).

2. Conventional 5-mm NMR probe heads are used for all extracted samples (Fig. 3).
3. All preserved tissues in HR-MAS rotors from Subheading 3.4 can be analyzed at the very same NMR spectrometer; however, the spectrometer needs to be equipped with a HR-MAS probe head (Fig. 3).
4. Insert the sample into the NMR spectrometer.
5. Choose the appropriate probe head in use.
6. Perform automatic matching and tuning (such as the "ATMA" command on Bruker systems).
7. Lock the field either to deuterium water (water-soluble extracts and HR-MAS tissues) or deuterium methanol (lipid extracts).
8. Begin gradient shimming (such as the "topspin3d" command on Bruker systems).
9. For all one-dimensional ^{1}H-NMR spectrum acquisition on extracted samples, a standard water presaturation pulse program "zgpr" can be used to suppress water residue signal (*see* **Note 7** for 2D-NMR for structure elucidation).
10. Confirm/optimize the parameters for the fully relaxed ^{1}H-NMR spectra. For example, with the operating frequency for proton channel: 500.24 MHz; power level pl1 = 3 dB; power level for water suppression pl9 = 55 dB; power angle p1 = 7.5 μs (90° pulse); power angle for water suppression p12 = 60 μs; water suppression at O1 = 4.76 ppm; relaxation delay d1 = 12.85 s (5*T1); delay for power switching d12 = 20 μs; short delay d13 = 3 μs; spectral width sw = 12 ppm; total number of scans ns = 40.
11. The total acquisition time of fully relaxed ^{1}H-NMR spectra using a standard water presaturation pulse program is usually 8–10 min (depending on the total number of scans).
12. Alternatively, water suppression can be achieved using NOESY or COSY pulse sequences.
13. For HR-MAS ^{1}H-NMR spectrum acquisition, usually a 4-mm HR-MAS dual inverse (^{1}H/^{13}C) probe head is used (Fig. 3).
14. HR-MAS spectra should be acquired using a 1D Carr-Purcell-Meiboom-Gill (CPMG) sequence with water presaturation (cpmgppr1d).
15. The sample should be spinning at 4–6 kHz at 285°K.
16. Each 1D ^{1}H-HR-MAS NMR spectrum will be acquired applying 400–512 transients (scans), a spectral width of 6000 Hz, a data size of 64 K points, and acquisition time of 5–200 s, a relaxation delay of 4–12 s an echo time of 74 ms, and inter-echo spacing of 2 ms.

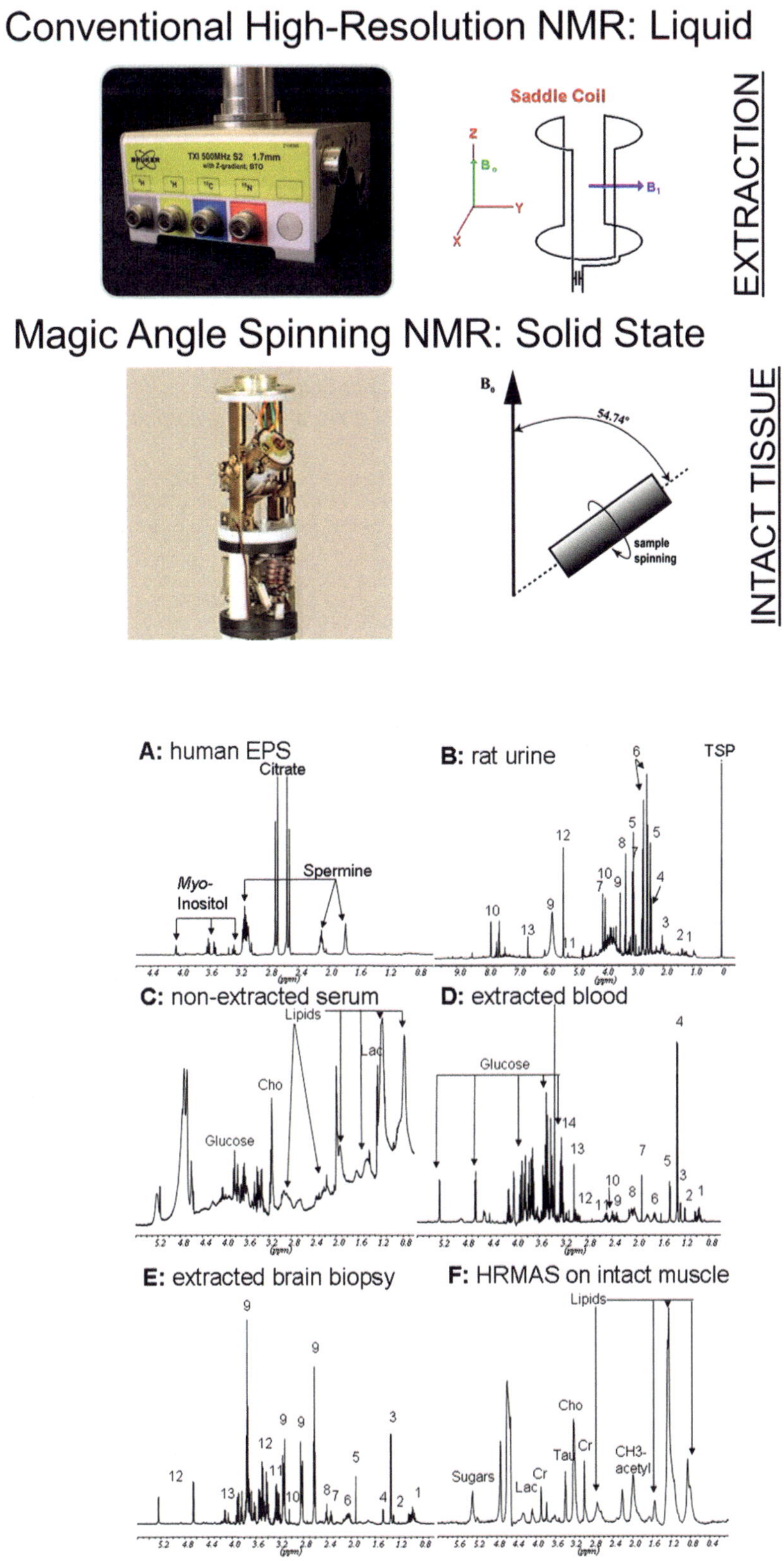

Fig. 3 Conventional (liquid state) and high-resolution magic angle spinning (solid state, non-destructed tissues) NMR: NMR probe heads and ^{1}H-NMR spectra from blood and tissue extracts and non-extracted tissue by HR-MAS

17. Finally, in vivo ^{1}H-MRS can be considered for noninvasive major metabolite detection and comparison with the ex vivo data (Fig. 4a, b).
18. For MRS, an anatomical MRI should first be obtained (usually T1 or T2 weighted, Fig. 4b).

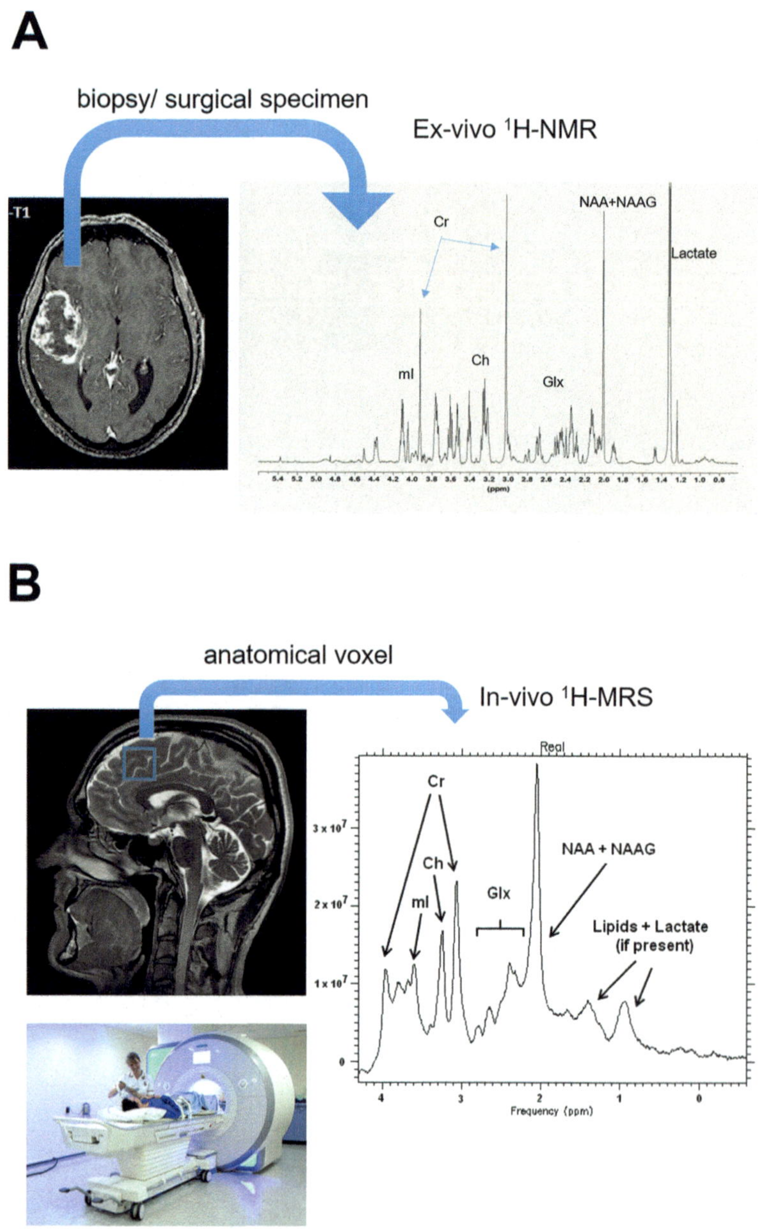

Fig. 4 Comparison between ex vivo ^{1}H-NMR and in vivo ^{1}H-MRS of human brain: (**a**) ex vivo ^{1}H-NMR on human brain glioma acid extracts; (**b**) in vivo MRS on a healthy human brain. *Ch* total choline, *Cr* total creatine, *Glx* glutamine and glutamate, *mI* myoinositol, *NAA + NAAG* N-acetylaspartate and N-acetylaspartylglutamate

19. Then, an anatomical voxel (usually $2 \times 2 \times 2$ cm^3 for humans) will be placed at the region of interest (preferably a homogenous tissue region), from which a water-suppressed ^{1}H-NMR spectrum will be obtained.
20. Based on the organ and metabolites of interest, the choice of data acquisition (PRESS, STEAM, single-voxel, 2D-3D CSI) and editing techniques (echo time, TE) should be made [29, 30].
21. The different metabolites have different T2 relaxation times; therefore, their detection and spectral appearance should be improved by TE filtering.
22. As a general rule, use long echo-time (135–270 ms) techniques for less noise but fewer metabolites.
23. Short echo times (20–30 ms) will demonstrate more metabolites but higher background noise.
24. NAA and NAAG, choline, creatine, and lactate can be detected on both short and long echo-time spectra (Fig. 4b).
25. Myoinositol, glutamine, and glutamate (Glx, Fig. 4b) are best achieved by short echo times.
26. Glycine in tumors is best identified on the 135 ms spectrum.

3.6 Spectral Analysis and Metabolite Quantification

1. Each spectrum will undergo a post-processing logarithm which is comprised of Fourier transformation, line broadening (0.1–1 Hz), phasing (manual), baseline correction, and chemical shift calibration to TMSP at o ppm.
2. Tissue metabolites will be identified based on the results from our chemical shift database [6].
3. The quantitative inaccuracy of NMR (^{1}H- or multinuclear) for a "pure" compound (Fig. 2a, b) has been reported less than 2.0% which is an excellent limit for precise, accurate quantification (*see* **Note 8**).
4. For metabolite quantification, an external standard substance, trimethylsilyl propionic-2,2,3,3,-d_4 acid (TMSP, 20 and 50 mM in D_2O) will be added into a thin sealed glass capillary.
5. The final TMSP concentration (0.5 mmol/L and 1.10 mmol/L) in the capillary will be calculated prior to sample analysis using a standard amino acid solution (Fig. 2b).
6. The TMSP capillary will be placed into the NMR tube during the experiment (0.5 mmol/L for water-soluble extracts and 1.1 mmol/L for lipid extracts) and served as an external standard which allowed for absolute metabolite quantification in each study extract.
7. ^{1}H chemical shifts will be referred to TMSP signal at 0 ppm.

8. After performing Fourier transformation and making phase and baseline corrections, each ^{1}H peak will be integrated using 1D WINNMR or TopSpin program (Bruker Biospin Inc., Fremont, CA).
9. The absolute concentrations of single metabolites (Table 1) will be then referred to the TMSP integral and calculated according to the equation at the very end of this section (Eq. 2).
10. For in vivo MRS (Fig. 4b), usually metabolite ratios are reported (such as NAA/Cr or NAA/Ch) (*see* **Note 9**).
11. Alternatively, tissue water (in this case unsuppressed spectra need to be obtained) can be used as an internal standard.
12. For multivariate analysis, usually Principal Component Analysis (PCA) or Partial Least Squares Discriminant Analysis (PLS-DA) will be applied for clustering and biomarker identification (*see* **Note 10**).

3.7 NMR Quantification

$$\mathrm{Cx} = \frac{\mathrm{Ix} : \mathrm{Nx} \times \mathrm{C}}{\mathrm{I} : 9} \times V : M \qquad (2)$$

where Cx = metabolite concentration

Ix = integral of metabolite ^{1}H peak

Nx = number of protons in metabolite ^{1}H peak (from CH, CH_2, CH_3, etc.)

C = TMSP concentration in the glass capillary

I = integral of TMSP ^{1}H peak at 0 ppm (9 since TMSP has 9 protons)

V = volume of the extract

M = weight of tissue sample or volume of blood sample prior to extraction

4 Notes

1. While working with liquid nitrogen (sample collection), it is recommended to wear safety goggles.
2. Optimal sample size is 100–200 mg of tissue (low limit is 5 mg for HR-MAS NMR, high limit for liquid-phase extraction is 500 mg).
3. *For tissue sampling, no formalin or other cold preservatives* must be used except of liquid nitrogen.
4. *For whole blood and plasma collection, only heparin-preserved blood* should be used.
5. Please avoid direct contact of local anesthetics/analgesics and disinfectant with the biopsy to minimize NMR contamination and peak overlaps.

6. Most metabolomics studies are performed on 400–800 MHz NMR spectrometers. According to our PubMed search (as of May 2018), from 3238 articles, 500 MHz (11.7 T) is the mostly used proton frequency for high-resolution quantitative ^{1}H-NMR and structural two-dimensional experiments.
7. For metabolite chemical shift identification, a two-dimensional (2D)-H, C-HSQC (heteronuclear single-quantum correlation, "hsqcetgp") technique is used [31]. The experiments will be acquired with 512 increments and 256 scans per increment, using 90 °C pulse and a recovery delay of 1 s. The spectral width will be 10 ppm in the proton dimension and 140 ppm in the ^{13}C dimension. Lactate (Lac3, CH_3) will be used as a chemical shift reference for both carbon (21 ppm) and proton (1.32 ppm) axes. Tissue metabolites will be identified based on the results from our chemical shift database.
8. For quantitative NMR, a validation/QC process should be developed to assess precision, accuracy, linearity, day-to-day reproducibility, robustness, selectivity, specificity, metabolite stability, and, in case of extraction, metabolite recovery [32].
9. For in vivo magnetic resonance spectroscopy (MRS), human MRI scanner with the field strength of 1.5 and 3 T is usually used (FDA approved); for research, human 7 T MRI is available in some institutions. For small animal research, 4.7–11.7 T MRI is usually applied.
10. Software: TopSpin, XWINNMR, 1DWINNMR, 2DWINNMR, Amix-Viewer, Aurelia, ParaVision, as well as some clinical software are available.

Acknowledgment

This work was supported by the NIH grants S10 OD023485, R01 CA140368, R01 CA195708, P30 CA046934, and UL1 TR001082, the University of Colorado Cancer Center (UCCC), and the Colorado Clinical Translational Sciences Institute (CCTSI) Animal Imaging Shared Resource (AISR).

References

1. Nevedomskaya E, Pacchiarotta T, Artemov A, Meissner A, van Nieuwkoop C, van Dissel JT et al (2012) (1)H NMR-based metabolic profiling of urinary tract infection: combining multiple statistical models and clinical data. Metabolomics 8(6):1227–1235
2. Bouatra S, Aziat F, Mandal R, Guo AC, Wilson MR, Knox C et al (2013) The human urine metabolome. PLoS One 8(9):e73076
3. Schicho R, Shaykhutdinov R, Ngo J, Nazyrova A, Schneider C, Panaccione R et al (2012) Quantitative metabolomic profiling of serum, plasma, and urine by (1)H NMR spectroscopy discriminates between patients with inflammatory bowel disease and healthy individuals. J Proteome Res 11:3344
4. Bogren LK, Murphy CJ, Johnston EL, Sinha N, Serkova NJ, Drew KL (2014) 1H-NMR

metabolomic biomarkers of poor outcome after hemorrhagic shock are absent in hibernators. PLoS One 9(9):e107493

5. Serkova NJ, Brown MS (2012) Quantitative analysis in magnetic resonance spectroscopy: from metabolic profiling to in vivo biomarkers. Bioanalysis 4(3):321–341
6. Wishart DS, Jewison T, Guo AC, Wilson M, Knox C, Liu Y et al (2013) HMDB 3.0—The Human Metabolome Database in 2013. Nucleic Acids Res 41(Database issue):D801–D807
7. El-Sayed S, Bezabeh T, Odlum O, Patel R, Ahing S, MacDonald K et al (2002) An ex vivo study exploring the diagnostic potential of 1H magnetic resonance spectroscopy in squamous cell carcinoma of the head and neck region. Head Neck 24(8):766–772
8. Tzika AA, Cheng LL, Goumnerova L, Madsen JR, Zurakowski D, Astrakas LG et al (2002) Biochemical characterization of pediatric brain tumors by using in vivo and ex vivo magnetic resonance spectroscopy. J Neurosurg 96(6):1023–1031
9. Kebir S, Rauschenbach L, Galldiks N, Schlaak M, Hattingen E, Landsberg J et al (2016) Dynamic O-(2-[18F]fluoroethyl)-L-tyrosine PET imaging for the detection of checkpoint inhibitor-related pseudoprogression in melanoma brain metastases. Neuro-Oncology 18(10):1462–1464
10. Serkova N, Bendrick-Peart J, Alexander B, Tissot van Patot MC (2003) Metabolite concentrations in human term placentae and their changes due to delayed collection after delivery. Placenta 24(2–3):227–235
11. Serkova NJ, Glunde K (2009) Metabolomics of cancer. Methods Mol Biol 520:273–295
12. Rudolph MC, Wellberg EA, Lewis AS, Terrell KL, Merz AL, Maluf NK et al (2014) Thyroid hormone responsive protein Spot14 enhances catalysis of fatty acid synthase in lactating mammary epithelium. J Lipid Res 55(6):1052–1065
13. Serkova NJ, Rose JC, Epperson LE, Carey HV, Martin SL (2007) Quantitative analysis of liver metabolites in three stages of the circannual hibernation cycle in 13-lined ground squirrels by NMR. Physiol Genomics 31(1):15–24
14. Serkova N, Klawitter J, Niemann CU (2003) Organ-specific response to inhibition of mitochondrial metabolism by cyclosporine in the rat. Transpl Int 16(10):748–755
15. Niemann CU, Saeed M, Akbari H, Jacobsen W, Benet LZ, Christians U et al (2002) Close association between the reduction in myocardial energy metabolism and infarct size: dose-response assessment of cyclosporine. J Pharmacol Exp Ther 302(3):1123–1128
16. Serkova N, Fuller TF, Klawitter J, Freise CE, Niemann CU (2005) H-NMR-based metabolic signatures of mild and severe ischemia/reperfusion injury in rat kidney transplants. Kidney Int 67(3):1142–1151
17. Niemann CU, Choi S, Behrends M, Hirose R, Noh J, Coatney JL et al (2006) Mild hypothermia protects obese rats from fulminant hepatic necrosis induced by ischemia-reperfusion. Surgery 140(3):404–412
18. Serkova NJ, Jackman M, Brown JL, Liu T, Hirose R, Roberts JP et al (2006) Metabolic profiling of livers and blood from obese Zucker rats. J Hepatol 44(5):956–962
19. Raina K, Serkova NJ, Agarwal R (2009) Silibinin feeding alters the metabolic profile in TRAMP prostatic tumors: 1H-NMRS-based metabolomics study. Cancer Res 69(9):3731–3735
20. Klawitter J, Gottschalk S, Hainz C, Leibfritz D, Christians U, Serkova NJ (2010) Immunosuppressant neurotoxicity in rat brain models: oxidative stress and cellular metabolism. Chem Res Toxicol 23(3):608–619
21. Gottschalk S, Cummins CL, Leibfritz D, Christians U, Benet LZ, Serkova NJ (2011) Age and sex differences in the effects of the immunosuppressants cyclosporine, sirolimus and everolimus on rat brain metabolism. Neurotoxicology 32(1):50–57
22. Serkova NJ, Van Rheen Z, Tobias M, Pitzer JE, Wilkinson JE, Stringer KA (2008) Utility of magnetic resonance imaging and nuclear magnetic resonance-based metabolomics for quantification of inflammatory lung injury. Am J Physiol Lung Cell Mol Physiol 295(1):L152–L161
23. Raina K, Ravichandran K, Rajamanickam S, Huber KM, Serkova NJ, Agarwal R (2013) Inositol hexaphosphate inhibits tumor growth, vascularity, and metabolism in TRAMP mice: a multiparametric magnetic resonance study. Cancer Prev Res 6(1):40–50
24. Cheng LL, Wu C, Smith MR, Gonzalez RG (2001) Non-destructive quantitation of spermine in human prostate tissue samples using HRMAS 1H NMR spectroscopy at 9.4 T. FEBS Lett 494(1–2):112–116
25. Martinez-Granados B, Monleon D, Martinez-Bisbal MC, Rodrigo JM, del Olmo J, Lluch P et al (2006) Metabolite identification in human liver needle biopsies by high-resolution magic angle spinning 1H NMR spectroscopy. NMR Biomed 19(1):90–100

26. Tessem MB, Swanson MG, Keshari KR, Albers MJ, Joun D, Tabatabai ZL et al (2008) Evaluation of lactate and alanine as metabolic biomarkers of prostate cancer using 1H HR-MAS spectroscopy of biopsy tissues. Magn Reson Med 60(3):510–516
27. Chan EC, Koh PK, Mal M, Cheah PY, Eu KW, Backshall A et al (2009) Metabolic profiling of human colorectal cancer using high-resolution magic angle spinning nuclear magnetic resonance (HR-MAS NMR) spectroscopy and gas chromatography mass spectrometry (GC/MS). J Proteome Res 8(1):352–361
28. Li M, Song Y, Cho N, Chang JM, Koo HR, Yi A et al (2011) An HR-MAS MR metabolomics study on breast tissues obtained with core needle biopsy. PLoS One 6(10):e25563
29. Meisamy S, Bolan PJ, Baker EH, Pollema MG, Le CT, Kelcz F et al (2005) Adding in vivo quantitative 1H MR spectroscopy to improve diagnostic accuracy of breast MR imaging: preliminary results of observer performance study at 4.0 T. Radiology 236(2):465–475
30. Shin HJ, Baek HM, Cha JH, Kim HH (2012) Evaluation of breast cancer using proton MR spectroscopy: total choline peak integral and signal-to-noise ratio as prognostic indicators. AJR Am J Roentgenol 198(5):W488–W497
31. Rudolph MC, McManaman JL, Phang T, Russell T, Kominsky DJ, Serkova NJ et al (2007) Metabolic regulation in the lactating mammary gland: a lipid synthesizing machine. Physiol Genomics 28(3):323–336
32. Serkova NJ, Gamito EJ, Jones RH, O'Donnell C, Brown JL, Green S et al (2008) The metabolites citrate, myo-inositol, and spermine are potential age-independent markers of prostate cancer in human expressed prostatic secretions. Prostate 68(6):620–628

Part VII

Towards Personalized Metabolomics

Chapter 24

Metabolomic Applications in Radiation Biodosimetry

Evagelia C. Laiakis

Abstract

Metabolomics, the collective assessment and quantification of small molecules in a given biofluid or tissue sample, has provided new ways in evaluating an individual's exposure level to ionizing radiation or other genotoxic stressors. Protocols that are routinely utilized for the preparation of samples from rodents to patients are presented here in order to be analyzed by high-throughput liquid chromatography-mass spectrometry techniques. These protocols are based on established methods in our laboratory that have been used extensively in radiation biodosimetry through metabolomics. These protocols are focused on general profiling of samples and therefore do not concentrate on extraction of specific classes on metabolites (e.g., eicosanoids).

Key words Metabolomics, Lipidomics, Radiation, Tissues, Biofluids, Rodents, Primates

1 Introduction

Metabolomics and lipidomics have provided new insights into the metabolic status of an individual following specific injuries and exposure to genotoxic agents, such as ionizing radiation. While targeted approaches utilize initial steps for sample cleanup and elimination of a large number of metabolites that can be time-consuming and laborious [1], high-throughput metabolomics eliminates this step and takes advantage of simple cleanup techniques based primarily on protein precipitation. These simplified cleanup methods have been applied to multiple biofluids (urine, blood, saliva) [2–12] and other materials (feces, cells, tissues) [13, 14] from rodent to human samples (comprehensively reviewed in [15, 16]) (Fig. 1).

The methods presented here are suitable for analysis with liquid chromatography-mass spectrometry. Sensitivity, mass accuracy, reliability, and most importantly high-throughput processing have been integral components of our studies. Therefore, this chapter does not present methods for preparation of samples for nuclear magnetic resonance or gas chromatography-mass spectrometry

Angelo D'Alessandro (ed.), *High-Throughput Metabolomics: Methods and Protocols*, Methods in Molecular Biology, vol. 1978, https://doi.org/10.1007/978-1-4939-9236-2_24, © Springer Science+Business Media, LLC, part of Springer Nature 2019

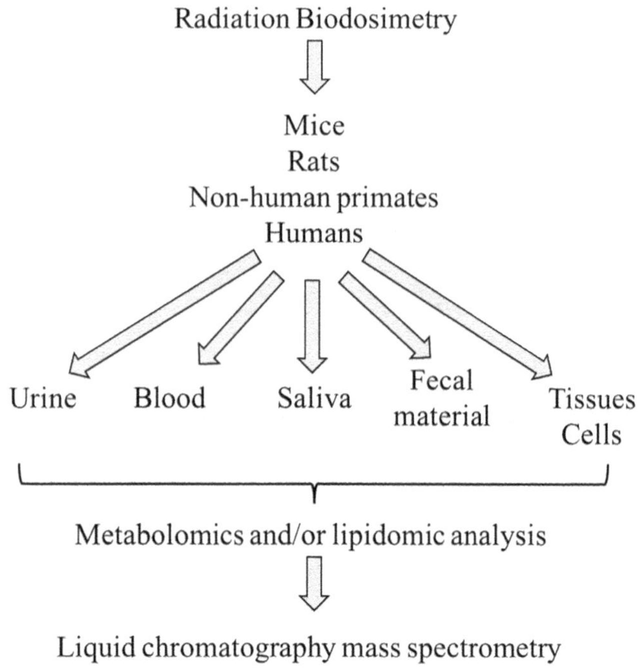

Fig. 1 General schematic of the materials that have been analyzed with liquid chromatography-mass spectrometry, for which detailed preparation protocols are provided

analyses. We are able to routinely detect >2000 spectral features with each protocol with an *m/z* between 50 and 850, in both positive and negative ionization modes. These protocols have provided reliable results in high-throughput radiation biodosimetry, with cross-species validation of metabolites identified as radiation biomarkers [16], and identification of biomarkers that can be potentially used as a signature to design targeted and clinically relevant approaches. Finally, the majority of these protocols use reagents that are common in every mass spectrometry laboratory, with the exception of the specific internal standards.

2 Materials

1. Prepare all solutions with ultrapure water (with sensitivity of 18.2 MΩ-cm at 25 °C or HPLC grade and higher). All other solvents, such as acetonitrile, methanol, and chloroform, should also be of the highest purity possible. Prepare all reagents before every experiment and store according to each method. Prepare enough volume of the solutions to complete preparation of all samples in one batch to minimize variability within the experiment. All protocols (Subheadings 2.1–2.5 and 2.8) should include the internal standards debrisoquine

sulfate (final concentration 4 μM), 4-nitrobenzoic acid (final concentration 30 μM), and chlorpropamide (final concentration 5 μM), unless indicated otherwise. However, if certain laboratories have established internal standards in their studies, they can use those instead. Lipidomic protocols (Subheadings 2.6 and 2.7) should incorporate the following internal standards: MG(17:1), DG(20:1/20:1), TG(19:1/19:1/19:1), PA(14:0/14:0), PC(14:0/14:0), PE(14:0/14:0), phosphatidylserine PS(14:0/14:0), and d8 arachidonic acid. Alternatively, a commercially available premixed lipid cocktail can be used, specific for lipidomic applications.

2. Quality controls: Create pooled samples by pooling 2–5 μL of each sample in a single microcentrifuge tube. Treat as an individual sample when preparing your samples and incorporate in your experimental analysis at least once every 10–15 samples.
3. Equipment needed: Microcentrifuge, tissue homogenizer, speed vacuum, nitrogen stream, water bath, and sonicator.
4. Use borosilicate glass disposable serological pipettes and dedicated glassware (bottles, graduated cylinders, beakers) for solvents and solutions. Plastic (siliconized preferred) microcentrifuge tubes can be used, unless otherwise indicated.

2.1 Urine Metabolomics

1. 50%:50% acetonitrile/water and chill on ice.

2.2 Serum/Plasma Metabolomics

1. *Method A:* 66%:34% acetonitrile/water with internal standards. Chill on ice.
2. *Method B:* 40%:25%:35% acetonitrile/methanol/water (premix A with internal standards) and 5%:1%:94% methanol/acetonitrile/water (premix B). Chill both on ice.

2.3 Cell Extracts for Metabolomics

1. *Method A:* Maintain chloroform at room temperature. All other solvents [phosphate-buffered saline (PBS), methanol, acetonitrile] should be chilled either at 4 °C or on ice. Chilled methanol should contain internal standards.
2. *Method B:* Maintain water and 50%:50% acetonitrile/water (premix containing internal standards) at room temperature. This method is primarily for polar molecules.

2.4 Tissue Extraction Protocol for Metabolomics

1. 50%:50% methanol/water (premix A). Make two batches, one containing internal standards and one without. Chill on ice.
2. 50%:50% acetonitrile/water (premix B with internal standards). Chill on ice.

2.5 Saliva Metabolomics

1. 50%:50% acetonitrile/water with internal standards. Chill on ice.

2.6 Tissue Lipidomics

1. 50%:25%:25% isopropanol/acetonitrile/water with lipidomic internal standards.

2.7 Serum/Plasma Lipidomics

1. Chill 2:1 v/v chloroform/methanol at −20 °C, containing the lipidomic internal standards (premix A).
2. 50%:25%:25% isopropanol/acetonitrile/water (premix B).

2.8 Microbiome Metabolomics from Fecal Material

1. 100% acetonitrile with internal standards.
2. 98%:2% acetonitrile/water with 0.1% formic acid.

3 Methods

All procedures can be carried out in room temperature, unless otherwise specified.

3.1 Urine Metabolomics

1. For rodent urine: Mix 20 μL urine with 80 μL premix (*see* Subheading 2.1) (1:4 v/v) in a 1.5 mL microcentrifuge tube, and vortex briefly. For more dilute urine (e.g., from nonhuman primates and humans): Mix 50 μL urine with 50 μL premix (*see* Subheading 2.1) (1:1 v/v) in a 1.5 mL microcentrifuge tube, and vortex briefly.
2. Centrifuge for 20 min at 13,000 × *g* at 4 °C to precipitate the protein.
3. Recover ~90 μL and place in a mass spectrometry sample vial with a pre-slit cover.
4. Samples can be stored at 4 °C overnight or −20 °C for longer.

3.2 Serum/Plasma Metabolomics (Method A)

1. For rodent samples: Mix 5 μL sample with 195 μL premix (*see* Subheading 2.2, **item 1**) (1:40 dilution) in a 1.5 mL microcentrifuge tube, and vortex briefly. For nonhuman primate and human samples: Mix 40 μL with 160 μL premix (*see* Subheading 2.2, **item 1**) (1:5 dilution) in a 1.5 mL microcentrifuge tube, and vortex briefly.
2. Incubate on ice for 10 min.
3. Centrifuge for 10 min at 13,000 × *g* at 4 °C to precipitate the protein.
4. Recover >100 μL and place in a mass spectrometry sample vial with a pre-slit cover.
5. Samples can be stored at −20 °C overnight or −80 °C for longer.

3.3 Serum/Plasma Metabolomics (Method B)

1. For rodent samples: Mix 5 μL sample with 195 μL premix (*see* Subheading 2.2, **item 2**, premix A) (1:40 dilution) in a 1.5 mL microcentrifuge tube, and vortex briefly. For nonhuman primate and human samples: Mix 25 μL with 175 μL premix (40:25:35) (*see* Subheading 2.2, **item 2**) (1:8 dilution) in a 1.5 mL microcentrifuge tube, and vortex briefly.
2. Incubate on ice for 15 min.
3. Centrifuge for 20 min at 13,000 × *g* at 4 °C to precipitate the protein.
4. Recover the supernatant in a clean microcentrifuge tube, and discard the protein precipitate.
5. Add 200 μL of 100% acetonitrile, followed by another incubation on ice for 15 min.
6. Centrifuge for 20 min at 13,000 × *g* at 4 °C to precipitate the protein.
7. Recover the supernatant in a clean microcentrifuge tube.
8. Vacuum dry the sample using a speed vac with no heat.
9. Reconstitute in chilled premix B (*see* Subheading 2.2, **item 2**).
10. Centrifuge for 20 min at 13,000 × *g* at 4 °C.
11. Transfer the supernatant into a mass spectrometry sample vial with a pre-slit cover.
12. Samples can be stored at −20 °C overnight or −80 °C for longer.

3.4 Cell Extracts (Method A)

1. For adherent cell cultures: Wash exponentially growing cells (total number $\geq 10^7$) 3× with chilled PBS, and scrape into a small PBS volume. For cells in suspension: Wash pellets of exponentially growing cells (total number $\geq 10^7$) 3× with chilled PBS.
2. Centrifuge for 10 min at 200 × *g* at 4 °C.
3. Discard supernatant. Pellet can either be saved at −80 °C until extraction or can be processed immediately.
4. Resuspend the pellet in 150 μL of water and pipette gently to loosen the pellet.
5. Cell disruption can be achieved by quick freeze of cells on dry ice, followed by a 90 s heat shock bath in 37 °C. Repeat three times, followed by a 30 s sonication.
6. Remove 2–5 μL of the mix, and assess protein levels in each sample with a protein assay capable of quantifying low protein levels on a 96-well plate (*see* **Note 1**).
7. Add 600 μL of chilled methanol containing internal standards to each sample, and vortex slightly.
8. Incubate on ice for 15 min.

9. Add 600 μL of 100% room temperature chloroform to each sample, and vortex slightly.
10. Centrifuge for 10 min at 16,200 × *g* at room temperature.
11. Transfer each phase to a separate clean 1.5 mL microcentrifuge tube, avoiding contamination from the interphase. The lower phase is the organic containing primarily nonpolar molecules. The upper aqueous phase contains primarily polar molecules. The interphase contains primarily precipitated proteins (*see* **Note 2**).
12. Add 600 μL of chilled 100% acetonitrile to each tube (organic and aqueous phases), and vortex.
13. Incubate at −20 °C overnight.
14. Centrifuge for 10 min at 16,200 × *g* at room temperature.
15. Carefully transfer the supernatant in clean tubes and discard the pellets.
16. Vacuum dry the sample using a speed vac with no heat.
17. Resuspend in 100 μL of 50%:50% water/acetonitrile.
18. The two phases can be combined into one mass spectrometry sample vial or processed separately.

3.5 Cell Extracts (Method B)

1. For adherent cell cultures: Wash exponentially growing cells (total number $\geq 10^7$) 3× with chilled PBS, and scrape into a small PBS volume. For cells in suspension: Wash pellets of exponentially growing cells (total number $\geq 10^7$) 3× with chilled PBS.
2. Centrifuge for 10 min at 200 × *g* at 4 °C.
3. Discard supernatant. Pellet can either be saved at −80 °C until extraction or can be processed immediately.
4. Resuspend the pellet in 150 μL of water and pipette gently to loosen the pellet.
5. Sonicate for 30 s to disrupt the cell membranes.
6. Remove 2–5 μL of the mix, and assess protein levels in each sample with a protein assay capable of quantifying low protein levels on a 96-well plate (*see* **Note 1**). Alternatively, a 20 μL aliquot can be assessed with a regular BCA assay.
7. Mix each pellet with premix (*see* Subheading 2.3, **item 2**) in a 1:1 v/v ratio.
8. Centrifuge for 20 min at 13,000 × *g* at 4 °C.
9. Carefully transfer the supernatant in clean tubes and discard the pellets.
10. Vacuum dry the sample using a speed vac with no heat.
11. Resuspend the pellet in 300 μL of water.

12. Filter the supernatant with a 0.2 μm column to remove particulates (*see* **Note 3**).
13. Transfer the eluent into a mass spectrometry sample vial with a pre-slit top.

3.6 Tissue Extraction

1. Add chilled premix A with internal standards (*see* Subheading 2.4) to the tissue sample (3 volumes or 300 μL) (*see* **Note 4**).
2. Homogenize each sample with a tissue homogenizer, thoroughly cleaning the probe between each homogenization with PBS.
3. Remove 2–5 μL of the mix, and assess protein levels in each sample with a protein assay capable of quantifying low protein levels on a 96-well plate (*see* **Note 1**). Alternatively, a 20 μL aliquot can be assessed with a regular BCA assay.
4. Centrifuge for 15 min at 13,000 × *g* at 4 °C.
5. Carefully transfer the supernatant to a clean 1.5 mL microcentrifuge, without disrupting the pellet.
6. Add an equal volume of chilled premix B (*see* Subheading 2.4) (300 μL) to the pellet, and vortex thoroughly.
7. Incubate on ice for 15 min.
8. Centrifuge for 15 min at 13,000 × *g* at 4 °C.
9. Transfer the supernatant to the previously collected sample (*see* **step 5**) to combine the two.
10. Vacuum dry the sample using a speed vac with no heat.
11. Resuspend the sample in 150 μL of premix A without internal standards (*see* Subheading 2.4), and vortex thoroughly.
12. Transfer the sample into a mass spectrometry sample vial with a pre-slit top.
13. Store either at −20 °C for short-term storage or −80 °C for longer-term storage.

3.7 Saliva Metabolomics

1. Thaw saliva samples on ice and maintain on ice.
2. Remove 2–5 μL of the mix, and assess protein levels in each sample with a protein assay capable of quantifying low protein levels on a 96-well plate (*see* **Note 1**).
3. Mouse: Add 95 μL of chilled premix (*see* Subheading 2.5) to 5 μL of saliva. Human or nonhuman primates: Add chilled premix (*see* Subheading 2.5) to each saliva sample in a 1:1 (v:v) ratio.
4. Centrifuge for 15 min at 13,000 × *g* at 4 °C.
5. Transfer supernatant into a mass spectrometry sample vial with a pre-slit top.
6. Store either at −20 °C for short-term storage or −80 °C for longer-term storage.

3.8 Tissue Lipidomics

1. Label glass test tubes, and add 1 mL chilled methanol with internal lipidomic standards in each.
2. Maintain the tubes on ice.
3. Add a tissue sample on each tube (*see* **Note 4**).
4. Homogenize each sample with a tissue homogenizer on an ice bath, thoroughly cleaning the probe between each homogenization with PBS.
5. Return the tubes on ice.
6. Add 2 mL of 100% chloroform to each sample (*see* **Note 5**).
7. Vortex each sample for 10 s.
8. Add 1 mL of water to each sample.
9. Vortex each sample for 10 s.
10. Centrifuge for 15 min at 1000 × *g* at 4 °C (*see* **Note 6**).
11. Label 1.5 mL siliconized microcentrifuge tubes and with a glass Pasteur pipette transfer the lower organic phase. Save the upper aqueous phase in another set of tubes for metabolomic analysis.
12. Vacuum dry the sample using a speed vac with no heat. Alternatively, evaporate the sample under a gentle stream of nitrogen gas (*see* **Note** 7).
13. Dissolve the dried lipid extract in 100–200 μL of premix (*see* Subheading 2.6), and transfer to a glass mass spectrometry sample vial with a pre-slit top.
14. Store at −80 °C if the samples will not be processed immediately (*see* **Note 8**).

3.9 Serum/Plasma Lipidomics (Fig. 2)

1. Add 25 μL of sample (*see* **Note 9**) in a 1.5 mL siliconized microcentrifuge tube (*see* **Notes 10** and **11**).
2. Add chilled premix A (*see* Subheading 2.7) standards to the sample in a 1:4 v/v.

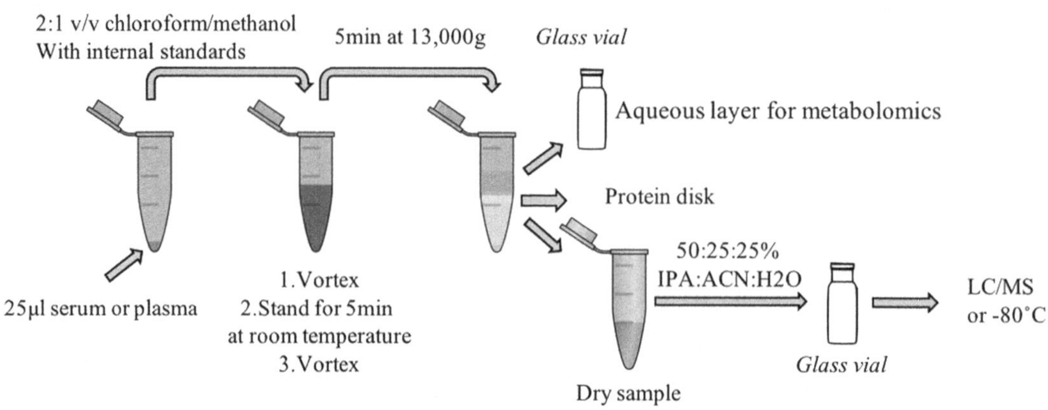

Fig. 2 Diagram of lipidomic preparation for serum or plasma

3. Vortex each sample for 30 s at room temperature and allow to stand for 5 min.
4. Repeat vortexing for 30 s.
5. Centrifuge for 5 min at 13,000 × *g* at room temperature to separate the organic from the aqueous phases. The middle disk is the precipitated protein.
6. Utilizing a glass Pasteur pipette, carefully collect the top aqueous layer, and save into a glass tube with Teflon-lined screw caps (*see* **Note 2**).
7. Save at −80 °C for metabolomic analysis.
8. Utilizing glass Pasteur pipettes, carefully pierce the protein disk (*see* **Note 12**), and extract the bottom organic phase that contains all the lipids (*see* **Note 2**).
9. Transfer the organic phase in a clean 1.5 mL siliconized microcentrifuge tube.
10. Vacuum dry the sample using a speed vac with no heat. Alternatively, evaporate the sample under a gentle stream of nitrogen gas (*see* **Note 7**).
11. Store at −80 °C if the sample will not be processed immediately.
12. Dissolve the dried lipid extract in premix B (*see* Subheading 2.7), and transfer to a glass mass spectrometry sample vial with pre-slit top.
13. Store at −80 °C if the samples will not be processed immediately (*see* **Note 8**).

3.10 Microbiome Metabolomics from Fecal Material

1. Add 50–100 mg dried fecal material in a 1.5 mL microcentrifuge tube.
2. Add 150 μL of water in the tube.
3. Homogenize each sample with a tissue homogenizer, thoroughly cleaning the probe between each homogenization with PBS.
4. Heat the samples to 37 °C in a water bath for 1.5 min and immediately place on ice.
5. Remove 2–5 μL of the mix, and assess protein levels in each sample with a protein assay capable of quantifying low protein levels on a 96-well plate (*see* **Note 1**).
6. Add 150 μL of 100% acetonitrile in each sample, containing metabolomic internal standards.
7. Vortex briefly.
8. Centrifuge for 20 min at 13,000 × *g* at 4 °C.
9. Transfer the supernatant to a clean 1.5 mL microcentrifuge tube.

10. Vacuum dry the sample using a speed vac with no heat. Alternatively, evaporate the sample under a gentle stream of nitrogen gas (*see* **Note 7**).
11. Resuspend the sample in 300 μL of 98%:2% water/acetonitrile with 0.1% formic acid.
12. Transfer the supernatant into a mass spectrometry sample vial with a pre-slit cover.

4 Notes

1. Protein levels can be used as a normalization method of the deconvoluted mass spectrometry data as an alternate to TIC normalization. Protein normalization is the preferred method when cell extracts, tissue samples, saliva, or fecal samples are assessed. Kits that can assess levels of protein <200 μg/mL are commercially available.
2. It is far preferable to sacrifice a small amount of sample from the top and bottom phase in order to avoid contamination from the protein disk. Therefore, recovery will not be 100%. However, consistency on sample removal can be achieved by setting a predetermined setting on your pipette.
3. We have used filters for samples other than cell extracts. In rare occasions, the sample may contain particulates, mostly due to accidental contamination from the pellet or due to high lipid content, which can create problems with the chromatography and even clog the column. In such occasions, we have filtered all samples, which has eliminated the issue.
4. A small 10–30 μg piece of tissue is sufficient for metabolomic analysis. To extract the frozen tissue from a cryovial, gently and briefly cover the cryovial in your palm for a few seconds to dislodge from the walls. To cut a piece of frozen tissue, place a plastic dish (could be a Petri dish or a pipette tip box cover) covered with aluminum foil over dry ice, and utilizing a sharp scalpel, cut a small tissue sample, and place in a 1.5 mL microcentrifuge tube on ice. Return the rest of the tissue to −80 °C. Clean the scalpel thoroughly between each tissue.
5. It is best to do this in a fume hood to minimize exposure to harmful chloroform.
6. It will be difficult to actually put glass test tubes in a centrifuge; therefore, we recommend splitting the sample into two in 1.5 mL siliconized microcentrifuge tubes and in the end recombining the two parts.
7. Nitrogen gas is the preferred method anytime you are analyzing lipids. However, if you lack this capability in your lab, a

carefully observed drying down with a speed vac is perfectly acceptable.

8. For lipidomic analysis it is ideal to process the samples on the LC-MS shortly after sample preparation.
9. The same volume can be used from rodents to larger mammals (e.g., nonhuman primates, humans).
10. Be careful not to use regular microcentrifuge tubes, as the siliconized tubes will minimize the sample adhesion to the walls and therefore sample loss.
11. With lipidomic analysis, it is strongly urged to use glass Pasteur pipettes and pipette tips. Although cumbersome to thoroughly clean the tips after each use or expensive to purchase new ones, for low abundance molecules it can make a significant difference.
12. The protein disk can be saved for proteomic analysis.

Acknowledgments

This work was supported by the National Institutes of Health National Institute of Allergy and Infectious Diseases (NIAID) grant U19 AI067773 (P.I. David J. Brenner, performed as part of Columbia University Center for Medical Countermeasures against Radiation), NIAID R01AI101798 (P.I. Albert J. Fornace, Jr.), National Cancer Institute R01CA184168 (P.I. Albert J. Fornace, Jr.), and a pilot grant from the Opportunity Funds Management Core of the Centers for Medical Countermeasures against Radiation (NIAID, U19 AI067773, P.I. Evagelia C. Laiakis).

References

1. Roberts LD, Souza AL, Gerszten RE, Clish CB (2012) Targeted metabolomics. Curr Protoc Mol Biol Chapter 30:Unit 30.2.1-24
2. Tyburski JB, Patterson AD, Krausz KW, Slavík J, Fornace AJ, Gonzalez FJ et al (2008) Radiation metabolomics. 1. Identification of minimally invasive urine biomarkers for gamma-radiation exposure in mice. Radiat Res 170:1–14
3. Laiakis EC, Morris GA, Fornace AJ, Howie SR (2010) Metabolomic analysis in severe childhood pneumonia in the Gambia, West Africa: findings from a pilot study. PLoS One 5:9
4. Laiakis EC, Mak TD, Anizan S, Amundson SA, Barker CA, Wolden SL et al (2014) Development of a metabolomic radiation signature in urine from patients undergoing total body irradiation. Radiat Res 181:350–361
5. Laiakis EC, Strassburg K, Bogumil R, Lai S, Vreeken RJ, Hankemeier T et al (2014) Metabolic phenotyping reveals a lipid mediator response to ionizing radiation. J Proteome Res 13:4143–4154
6. Mak TD, Tyburski JB, Krausz KW, Kalinich JF, Gonzalez FJ, Fornace AJ (2015) Exposure to ionizing radiation reveals global dose- and time-dependent changes in the urinary metabolome of rat. Metabolomics 11: 1082–1094
7. Pannkuk EL, Laiakis EC, Mak TD, Astarita G, Authier S, Wong K et al (2016) A lipidomic and metabolomic serum signature from nonhuman primates exposed to ionizing radiation. Metabolomics 12:5
8. Laiakis EC, Strawn SJ, Brenner DJ, Fornace AJ (2016) Assessment of saliva as a potential

biofluid for biodosimetry: a pilot metabolomics study in mice. Radiat Res 186:92–97

9. Laiakis EC, Pannkuk EL, Diaz-Rubio ME, Wang YW, Mak TD, Simbulan-Rosenthal CM et al (2016) Implications of genotypic differences in the generation of a urinary metabolomics radiation signature. Mutat Res 788:41–49
10. Laiakis EC, Pannkuk EL, Chauthe SK, Wang YW, Lian M, Mak TD et al (2017) A serum small molecule biosignature of radiation exposure from total body irradiated patients. J Proteome Res 16:3805–3815
11. Pannkuk EL, Laiakis EC, Singh VK, Fornace AJ (2017) Lipidomic signatures of nonhuman primates with radiation-induced hematopoietic syndrome. Sci Rep 7:9777
12. Laiakis EC, Wang YW, Young EF, Harken AD, Xu Y, Smilenov L et al (2017) Metabolic dysregulation after neutron exposures expected from an improvised nuclear device. Radiat Res 188:21–34
13. Patterson AD, Li H, Eichler GS, Krausz KW, Weinstein JN, Fornace AJ et al (2008) UPLC-ESI-TOFMS-based metabolomics and gene expression dynamics inspector self-organizing metabolomic maps as tools for understanding the cellular response to ionizing radiation. Anal Chem 80:665–674
14. Goudarzi M, Mak TD, Jacobs JP, Moon BH, Strawn SJ, Braun J et al (2016) An integrated multi-omic approach to assess radiation injury on the host-microbiome axis. Radiat Res 186:219–234
15. Coy SL, Cheema AK, Tyburski JB, Laiakis EC, Collins SP, Fornace A (2011) Radiation metabolomics and its potential in biodosimetry. Int J Radiat Biol 87:802–823
16. Pannkuk EL, Fornace AJ, Laiakis EC (2017) Metabolomic applications in radiation biodosimetry: exploring radiation effects through small molecules. Int J Radiat Biol 93:1151–1176

Chapter 25

Metabolomics Analyses to Investigate the Role of Diet and Physical Training

Pol Herrero, Miguel Ángel Rodríguez, Maria Rosa Ras, Antoni del Pino, Lluís Arola, and Núria Canela

Abstract

The metabolic and physiologic responses to healthy dietary habits and physical exercise have become an increasingly interesting research area, since equilibrated diet and regular physical activity are commonly recommended for their antioxidant capacity and for the prevention and treatment of several disorders as insulin resistance, dyslipidemia, obesity, and hypertension that may result in cardiovascular disease and type II diabetes.

Nutritional and exercise-induced responses and the biological mechanisms that explain these associations have been tackled by several researchers using metabolomic approaches that have emerged as a powerful tool to comprehensively evaluate individual metabolic signatures, analyzing metabolome composition in serum, urine, stool, or tissue samples.

The overview of the wide range of metabolites related to dietary and to physical training interventions reported from numerous human or animal studies endorses the complexity for assessing metabolic changes and compound identification, and a combination of targeted and non-targeted global profiling studies is recommended for increasing the understanding of nutrition and exercise metabolic mechanisms.

The present protocol attempt to identify variations in human blood circulating metabolites with a multiplatform global analysis based on LC-MS, GC-MS, and NMR, combining targeted and untargeted strategies, to complete the holistic understanding of the serum metabolome.

Key words Metabolomics, Diet, Nutrition, Physical training, Liquid chromatography, Gas chromatography, Mass spectrometry, Nuclear magnetic resonance

1 Introduction

Nutritional studies have significantly increased in the last decade, demonstrating a trend in promoting healthy properties of smart foods, phytochemicals, nutraceuticals, and dietary patterns [1–5]. Otherwise, it is well known that physical activity confers a wide range of health benefits associated with reduced risks of cardiovascular fail, diabetes, osteoporosis, cancer, and other disorders [6–11].

Metabolomics has emerged as a useful high-throughput and sensitive approach to identify and asses the alteration in metabolic

Angelo D'Alessandro (ed.), *High-Throughput Metabolomics: Methods and Protocols*, Methods in Molecular Biology, vol. 1978, https://doi.org/10.1007/978-1-4939-9236-2_25, © Springer Science+Business Media, LLC, part of Springer Nature 2019

phenotypes that links complex relationships between dietary exposures and physical training to metabolic disorders that lead to chronic diseases [12–14]. Metabolomic signature could elucidate further changes in the metabolome that may help us understand the mechanisms whereby diet and periodic training exert its impact in subjects.

Related to dietary-physical activity, several metabolomic strategies can be distinguished, being the principal trials: (1) the metabolome comparative between control groups and dietary-exercise interventions and (2) to determine the metabolic response with respect to time.

Nutritional and physical training-related metabolomic investigations should include both targeted (focused to a select group of metabolites) and non-targeted (all detectable metabolites) experimental approaches in a variety of noninvasively collected biological samples.

Literature reveals a breadth of metabolites influenced in dietary and physical training intervention studies; therefore researchers applied comprehensive metabolomic methodologies to uncover metabolite profile and monitored the changes of specific compounds. Some studies have attempted to identify distinctive metabolite profiles associated with diet effect on obesity in adults [15] and others with youths [16], pointing to the importance of biomarker discovery for early diagnosis, treatment, and assessment of lifestyle-related diseases and demonstrating that the major changes are on amino acids, lipid metabolism (fatty acids, acylcarnitines, sphingolipids, etc.), vitamins, and catabolism [17–19]. On the other hand, higher level of physical activity is associated with human metabolomic signatures that are consistent with better cardiometabolic health, with lower levels of branched chain amino acids and monosaccharides [20] and altered circulating bile acids and glycerophospholipids [11].

The present protocol is focused on the methodology to evaluate and detect diet and physical exercise responsive metabolites from blood, as well as specific experimental considerations in the design of multiplatform metabolomic studies. Our gold standard methodology is based on a global targeted approach with the potential for novel discoveries and was accomplished combining three different technological platforms, using nuclear magnetic resonance (NMR), gas chromatography coupled to mass spectrometry (GC-MS), and liquid chromatography coupled to mass spectrometry (LC-MS): all powerful techniques very suitable for metabolomic studies that allows to achieve a more complete vision of the metabolome identifying a wide range of metabolites altered among the experimental groups. This targeted metabolic approach visualizes the breadth of metabolism that is influenced by the diet and exercise and enables increased sensitivity and absolute quantitation; otherwise it requires a priori knowledge of the metabolites of interest. In metabolomic studies it is recommended to combine targeted and non-targeted approaches to enable unbiased metabolite profiling. For that reason a key characteristic of our strategy is that MS instruments are operated in scan mode; therefore the raw data generated could be used simultaneously for targeted and untargeted metabolomics analysis that permits the identification of potential biomarkers and the characterization of the metabolic signatures that may be predictive of the health status.

The aim of the protocol presented in this chapter is to perform a global metabolic profiling on human plasma samples (GlobalMet-p profiling Method) measuring over 200 compounds covering a wide range of metabolomic pathways: steroids, glycerophospholipids, sphingolipids, fatty acyls, cholesteryl esters, cholesterol, glycerolipids, amino acids and amino acid derivatives, central carbon metabolism, carboxylic acids, vitamins, acylcarnitines, and other metabolites.

This procedure includes:

- Experimental design.
- Sample preparation and metabolite extraction from plasma.
- Multiplatform analysis using three different infrastructures:
  - GlobalMet polar metabolites method by gas chromatography coupled to high-resolution mass spectrometry (GC-HRMS).
  - GlobalMet lipidic metabolites method by liquid chromatography coupled to high-resolution mass spectrometry (LC-HRMS) and proton nuclear magnetic resonance (^{1}H-NMR).
  - GlobalMet acylcarnitines analysis by liquid chromatography coupled to tandem mass spectrometry (LC-MS/MS).
- Data analysis.

2 Materials

Some reagents and instruments are required for several methods englobed in this protocol and are listed as general materials.

General Reagents

- Ultrapure water.
- Methanol LC-MS grade.
- Formic acid LC-MS grade.

General Instruments and Materials

- 10 mL Vacutainer® (Not SST tubes).
- 15 mL polypropylene conical tubes.
- Sterile cryovials with writing surface.

- Eppendorf tubes of 1.5 or 2 mL capacity.
- 1.5 mL chromatographic vials equipped with 200 μL inserts.
- A set of pipettes and corresponding disposable tips (0.5–10; 5–50, 20–200; 100–1000 μL).
- Laboratory benchtop centrifuge.
- SpeedVac or equivalent dryer system (*see* **Note 1**).
- Vortex and orbital mixers.
- Incubator or oven.

2.1 Materials for GlobalMet-p Polar Metabolites Method

2.1.1 Standards and Reagents

- Lactic acid, glycolic acid, alanine, glycine, 2-hydroxybutanoic acid, 3-hydroxybutanoic acid, valine, leucine, glycerol, isoleucine, proline, glyceric acid, serine, threonine, methionine, oxoproline, glutamic acid, phenylalanine, lysine, glutamine, citric acid, fructose, glucose, tyrosine, palmitic acid, linoleic acid, oleic acid, stearic acid, tryptophan, sucrose, alpha-tocopherol, and cholesterol. Stock solution of 2 g/L of each standard is prepared by dissolving them in pure methanol or water depending on their solubility and stored at −20 °C (*see* **Note 2**). Working calibration solution at a concentration of 50 μg/mL is prepared by mixing equal amounts of each standard in MeOH/water 8:1.
- Internal standards: succinic acid d_4, norvaline, $[^{13}C_3]$ glycerol, $[^{13}C_6]$ glucose, $[^{13}C,d_3]$L-methionine, and myristic acid d_{27}. Stock solution of 2 g/L of each standard is prepared by dissolving them in pure methanol or water depending on their solubility and stored at −20 °C (*see* **Note 2**). Working calibration solution at a concentration of 10 μg/mL is prepared by mixing equal amounts of each internal standard in MeOH/water 8:1.
- 2,2,2-Trifluoro-*N*-methyl-*N*-(trimethylsilyl)-acetamide (MSTFA) plus 1% chlorotrimethylsilane (TMCS).
- Methoxyamine hydrochloride.
- Pyridine.

2.1.2 Equipment and Other Materials

- Gas chromatograph coupled to a quadrupole time-of-flight mass spectrometer (e.g., GC-QTOF 7200 from Agilent Technologies, Santa Clara, CA, USA) equipped with an electron impact interface and split/splitless injection port. Equivalent systems from other vendors are acceptable.
- Column: HP-5ms (30 m × 0.25 mm i.d., 0.25 μm) from Agilent Technologies or equivalent.

2.2 Materials for GlobalMet-p Lipidic Metabolites Method

2.2.1 Standards and Reagents

- 18:0–18:1 PC, 18:0–18:1 PE, 18:1 LPC, 18:1 LPE, 18:0–18:1 DG, 18:0–18:1–18:0 TG, and 18:1 SM unlabeled lipid standards. Stock solution of 1 mg/L of each standard is prepared by dissolving them in pure methanol and stored at −20 °C. Working calibration solution is prepared by mixing equal amounts of each standard at a concentration of 10 μg/mL.
- SPLASH® Lipidomix® Mass Spec Standard (Cat. Number 330707) from Avanti Polar Lipids (Alabaster, AL, USA) containing 15:0–18:1(d7) PC (160 μg/mL), 15:0–18:1(d7) PE (5 μg/mL), 15:0–18:1(d7) PS (5 μg/mL), 15:0–18:1(d7) PG (30 μg/mL),15:0–18:1(d7) PI (10 μg/mL), 15:0–18:1(d7) PA (7 μg/mL), 18:1(d7) LPC (25 μg/mL), 18:1(d7) LPE (5 μg/mL), 18:1(d7) Chol Ester (350 μg/mL), 18:1(d7) MG (2 μg/mL), 15:0–18:1(d7) DG (10 μg/mL), 15:0–18:1(d7)–15:0 TG (55 μg/mL), 18:1(d9) SM (30 μg/mL), and cholesterol (d7) (100 μg/mL). The working solution is made by preparing a 1:50 dilution of SPLASH with pure methanol. The solutions should be stored at −20 °C
- 2-Propanol LC-MS grade.
- Ammonium formate 10 M solution in water.
- Methyl tert-butyl ether HPLC grade.
- Mobile phase A (ultrapure water). It should be changed daily.
- Mobile phase B (methanol).
- Mobile phase C (2-propanol).
- Mobile phase D (200 mM ammonium formate, 1% formic acid in ultrapure water). This should be prepared freshly.
- Tetramethylsilane (TMS).
- Deuterated chloroform $CDCl_3$NMR grade.
- Deuterated methanol CD_3OD NMR grade.

2.2.2 Equipment and Other Materials

- Quadrupole time-of-flight mass spectrometer (e.g., QTOF 6550 series from Agilent Technologies) equipped with an electrospray interface. Equivalent systems from other vendors are acceptable
- Ultrahigh-performance liquid chromatograph equipped with solvents reservoir, degasser, 1200 bar quaternary pump, autosampler, and column oven (e.g., 1290 Infinity series systems from Agilent Technologies). Equivalent systems from other vendors are acceptable.
- Analytical column: Kinetex C18-EVO 2.7 μm (100 × 2.1 mm) from Phenomenex (Torrance, CA, USA).
- 600 MHz NMR spectrometer (e.g., Avance III 600 from Bruker, Germany) operating at a proton frequency of 600.20 MHz using a 5 mm CPTCI triple resonance (^{1}H, ^{13}C, ^{31}P) gradient cryoprobe.

2.3 Materials for GlobalMet-p Acylcarnitines Method

2.3.1 Standards and Reagents

- Unlabeled carnitine standards containing carnitine (free carnitine, CN) 152.0 nmol/mL, acetylcarnitine (C2) 38.0 nmol/mL, propionylcarnitine (C3) 7.6 nmol/mL, butyrylcarnitine (C4) 7.6 nmol/mL, isovalerylcarnitine (C5) 7.6 nmol/mL, octanoylcarnitine (C8) 7.6 nmol/mL, myristoylcarnitine (C14) 7.6 nmol/mL, and palmitoylcarnitine (C16) 15.2 nmol/mL. The dry standards are reconstituted in 1 mL of methanol-water (1:1, v,v) and stored in a tightly sealed vial at 4 °C (*see* **Note 3**).
- Labeled carnitine internal standards set B from Cambridge Isotopes (NSK-B-1) containing $^{2}H_{9}$ carnitine (free carnitine, CN) 152.0 nmol/mL, $^{2}H_{3}$ acetylcarnitine (C2) 38.0 nmol/mL, $^{2}H_{3}$ propionylcarnitine (C3) 7.6 nmol/mL, $^{2}H_{3}$ butyrylcarnitine (C4) 7.6 nmol/mL, $^{2}H_{9}$ isovalerylcarnitine (C5) 7.6 nmol/mL, $^{2}H_{3}$ octanoylcarnitine (C8) 7.6 nmol/mL, $^{2}H_{9}$ myristoylcarnitine (C14) 7.6 nmol/mL, and $^{2}H_{3}$ palmitoylcarnitine (C16) 15.2 nmol/mL. The dry standards are reconstituted in 1 mL of methanol-water (1:1, v,v) and stored in a tightly sealed vial at 4 °C (*see* **Note 3**).
- Anhydrous butanolic hydrogen chloride solution ~3 M for GC derivatization.
- Heptafluorobutyric acid for ion chromatography.
- Acetic acid LC-MS grade.
- Acetonitrile LC-MS grade.
- Mobile phase A (0.1% acetic acid and 0.01% heptafluorobutyric acid in ultrapure water). This should be prepared freshly.
- Mobile phase B (0.1% acetic acid and 0.01% heptafluorobutyric acid in acetonitrile). This should be prepared freshly.

2.3.2 Equipment and Other Materials

- Triple quadrupole mass spectrometer (e.g., QqQ 6490 series from Agilent Technologies) equipped with an electrospray interface. Equivalent systems from other vendors are acceptable.
- Ultrahigh-performance liquid chromatograph equipped with solvents reservoir, degasser, 1200 bar binary pump, autosampler, and column oven (e.g., 1290 Infinity series systems from Agilent Technologies). Equivalent systems from other vendors are acceptable.
- Analytical column: XBridge BEH C18 Column, 3.5 μm (2.1 mm × 150 mm), from Waters (Millstone, NJ, USA) or equivalent.

3 Methods

3.1 Experimental Design in Nutritional and Physical Training Metabolomics

The study design is especially fundamental to dietary-exercise metabolomic investigations given the large number of measured metabolites and the natural variability intrinsic to biological samples [21]. Then, some considerations must be taken into account in the experimental design to guarantee reproducible and successful experiments that minimize erroneous variability. These considerations are related to external factors that can affect the experiment, like the nature of the sample matrix, collection and storage, study size, randomization, study controls, time points, and individual metabolite variation.

Sample matrix, collection, and storage: Before performing a metabolomic experiment, we firstly should know the sample type and amount. Samples that may reflect metabolome changes due to diet and exercise include urine, plasma or serum, saliva, feces, and extracts of cells from muscle and liver tissues [18]. Nonetheless, for the major number of dietary studies, serum is the selected matrix to follow metabolites transit from food intake to gut and other tissues, while for detoxification and degradation experiments, urine and feces are more suitable models.

For body fluids, as blood, a rapid pretreatment after sampling is performed, by adding anticoagulants and preservatives, and then samples must be stored properly at −80 °C until metabolite extraction. Pooling and repeated freezing and thawing must be avoided, to reduce serum/plasma hemolysis. In addition, in plasma samples, sodium heparin should be used as anticoagulant, instead of EDTA, because of its interfering effects in metabolite pattern. Moreover for NMR, samples should not contain alcohol or anesthetics (*see* **Notes 4** and **5**).

Study size: Number of specimens is another important consideration in experimental design and should be defined before beginning the investigation to guarantee that an appropriate number of biological replicates are included in each biological group, to confidently give results in a statistically robust manner.

The minimum number of biological replicates for statistically significant assessment of a metabolomic dataset could be decided from a pilot experiment or from a previously reported variation in a similar study. In metabolomic studies, there is no standard method for the determination of sample size, and few strategies are appearing in the literature for sample size selection in the high-dimensional data setting [22, 23].

In laboratory model animals, variability is highly controlled, and fewer replicates are necessary. Typically the minimum per biological group should be about six to ten biological replicates. Nevertheless, for human specimens, where variability is not well controlled, more replicates, up to hundreds, are required to assure statistically significant and robust results.

Randomization: The design must join randomization and technical replicates in a similar way as clinical trials. Randomizing the samples secures an experiment against fortuitous elements, such as temperature or humidity variations. Technical replicates, in which a sample is measured several times, could be performed to minimize variability induced by the metabolite detection platform. These types of variation are independent from biological variation and show the consistency of the analytical measurement. Nevertheless, in the present protocol several quality control samples are included, and technical replicates could be omitted.

Study controls: The principal diet-exercise studies are focused on a parallel design, where at least two groups are followed up along the time, and one or more groups get a treatment, and at least one group acts as control. The intervention effect is defined as the difference in the parameter of interest between the control and the treatment groups at the start and the end of the intervention [24, 25].

Time points: Some other really important issue is at what time point after the diet-exercise treatment samples will be collected. A literature study or pilot data will be indispensable, because the extent of the intervention effect may fluctuate from seconds to hours depending on the metabolite and treatment. Based on time duration the metabolomic studies can be distinguished between short-term to evaluate an acute or transient response and long-term trials to asses chronic response on metabolic phenotype [26, 27].

Individual metabolite variation: Animal models involve many sources of variation, related to the genotype, gender, environment, chronic diseases, medications, food intake, and many more. All these factors could affect metabolite profile, and it is difficult to control these sources of variation, and so we need to ensure that the variability is equivalent and balanced between different groups of subjects. Thus, nutritional and physical training trials are often designed as crossover studies with the benefit that each individual serves as its own baseline control, being particularly efficient to identify differences in the presence of large interindividual variation.

3.2 Sample Preparation

For GlobalMet-p method, total blood is collected and processed to plasma as follows:

3.2.1 Collection of Blood Plasma for Metabolomic Analysis

1. Total blood is collected in Vacutainer® tubes that contain lithium heparin (*see* **Notes 6–8**).
2. Spin the blood samples for 10–15 min at 2000 × *g* at 4 °C in a refrigerated centrifuge to remove the blood cells to the bottom of the tube (*see* **Note 9**).
3. Immediately transfer 500 μL aliquots of the plasma solution into fresh 2 mL labeled cryovials using a pipette with appropriate filter tip. Take care to avoid taking any red blood cells over into the plasma.
4. Samples should be stored at −80 °C until analysis and all transportation should be carried out under dry ice.

3.2.2 Metabolites Extraction from Plasma

The extraction of metabolites includes different methods depending on the analytical platform that is used to perform the metabolites determination (*see* **Note 10**).

GlobalMet-p Polar Metabolites Extraction

1. Allow all specimens, reagents, and internal standard solution to reach ambient laboratory temperature for 30 min.
2. Vortex each sample tube to ensure homogeneity.
3. Pipette 100 μL of each plasma sample to a 1.5 mL Eppendorf tube.
4. Add 25 μL of internal standard solution (prepared in methanol).
5. Add 800 μL of methanol-water (8:1, v,v) and vortex for 1 min.
6. Incubate the tubes at −20 °C for 30 min.
7. Centrifuge the tube at 21,380 × *g* for 10 min to pellet the protein precipitate.
8. Collect 750 μL of the supernatant into a new 1.5 mL microcentrifuge tube, and evaporate to dryness in the SpeedVac concentrator at ambient temperature.
9. Add 25 μL of methoxyamine hydrochloride solution prepared in pyridine to the dried samples to protect the reactive carbonyls.
10. Incubate at 37 °C for 1.5 h.
11. Allow the tube to cool for 5 min.
12. Add 20 μL of MSTFA+ 1% TMCS to obtain the metabolite O-trimethylsilyl derivatives and allow to react for 1 h in rest.
13. Transfer the extract to a 1.5 mL chromatographic vials equipped with 200 μL inserts.

The sample is ready for their analysis by GC-HRMS. Usual injection volume is 1 μL (*see* **Note 11**).

GlobalMet-p Lipidic Metabolites Extraction

The lipidic extraction protocol has some different procedure for samples to be analyzed by NMR or LC-MS.

1. For LC-MS analysis, allow all specimens, reagents, and internal standard solution to reach ambient laboratory temperature for 30 min.
2. Vortex each specimen tube to ensure sample homogeneity.
3. Pipette 100 μL of each plasma sample to a 1.5 mL Eppendorf tube.
4. Add 323 μL of methanol and vortex for 1 min.
5. Add 1077 μL of methyl tert-butyl ether and vortex for 1 min.
6. Add 180 μL of water and vortex for 1 min.
7. Keep the tubes to −20 °C for 30 min.

8. Centrifuge the tube at 21,380×*g* for 10 min to separate organic and aqueous phases (pellet protein precipitate remaining in the liquid-liquid interface).
9. Transfer 10 μL of the supernatant into a 1.5 mL chromatographic vials equipped with 200 μL inserts for UHPLC-HRMS analysis.
10. Add 25 μL of internal standard solution (prepared in methanol).
11. Add 165 μL of methanol.
12. Vortex the vial and the sample is ready for their analysis by UHPLC-HRMS. Usual injection volume is 2 μL.
13. For NMR samples, collect 900 μL of the supernatant from point 8 into a new 1.5 mL Eppendorf tube, and evaporate to dryness in the SpeedVac concentrator at ambient temperature.
14. Reconstitute the samples with 700 μL of 0.01% tetramethylsilane solution (0.067 mM) of deuterated chloroform-methanol (2:1, v,v) with a 4% of deuterated water. Calibrate electronic reference signal (ERETIC signal) against a standard concentration sample. Transfer ERETIC concentration to a blank measurement of solvent in order to quantify all the samples.
15. Vortex for 1 min and transfer to a NMR sample tube.
16. The sample is ready for their analysis by ^{1}H-NMR and could be stored in a refrigerated sample changer until NMR measurement (*see* **Note 12**).

GlobalMet-p Acylcarnitines Extraction

1. Allow all specimens, reagents, and internal standard solution to reach ambient laboratory temperature for 30 min.
2. Vortex each sample tube to ensure homogeneity
3. Pipette 50 μL of each plasma sample to a 1.5 mL Eppendorf tube.
4. Add 30 μL of internal standard solution (prepared in methanol)
5. Add 370 μL of methanol and vortex for 1 min.
6. Shake the tube for 10 min in an orbital shaker at RT.
7. Centrifuge the tube at 21,380×*g* for 10 min to pellet the protein precipitate.
8. Collect 200 μL of the supernatant into a new 1.5 mL microcentrifuge tube, and evaporate to dryness in the SpeedVac concentrator at ambient temperature.
9. Add 100 μL of butanolic hydrogen chloride solution to the dried samples to derivatizate the acylcarnitines to their butyl esters.
10. Incubate at 60 °C for 15 min.

11. Allow the tube to cool for 5 min, and evaporate to dryness in the SpeedVac concentrator at ambient temperature.
12. Reconstitute the samples with 200 μL methanol-water (1:1, v,v).
13. Centrifuge the tube at 21,380 × *g* for 10 min.
14. Transfer the supernatant to a 1.5 mL chromatographic vials equipped with 200 μL inserts.

The sample is ready for their analysis by UHPLC-MS/MS. Usual injection volume is 5 μL.

4 Analytical Methods

4.1 GlobalMet-p Polar Metabolites Analytical Method

4.1.1 Standard Calibration Curve Preparation

1. Allow reagents, standard solution, and internal standard solution to reach ambient laboratory temperature for 30 min.
2. Pipette 100 μL of ultrapure water to a 1.5 mL Eppendorf tube.
3. Prepare the different calibration levels desired by diluting the adequate amount of calibration mix working solution in 1.5 mL Eppendorf tube. Suggested calibration range is from 0 to 50 μg/mL.
4. Add 25 μL of 10 μg/mL internal standard solution.
5. Add methanol-water (1:1, v,v) up to 800 μL and vortex for 1 min.
6. Proceed as from **step 6** of sample preparation procedure.

4.1.2 GC-HRMS Analysis

1. Set up the method with parameters showed in Table 1.
2. Before analysis, tune/check and calibrate the mass spectrometer in positive ionization mode for an optimal method performance using vendor tuning solution.
3. Run a blank (hexane), calibration standard and QC sample (QC should be a pool of different samples from study) to check sensitivity and that retention times meet with the theoretical ones. If not, check GC and MS performance (*see* **step 2**).
4. Set up a sequence as follows: blank, cal. standard, blank, QC, 1–10 samples, cal. standard, blank, QC, 11–20 samples,... blank. Perform a mass calibration after each ten sample cycle.
5. Calibration curves for each compound (*see* Table 2 for corresponding standard and internal standard for each compound) are obtained by running standard calibrators in increasing concentrations.
6. The quantification of polar metabolites in biological samples is done by the interpolation of compound/internal standard response ratio with the corresponding calibration curve.

Table 1
GC-HRMS operating conditions

Parameter	Value
Carrier gas	He
Gas flow	1.2 mL/min
Injection mode	Split 1:50
Injection temperature	250 °C
Oven program	70 °C 0 min, increase at 15 °C/min to 100 °C, increase at 25 °C/min to 320 °C, and held for 3.2 min
Total run time	14 min
Ionization	EI+
Voltage	70 eV
Source temp.	250 °C
Acquisition range	35–700 *m/z*
Acquisition rate	5 spec/seg

Table 2
Exact measured ion mass, retention times, and calibration curve for the metabolites determined by GC-HRMS

Metabolite	Internal standard	Quantifier ion (*m/z*)	RT (min)
Lactic acid	Succinic acid d_4	191.0917	3.24
Glycolic acid	Succinic acid d_4	147.0656	3.34
Alanine	Succinic acid d_4	116.0902	3.53
Glycine	Succinic acid d_4	102.0744	3.63
2-Hydroxybutanoic acid	Succinic acid d_4	205.1074	3.70
3-Hydroxybutanoic acid	Succinic acid d_4	191.0900	3.91
Valine	Norvaline	144.1215	4.29
Leucine	Norvaline	158.1375	4.63
Glycerol	[$^{13}C_3$]glycerol	218.1161	4.64
Isoleucine	Norvaline	158.137	4.77
Proline	Norvaline	142.1059	4.80
Glyceric acid	Succinic acid d_4	292.1356	4.99
Serine	Norvaline	204.1247	5.16
Threonine	Norvaline	219.114	5.32

(continued)

Table 2 (continued)

Metabolite	Internal standard	Quantifier ion (*m/z*)	RT (min)
Methionine	[^{13}C,d_3]L-methionine	176.0936	6.02
Oxoproline	[^{13}C,d_3]L-methionine	156.0840	6.13
Glutamic acid	Norvaline	246.135	6.50
Phenylalanine	Norvaline	218.1036	6.57
Lysine	Norvaline	156.1199	6.91
Glutamine	Norvaline	156.0848	7.00
Citric acid	Succinic acid d_4	273.0991	7.44
Fructose	[$^{13}C_6$]glucose	307.1591	7.72
Glucose	[$^{13}C_6$]glucose	319.1596	7.83
Tyrosine	Norvaline	218.1041	7.94
Palmitic acid	Myristic acid d_{27}	313.2573	8.31
Linoleic acid	Myristic acid d_{27}	337.2500	8.93
Oleic acid	Myristic acid d_{27}	339.2719	8.94
Stearic acid	Myristic acid d_{27}	341.2869	9.03
Tryptophan	Norvaline	202.105	9.04
Sucrose	[$^{13}C_6$]glucose	361.1697	10.50
Alpha-tocopherol	Myristic acid d_{27}	502.4213	12.04
Cholesterol	Myristic acid d_{27}	458.3938	12.19

4.2 GlobalMet-p Lipidic Metabolites Analytical Method

4.2.1 Standard Calibration Curve Preparation for UHPLC-HRMS

1. Allow reagents, standard solution, and internal standard solution to reach ambient laboratory temperature for 30 min.
2. Prepare the different calibration levels desired by diluting the adequate amount of calibration mix working solution in 1.5 mL Eppendorf tube. Suggested range is from 25 to 5000 μg/L for each family lipid standards.
3. Add 25 μL of internal standard solution
4. Add methanol up to reach 200 μL and vortex for 1 min.
5. Transfer to 1.5 mL chromatographic vials equipped with 200 μL inserts for UHPLC-HRMS analysis.

4.2.2 UHPLC-HRMS Analysis

1. Set up the method with parameters showed in Table 3 for UHPLC and in Table 4 for MS.
2. Before analysis, tune/check and calibrate the mass spectrometer in positive ionization mode for an optimal method performance using vendor tuning solution.

Table 3
UHPLC operating conditions

Time (min)	%A	%B	%C	%D	Flow (mL/min)
0.5	40.00	10.00	45.00	5.00	0.6
1.5	37.80	9.50	47.70	5.00	0.6
1.6	29.00	7.50	58.50	5.00	0.6
5.0	26.80	7.00	61.20	5.00	0.6
5.1	13.60	4.00	77.40	5.00	0.6
7.5	11.40	3.50	80.10	5.00	0.6
9.0	11.40	3.50	80.10	5.00	0.6
9.5	0.00	0.00	100.00	0.00	0.6
11.5	0.00	0.00	100.00	0.00	0.6
11.6	40.00	10.00	45.00	5.00	0.6
14.0	40.00	10.00	45.00	5.00	0.6

Table 4
MS operating parameters

Parameter	Value
Ionization	ESI+
Gas flow	11 L/min
Gas temperature	180 °C
Sheath gas flow	12 L/min
Sheath gas temperature	300 °C
Nebulizer	35 psi
Capillary voltage	3500 V
Nozzle voltage	500 V
Acquisition range	300–1000 *m/z*
Acquisition rate	3 spec/seg
Reference mass	922.0098 *m/z*

3. Prepare the chromatographic system by purging the solvents for at least 5 min (5 mL/min) with freshly prepared mobile phase. Be sure that no air pumps are in the system.
4. Turn on column oven and set to 60 °C.
5. Allow to equilibrate column by flushing 35%A, 30%B, 30%C, 5%D for at least 20 min. Monitor the system pressure to be sure that the system is equilibrated.

6. Run a blank (made with methanol), calibration standard and QC sample (QC should be a pool of different samples from study) to check sensitivity and that retention times meet with the theoretical ones (*see* Table 6). If not, adjust retention times or check MS performance again (*see* **step 2**).
7. Set up a sequence as follows: blank, cal. standard, blank, QC, 1–10 samples, cal. standard, blank, QC, 11–20 samples,..., blank. To avoid carryover, methanol/2-propanol (1:1, v,v) should be used as needle clean solvent.
8. Calibration curves for each compound (each lipid is quantified according to their family-specific standard) are obtained by running standard calibrators in increasing concentrations.
9. After the last chromatographic run, flush the column with acetonitrile for 30 min.
10. The quantification of lipids in biological samples is done by the interpolation of compound/internal standard response ratio with the corresponding calibration curve. (*See* Table 5 for information about pseudo-molecular ion of lipids, adduct measured, retention time, and internal standard used.)

Table 5
Pseudo-molecular ion of lipids, adduct measured, retention time, and internal standard used

Metabolite	Internal standard	Quantifier ion (*m/z*)	Adduct	RT (min)
LPC 14:0	LPC 18:1-d7	468.3085	$[M^+H]^+$	1.31
LPC 18:3	LPC 18:1-d7	518.3241	$[M^+H]^+$	1.41
LPC 20:5	LPC 18:1-d7	542.3241	$[M^+H]^+$	1.42
LPC 16:1	LPC 18:1-d7	494.3241	$[M^+H]^+$	1.45
LPC 18:4	LPC 18:1-d7	516.3085	$[M^+H]^+$	1.45
LPC 15:0	LPC 18:1-d7	482.3241	$[M^+H]^+$	1.52
LPC 18:2	LPC 18:1-d7	520.3398	$[M^+H]^+$	1.60
LPC 20:4	LPC 18:1-d7	544.3398	$[M^+H]^+$	1.62
LPC 22:6	LPC 18:1-d7	568.3398	$[M^+H]^+$	1.62
LPC 16:0	LPC 18:1-d7	496.3398	$[M^+H]^+$	1.72
LPC 22:5	LPC 18:1-d7	570.3554	$[M^+H]^+$	1.74
LPC 20:3	LPC 18:1-d7	546.3554	$[M^+H]^+$	1.77
LPC 18:1	LPC 18:1-d7	522.3554	$[M^+H]^+$	1.84
LPC 16:0 e	LPC 18:1-d7	482.3605	$[M^+H]^+$	1.91
LPC 16:1 e	LPC 18:1-d7	480.3449	$[M^+H]^+$	1.91
LPC 17:0	LPC 18:1-d7	510.3554	$[M^+H]^+$	1.94

(continued)

Table 5
(continued)

Metabolite	Internal standard	Quantifier ion (*m/z*)	Adduct	RT (min)
LPC 20:2	LPC 18:1-d7	548.3711	$[M^{+}H]^{+}$	2.00
LPC 18:0	LPC 18:1-d7	524.3711	$[M^{+}H]^{+}$	2.19
LPC 20:1	LPC 18:1-d7	550.3867	$[M^{+}H]^{+}$	2.30
LPC 18:0 e	LPC 18:1-d7	510.3918	$[M^{+}H]^{+}$	2.43
LPC 20:0	LPC 18:1-d7	552.4024	$[M^{+}H]^{+}$	2.69
SM 32:2	SM 36:2-d7	673.5279	$[M^{+}H]^{+}$	3.14
SM 32:1	SM 36:2-d7	675.5436	$[M^{+}H]^{+}$	3.38
SM 34:2	SM 36:2-d7	701.5592	$[M^{+}H]^{+}$	3.51
SM 32:0	SM 36:2-d7	677.5592	$[M^{+}H]^{+}$	3.53
SM 33:1	SM 36:2-d7	689.5592	$[M^{+}H]^{+}$	3.57
SM 35:0	SM 36:2-d7	719.5698	$[M^{+}H]^{+}$	3.58
PC 32:2	PC 33:1-d7	730.5381	$[M^{+}H]^{+}$	3.78
PC 33:2	PC 33:1-d7	730.5381	$[M^{+}H]^{+}$	3.78
SM 34:1	SM 36:2-d7	703.5749	$[M^{+}H]^{+}$	3.80
PC 34:4	PC 33:1-d7	754.5381	$[M^{+}H]^{+}$	3.81
PC 30:0	PC 33:1-d7	706.5381	$[M^{+}H]^{+}$	3.84
PC 34:3	PC 33:1-d7	756.5538	$[M^{+}H]^{+}$	3.93
SM 36:2	SM 36:2-d7	729.5905	$[M^{+}H]^{+}$	3.98
SM 35:1	SM 36:2-d7	717.5905	$[M^{+}H]^{+}$	4.05
PC 36:5	PC 33:1-d7	780.5538	$[M^{+}H]^{+}$	4.06
PC 37:6	PC 33:1-d7	792.5538	$[M^{+}H]^{+}$	4.06
PC 32:1	PC 33:1-d7	732.5538	$[M^{+}H]^{+}$	4.07
PC 31:0	PC 33:1-d7	720.5538	$[M^{+}H]^{+}$	4.07
PC 34:2	PC 33:1-d7	758.5694	$[M^{+}H]^{+}$	4.27
PC 33:1	PC 33:1-d7	746.5694	$[M^{+}H]^{+}$	4.35
PC 35:4	PC 33:1-d7	768.5538	$[M^{+}H]^{+}$	4.35
SM 36:1	SM 36:2-d7	731.6062	$[M^{+}H]^{+}$	4.36
PC 36:4	PC 33:1-d7	782.5694	$[M^{+}H]^{+}$	4.36
PC 38:6	PC 33:1-d7	806.5694	$[M^{+}H]^{+}$	4.36
PC 32:0	PC 33:1-d7	734.5694	$[M^{+}H]^{+}$	4.46
PC 38:5	PC 33:1-d7	808.5851	$[M^{+}H]^{+}$	4.51
PC 36:3	PC 33:1-d7	784.5851	$[M^{+}H]^{+}$	4.54

(continued)

Table 5 (continued)

Metabolite	Internal standard	Quantifier ion (*m/z*)	Adduct	RT (min)
PE 36:4	PE 33:1-d7	740.5225	$[M^+H]^+$	4.55
SM 36:0	SM 36:2-d7	733.6218	$[M^+H]^+$	4.57
SM 38:2	SM 36:2-d7	757.6218	$[M^+H]^+$	4.59
PC 34:3 e	PC 33:1-d7	742.5745	$[M^+H]^+$	4.59
PE 32:0	PE 33:1-d7	692.5225	$[M^+H]^+$	4.60
PC 35:2	PC 33:1-d7	772.5851	$[M^+H]^+$	4.61
PC 34:1	PC 33:1-d7	760.5851	$[M^+H]^+$	4.65
PC 36:5 e	PC 33:1-d7	766.5745	$[M^+H]^+$	4.65
PC 33:0	PC 33:1-d7	748.5851	$[M^+H]^+$	4.66
PC 37:4	PC 33:1-d7	796.5851	$[M^+H]^+$	4.68
PC 36:4 e	PC 33:1-d7	768.5902	$[M^+H]^+$	4.74
PC 38:6 e	PC 33:1-d7	792.5902	$[M^+H]^+$	4.75
PC 32:1 e	PC 33:1-d7	718.5745	$[M^+H]^+$	4.80
PC 36:3 e	PC 33:1-d7	770.6058	$[M^+H]^+$	4.82
PC 38:5 e	PC 33:1-d7	795.6136	$[M^+H]^+$	4.91
PE 36:5 e	PE 33:1-d7	724.5276	$[M^+H]^+$	4.92
PC 36:2	PC 33:1-d7	786.6007	$[M^+H]^+$	4.96
PC 34:2 e	PC 33:1-d7	744.5902	$[M^+H]^+$	5.01
PC 35:1	PC 33:1-d7	774.6007	$[M^+H]^+$	5.04
PC 38:4	PC 33:1-d7	810.6007	$[M^+H]^+$	5.07
PC 40:6	PC 33:1-d7	834.6007	$[M^+H]^+$	5.07
PC 34:1 e	PC 33:1-d7	746.6058	$[M^+H]^+$	5.08
PE 38:6 e	PE 33:1-d7	750.5432	$[M^+H]^+$	5.09
SM 38:1	SM 36:2-d7	759.6375	$[M^+H]^+$	5.12
PC 34:0	PC 33:1-d7	762.6008	$[M^+H]^+$	5.17
PC 40:5	PC 33:1-d7	836.6164	$[M^+H]^+$	5.24
PC 38:3	PC 33:1-d7	812.6164	$[M^+H]^+$	5.30
PC 36:2 e	PC 33:1-d7	772.6215	$[M^+H]^+$	5.31
SM 40:2	SM 36:2-d7	785.6531	$[M^+H]^+$	5.38
PC 36:1	PC 33:1-d7	788.6164	$[M^+H]^+$	5.44
SM 42:3	SM 36:2-d7	811.6688	$[M^+H]^+$	5.46
SM 39:1	SM 36:2-d7	773.6531	$[M^+H]^+$	5.56

(continued)

Table 5
(continued)

Metabolite	Internal standard	Quantifier ion (*m/z*)	Adduct	RT (min)
PC 38:4 e	PC 33:1-d7	796.6215	$[M^{+}H]^{+}$	5.58
PC 40:4	PC 33:1-d7	838.6320	$[M^{+}H]^{+}$	5.68
PC 38:2	PC 33:1-d7	814.6320	$[M^{+}H]^{+}$	5.71
PC 40:5 e	PC 33:1-d7	822.6371	$[M^{+}H]^{+}$	5.72
PE 38:5 e	PE 33:1-d7	752.5589	$[M^{+}H]^{+}$	5.80
PC 38:3 e	PC 33:1-d7	798.6371	$[M^{+}H]^{+}$	5.82
SM 41:2	SM 36:2-d7	799.6688	$[M^{+}H]^{+}$	5.85
SM 40:1	SM 36:2-d7	787.6688	$[M^{+}H]^{+}$	5.97
SM 42:2	SM 36:2-d7	813.6844	$[M^{+}H]^{+}$	6.02
PC 36:0	PC 33:1-d7	790.6320	$[M^{+}H]^{+}$	6.05
SM 40:0	SM 36:2-d7	789.6844	$[M^{+}H]^{+}$	6.16
SM 43:2	SM 36:2-d7	827.7001	$[M^{+}H]^{+}$	6.16
SM 41:1	SM 36:2-d7	801.6844	$[M^{+}H]^{+}$	6.21
PC 40:4 e	PC 33:1-d7	824.6528	$[M^{+}H]^{+}$	6.24
PC 42:5 e	PC 33:1-d7	850.6684	$[M^{+}H]^{+}$	6.26
SM 42:1	SM 36:2-d7	815.7001	$[M^{+}H]^{+}$	6.37
SM 43:1	SM 36:2-d7	829.7157	$[M^{+}H]^{+}$	6.41
PC 42:4 e	PC 33:1-d7	852.6841	$[M^{+}H]^{+}$ and $[M^{+}NH_4]^{+}$	6.51
TG 46:2	TG 48:1-d7	792.7076	$[M^{+}NH_4]^{+}$	8.13
TG 48:3	TG 48:1-d7	818.7232	$[M^{+}NH_4]^{+}$	8.24
TG 52:6	TG 48:1-d7	868.7389	$[M^{+}NH_4]^{+}$	8.28
TG 50:4	TG 48:1-d7	844.7389	$[M^{+}NH_4]^{+}$	8.36
TG 54:7	TG 48:1-d7	894.7545	$[M^{+}NH_4]^{+}$	8.38
TG 46:1	TG 48:1-d7	794.7232	$[M^{+}NH_4]^{+}$	8.39
TG 48:2	TG 48:1-d7	820.7389	$[M^{+}NH_4]^{+}$	8.51
TG 52:5	TG 48:1-d7	870.7545	$[M^{+}NH_4]^{+}$	8.51
TG 54:6	TG 48:1-d7	896.7702	$[M^{+}NH_4]^{+}$	8.63
TG 50:3	TG 48:1-d7	846.7545	$[M^{+}NH_4]^{+}$	8.63
TG 46:0	TG 48:1-d7	796.7389	$[M^{+}NH_4]^{+}$	8.69
TG 52:4	TG 48:1-d7	872.7702	$[M^{+}NH_4]^{+}$	8.79
TG 48:1	TG 48:1-d7	822.7545	$[M^{+}NH_4]^{+}$	8.80
TG 54:5	TG 48:1-d7	898.7858	$[M^{+}NH_4]^{+}$	8.92

(continued)

Table 5 (continued)

Metabolite	Internal standard	Quantifier ion (*m/z*)	Adduct	RT (min)
TG 50:2	TG 48:1-d7	848.7702	$[M^{+}NH_4]^{+}$	8.95
TG 52:3	TG 48:1-d7	874.7858	$[M^{+}NH_4]^{+}$	9.10
TG 51:2	TG 48:1-d7	862.7858	$[M^{+}NH_4]^{+}$	9.16
TG 48:0	TG 48:1-d7	824.7702	$[M^{+}NH_4]^{+}$	9.17
TG 54:4	TG 48:1-d7	900.8015	$[M^{+}NH_4]^{+}$	9.21
TG 50:1	TG 48:1-d7	850.7858	$[M^{+}NH_4]^{+}$	9.31
TG 52:2	TG 48:1-d7	876.8015	$[M^{+}NH_4]^{+}$	9.44
TG 54:3	TG 48:1-d7	902.8171	$[M^{+}NH_4]^{+}$	9.59
TG 50:0	TG 48:1-d7	852.8015	$[M^{+}NH_4]^{+}$	9.76
TG 52:1	TG 48:1-d7	878.8171	$[M^{+}NH_4]^{+}$	9.92
TG 54:2	TG 48:1-d7	904.8328	$[M^{+}NH_4]^{+}$	10.07

4.2.3 ^{1}H-NMR Analysis

1. Reference lipidic extract into 5 mm o.d. NMR tubes is measured and recorded using a simple presaturation sequence (RD-90°-ACQ) to eliminate the residual water moisture of deuterated methanol. Solvent presaturation with irradiation power of 30 Hz is applied during recycling delay (RD = 5 s) and mixing time.
2. Calibrate the 90° pulse length using nutation sequences to ensure accurate quantification. Set the spectral width around 12 kHz (20 ppm), and acquire at least a total of 64 transients to obtain an adequate signal-to-noise (s/n) ratio into 64 k data points for each ^{1}H spectrum. The exponential line broadening is applied before Fourier transformation with a value of 0.3 Hz.
3. Select suitable recycle delay (RD bigger than five times the longer T1 relaxation time) that helps in suitable water suppression. Typical parameters might be an RD of 5 s and a total acquisition time (AQ) of 2.73 s for a total recycling time of 7.7 s.
4. The frequency domain spectrum is manually phased and baseline-corrected using TopSpin software (Bruker) or equivalent.
5. Set up the sequence of samples as follows: QC, 1–10 samples, QC, 11–20 samples…, QC in order to check reproducibility into the dataset.

6. Once the parameter set for lipid samples is optimized, set up the automation method as follows: Adjust the temperature of the probe (e.g., 300 K), insertion and transport of sample, and delay period for temperature equilibration (~5 min) with the values obtained for reference lipidic extract. Lock the instrument to the deuterated methanol signal, tune and match the probe, and adjust shims to optimize spectral line shape with automatic shimming routine using the field map previously measured. Half-height linewidth of ~1.5 Hz should be readily obtainable on lipidic samples extracted from plasma.
7. Process spectra in automatic way with the parameters optimized for the lipidic reference. These instructions assume the use of a Bruker® AVANCE III 600-MHz NMR spectrometer with a refrigerated sample changer; however, they should be easily adapted to other instruments.
8. After preprocessing and visual checking of NMR dataset, specific ^1H regions of lipids and phospholipids based on moieties with specific signals were identified by in the spectra using a comparison with lipids standards into AMIX 3.9 software. Curated identified regions across the spectra were integrated using the same AMIX 3.9 software package and exported to spreadsheet software in order to calculate relative concentrations. In addition, we assigned metabolites by ^1H–^1H homonuclear correlation (COSY) and ^1H–^{13}C heteronuclear (HSQC) 2D NMR experiments and by correlation with pure compounds run in-house. The regions used for quantification are reported in Table 6 [28].

4.3 GlobalMet-p Acylcarnitines Analytical Method

4.3.1 Standard Calibration Curve Preparation for UHPLC-HRMS

1. Allow reagents, standard solution, and internal standard solution to reach ambient laboratory temperature for 30 min.
2. Pipette 50 μL of ultrapure water to a 1.5 mL Eppendorf tube.
3. Add adequate amount of standard solution to create the different calibration levels desired. *See* Table 7 for suggested calibration range.
4. Add 30 μL of internal standard solution
5. Add methanol up to reach 400 μL and vortex for 1 min.
6. Procced as GlobalMet-p acylcarnitines extraction protocol starting from **step 6**.

4.3.2 UHPLC-MS/MS Analysis

1. Set up the method with parameters showed in Table 8 for UHPLC and in Tables 9 and 10 for MS [29, 30].
2. Before analysis, tune/check the mass spectrometer in positive ionization mode for an optimal method performance using vendor tuning solution.

Table 6
^{1}H-NMR spectral regions for quantification of lipids

Lipid molecule	ppm range
TMS reference	0
Total cholesterol 3H (C3–H)	0.70–0.67
Methyl–CH_3 fatty acyls	0.91–0.85
Omega-3 (methyl)	0.98–0.94
Free cholesterol (C19–3H)	1.02–1.00
Esterified cholesterol (C19–3H)	1.05–1.02
Methylene CH_2 of fatty acyls	1.37–1.22
bCH_2 of fatty acyls	1.66–1.58
ARA+EPA bCH_2	1.72–1.67
ARA+EPA gCH_2	2.15–2.10
aCH_2 of fatty acyls	2.38–2.26
DHA	2.43–2.39
Linoleic	2.80–2.76
Polyunsaturated fatty acids (PUFA)	2.88–2.80
Sphingomyelin-$(CH_3)_3$	3.20–3.18
Lysophosphatidylcholine-$(CH_3)_3$	3.21–3.19
Phosphatidylcholine-$(CH_3)_3$	3.24–3.21
Phosphatidylcholine-CH_2–N–$(CH_3)_3$	3.62–3.59
Total phospholipids (glycerol C3–2H)	4.00–3.96
Triglycerides (glycerol C1,3-H)	4.20–4.13
Total phospholipids (glycerol C2–H)	5.26–5.21
Triglycerides (glycerol C2–H)	5.31–5.26
Monounsaturated fatty acids (MUFA)	5.44–5.31
Sphingomyelin-CH=CH–	5.74–5.67
Electronic reference (ERETIC)	11.05–10.95

3. Prepare the chromatographic system by purging the solvents for at least 5 min (5 mL/min) with freshly prepared mobile phase. Be sure that no air pumps are in the system.
4. Turn on column oven and set to 35 °C.
5. Allow to equilibrate column by flushing 50% B for at least 20 min. Monitor the system pressure to be sure that system is equilibrated.

Table 7
Acylcarnitine calibration range

Acylcarnitine	Concentration range (nmol/mL)
C0	0–152
C2	0–19
C3	0–0.76
C4	0–0.76
C5	0–0.23
C8	0–0.76
C14	0–0.23
C16	0–0.15

Table 8
UHPLC operating conditions

Time (min)	%B	Flow (mL/min)
0.00	1	0.4
0.10	17	0.4
0.28	24	0.4
0.35	26	0.4
0.80	29	0.4
1.71	31	0.4
2.96	34	0.4
4.50	35	0.4
5.44	37	0.4
8.01	40	0.4
11.30	70	0.4
12.50	80	0.4
13.50	90	0.4
14.50	95	0.4
15.50	95	0.4
16.50	1	0.4

Table 9
MS operating parameters

Parameter	Value
Ionization	ESI+
Gas flow	17 L/min
Gas temperature	180 °C
Sheath gas flow	10 L/min
Sheath gas temperature	400 °C
Nebulizer	50 psi
Capillary voltage	1500 V
Nozzle voltage	0 V

6. Run a blank (made with methanol-water (1:1, v,v)), calibration standard and QC sample (QC should be a pool of different samples from study) to check sensitivity and that retention times meet with the theoretical ones. If not, adjust retention times or check MS performance again (*see* **step 2**).
7. Set up a sequence as follows: blank, cal. standard, blank, QC, 1–10 samples, cal. standard, blank, QC, 11–20 samples,..., blank. To avoid carryover, methanol-water (1:1, v,v) should be used as needle clean solvent.
8. Calibration curves for each compound are obtained by running standard calibrators in increasing concentrations.
9. After the last chromatographic run, flush the column with acetonitrile for 30 min.
10. The quantification of acylcarnitines in biological samples is done by the interpolation of compound/internal standard response ratio with the corresponding calibration curve.

4.4 Statistical Data Analysis

4.4.1 Multivariate Analysis

Multivariate analysis based on a combination of hierarchical clustering analysis (HCA), principal components analysis (PCA), and partial least squares for discriminant analysis (PLS-DA) can be performed to evaluate the influence of diet and physical exercise on the metabolome, to identify those biologically relevant metabolites for further analyses. Simply, the essence of PCA and PLS is to differentiate between classes in highly complex datasets, despite within class variability. A class can refer to any biologically relevant classification, such as humans treated with a specific diet or exposed to particular physical training.

1. The HCA is conducted on both metabolites and conditions. The distance metric selected for the analysis is euclidean, and the linkage rule is centroid.

Table 10
Acylcarnitine MRM transitions, retention times, collision energy, and calibration curve used

Carnitine	Calibration curve	Transition	CE (V)	Retention time (min)
C0	C0	218.1 → 85.0	30	4.20
C2:0	C2	260.1 → 85.0	30	4.56
C3:0	C3	274.1 → 85.0	30	5.03
C4:1/2-M-C3:1	C3	286.1 → 85.0	30	5.20
C4:0	C4	288.2 → 85.0	30	5.65
2-M-C4:1/3-M-C4:1	C5	300.2 → 85.0	30	5.97
2-M-C4:0/3-M-C4:0/C5:0	C5	302.1 → 85.0	30	6.59
C4-OH a and b/2-M-C3-OH	C5	304.1 → 85.0	30	4.52
C6:0	C5	316.4 → 85.0	30	8.14
C5-OH	C5	318.1 → 85.0	30	4.74
C7:0	C5	330.2 → 85.0	30	10.45
C6-OH a and b	C5	332.2 → 85.0	30	5.50
C8:1	C8	342.0 → 85.0	30	9.50
C8:0	C8	344.1 → 85.0	30	11.36
C9:0	C8	358.2 → 85.0	30	12.02
C3-DC	C8	360.2 → 85.0	30	7.58
C10:2	C8	368.1 → 85.0	30	11.80
C10:1	C8	370.2 → 85.0	30	12.48
C10:0/C4:1-DC	C8	372.2 → 85.0	30	12.70
C4-DC/C3-DC-M	C8	374.2 → 85.0	30	8.20
C11:0/C5:1-DC/iso-C11:0	C8	386.3 → 85.0	30	13.24
C5-DC	C8	388.3 → 85.0	30	8.76
C12:1	C8	398.1 → 85.0	30	13.33
C12:0	C8	400.3 → 85.0	30	13.45
C6-DC	C8	402.3 → 85.0	30	10.16
C5-M-DC	C8	402.3 → 85.0	30	9.88
C13:0	C14	414.3 → 85.0	30	14.08
C7-DC	C14	416.3 → 85.0	30	10.93
C14:2	C14	424.3 → 85.0	30	13.34
C14:1	C14	426.1 → 85.0	30	14.11
C14:0	C14	428.3 → 85.0	30	14.05

(continued)

2. PCA represents an unsupervised linear model, aiming an effective dimension reduction and exploratory visualization, to explain the variance within a dataset by a smaller number of mutually decorrelated principal components that in metabolomic studies are vectors of metabolite contribution.
3. Finally, a supervised linear model such us PLS-DA can be performed to unravel inherent metabolite profiles strongly associated with the predefined case-control design by a linear regression model.

4.4.2 Univariate Statistical Analysis

After this previous analysis, metabolites profiles can be studied more thoroughly. The differences among groups are assessed using one- and two-way ANOVA.

1. First, two-way ANOVA is carried out to evaluate the effects of diet, exercise, and their interactions.
2. When one or both main effects are statistically significant, one-way ANOVA is used to determine the differences between the means.
3. When only the interaction between diet and exercise is statistically significant according to the two-way ANOVA model, one-way ANOVA is carried out to compute pairwise comparisons between diet groups (i.e., the effect of exercise within diet groups).
4. The assumption of normality can be determined using the Shapiro-Wilk test, and the homoscedasticity between groups can be determined using Levene's test.
5. Tukey post hoc contrast is used when variances between groups are similar, and Dunnett's T3 test is used if this assumption is not fulfilled.
6. Student's t-test is used for single statistical comparisons. A two-tailed value of $p < 0.05$ is considered statistically significant for all tests.
7. To handle false discovery rates from multiple comparisons, the cutoff point for significance is calculated according to the Benjamini-Hochberg correction [31] at a level of 5%.

4.5 Untargeted Data Processing and Metabolite Identification (Optional Step)

From MS and NMR raw data obtained in scan mode or full NMR spectral window, respectively, an untargeted data analysis could be, optionally, performed. All the software programs used to process the data can be from a commercial vendor or open platforms [32].

Table 10 (continued)

Carnitine	Calibration curve	Transition	CE (V)	Retention time (min)
C15:0/iso-C15:0	C14	442.3 → 85.0	30	14.80
C16:2	C16	452.3 → 85.0	30	13.90
C16:1	C16	454.3 → 85.0	30	14.30
C16:0	C16	456.5 → 85.0	30	14.85
C16:0-OH a and b	C16	472.3 → 85.0	30	13.70
C18:2	C16	480.3 → 85.0	30	14.40
C18:1	C16	482.3 → 85.0	30	14.90
C18:0	C16	484.4 → 85.0	30	15.10

5 Notes

1. To dry samples, a N2 flow-based systems can be used instead of a SpeedVac.
2. All standards can be dissolved in water except palmitic acid, oleic acid, alpha-tocopherol, and cholesterol, which were dissolved in methanol. Otherwise, phenylalanine and tyrosine can be dissolved in water with 10% NH4OH commercial solution at 25–30%.
3. In order to maintain the integrity of the solution, we recommend storing the sealed vials in a second sealed container. Discard this solution after 1 month, according to supplier recommendations
4. If urine, feces, or intestinal contents samples are used, sodium azide should be added to eliminate microbial activity.
5. For cellular samples, after some PBS washes, a rapid quenching or inactivation of enzymatic activity are required, to keeping cells unaltered.
6. Gloves must be worn at all times when handling specimens. This includes during removal of the rubber stopper from the blood tubes, centrifugation, pipetting, disposal of contaminated tubes, and cleanup of any spills. Tubes, needles, and pipettes must be properly disposed of in biohazard containers, in accordance with institutional requirements.
7. All instruments and material (storage, shipping, and centrifuge) must be labeled as biohazard.
8. It is important to prevent hemolysis in these samples.

9. When removing the cryovial tube cap, care should be taken to avoid disturbing the pellet or causing flopping of the blood.
10. An automated liquid handling platform can be used to extract plasma samples in 96-well format plates maintaining the same steps and volumes.
11. Upon derivatized, samples from polar extractions should be analyzed within the next 24 h.
12. Upon resuspended in chloroform-methanol, lipidic extracts should be analyzed within the next 24 h.

References

1. Terra X, Pallarés V, Ardèvol A et al (2011) Modulatory effect of grape-seed procyanidins on local and systemic inflammation in diet-induced obesity rats. J Nutr Biochem 22:380–387
2. Sut S, Baldan V, Faggian M et al (2016) Nutraceuticals, a new challenge for medicinal chemistry. Curr Med Chem 23(28):3198–3223
3. Braconi D, Bernardini G, Millucci L et al (2018) Foodomics for human health: current status and perspectives. Expert Rev Proteomics 15:153–164
4. Catalán Ú, Barrubés L, Valls RM et al (2017) In vitro metabolomic approaches to investigating the potential biological effects of phenolic compounds: an update. Genomics Proteomics Bioinformatics 15:236–245
5. Khakimov B, Poulsen SK, Savorani F et al (2016) New nordic diet versus average danish diet: a randomized controlled trial revealed healthy long-term effects of the new nordic diet by GC–MS blood plasma metabolomics. J Proteome Res 15:1939–1954
6. Chorell E, Moritz T, Branth S et al (2009) Predictive metabolomics evaluation of nutrition-modulated metabolic stress responses in human blood serum during the early recovery phase of strenuous physical exercise. J Proteome Res 8:2966–2977
7. Gill JMR, Hardman AE (2003) Exercise and postprandial lipid metabolism: an update on potential mechanisms and interactions with high-carbohydrate diets (review). J Nutr Biochem 14:122–132
8. Thompson PD, Franklin BA, Balady GJ et al (2007) Exercise and acute cardiovascular events: placing the risks into perspective: a scientific statement from the American Heart Association Council on Nutrition, Physical Activity, and Metabolism and the Council on Clinical Cardiology. Circulation 115:2358–2368
9. Boccatonda A, Tripaldi R, Davì G et al (2016) Oxidative stress modulation through habitual physical activity. Curr Pharm Des 22:3648–3680
10. Yan BAJ, Wang G et al (2009) Metabolomic investigation into variation of endogenous metabolites in professional athletes subject to strength-endurance training. J Appl Physiol 106:531–538
11. Suárez-García S, Del Bas JM, Caimari A et al (2017) Impact of a cafeteria diet and daily physical training on the rat serum metabolome. PLoS One 12:1–19
12. Gami AS, Witt BJ, Howard DE et al (2007) Metabolic syndrome and risk of incident cardiovascular events and death. J Am Coll Cardiol 49:403–414
13. Kaur J (2014) A comprehensive review on metabolic syndrome. Cardiol Res Pract 2014:1–21
14. Alberti KGMM, Eckel RH, Grundy SM et al (2009) Harmonizing the metabolic syndrome: a joint interim statement of the International Diabetes Federation Task Force on Epidemiology and Prevention; National Heart, Lung, and Blood Institute; American Heart Association; World Heart Federation; International. Circulation 120:1640–1645
15. Wang Y, Liu D, Li Y et al (2016) Metabolomic analysis of serum from obese adults with hyperlipemia by UHPLC-Q-TOF MS/MS. Biomed Chromatogr 30:48–54
16. Mihalik SJ, Michaliszyn SF, de las Heras J et al (2012) Metabolomic profiling of fatty acid and amino acid metabolism in youth with obesity and type 2 diabetes: evidence for enhanced mitochondrial oxidation. Diabetes Care 35:605–611
17. Ryan EP, Heuberger AL, Broeckling CD et al (2013) Advances in nutritional metabolomics. Curr Metabolomics 1:109–120

18. Walsh MC, Brennan L, Malthouse JPG et al (2006) Effect of acute dietary standardization on the urinary, plasma, and salivary metabolomic profiles of healthy humans. Am J Clin Nutr 84:531–539
19. Winnike JH, Busby MG, Watkins PB et al (2009) Effects of a prolonged standardized diet on normalizing the human metabolome. Am J Clin Nutr 90:1496–1501
20. Xiao Q, Moore SC, Keadle SK et al (2016) Objectively measured physical activity and plasma metabolomics in the Shanghai Physical Activity Study. Int J Epidemiol 45:1433–1444
21. Xu J, Yang S, Cai S et al (2010) Identification of biochemical changes in lactovegetarian urine using ^{1}H NMR spectroscopy and pattern recognition. Anal Bioanal Chem 396:1451–1463
22. Nyamundanda G, Brennan L, Gormley IC (2010) Probabilistic principal component analysis for metabolomic data. BMC Bioinformatics 11:571
23. Garcia AL, Mohan R, Koebnick C, Bub A, Heuer T, Strassner C, Groeneveld MJ, Katz N, Elmadfa I, Leitzmann C, Hoffmann I (2010) Plasma β-carotene is not a suitable biomarker of fruit and vegetable intake in German subjects with a long-term high consumption of fruits and vegetables - Karger Publishers. Ann Nutr Metab 56:23–30
24. Lehmann R, Zhao X, Weigert C et al (2010) Medium chain acylcarnitines dominate the metabolite pattern in humans under moderate intensity exercise and support lipid oxidation. PLoS One 5:e11519
25. Zhao X, Peter A, Fritsche J et al (2009) Changes of the plasma metabolome during an oral glucose tolerance test: is there more than glucose to look at? Am J Physiol Metab 296:E384–E393
26. Jacobsen R, Lorenzen JK, Toubro S et al (2005) Effect of short-term high dietary calcium intake on 24-h energy expenditure, fat oxidation and fecal fat excretion. Int J Obes 29:292–301
27. Zheng H, Lorenzen J, Astrup A et al (2016) Metabolic effects of a 24-week energy-restricted intervention combined with low or high dairy intake in overweight women: an NMR-based metabolomics investigation. Nutrients 8:108
28. Vinaixa M, Ángel Rodríguez M, Rull A et al (2010) Metabolomic assessment of the effect of dietary cholesterol in the progressive development of fatty liver disease. J Proteome Res 9:2527–2538
29. Giesbertz P, Ecker J, Haag A et al (2015) An LC-MS/MS method to quantify acylcarnitine species including isomeric and odd-numbered forms in plasma and tissues. J Lipid Res 56:2029–2039
30. Millington DS, Stevens RD (2011) Acylcarnitines: analysis in plasma and whole blood using tandem mass spectrometry. Methods Mol Biol 708:55–72
31. Benjamini Y, Hochberg Y (1995) Controlling the false discovery rate: a practical and powerful approach to multiple testing. https://www.jstor.org/stable/2346101
32. Xia J, Wishart DS (2016) Using MetaboAnalyst 3.0 for comprehensive metabolomics data analysis. Curr Protoc Bioinformatics 55:14.10.1–14.10.91

Chapter 26

Blood Biomarkers in Sports Medicine and Performance and the Future of Metabolomics

Iñigo San-Millán

Abstract

The field of sports medicine and performance has undergone an important transformation in the past years where the scientific approach is becoming increasingly more important for teams and athletes. Physical and physiological fitness, nutrition, fatigue and recovery, as well as injury prevention are key elements of the scientific monitoring of athletes nowadays. Many different methods are used nowadays as part of the scientific monitoring and testing of the competitive athlete. Among them, physiological and metabolic testing, biomechanical and movement assessments, GPS-based tracking systems, heart rate monitors, power meters, and training software are an integrative part of the scientific monitor program of many teams and athletes.

Blood biomarkers through traditional blood analysis have been used for over three decades (mainly in Europe) to monitor athletic performance. In the same manner that different cells in the body respond to the stress of an infection or a disease, cells in athletes respond to the stress of competition and training. Nowadays, the area of blood biomarkers is an emerging field in the US offering important level of possibilities to monitor athletes. The field of metabolomics can offer a significantly higher level of blood biomarkers for sports medicine and performance monitoring.

Key words Sports medicine, Metabolism, Performance, Biomarkers

1 Introduction

The modern high-throughput bioanalytical platforms, in combination with computational capabilities for data analysis and interpretation, make it possible to quantify hundreds of metabolites in a single analysis.

The two main areas of utilization in blood biomarkers are for performance and fitness as well as to monitor for overtraining. The intention of this chapter is to show the current methodologies used to monitor performance and training of competitive athletes where the use of metabolomics holds great future. The rapid emergence of metabolomics has enabled system-wide measurements of metabolites in various organisms. However, advances in the mechanistic understanding of metabolic networks remain limited,

Angelo D'Alessandro (ed.), *High-Throughput Metabolomics: Methods and Protocols*, Methods in Molecular Biology, vol. 1978, https://doi.org/10.1007/978-1-4939-9236-2_26,

as most metabolomics studies cannot routinely provide accurate metabolite identification, absolute quantification and flux measurement. Stable isotope labeling offers opportunities to overcome these limitations. Here we describe current approaches in the area of sports medicine and performance and provide examples of the significant impact that these studies have had on our understanding of cellular metabolism. Furthermore, we discuss recently developed software solutions for the analysis of stable isotope-labeled metabolomics data and propose the bioinformatics solutions that will pave the way for the broader application and optimal interpretation of system-scale labeling studies in metabolomics.

1.1 The Use of Blood Biomarkers to Assess Metabolic Fitness and Performance

There are not currently many biomarkers to assess metabolic fitness and performance. Because of its easy analysis and accessibility and because it is one of the most important signatures of cellular metabolism, lactate has been a preferred biomarker to assess physical and metabolic fitness.

Lactate testing is probably the best method that we have to predict performance in endurance events [1, 2], it is probably the parameter that discriminates the most between different levels of athletic performance [3] (Fig. 1), and it can also be a very good parameter to prescribe exercise training zones and training programs for athletes [1, 4]. Lactate is the mandatory by-product of glucose oxidation (glycolysis). During glycolysis, lactic acid is formed, and it rapidly dissociates to lactate and a hydrogen ion (H+). For almost 50 years, the notion of lactate being a "waste product" as a result of anaerobic metabolism prevailed and was embedded in textbooks. However, the work done for the past 40 years by Dr. George Brooks from the University of California Berkeley has showed that lactate is not a waste product but a key regulator of different metabolic processes with hormone-like properties ("lacthormone") [5–7] and even a master regulator in cancer as Brooks and I have proposed [8]. During high-intensity exercise, ATP generation demands increase, and glucose becomes the predominant muscle substrate for ATP generation. As there is an increased in glycolysis leading to pyruvate formation, lactate production increases as well. When the rate of pyruvate formation due to excessive mass action glycolytic flux exceeds the rate of oxidation by the enzyme pyruvate dehydrogenase (PDH), lactate is then formed [9–11]. Pyruvate then will be reduced to lactate by lactate dehydrogenase A (LDHA) resulting in increased cytosolic lactate concentration. Lactate is oxidized in mitochondria via the mitochondrial lactate oxidation complex (mLOC) comprised of a mitochondrial LDH (mLDH), a monocarboxylate transporter (MCT1), its chaperone (CD147/Basigin), and cytochrome oxidase (COx) [12]. When lactate production exceeds mitochondrial clearance, lactate accumulates, and the excessive accumulation of H^+ ions derived from lactate results in an important decrease in

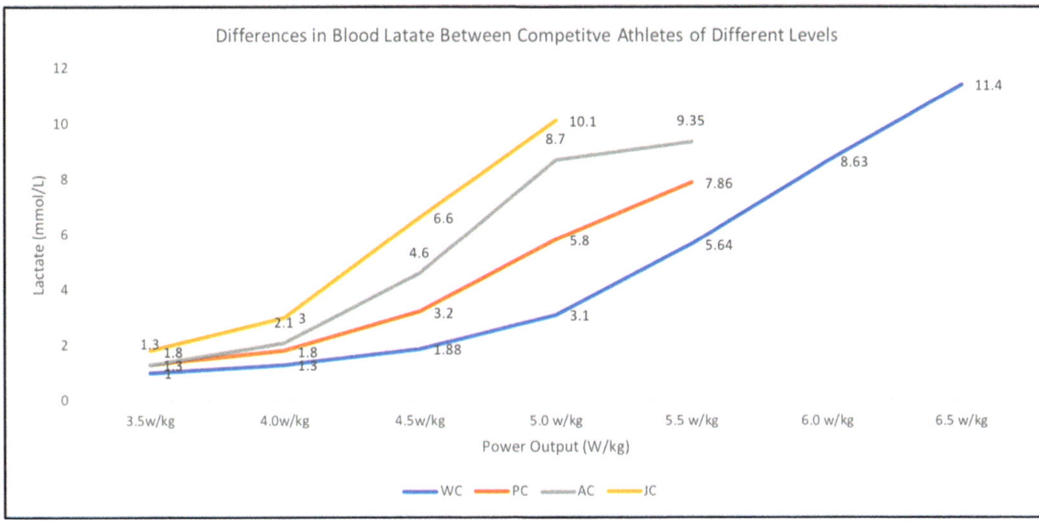

Fig. 1 Blood lactate levels during an incremental test in athletes of different categories. *JC* elite junior cyclists, *AC* elite amateur cyclists/domestic professionals, *PC* international level professional cyclists, *WPC* world-class professional cyclists and Grand Tours winners of podium finishers (Modified from San Millán et al. [3])

muscle pH to 6.2–7.0 [13–15]. The acidosis and H^+ ions associated with lactate interfere with muscle contraction and can elicit an important decrease in peak contractile force and a decrease in muscle shortening velocity [16]. Furthermore, H^+ ions compete with calcium (Ca^{++}) for the Troponin C binding site interfering with muscle contraction as well [13, 16–18]. Moreover, it has been shown that lactate-derived H^+ ions inhibit Ca^{++} release and uptake from sarcoplasmic reticulum [16, 17, 19, 20] further interfering with muscle contraction and athletic performance. Therefore, lactate clearance capacity during exercise is a key parameter in athletic performance.

1.2 Lactate as a Surrogate for Substrate Utilization During Exercise

The ability to oxidize substrates during exercise is very important for athletic performance. In endurance sports it is important to oxidize fatty acids in order to preserve skeletal and liver glycogen for the last part of the race when glycolytic efforts will determine the winner of the competition. Historically, fatty acid and carbohydrate oxidation (FATox and CHOox) have been calculated through stable-labeled isotope studies during exercise which is invasive and expensive to do. Recently, the use of stoichiometric analysis of FATox and CHOox through indirect calorimetry has allowed us to indirectly assess substrate oxidation. The measurement of blood lactate levels could be a good surrogate for substrate utilization and provide important information on muscle metabolism during exercise. We have recently shown that lactate is highly correlated with both fat and carbohydrate metabolism during exercise [21]

(Fig. 2), and although not possible through lactate measurement to quantify the amount of fat and CHO being oxidized, at least it is possible to estimate whether there is muscle oxidation of fat and CHO or just CHO. Further, lactate inhibits lipolysis by inhibiting GPR81 receptor in adipose tissue [22]. Moreover, we have recently shown that lactate downregulates cardiolipin activity which regulates both fatty acid transporters CPT-1 and CPT-2 across mitochondria and therefore can interfere with mitochondrial fatty acid transport and oxidation (San-Millan et al., unpublished data). Therefore, a poor lactate clearance capacity could also interfere with fatty acid oxidation, and the measurement of blood lactate could be an indirect manner to assess fat and carbohydrate oxidation during exercise. The use of novel metabolomics methodologies could substantially increase the amount of knowledge of muscle metabolism and substrate utilization during exercise with important applications in the area of sports performance.

1.3 Lactate Measurement as an Indirect Parameter to Assess Mitochondrial Function

During exercise, glycolytic activity is directly correlated to exercise intensity and ATP demand [21]. The mass action effect of rapid glycolysis increases cytosolic pyruvate levels which need to be oxidized in the mitochondria for continuation into Krebs cycle for ATP synthesis. The oxidation of pyruvate to Acetyl-CoA is mediated by the enzyme pyruvate dehydrogenase (PDH) which has a low Michaelis constant (K_m) and saturates rapidly as a result of the increased glycolytic flux leading to an increase in cytosolic pyruvate which in turn is reduced to lactate by lactate dehydrogenase A (LDHA). As discussed earlier, lactate is oxidized in the mitochondrial oxidation complex relieving the accumulation of cytosolic lactate. When there is a mitochondrial dysfunction or, simply, mitochondria are not very efficient to clear lactate, it accumulates in the cytosol and then is exported to the blood for further oxidation in most tissues of the body [5, 23, 24]. Therefore, an increased in blood lactate is a surrogate of mitochondrial function. Elite athletes have the highest mitochondrial function of all humans [25–28], and as seen in Fig. 1, their blood lactate levels are significantly lower compared to different levels of competitive cyclists and different populations (Fig. 3). The use of lactate as an indirect parameter of mitochondrial function is not new as in 1923, Nobel Laureate Otto Warburg discovered that cancer cells were characterized by an accelerated glycolysis with an exacerbated production of lactate (Warburg effect) which led him to believe that cancer was an injury to mitochondrial respiration [29].

The use of metabolomics holds important promise to discover novel biomarkers of mitochondrial function not just in sports performance but also in health and disease as it has been shown that mitochondrial dysfunction is a major player in the pathogenesis of different diseases like type 2 diabetes, metabolic syndrome, cardiovascular disease, or even Alzheimer's disease.

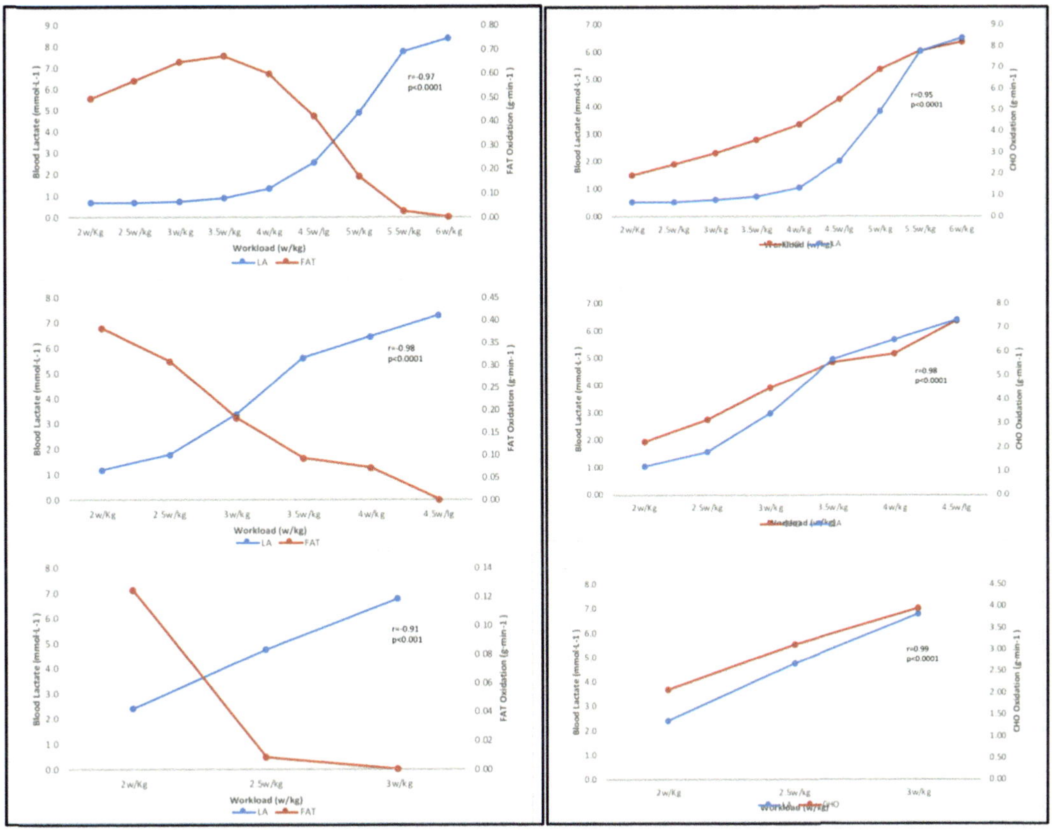

Fig. 2 Relationships between lactate and fat oxidation and carbohydrate oxidation during exercise between populations of different metabolic phenotypes. *A* international-level professional cyclists, *B* recreational athletes, *C* patients with metabolic syndrome (Modified from San-Millan and Brooks [21])

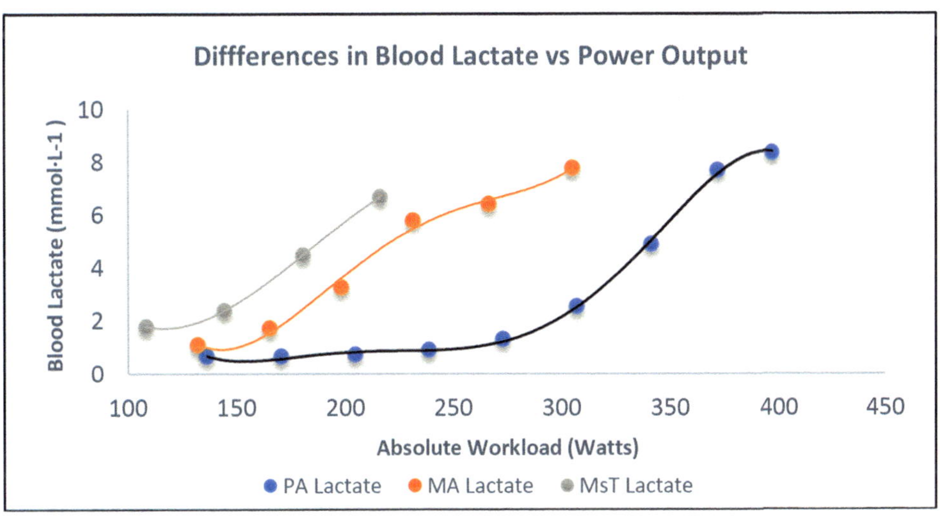

Fig. 3 Comparison of blood lactate levels between different populations. *PA* international-level professional cyclists, *MA* moderately active individuals, *MsT* patients with metabolic syndrome (Modified from San-Millan and Brooks [21])

2 The Use of Blood Biomarkers to Assess Overtraining in Sports

Overtraining has become a prevalent term and condition among athletes of all categories, affecting both international-level athletes and recreational ones. Among recreational athletes, a big number of them are not aware of the condition and can be suffering from overtraining for several months or even years. Elite athletes, however, have more experience as well as many references throughout the season in the form of high-caliber competitions where they can realize that their performance has deteriorated. Furthermore, elite athletes are usually part of a multidisciplinary team of experts (coaches, sports scientists, therapists, physicians, etc.) where they can detect overtraining and correct it. Monitoring for overtraining through blood analysis is a fundamental piece of a training program of top-level athletes and has been done (mainly in Europe) for over three decades.

2.1 Causes of Overtraining

There are different causes of overtraining, but the main causes are excessive training, poor recovery and assimilation, poor nutrition, and psychological stress.

2.1.1 Excessive Training

Many athletes train more than what they can assimilate. In many other instances, the exercise intensities and workload are not properly individualized and prescribed which can result in overtraining. The only manner to individualize training to an athlete is through different testing approaches, from performing both physiological and metabolic testing to strength and conditioning testing specific for each athlete and sport. Unfortunately, many athletes receive general "one-size-fits-all" training programs. This means that for some athletes, the training dose may not be enough to improve, while for others the dose may be appropriate, but for a good number of athletes, such a training dose may be too much leading to overtraining.

2.1.2 Poor Recovery

Although many times unrecognized, recovery is part of training. There are two main principles of training: overload and supercompensation. The overload principle pertains to the training itself which if individualized will elicit the right physiological and metabolic stress needed to improve performance. Supercompensation refers to the piece of a training program where an athlete has assimilated the prescribed workload and has made progress in performance. However, recovery is the key element to assimilate any training workload and elicit the pursued supercompensation. Many times, athletes do not take the enough time to recover as in general, there is a misconception among many athletes and coaches across all sports that a recovery day is a "lost" day. This misconception constantly leads to overtraining.

If after a few days of hard training or a block of a few weeks of hard training, an athlete does not take the proper recovery to assimilate the workload prescribed; he or she will fall into an overtraining situation which can be chronic and last months or even years.

2.2 Inadequate Nutrition

Nutrition is a key part of a training regime of any athlete. The right amount of total calories as well as macro- and micronutrients is essential to provide the body with the necessary elements to achieve recovery, increase muscle size, improve metabolic function, and overall contribute to the assimilation process of training. For decades, different diets and ideas have inundated the field of sports nutrition. Many of these ideas and philosophies have never been proven scientifically, and yet, many athletes and coaches keep falling for them. Carbohydrates have been a target of "novel" dietary approaches for several decades where they are "demonized" and there have been thousands of books, general media articles, and blogs discrediting the importance of carbohydrate for athletic performance and even blaming them for a good number of different diseases.

However, for at least the past 60 years, the science is solid about the importance of carbohydrates in athletic performance and recovery. Multiple studies show that fatigue and decrease in performance are associated with low carbohydrate diets due to decreased glycogen stores [30–36]. When glycogen storages are low or there is a glycogen depletion, there is an increase in proteolysis for energy purposes, especially through gluconeogenesis [37, 38] which may lead to increased catabolism and decreased anabolism as well as overtraining [39–41]. As discussed later, an increase in protein breakdown for gluconeogenesis purposes can lead to skeletal muscle microtears which in turn, as it has been shown, lead to a decreased capacity to store glycogen for several days until muscle is repaired [42, 43] posing extra metabolic and energetic challenges for performance.

2.3 Psychological Stress

Psychological stress and overtraining have been well established for years [44]. Both competitive and recreational athletes may experience psychological stress from different etiologies. Competitive athletes typically develop a psychological stress derived from the high demands of the competition, results, or necessity to renew contracts. Recreational athletes tend to develop a different kind of stress derived from their busy lives and schedules which are mainly around work and the difficulty to balance work, family, training, and recovery properly. Psychological stress may elicit important response of catabolic hormones like cortisol [45] which may increase catabolic activity and impede recovery leading to overtraining.

2.4 Monitoring Overtraining Through Blood Biomarkers

There are different biomarkers that have been used for years to monitor for overtraining in athletes. Different hematological, biochemical, hormonal, and serological biomarkers can be found in the blood. The intention of this chapter is to summarize the main biomarkers used to monitor for overtraining, specifically those ones related to oxygen transport, muscle damage, and physiological stress.

3 Biomarkers for Oxygen Transport: Hemoglobin

Hemoglobin (Hgb) is a tetramer protein consisting of two dimers that bind oxygen (O_2). Hemoglobin resides inside red blood cells and is the protein in charge of transporting oxygen in the blood from the lungs to the rest of tissues in the body. Each Hgb molecule is made up of four heme groups containing iron to which O_2 ultimately binds. We could consider red blood cells as the taxi for O_2, and Hgb is the seat where O_2 sits. Each gram of Hgb transports 1.34 ml of O_2. Therefore, a normal male with a Hgb of 15 g/dl in the blood can transport 20.1 g/dl of O_2 which is called the O_2-carrying capacity. The measurement of Hgb has been used for many years in sports to monitor for overtraining and performance [46–50]. A decrease in the Hgb content in the blood would lead to a decrease in the O_2-carrying capacity and therefore athletic performance. As an example, if a normal athlete with 15 g/dl of Hgb and therefore a 20.1 g/dl of O_2-carrying capacity decreases the Hgb content to 14 g/dl, the O_2 capacity would decrease to 18.76 g/dl of O_2, which would represent close to a 7% decrease (Fig. 4). Although not a significant decrease in O_2-carrying capacity, it could have a decrease in athletic performance. However, it is not rare to observe athletes with much larger decreases in Hgb. An athlete whose baseline Hgb is 15 g/dl and decreases it to 12 g/dl would represent a subclinical anemia with significant decrease in O_2-carrying capacity of ~20% and therefore an important decrease in athletic performance.

There are several reasons why athletes decrease their Hgb. The process of red blood cell formation is called erythropoiesis. An average person produces around 2.4 million red blood cells per second [51] or around 200,000,000,000 per day in order to replace the amount of red blood cells destroyed per day. Therefore, erythropoiesis is not just an amazing anabolic process but is also an incredibly well-balanced process. The main reason for a decrease in O_2-carrying capacity is a decreased production of red blood cells and Hgb due to different reasons. When an athlete is tired or fatigued, many anabolic processes in the body are decreased, and erythropoiesis can be affected, and therefore, a decrease in the production of Hgb- and O_2-carrying capacity occurs. Another

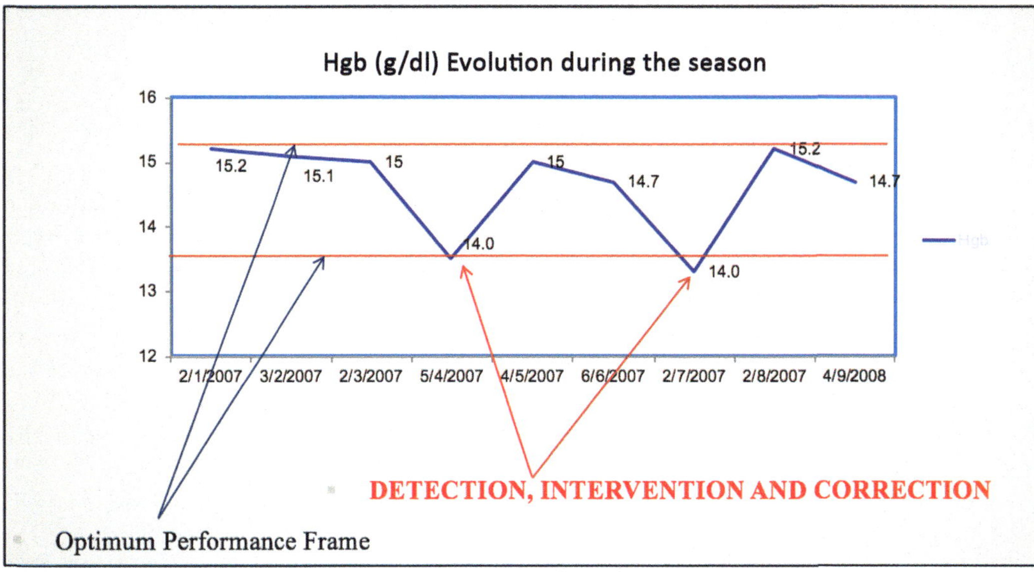

Fig. 4 Evolution of hemoglobin (g/dl) throughout the season in a professional athlete. When a drop of 1 g/dl was found, training was modified in order to allow increased recovery to normalize values (Image from San-Millan's lab)

common cause for the decrease in Hgb may be due to iron deficiency due to poor nutrition and poor intestinal absorption, or in case of women, it can also be worsen by menstruation. Iron is required for hemoglobin synthesis and is also where O_2 binds and therefore iron deficiency state will reduce the content of hemoglobin and O_2-carrying capacity. Further, intestinal iron absorption is regulated by the hepatic protein hepcidin [52, 53]. When there is inflammation as a result of high-intensity exercise or chronic overtraining, hepcidin release from the liver can be increased and reduce intestinal absorption of iron leading to iron deficiency [54–58]. Hemolysis as a result of repetitive foot striking was thought for many years to underlie decreases in Hgb in athletes. However, experience shows us that thousands of runners do not necessarily show a decrease in Hgb simply due to the action of running, and, in fact, there are many other non-weight-bearing sports where "sports anemia" occurs like cycling, swimming, or hockey.

Monitoring Hgb is an effective manner to assess the O_2-carrying capacity throughout the season or when any decrease in performance is suspected. Figure 4 illustrates the example of a Hgb monitoring throughout the season which was helpful to modify the training of an athlete, increasing recovery period, whenever a drop in Hgb was detected.

4 Biomarkers for Muscle Damage

Muscle damage is quite prevalent in many sports due to the excessive stress of competition or even training. Those exercises with a prevalent component of eccentric contraction are more prone to muscle damage [59–63]. Weight lifting is the most prevalent elicitor of muscle damage in sports. In many instances, the incorrect balance of excessive training, improper training prescription and programing, poor recovery, and nutrition are the main responsible for muscle damage. In a first phase, microtears occur, and the contents of several components of muscle fibers leak out to the blood where they can be quantified. This cellular content is mainly made up of enzymes and—if muscle damage is more severe—can also involve some contractile proteins. There are different muscle enzymes that can be used for the assessment of muscle damage. Creatine kinase (CK) and lactate dehydrogenase (LDH) are the enzymes most used as biomarkers for muscle damage [59, 62–66]. Other enzymes that can be altered are alanine aminotransferase (ALT/GPT) and aspartate aminotransferase (AST/GOT). A contractile protein, Troponin I is also used to evaluate muscle damage [67]. Several inflammation mediators like the cytokines [68], free radicals or, more appropriate, reactive oxidative species (ROS) [69], and hormonal responses [70] are also used for muscle damage assessment. A combination of different biomarkers could also shed more appropriate light into muscle damage status (Fig. 5). The use of metabolomics could provide significantly more insight and precision to monitor anabolic and catabolic states which can contribute to diagnose muscle damage at an earlier state or even before it happens as well as to improve the monitoring of recovery by assessing different anabolic metabolites and pathways.

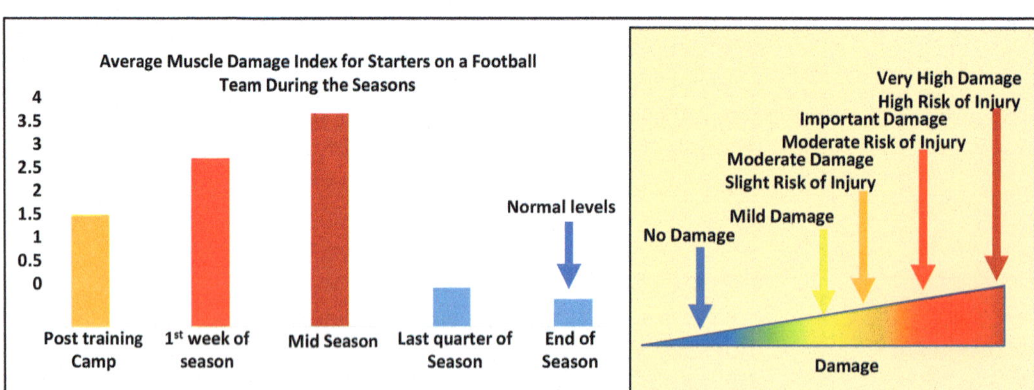

Fig. 5 Evolution of average muscle damage index (combination of different biomarkers) throughout the season among the starting players of a Division 1 American Football team. At mid-season there was observe an important degree of muscle damage which was then corrected by adjusting working load accordingly (Data from San-Millan's Lab)

5 Endocrinological Biomarkers

When overtraining occurs, different hormones can be affected and therefore disrupt different metabolic processes. In an acute phase, different endocrinological responses react to a physiological stress like excessive training, poor recovery, or poor nutrition eliciting low-energy availability. In order to increase energy availability and as an evolutionary mechanism in humans, different hormones are mobilized to increase energy availability mainly by breaking down protein for gluconeogenic purposes which eventually will lead to a catabolic state if the overtraining state is not corrected. In a chronic phase, different endocrinological responses tend to spare the body from incurring in more damage by decreasing exercise capacity as a self-protective mechanism to decrease further body damage.

Cortisol is the most immediate hormone acting upon excessive physiological (and also psychological stress). Cortisol is a hormone which is released from the adrenal cortex within the adrenal gland. Cortisol is a catabolic hormone involved in lipolysis, glycogenolysis, and proteolysis to mobilize fatty acids, glucose, and amino acids for energy purposes. In the acute phase of overtraining, energy availability needs to be augmented to overcome excessive training, poor recovery, or poor nutrition. Cortisol release is augmented to increase energy availability via lipolysis, glycogenolysis, and proteolysis [71]. The level of exercise intensity is a major driver of cortisol release, and its release is directly dependent of exercise intensity [72]. Coaches should be aware of the effects of cortisol and exercise intensities when creating training programs for athletes. The cortisol response and therefore increase of chances of overtraining are always higher in athletes who constantly train at high exercise intensities.

Cortisol responds to adrenocorticotropic hormone (ACTH) released from the anterior pituitary gland which in turn is stimulated by corticotropin-releasing hormone (CRH) released from the hypothalamus. This complex constitutes the hypothalamic-pituitary-adrenal axis (HPA axis) which acts in conjunction when stimulated by physiological stress in a manner dependent on exercise intensity and physiological stress [72]. This catabolism can be counteracted by anabolic hormones like growth hormone and testosterone. The testosterone/cortisol ratio is an effective manner to assess the catabolic vs anabolic balance in athletes.

If an athlete stays under stressful physiological conditions over time, the catabolic effects of cortisol will be important. Furthermore, the anabolic capacity from the body will be reduced, especially the testosterone responses will be blunted leading to an increased catabolic state, overtraining, and decrease in performance [73].

When the acute phase of overtraining is not corrected, an athlete enters the chronic phase, characterized by a decreased capacity to perform work at high exercise intensities and by chronic fatigue. In these instances, a decreased response of the HPA axis and decreased cortisol levels are observed [74, 75]. Catecholamine activation and prevalence of parasympathetic activity are also decreased [74, 76]. This decrease of HPA axis and adrenergic activity has gained popularity and named "adrenal fatigue" by many. This term has not been scientifically demonstrated, and it could be possible that the hormonal response to chronic overtraining could just be an evolutionary self-defensive and protective mechanism that the body has developed to "adapt" to chronic overtraining and preserve resources.

A big number of endurance athletes are being diagnosed nowadays with hypothyroidism which seems to have become a plague, especially among endurance athletes. It is important to pay attention to this phenomenon as it is quite possible that most athletes diagnosed with hypothyroidism actually may be misdiagnosed and they are simply chronically overtrained. The typical symptom of a person with hypothyroidism is chronic fatigue which overlaps with the symptom of chronic overtraining. Many athletes decide to finally go and see clinician to perform some blood tests. Thyroid-stimulating hormone (TSH), which stimulates the release of thyroid hormones T3 and T4 from the thyroid gland, is a routine parameter explored as part of most comprehensive metabolic panel tests. Hypothyroidism is characterized by increased TSH levels, and these elevated levels can also be found in athletes after several weeks of endurance training [77]. Although it is not clear why so many endurance athletes are currently "diagnosed" with hypothyroidism, there is a need to research these situations to make sure that athletes are not misdiagnosed. The field of metabolomics cannot only shed new light into this hypothyroidism epidemic but also increase the accuracy and monitoring capabilities to measure the different hormonal changes in the athletic population.

6 Serology/Infectious Markers

One of the observed features in overtraining athletes is a decreased immune system [78, 79] especially susceptible for upper respiratory tract infections (URTIs). There is even a time "window" of 3–72 h after intense exercise when athletes' immune systems are blunted and more susceptible to bacterial and viral infections [78–80]. Furthermore, it is well known that hypercortisolemia decreases the immune system [81]. Therefore, athletes that train with high cortisol levels due to excessive high-intensity sessions can also be more susceptible to infections.

7 Conclusion

In summary, the use of blood biomarkers in the area of sports medicine and performance has been successfully utilized for over 30 years to monitor performance, overtraining, and overall health in competitive athletes. Recently, the area of blood biomarkers has grown strongly among recreational athletes and fitness enthusiasts. However, although the understanding and expertise in interpreting blood biomarkers in athletes have increased and been mastered by many, the developing of novel biomarkers has been quite scarce in the last 30 years. The relatively new field of metabolomics, not applied in the world of sports yet, offers tremendous opportunities to improve the understanding of human performance and adaptations to exercise. The significantly higher amounts of metabolites that can be analyzed, the wider perspective of novel metabolic signatures, and the significantly lower amount of blood needed (~10 μl) as well as the large throughput of samples through isotope-ratio mass spectrometry (IR/MS), gas chromatography-mass spectrometry (GCMS), and LC-tandem mass spectrometry will set the path for metabolomics to be the future of monitoring and diagnostics in sports medicine and performance.

References

1. Billat LV (1996) Use of blood lactate measurements for prediction of exercise performance and for control of training. Recommendations for long-distance running. Sports Med 22:157–175
2. Jacobs I, Sjodin B, Schele R (1983) A Single Blood Lactate Determination as an Indicator of Cycle Ergometer Endurance Capacity. Eur J Appl Physiol Occup Physiol 50:355–364
3. San Millan I, Gonzalez-Haro C, Sagasti M (2009) Physiological differences between road cyclists of different categories. A new approach. Med Sci Sports Exerc 41:48
4. Jacobs I (1986) Blood lactate. Implications for training and sports performance. Sports Med 3:10–25
5. Brooks GA (2000) Intra- and extra-cellular lactate shuttles. Med Sci Sports Exerc 32:790–799
6. Brooks GA (2002) Lactate shuttles in nature. Biochem Soc Trans 30:258–264
7. Brooks GA (2009) Cell-cell and intracellular lactate shuttles. J Physiol 587:5591–5600
8. San-Millan I, Brooks GA (2017) Reexamining cancer metabolism: lactate production for carcinogenesis could be the purpose and explanation of the Warburg Effect. Carcinogenesis 38:119–133
9. Brooks GA, Dubouchaud H, Brown M, Sicurello JP, Butz CE (1999) Role of mitochondrial lactate dehydrogenase and lactate oxidation in the intracellular lactate shuttle. Proc Natl Acad Sci U S A 96:1129–1134
10. Gollnick PD, Hodgson DR (1986) The identification of fiber types in skeletal muscle: a continual dilemma. Exerc Sport Sci Rev 14:81–104
11. Heigenhauser GJ, Parolin ML (1999) Role of pyruvate dehydrogenase in lactate production in exercising human skeletal muscle. Adv Exp Med Biol 474:205–218
12. Hashimoto T, Hussien R, Cho HS, Kaufer D, Brooks GA (2008) Evidence for the mitochondrial lactate oxidation complex in rat neurons: demonstration of an essential component of brain lactate shuttles. PLoS One 3:e2915
13. Donaldson SK, Hermansen L, Bolles L (1978) Differential, direct effects of H+ on Ca2+ – activated force of skinned fibers from the soleus, cardiac and adductor magnus muscles of rabbits. Pflugers Arch 376:55–65
14. Sahlin K (1978) Intracellular pH and energy metabolism in skeletal muscle of man. With

special reference to exercise. Acta Physiol Scand Suppl 455:1–56
15. Sahlin K, Harris RC, Nylind B, Hultman E (1976) Lactate content and pH in muscle obtained after dynamic exercise. Pflugers Arch 367:143–149
16. Fitts R (2004) Mechanisms of muscular fatigue. Karger, Basel
17. Fabiato A, Fabiato F (1978) Effects of pH on the myofilaments and the sarcoplasmic reticulum of skinned cells from cardiac and skeletal muscles. J Physiol 276:233–255
18. Miyake S, Ishii Y, Watari T, Huang Z, Tsuchiya T (2003) The influences of L(+)-lactate and pH on contractile performance in rabbit glycerinated skeletal muscle. Jpn J Physiol 53:401–409
19. Kristensen M, Albertsen J, Rentsch M, Juel C (2005) Lactate and force production in skeletal muscle. J Physiol 562:521–526
20. Messonnier L, Kristensen M, Juel C, Denis C (2007) Importance of pH regulation and lactate/H+ transport capacity for work production during supramaximal exercise in humans. J Appl Physiol (1985) 102:1936–1944
21. San-Millan I, Brooks GA (2018) Assessment of metabolic flexibility by means of measuring blood lactate, fat, and carbohydrate oxidation responses to exercise in professional endurance athletes and less-fit individuals. Sports Med 48:467–479
22. Cai TQ, Ren N, Jin L, Cheng K, Kash S, Chen R, Wright SD, Taggart AK, Waters MG (2008) Role of GPR81 in lactate-mediated reduction of adipose lipolysis. Biochem Biophys Res Commun 377:987–991
23. Brooks GA (1991) Current concepts in lactate exchange. Med Sci Sports Exerc 23:895–906
24. Brooks GA (2018) The science and translation of lactate shuttle theory. Cell Metab 27:757–785
25. Hoppeler H, Luthi P, Claassen H, Weibel ER, Howald H (1973) The ultrastructure of the normal human skeletal muscle. A morphometric analysis on untrained men, women and well-trained orienteers. Pflugers Arch 344:217–232
26. Jacobs RA, Lundby C (2013) Mitochondria express enhanced quality as well as quantity in association with aerobic fitness across recreationally active individuals up to elite athletes. J Appl Physiol (1985) 114:344–350
27. Jacobs RA, Rasmussen P, Siebenmann C, Diaz V, Gassmann M, Pesta D, Gnaiger E, Nordsborg NB, Robach P, Lundby C (2011) Determinants of time trial performance and maximal incremental exercise in highly trained endurance athletes. J Appl Physiol (1985) 111:1422–1430
28. Seto JT, Quinlan KG, Lek M, Zheng XF, Garton F, MacArthur DG, Hogarth MW, Houweling PJ, Gregorevic P, Turner N, Cooney GJ, Yang N, North KN (2013) ACTN3 genotype influences muscle performance through the regulation of calcineurin signaling. J Clin Invest 123:4255–4263
29. Warburg OA, Minami S (1923) Versuche an Überlebendem Carcinom-gewebe. Klin Wochenschr 2:776–777
30. Coggan AR, Coyle EF (1987) Reversal of fatigue during prolonged exercise by carbohydrate infusion or ingestion. J Appl Physiol (1985) 63:2388–2395
31. Coyle EF, Coggan AR, Hemmert MK, Ivy JL (1986) Muscle glycogen utilization during prolonged strenuous exercise when fed carbohydrate. J Appl Physiol (1985) 61:165–172
32. Coyle EF, Hagberg JM, Hurley BF, Martin WH, Ehsani AA, Holloszy JO (1983) Carbohydrate feeding during prolonged strenuous exercise can delay fatigue. J Appl Physiol Respir Environ Exerc Physiol 55:230–235
33. Hermansen L, Hultman E, Saltin B (1967) Muscle glycogen during prolonged severe exercise. Acta Physiol Scand 71:129–139
34. Maughan RJ, Greenhaff PL, Leiper JB, Ball D, Lambert CP, Gleeson M (1997) Diet composition and the performance of high-intensity exercise. J Sports Sci 15:265–275
35. McConell G, Snow RJ, Proietto J, Hargreaves M (1999) Muscle metabolism during prolonged exercise in humans: influence of carbohydrate availability. J Appl Physiol (1985) 87:1083–1086
36. Ren JM, Broberg S, Sahlin K, Hultman E (1990) Influence of reduced glycogen level on glycogenolysis during short-term stimulation in man. Acta Physiol Scand 139:467–474
37. Lemon PW, Mullin JP (1980) Effect of initial muscle glycogen levels on protein catabolism during exercise. J Appl Physiol Respir Environ Exerc Physiol 48:624–629
38. Tarnopolsky MA, Atkinson SA, Phillips SM, MacDougall JD (1995) Carbohydrate loading and metabolism during exercise in men and women. J Appl Physiol (1985) 78:1360–1368
39. Sherman WM (1995) Metabolism of sugars and physical performance. Am J Clin Nutr 62:228S–241S
40. Sherman WM, Wimer GS (1991) Insufficient dietary carbohydrate during training: does it impair athletic performance? Int J Sport Nutr 1:28–44

41. Snyder AC, Kuipers H, Cheng B, Servais R, Fransen E (1995) Overtraining following intensified training with normal muscle glycogen. Med Sci Sports Exerc 27:1063–1070
42. Costill DL, Pascoe DD, Fink WJ, Robergs RA, Barr SI, Pearson D (1990) Impaired muscle glycogen resynthesis after eccentric exercise. J Appl Physiol (1985) 69:46–50
43. O'reilly KP, Warhol MJ, Fielding RA, Frontera WR, Meredith CN, Evans WJ (1987) Eccentric exercise-induced muscle damage impairs muscle glycogen repletion. J Appl Physiol (1985) 63:252–256
44. Meyers AW, Whelan JP (1988) Systematic model for understanding psychosocial influences in overtraining. In: Kreider RB, Fry AC, O'Toole ML (eds) Overtraining in sport. Human Kinetics, Champaign, IL
45. Alberti KG, Johnston DG (1977) Cortisol and catabolism: a new perspective. Clin Sci Mol Med 52:333–336
46. Borges GF, Rama LM, Pedreiro S, Rosado F, Alves F, Santos AM, Paiva A, Teixeira AM (2012) Haematological changes in elite kayakers during a training season. Appl Physiol Nutr Metab 37:1140–1146
47. Celsing F, Svedenhag J, Pihlstedt P, Ekblom B (1987) Effects of anaemia and stepwise-induced polycythaemia on maximal aerobic power in individuals with high and low haemoglobin concentrations. Acta Physiol Scand 129:47–54
48. Dressendorfer RH, Wade CE, Amsterdam EA (1981) Development of pseudoanemia in marathon runners during a 20-day road race. JAMA 246:1215–1218
49. Rushall BS, Busch JD (1980) Hematological responses to training in elite swimmers. Can J Appl Sport Sci 5:164–169
50. Woodson RD (1984) Hemoglobin concentration and exercise capacity. Am Rev Respir Dis 129:S72–S75
51. Sackmann E (1995) Biological membranes architecture and function. Elsevier Science, Amsterdam
52. Delaby C, Deybach JC, Beaumont C (2007) Hepcidin and iron metabolism. Rev Med Interne 28:510–512
53. Eleftheriadis T, Liakopoulos V, Antoniadi G, Kartsios C, Stefanidis I (2009) The role of hepcidin in iron homeostasis and anemia in hemodialysis patients. Semin Dial 22:70–77
54. Auersperger I, Knap B, Jerin A, Blagus R, Lainscak M, Skitek M, Skof B (2012) The effects of 8 weeks of endurance running on hepcidin concentrations, inflammatory parameters, and iron status in female runners. Int J Sport Nutr Exerc Metab 22:55–63
55. Kong WN, Gao G, Chang YZ (2014) Hepcidin and sports anemia. Cell Biosci 4:19
56. Liu YQ, Chang YZ, Zhao B, Wang HT, Duan XL (2011) Does hepatic hepcidin play an important role in exercise-associated anemia in rats? Int J Sport Nutr Exerc Metab 21:19–26
57. Peeling P (2010) Exercise as a mediator of hepcidin activity in athletes. Eur J Appl Physiol 110:877–883
58. Peeling P, Dawson B, Goodman C, Landers G, Wiegerinck ET, Swinkels DW, Trinder D (2009) Effects of exercise on hepcidin response and iron metabolism during recovery. Int J Sport Nutr Exerc Metab 19:583–597
59. Brancaccio P, Maffulli N, Limongelli FM (2007) Creatine kinase monitoring in sport medicine. Br Med Bull 81-82:209–230
60. Clarkson PM, Hubal MJ (2002) Exercise-induced muscle damage in humans. Am J Phys Med Rehabil 81:S52–S69
61. Coratella G, Chemello A, Schena F (2016) Muscle damage and repeated bout effect induced by enhanced eccentric squats. J Sports Med Phys Fitness 56:1540–1546
62. Nosaka K, Clarkson PM (1996) Variability in serum creatine kinase response after eccentric exercise of the elbow flexors. Int J Sports Med 17:120–127
63. Son HJ, Lee YH, Chae JH, Kim CK (2015) Creatine kinase isoenzyme activity during and after an ultra-distance (200 km) run. Biol Sport 32:357–361
64. Clarkson PM, Ebbeling C (1988) Investigation of serum creatine kinase variability after muscle-damaging exercise. Clin Sci (Lond) 75:257–261
65. Sorichter S, Koller A, Haid C, Wicke K, Judmaier W, Werner P, Raas E (1995) Light concentric exercise and heavy eccentric muscle loading: effects on CK, MRI and markers of inflammation. Int J Sports Med 16:288–292
66. Wolf PL, Lott JA, Nitti GJ, Bookstein R (1987) Changes in serum enzymes, lactate, and haptoglobin following acute physical stress in international-class athletes. Clin Biochem 20:73–77
67. Sorichter S, Mair J, Koller A, Gebert W, Rama D, Calzolari C, Artner-Dworzak E, Puschendorf B (1997) Skeletal troponin I as a marker of exercise-induced muscle damage. J Appl Physiol (1985) 83:1076–1082
68. Moldoveanu AI, Shephard RJ, Shek PN (2001) The cytokine response to physical activity and training. Sports Med 31:115–144

69. Powers SK, Jackson MJ (2008) Exercise-induced oxidative stress: cellular mechanisms and impact on muscle force production. Physiol Rev 88:1243–1276
70. Kraemer WJ, Ratamess NA (2005) Hormonal responses and adaptations to resistance exercise and training. Sports Med 35:339–361
71. Budgett R (1990) Overtraining syndrome. Br J Sports Med 24:231–236
72. Hill EE, Zack E, Battaglini C, Viru M, Viru A, Hackney AC (2008) Exercise and circulating cortisol levels: the intensity threshold effect. J Endocrinol Investig 31:587–591
73. Urhausen A, Kullmer T, Kindermann W (1987) A 7-week follow-up study of the behaviour of testosterone and cortisol during the competition period in rowers. Eur J Appl Physiol Occup Physiol 56:528–533
74. Brooks K, Carter J (2013) Overtraining, exercise, and adrenal insufficiency. J Nov Physiother 3(125)
75. Lehmann M, Gastmann U, Petersen KG, Bachl N, Seidel A, Khalaf AN, Fischer S, Keul J (1992) Training-overtraining: performance, and hormone levels, after a defined increase in training volume versus intensity in experienced middle- and long-distance runners. Br J Sports Med 26:233–242
76. Urhausen A, Gabriel H, Kindermann W (1995) Blood hormones as markers of training stress and overtraining. Sports Med 20:251–276
77. Simsch C, Lormes W, Petersen KG, Baur S, Liu Y, Hackney AC, Lehmann M, Steinacker JM (2002) Training intensity influences leptin and thyroid hormones in highly trained rowers. Int J Sports Med 23:422–427
78. Hackney AC, Koltun KJ (2012) The immune system and overtraining in athletes: clinical implications. Acta Clin Croat 51:633–641
79. Nieman DC (2000) Special feature for the Olympics: effects of exercise on the immune system: exercise effects on systemic immunity. Immunol Cell Biol 78:496–501
80. Mackinnon LT (2000) Special feature for the Olympics: effects of exercise on the immune system: overtraining effects on immunity and performance in athletes. Immunol Cell Biol 78:502–509
81. Fareau GG, Vassilopoulou-Sellin R (2007) Hypercortisolemia and infection. Infect Dis Clin North Am 21:639–657, viii

Chapter 27

Personalized Metabolomics

David P. Marciano and Michael P. Snyder

Abstract

The human metabolome is the cumulative product of ingested metabolites and those produced by the body and its microbiota. Together these metabolites can dynamically report on the health and disease state of an individual, as well as their response to drug treatments and other external perturbations. Profiling metabolites in human body fluids provides an opportunity to identify biomarkers and stratify patients for personalized treatments but requires the development of high-throughput approaches compatible with large cohort and longitudinal studies. Here we review in detail sample preparation and analytical liquid chromatography-mass spectrometry (LC-MS) methods to measure the broad chemical diversity of metabolites found in human plasma and urine.

Key words Personalized metabolomics, Plasma, Urine, Metabolites, HILIC, Reverse phase, Liquid chromatography-mass spectrometry, LC-MS

1 Introduction

Understanding health and disease requires the detailed analysis of both an individual's genome and the molecular events that determine human physiology [1]. The advent of accessible genomic sequencing has revolutionized the diagnosis of monogenetic disorders and enabled the expanded clinical application of personalized disease risk [2] and drug response prediction [3]. However, despite these advances the static nature of the genome limits the prognostic power for many of the most common diseases with origins that are complex and strongly influenced by the environment. As such, significant efforts have focused on capturing the dynamic aspects of human health by integrating the measurement of the microbiome, transcriptome, proteome, and metabolome from clinical samples with patient genomes [4]. This approach has been used to study the health of individuals longitudinally [5, 6] and identify biomarkers in large disease cohorts [7]. It has also revealed opportunities to therapeutically target the microbiome to treat human disease [8].

Angelo D'Alessandro (ed.), *High-Throughput Metabolomics: Methods and Protocols*, Methods in Molecular Biology, vol. 1978, https://doi.org/10.1007/978-1-4939-9236-2_27,

The metabolome in particular has long been leveraged for its clinical utility, with dozens of metabolites routinely measured as FDA-approved diagnostics. The goal of modern clinical metabolomics is to expand these measurements to include thousands of molecules [9]. To this end, robust LC-MS-based methods have been developed that approach the coverage, throughput, and cost requirements of routine clinical application. These advances have enabled large-scale studies to identify common genetic polymorphism that influence metabolism [10] and are central to the widespread adoption of personalized metabolomics to stratify patients and inform medical decisions. In this chapter we describe the sample preparation, LC-MS methods, and data analysis workflow to enable the untargeted profiling of blood plasma and urine metabolites.

2 Materials

All solutions should be prepared with >99% ultrapure HPLC-MS grade reagents in glass bottles to minimize background signal. Standard deionized water contains significant amounts of organic contaminants. Powder-free nitrile gloves should be worn during sample preparation.

2.1 Blood Plasma and Urine Preparation

1. EDTA-coated whole blood collection tube (e.g., BD Biosciences K2 EDTA #366643).
2. Sterile urine collection cup.

2.2 Metabolite Extraction

1. Plasma precipitation solvent: HPLC-MS grade acetone-acetonitrile-methanol (1:1:1, v/v) (*see* **Note 1**).
2. Urine precipitation solvent: HPLC-MS grade methanol.
3. 2 mL polypropylene microcentrifuge tubes (e.g., Eppendorf LoBind) (*see* **Note 2**).
4. Sample mixer (e.g., HulaMixer).
5. Nitrogen evaporator (e.g., Biotage TurboVap LV or equivalent).
6. LC compatible glass vials and low volume inserts (e.g., WAT094171DV).

2.3 Mass Spectrometer Analysis

The recommended chromatographic columns and mobile phase buffers were determined to provide optimal metabolite coverage and peak shapes across a broad chemical diversity in a systematic evaluation using standard mixtures and biological fluid samples [11]. To maximize the diversity of metabolite coverage, we recommend parallel use of hydrophilic interaction liquid chromatography (HILIC) and reversed-phase liquid chromatography (RPLC) in both positive and negative electrospray ionization (ESI) modes. The availability of MS instrument time is an important consideration when designing a metabolomic experiment along with sample number, power, and downstream analyses [12].

1. Mobile phase buffers

Chromatographic mode	Mobile phase buffer A	Mobile phase buffer B
HILIC	Acetonitrile/water (5/95, v/v), 10 mM ammonium acetate pH 6.9	Acetonitrile/water (95/5, v/v), 10 mM ammonium acetate pH 6.9
RPLC	0.06% acetic acid in 100% Water	0.06% acetic acid in 100% methanol

2. Chromatographic columns

Chromatographic mode	Body fluid	Column name	Dimensions, particle, and pore size
HILIC	Plasma and urine	ZIC-HILIC (Merck Millipore)	2.1 × 100 mm, 1.7 μm, 100 Å
RPLC	Urine	Hypersil GOLD (Thermo)	2.1 × 150 mm, 1.9 μm, 175 Å
RPLC	Plasma	Zorbax SB aq (Agilent)	2.1 × 100 mm, 1.8 μm, 80 Å

2.4 Validation Reagents

1. The NIST SRM 1950 is a reference sample of metabolites in human plasma. It is recommended for validating methods and testing interlab variability (srm1950.nist.gov).
2. Isotopically labeled standards for metabolite identification. The gold standard for definitive metabolite identification is to spike the matrix (e.g., extracted plasma or urine metabolites) with isotopically labeled standards to confirm the retention time and MS-MS fragmentation patterns match.

2.5 Electrospray Ionization Mass Spectrometer (ESI-MS)

1. The methods described have been applied across several different ESI-MS platforms. The sample preparation, columns, and LC methods are intended to be broadly compatible with all platforms. Data acquisition settings will be specific to the mass spectrometer instrumentation.

3 Methods

The collection of blood and urine samples for large studies generally occurs over several months and involves multiple laboratory personnel. As such, it is strongly advised that the initial processing steps are standardized to avoid introducing handling variability. It is also recommended that whenever possible, patients are fasted before sample collection and instructed to minimize exercise, stress, and other behaviors that can alter the composition of the metabolome [13]. All biological samples should be randomized after collection to limit systematic batch effects during processing and data collection.

3.1 Blood Plasma Sample Preparation

1. Collect blood in EDTA-coated tubes, and store at 4 °C before centrifugation (*see* **Note 3**).
2. Centrifuge whole blood samples for 15 min at 2000 × *g* and 4 °C to deplete cells.
3. Immediately transfer supernatant (plasma) to polypropylene tubes in 100 μL aliquots and freeze at −80 °C. It is important to minimize freeze-thaw cycles.

3.2 Blood Plasma Metabolite Extraction

1. Thaw 100 μL plasma aliquot on ice (*see* **Note 4**).
2. Add 400 μL of cold precipitation solvent (Subheading 2.2, **item 1**) to 100 μL of plasma or urine.
3. Shake samples for 15 min in 4 °C cold room.
4. Transfer samples to −20 °C for 2 h to allow for protein precipitation.
5. Centrifuge samples at 20,000 × *g* for 15 min at 4 °C to pellet proteins.
6. Transfer supernatant to a clean tube and discard pellet.
7. Dry metabolite extracts with nitrogen evaporator (*see* **Note 5**).
8. Just prior to data acquisition, dissolve dried metabolites in 100 μL of cold methanol/water (1:1, v/v), and shake for 15 min in 4 °C cold room.
9. Centrifuge samples at 20,000 × *g* for 15 min at 4 °C.
10. Split supernatant into 3 × 30 μL aliquots: (1) HILIC aliquot, (2) RPLC aliquot, and (3) pooled QC sample aliquot (*see* **Note 6**).

3.3 Urine Sample Preparation

1. After collection, immediately centrifuge urine samples for 10 min at 21,000 × *g* and 4 °C.
2. Transfer supernatant into 100 μL aliquots to avoid freeze-thaw cycles, and store at −80 °C.

3.4 Urine Metabolite Extraction

1. Thaw 100 μL urine aliquot on ice (*see* **Note 4**).
2. Add 400 μL cold HPLC-MS grade methanol to 100 μL of urine.
3. Follow **steps 3–9** of Subheading 3.2.

3.5 LC-MS Methods

It is important to consider that metabolites are not stable at 4 °C indefinitely, particularly when collecting data for experiments that can span several weeks to months. It is recommended that batches of no more than 30 samples for HILIC and 50 samples for RPLC are run to limit the time samples are maintained at 4 °C to less than 48 h. Metabolomic sample data is collected in MS1 full scan mode for optimal peak quantification. MS2 data is collected on the QC sample to generate fragmentation patterns for subsequent metabolite identification. Data acquisition settings will be specific to the mass spectrometer instrumentation, but as a reference, the detailed parameters for the Agilent 6538 UHD Q-TOF were previously described [11].

1. Prepare LC-MS worklist (Fig. 1) with the appropriate LC method (gradient, flow rate, column temperature) described below. Sample injection volume is 5 μL. Before each injection, the column should be equilibrated for 5 min with 99% (HILIC) or 1% (RPLC) mobile phase buffer B.

Column name	Gradient	Flow rate	Column temperature (°C)
ZIC-HILIC	99% to 50% B 15 min	0.5 mL/min	40
Hypersil Gold	1% to 80% B in 10 min	0.25 mL/min	60
Zorbax SB aq	1% to 80% B in 9 min	0.6 mL/min	60

2. Before starting experimental data collection, new HILIC columns should be conditioned with eight QC sample injections. New RPLC columns should be conditioned with six QC sample injections (*see* **Note 7**).
3. Samples should be randomly split into batches and ordered for injection to avoid introducing analytical bias. For each batch of samples, positive and negative mode data should be collected in succession (Fig. 1).
4. QC sample should be collected in MS-MS mode to capture fragmentation patterns of metabolite features across all samples (*see* **Note 8**).
5. Each batch should end with a blank sample injection of 50/50 methanol/water to measure background signal that will be removed from biological sample analysis.

3.6 Data Processing and Analysis

The data acquired using the untargeted metabolomic methodologies described above is extremely complex, consisting of thousands of metabolite feature measurements from a large number of biological samples. This data is generally collected over long periods of time and in multiple batches, allowing for the introduction of random and systematic variability in retention time, mass accuracy, and signal sensitivity between samples. Fortunately, several user-friendly tools that require only a cursory knowledge of statistics are available to process and analyze this data. We recommend that novice users begin with XCMS online (www.xcmsonline/scripps.edu)

Injection #	Sample	Method	Injection #	Sample	Method
1	QC MS2	MS2 HILIC negative	37	QC MS2	MS2 HILIC positive
2	QC1	MS1 HILIC negative	38	QC1	MS1 HILIC positive
3	Sample 21	MS1 HILIC negative	39	Sample 21	MS1 HILIC positive
4	Sample 4	MS1 HILIC negative	40	Sample 4	MS1 HILIC positive
5	Sample 18	MS1 HILIC negative	41	Sample 18	MS1 HILIC positive
6	Sample 30	MS1 HILIC negative	42	Sample 30	MS1 HILIC positive
7	Sample 5	MS1 HILIC negative	43	Sample 5	MS1 HILIC positive
8	Sample 13	MS1 HILIC negative	44	Sample 13	MS1 HILIC positive
9	Sample 7	MS1 HILIC negative	45	Sample 7	MS1 HILIC positive
10	Sample 17	MS1 HILIC negative	46	Sample 17	MS1 HILIC positive
11	Sample 9	MS1 HILIC negative	47	Sample 9	MS1 HILIC positive
12	Sample 25	MS1 HILIC negative	48	Sample 25	MS1 HILIC positive
13	QC2	MS1 HILIC negative	49	QC2	MS1 HILIC positive
14	Sample 1	MS1 HILIC negative	50	Sample 1	MS1 HILIC positive
15	Sample 14	MS1 HILIC negative	51	Sample 14	MS1 HILIC positive
16	Sample 19	MS1 HILIC negative	52	Sample 19	MS1 HILIC positive
17	Sample 27	MS1 HILIC negative	53	Sample 27	MS1 HILIC positive
18	Sample 6	MS1 HILIC negative	54	Sample 6	MS1 HILIC positive
19	Sample 23	MS1 HILIC negative	55	Sample 23	MS1 HILIC positive
20	Sample 10	MS1 HILIC negative	56	Sample 10	MS1 HILIC positive
21	Sample 24	MS1 HILIC negative	57	Sample 24	MS1 HILIC positive
22	Sample 11	MS1 HILIC negative	58	Sample 11	MS1 HILIC positive
23	Sample 26	MS1 HILIC negative	59	Sample 26	MS1 HILIC positive
24	QC3	MS1 HILIC negative	60	QC3	MS1 HILIC positive
25	Sample 15	MS1 HILIC negative	61	Sample 15	MS1 HILIC positive
26	Sample 20	MS1 HILIC negative	62	Sample 20	MS1 HILIC positive
27	Sample 22	MS1 HILIC negative	63	Sample 22	MS1 HILIC positive
28	Sample 2	MS1 HILIC negative	64	Sample 2	MS1 HILIC positive
29	Sample 16	MS1 HILIC negative	65	Sample 16	MS1 HILIC positive
30	Sample 28	MS1 HILIC negative	66	Sample 28	MS1 HILIC positive
31	Sample 12	MS1 HILIC negative	67	Sample 12	MS1 HILIC positive
32	Sample 3	MS1 HILIC negative	68	Sample 3	MS1 HILIC positive
33	Sample 29	MS1 HILIC negative	69	Sample 29	MS1 HILIC positive
34	Sample 8	MS1 HILIC negative	70	Sample 8	MS1 HILIC positive
35	QC4	MS1 HILIC negative	71	QC4	MS1 HILIC positive
36	Blank	MS1 HILIC negative	72	Blank	MS1 HILIC positive

Fig. 1 Example LC-MS worklist for 30 sample experiments. Samples are loaded in randomized order with QC sample injected first and after every ten samples. MS-MS data is also collected on the QC sample with fragmentation patterns used for metabolite identification and isomer deconvolution

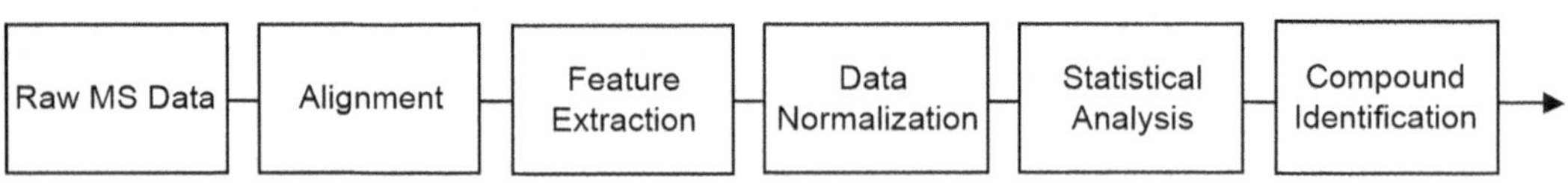

Fig. 2 Metabolomics data analysis workflow

for data processing, analysis, and visualization [14]. Further analyses and interpretation can be performed using MetaboAnalyst (www.metaboanalyst.ca) [15]. Tools to integrate metabolomic data with other omic information continue to be developed [16]. Below the recommended data analysis workflow is outlined and important considerations discussed (Fig. 2).

1. Spectral alignment adjusts for variability in metabolite retention times between sample injections. It is a particularly important processing step for studies that include large numbers of samples, as chromatographic retention times are known to drift over many injections and when columns are exchanged (Fig. 3) (*see* **Note 9**).
2. Feature extraction involves identifying and integrating all peaks and annotating their relevant spectral and chromatographic information. This includes identifying and merging adduct peaks that elute with the same retention time.
3. Data normalization is used to remove sources of systematic variability between sample injections due to factors such as sample dilution and technical variation (column aging, source contamination). This is particularly important when analyzing urine samples in which metabolite concentrations can vary greatly based on a sample dilution. In large-scale studies, the most basic data normalization methods are based on total ion intensity; however, more sophisticated methods have been developed that account for specific feature of signal intensity distributions [17, 18].
4. Statistical analysis will be specific to each experimental design but generally involves identifying significant differences in metabolite abundances between groups. Metabolomic datasets are high dimensional and require adjustment for false discovery rates (FDR) when conducting multiple comparisons. Many of these analysis methodologies are now standardized and integrated into XCMS Online and MetaboAnalyst but are important to be aware of. It is also recommended that metabolite features that are enriched less than fivefold over blanks be removed from the analysis, as well as those with a coefficient of variation (CV) greater than ~30% across QC samples.
5. Compound identification remains the most challenging and time-consuming analysis step and, for this reason, is generally only performed on metabolite features identified as significantly different after statistical analysis. The contribution of the microbiome to the blood plasma and urine metabolomes significantly increases their complexity and the challenge of obtaining definitive identifications. A putative metabolite identification can be made based on mass, retention time, and fragmentation pattern. These can be compared with in-house libraries and metabolomic spectra repositories (e.g., METLIN [19]). While many studies do not extend beyond these approaches, the gold standard for definitive metabolite identification and quantification remains spiking isotopically labeled standards into the QC sample before data acquisition. This enables direct comparison of the metabolite and standard in the same matrix, along with acquisition of standard curves for absolute quantification.

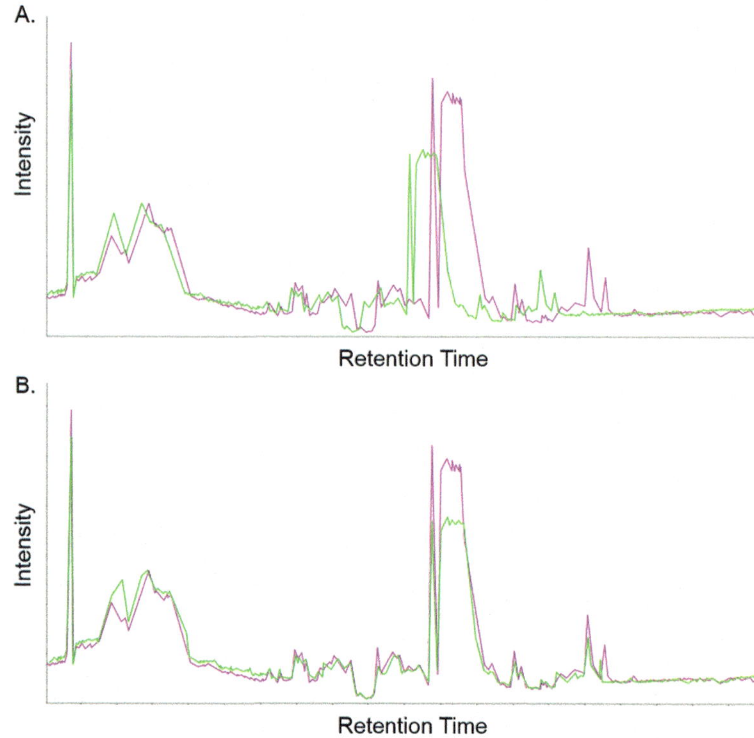

Fig. 3 Spectral alignment to correct retention time drift. (**a**) Comparison of total ion chromatograms from two sample injections demonstrates how retention times can change over the course of many injections or between columns. (**b**) Spectral alignment corrects retention time drift enabling direct comparison of features

4 Notes

1. Isotopically labeled internal standards can be included in the plasma precipitation solvent to allow for correction of downstream sample processing errors and to monitor drift in retention times and exact mass. It is recommended that ~10 standards with retention times that span the LC gradient be used for each mode.
2. The time solvents stored in polypropylene tubes should be minimized to avoid leaching of polymers. Glass vials should be used if long-term storage is required.
3. Blood plasma contains metabolically active cells that need to be removed quickly by centrifugation to preserve the metabolic state at sampling time.
4. It is recommended that no more than 24 samples are processed simultaneously.
5. Dried samples can be stored at −80 °C for several weeks or longer. For experiments with large numbers of samples, stop here until all samples are processed and are ready to proceed for data collection.

6. QC sample consists of 30 μL from all biological samples included in the study and is representative of both the sample matrix and metabolite composition. The QC sample will be analyzed between every ~10 injections to assess the variance across the analytical study.
7. Despite considerable advancement in HILIC methodologies, technical challenges still remain, and column performance should be closely monitored before and during biological sample injections.
8. The term feature refers to a mass spectral peak with a unique m/z and retention time that can represent a metabolite ion, isotope, adduct, fragment, or noise.
9. HILIC columns should be replaced every ~500 injections and RPLC column replaced every ~1000 injections.

Acknowledgment

This work was supported by NIH grant F32 HL 132452 to D.P.M.

References

1. Snyder M (2016) Genomics and personalized medicine: what everyone needs to know, 1st edn. Oxford University Press, New York, 166 pp
2. Esplin ED, Oei L, Snyder MP (2014) Personalized sequencing and the future of medicine: discovery, diagnosis and defeat of disease. Pharmacogenomics 15(14):1771–1790
3. Madhukar NS, Elemento O (2018) Bioinformatics approaches to predict drug responses from genomic sequencing. Methods Mol Biol 1711:277–296
4. Karczewski KJ, Snyder MP (2018) Integrative omics for health and disease. Nat Rev Genet 19(5):299–310
5. Chen R et al (2012) Personal omics profiling reveals dynamic molecular and medical phenotypes. Cell 148(6):1293–1307
6. Piening BD et al (2018) Integrative personal omics profiles during periods of weight gain and loss. Cell Syst 6(2):157–170.e8
7. Lee S et al (2016) Integrated network analysis reveals an association between plasma mannose levels and insulin resistance. Cell Metab 24(1):172–184
8. Pedersen HK et al (2016) Human gut microbes impact host serum metabolome and insulin sensitivity. Nature 535(7612):376–381
9. Beger RD et al (2016) Metabolomics enables precision medicine: "A White Paper, Community Perspective". Metabolomics 12(10):149
10. Gieger C et al (2008) Genetics meets metabolomics: a genome-wide association study of metabolite profiles in human serum. PLoS Genet 4(11):e1000282
11. Contrepois K, Jiang L, Snyder M (2015) Optimized analytical procedures for the untargeted metabolomic profiling of human urine and plasma by combining hydrophilic interaction (HILIC) and reverse-phase liquid chromatography (RPLC)-mass spectrometry. Mol Cell Proteomics 14(6):1684–1695
12. Hasin Y, Seldin M, Lusis A (2017) Multi-omics approaches to disease. Genome Biol 18(1):83
13. Yin P, Lehmann R, Xu G (2015) Effects of pre-analytical processes on blood samples used in metabolomics studies. Anal Bioanal Chem 407(17):4879–4892
14. Tautenhahn R et al (2012) XCMS Online: a web-based platform to process untargeted metabolomic data. Anal Chem 84(11):5035–5039
15. Xia J, Wishart DS (2016) Using MetaboAnalyst 3.0 for comprehensive metabolomics data analysis. Curr Protoc Bioinformatics 55:14.10.1–14.10.91

16. Cambiaghi A, Ferrario M, Masseroli M (2017) Analysis of metabolomic data: tools, current strategies and future challenges for omics data integration. Brief Bioinform 18(3): 498–510
17. Veselkov KA et al (2011) Optimized preprocessing of ultra-performance liquid chromatography/mass spectrometry urinary metabolic profiles for improved information recovery. Anal Chem 83(15):5864–5872
18. Wu Y, Li L (2016) Sample normalization methods in quantitative metabolomics. J Chromatogr A 1430:80–95
19. Guijas C et al (2018) METLIN: a technology platform for identifying knowns and unknowns. Anal Chem 90(5):3156–3164

Index

Angelo D'Alessandro (ed.), *High-Throughput Metabolomics: Methods and Protocols*, Methods in Molecular Biology, vol. 1978, https://doi.org/10.1007/978-1-4939-9236-2, © Springer Science+Business Media, LLC, part of Springer Nature 2019

Q

R

S

T

W

X

Printed by Printforce, the Netherlands